De Gruyter Textbook

Deuflhard / Weiser · Adaptive Numerical Solution of PDEs

Peter Deuflhard
Martin Weiser

Adaptive Numerical Solution of PDEs

De Gruyter

Mathematics Subject Classification 2010: Primary: 65-01; Secondary: 65F05, 65F08, 65F10, 65N06, 65N22, 65N25, 65N30, 65N35, 65N50, 65N55, 65M20, 65M55, 65M60.

ISBN 978-3-11-028310-5
e-ISBN 978-3-11-028311-2

Library of Congress Cataloging-in-Publication Data

A CIP catalog record for this book has been applied for at the Library of Congress.

Bibliographic information published by the Deutsche Nationalbibliothek

The Deutsche Nationalbibliothek lists this publication in the Deutsche Nationalbibliografie; detailed bibliographic data are available in the internet at http://dnb.dnb.de.

© 2012 Walter de Gruyter GmbH & Co. KG, Berlin/Boston

Typesetting: PTP-Berlin Protago-TeX-Production GmbH, www.ptp-berlin.eu

♾ Printed on acid-free paper

Printed in Germany

www.degruyter.com

Preface

The present volume deals with the *numerical solution of partial differential equations (PDEs)*. The focus is on elliptic and parabolic systems; hyperbolic conservation laws are only treated on an elementary level below turbulence. The book is written in the style of the two preceding volumes,

* *Numerical Analysis in Modern Scientific Computing. An Introduction*
  by Peter Deuflhard and Andreas Hohmann (to be named Volume 1), and

* *Scientific Computing with Ordinary Differential Equations*
  by Peter Deuflhard and Folkmar Bornemann (to be named Volume 2).

By and large, Volume 1 or any other introductory textbook on Numerical Analysis should suffice as a prerequisite. Volume 2 will be widely dispensable, since we included a shortcut presentation on stiff ODEs in Section 9.1. More sophisticated analytical tools needed in the text are comprehensively worked out in an Appendix.

Three main threads run through the entire book:

1. detailed derivation and analysis of *efficient adaptive algorithms*, a corresponding software list is at the end of the book;

2. clear orientation towards *scientific computing*, i.e., to complex problems from science, technology, and medicine;

3. as far as possible elementary, but not too elementary, *mathematical theory*, where the traditional field has been thinned in view of newer topics that are important with respect to (1) and (2).

This book is addressed to students of applied mathematics as well as to mathematicians, physicists, chemists, and engineers already on the job who are confronted with an *efficient* numerical solution of complex application problems. Conceptually, it is written as a textbook, but is also suited for private study and as background material. In our presentation we deliberately followed the famous dictum by Carl Friedrich Gauss that mathematics be a "science for the eye".[1] Consequently, we have inserted quite a number of figures and illustrating examples to explain complex relations. Moreover, we have worked out four nontrivial problems from application (in order of appearance:

---

[1] "Mein hochgeehrter Lehrer, der vor wenigen Jahren verstorbene Geheime Hofrath *Gauss* in Göttingen, pflegte in vertraulichem Gespräche häufig zu äussern, die Mathematik sei weit mehr eine Wissenschaft für das Auge als eine für das Ohr. ..."   published in: Kronecker's Werke, Volume 5, p. 391.

regenerative energy, nanotechnology, surgery, physiology) to demonstrate how far the concepts presented here would carry over into practice.

Acknowledgments. The contents of this book have profited in multiple ways from intensive professional discussions with my colleagues Ralf Kornhuber (FU Berlin), Harry Yserentant (TU Berlin), and Carsten Gräser (FU Berlin) about adaptive multigrid methods, Rupert Klein (FU Berlin) on computational fluid dynamics, Jens Lang (TU Darmstadt) on the adaptive time layer method, Ulrich Maas (KIT Karlsruhe) on the adaptive method of lines, Alexander Ostermann (U Innsbruck) on order reduction in one-step methods, Iain Duff (CERFACS) on direct sparse solvers, Volker Mehrmann (TU Berlin) on methods of numerical linear algebra, Wolfgang Dahmen (RWTH Aachen) and Angela Kunoth (U Paderborn) on proofs for adaptive finite element methods, Gabriel Wittum (U Frankfurt) on the comparative behavior of additive versus multiplicative multigrid methods, and Michael Wulkow (CIT) on the adaptive time layer method. We gratefully acknowledge contributions by and discussions with our ZIB colleagues Frank Schmidt on the modelling and numerics of the optical equations, Lin Zschiedrich on Rayleigh quotient minimization, Anton Schiela on many topics of the theoretical background, Bodo Erdmann on electrocardiology, Stefan Zachow on cranio-maxillo-facial surgery, Ulrich Nowak[†] on adaptivity in the method of lines, and Felix Lehman on solver comparisons. For valuable hints on the literature we wish to thank Christoph Schwab (ETH Zürich), Nick Trefethen (U Oxford), Zhiming Chen (Academia Sinica Beijing), Kunibert Siebert (U Duisburg-Essen), Ernst Hairer (U Geneva), and Gerhard Wanner (U Geneva).

Special thanks go to our former ZIB colleague Rainer Roitzsch, without whose profound knowledge concerning a wide range of tricky LaTeX questions, his continuous help with the subtle aspects of the figures as well as – down to the deadline – untiring watchdog activity regarding inconsistencies of the manuscript, this book would never have reached its final form. Last, but not least, we cordially thank Erlinda Cadano-Körnig and Regine Kossick for manifold behind-the-scene assistance, as well as the many colleagues who have supported us by reading earlier drafts of the manuscript.

Berlin, April 2012 Peter Deuflhard and Martin Weiser

[†] passed away in June 2011

Contents

Outline

This book deals with the numerical solution of elliptic and parabolic systems, while the numerical solution of hyperbolic conservation laws is only treated excluding turbulence. The emphasis lies on the derivation of *efficient* algorithms, which realize *adaptivity* with respect to *space and time*, with further modes of adaptivity included. The presentation of the mathematical theory lays special attention on simplicity, which does not mean that more difficult theoretical problems are circumvented. In order to elucidate complex relations, we have inserted numerous figures. Wherever appropriate, further reading is recommended at the end of a section. Just like in the two preceding volumes the historical evolution is interlaced. Each chapter finishes with an exercise section; programming exercises have been deliberately skipped, since they are too closely connected with the available software environment.

The book contains nine general methodological chapters, an appendix on theoretical background material, a software list, and a carefully prepared index.

The first two chapters give an introduction into *models* of partial differential equations (PDEs).

In **Chapter 1**, for readers to become acquainted with the field, we begin with a series of elementary PDEs. We start with the classics: Laplace and Poisson equations, including Robin boundary conditions, which run as a recurrent theme throughout the whole book. We also discuss the "ill-posed" elliptic initial value problem and give an example from medicine, where such a problem nevertheless arises. The early introduction of Laplacian eigenvalue problems pays off in subsequent chapters. Next, we treat the diffusion and the wave equation. Finally, we present the Schrödinger and Helmholtz equations, whereby we already leave the traditional classification into elliptic, parabolic, and hyperbolic PDEs.

In **Chapter 2** we show that the PDE examples of Chapter 1 actually arise in science and technology. Among the numerous possible application areas we selected electrodynamics, fluid dynamics, and elastomechanics. In these three areas we survey the established model hierarchies. As for fluid dynamics, turbulence models are deliberately excluded, since this would be a rather different mathematical world. However, for all three application fields we present nontrivial examples in subsequent chapters.

The next two chapters deal with *discretization methods* for elliptic problems.

In **Chapter 3** we start with elaborating *finite difference methods* on uniform meshes. For the Poisson model problems we derive the traditional approximation theory, first in the L^2-norm and the energy norm, then in the L^∞-norm. Next, we work out the various algorithmic strategies for nonuniform meshes and indicate their respective weak points. In particular, curved boundary geometries and strongly localized phenomena cause serious difficulties.

In **Chapter 4** we treat *Galerkin methods* as an alternative, both global *spectral methods* and *finite element methods* (FEM). We begin with an elaboration of the abstract theory from simple geometric concepts in Hilbert spaces. On this basis, we derive approximation properties for both types of methods, in FEM for the case of uniform meshes giving attention to boundary as well as eigenvalue problems. For the Fourier–Galerkin methods we present an adaptive variant that automatically determines a necessary dimension of the basis. For FEM we go deeper into the details of algorithmic realization, both in 2D and in 3D as well as for higher orders of local polynomials. In nonuniform FE meshes some angle condition plays an important role, which we derive in the simplest case and which will be useful in Section 6.2.2.

The next three chapters describe *algorithms* for the efficient numerical solution of linear elliptic boundary and eigenvalue problems.

In **Chapter 5** we deal with *discrete elliptic* boundary value problems where an a priori given mesh exists so that only elliptic grid equations of fixed dimension remain to be solved. For this kind of problem the algorithmic alternatives are either *direct sparse solvers* or *iterative methods*, in each case for symmetric positive definite matrices. For sparse solvers, we discuss the underlying principles of symbolic factorization and frontal techniques. Among the iterative solvers we introduce the classical matrix decomposition algorithms (Jacobi and Gauss–Seidel methods) as well as the method of conjugate gradients (cg-method), the latter including preconditioning and adaptive termination criteria in the *energy norm* – a technique not generally known, to which we will repeatedly recur in later chapters. Moreover, we present a variant of the cg-method for the minimization of the Rayleigh quotient. After that we discuss the selected iterative methods under the aspect of error smoothing, thus leading over to an elementary presentation of multigrid methods and of hierarchical basis methods, here for prescribed hierarchical meshes. For multigrid methods we offer the classical convergence proof, which, however, only covers W-cycles in the quasi-uniform grid case. More powerful proof techniques are postponed until Chapter 7. Finally, we compare these techniques with direct solvers when applied to each of the meshes from the hierarchy. As an example we give the power optimization of a Darrieus wind generator, where we use a potential flow model that is a slight extension of Section 2.2.1.

In **Chapter 6** we lay the basis for the efficient construction of adaptive hierarchical meshes. First we derive all established a posteriori error estimators, including the often neglected hierarchical error estimator. We present both a unified theory and details of algorithmic realization, especially in the context of finite elements. We incur, in sufficient detail, on mesh refinement. However, we circumvent techniques of *hp*-refinement, as we essentially confine ourselves mainly to finite elements of lower order. For a model refinement strategy we present a convergence theorem. In what follows the application of error estimators is worked out in view of an approximate equilibration of disctretization errors, both in theory and in algorithmic detail (local extrapolation). The theory is illustrated at an elementary example defined over a domain with a reentrant corner. On the basis of adaptive grids efficient direct or iterative solvers from numerical linear algebra as well as iterative multigrid methods can be realized. As a nontrivial example for the first case, we give a quadratic eigenvalue problem originating from the design of a plasmon-polariton wave guide, where we use our preknowledge on optical wave guides from Section 2.1.2.

In **Chapter 7** we approach *continuous* boundary value problems directly as problems in function space, i.e., via a combination of finite element discretization, multigrid method, and adaptive hierarchical grids. As a basis for a convergence theory for multigrid methods we first prepare the abstract tools of sequential and parallel subspace correction methods. In passing we also apply these tools to domain decomposition methods and to FEM of higher order. On this theoretical background we are then able to comparatively analyze multiplicative and additive multigrid approaches (such as HB and BPX preconditioners) for the adaptive case, too. As an algorithmically especially simple compromise between multiplicative and additive multigrid methods we also work out cascadic multigrid method. Finally, we present two types of adaptive multigrid methods for the solution of linear eigenvalue problems, a linear multigrid method and a multigrid variant of the Rayleigh quotient minimization.

The last two chapters leave the restriction to stationary linear problems and extend the presentation both to *nonlinear elliptic* and to *nonlinear parabolic* problems.

In **Chapter 8** we present *affine conjugate* Newton methods for *nonlinear elliptic boundary value problems*. Here, too, we first consider methods on a fixed spatial grid, i.e., discrete Newton methods for elliptic grid equations, where adaptivity is solely related to steplength strategies and truncation criteria for inner PCG iterations. Next, we introduce *fully adaptive Newton-multigrid methods*, where additionally the construction of suitable adaptive hierarchical grids with respect to the nonlinearity comes into play. A problem without a solution serves as illustration of the difference between discrete and continuous Newton methods. As a nontrivial example we finally add a method for operation planning in maxillo-facial surgery. Here we use previous knowledge about nonlinear elastomechanics from Section 2.3. As it turns out, this example forces one to leave the safe ground of convex minimization. For the special case we sketch an extended algorithmic approach.

In **Chapter 9** we treat *time dependent PDE systems of parabolic type*. We skip the presently en vogue discontinuous Galerkin methods, as they seem to be better suited for hyperbolic than for parabolic problems. In a kind of "quick run" through Volume 2 we put a review section on stiff ordinary differential equations at the beginning. Beyond Volume 2, we focus on the phenomenon of *order reduction* that arises in the discretization of parabolic differential equations. As in stationary problems in Chapter 5 and Section 8.1, we first treat the case of fixed spatial grids, the classical method of lines with adaptive timestep control. We then discuss its nontrivial extension to moving spatial grids, a method promising mainly for problems in one space dimension. As the alternative, we introduce the *adaptive method of time layers* (also: *Rothe method*), where spatial and temporal adaptivity can be coupled in some rather natural way. For this purpose, we derive an adaptive strategy. As a nontrivial example we finally give results on the numerical simulation of a model for the electrical excitation of the heart muscle, both for a healthy heart beat and for ventricular fibrillation.

Chapter 1

Elementary Partial Differential Equations

Ordinary differential equations (often abbreviated as ODEs) are equations connect-
ing functions and their derivatives with respect to *one* independent variable; in problems from science and technology the independent variable will be a time or a space
variable. ODEs appear as initial value problems or as boundary value problems, the
latter as "timelike" or as "spacelike", as distinguished and worked out in detail in
Volume 2. Partial differential equations (often referred to as PDEs) are equations con-
necting functions and their derivatives with respect to *several* independent variables;
in problems from science and technology these variables will be time and/or several
space variables. Here, too, initial value problems as well as boundary value problems
exist. However, as we will see in the following, in contrast to ODEs there exists a close
coupling between the structure of the PDE and corresponding well-defined initial or
boundary value problems.

The mathematical modelling of our real world by PDEs is extremely rich and poly-
morphic: They describe phenomena on the microlevel (e.g., materials science, chemistry, systems biology) as well as on the macrolevel (e.g., meteorology, climate science, medical technology), of engineering (e.g., structural mechanics, airplane design,
cosmology), of image processing, or of economy. If we attempted to treat this rich
world by a single general approach, its essentials would drop "through the conceptual
grate". For this reason, this text will not follow the usual mathematical derivation top-
down from the general to the specific. Rather, we sharpen our mathematical intuition
by starting with some important special cases classified as *scalar partial differential
equations of second order* (see, e.g., [56]).

1.1 Laplace and Poisson Equation

In some sense the heart and certainly the prototype of *elliptic* PDEs is the one named after the French mathematician and physicist (these professions were not separate then)
Pierre-Simon Marquis de Laplace (1749–1827).

Laplace Equation.   This equation reads

$$\Delta u(x) = 0, \quad x \in \Omega \subset \mathbb{R}^d \tag{1.1}$$

for a function $u : \Omega \to \mathbb{R}$ defined over some *domain*, i.e., an open, nonempty, and connected set. Therein Δ is the Laplace operator; for $d = 2$ and in Euclidean coordinates

$(x, y) \in \mathbb{R}^2$, it is defined by

$$\Delta u = u_{xx} + u_{yy} = \frac{\partial^2 u}{\partial x^2} + \frac{\partial^2 u}{\partial y^2}.$$

The transfer to the more general case $d > 2$ is self-explanatory. In applications, one often needs the representation of this operator in cylindrical or spherical coordinates, which we postpone until Exercises 1.1 and 1.2.

Poisson Equation. An inhomogeneous variant of the Laplace equation is named after the French mathematician and physicist Siméon Denis Poisson (1781–1840):

$$\Delta u = f \quad \text{in } \Omega \subset \mathbb{R}^d. \tag{1.2}$$

This PDE will accompany us as model problem over a wide range of the book.

1.1.1   Boundary Value Problems

The PDEs (1.1) and (1.2) appear in combination with the following *boundary conditions*:

- Dirichlet boundary conditions, named after the German mathematician Peter Gustav Lejeune Dirichlet (1805–1859), in *inhomogeneous* form[1]:

$$u|_{\partial\Omega} = \phi;$$

- *homogeneous* Neumann boundary conditions, named after the Leipzig mathematician Carl Gottfried Neumann[2] (1832–1925)

$$\frac{\partial u}{\partial n}\Big|_{\partial\Omega} = n^T \nabla u = 0;$$

- Robin boundary conditions, named after the French mathematician Victor Gustave Robin (1855–1897)

$$n^T \nabla u + \alpha(x)u = \beta(x), \quad x \in \partial\Omega,$$

where n is the *unit normal vector*, i.e., the vector orthogonal to the domain boundary $\partial\Omega$, pointing outward and normalized to length 1 (see also Definition (A.1) in the Appendix A.2).

Remark 1.1. When the Dirichlet boundary condition is inhomogeneous, as above, it can be homogenized exploiting the linearity of the whole problem. For this purpose

---

[1] Despite the name, these boundary conditions were never published by Dirichlet, but were only presented in his lectures in Göttingen.
[2] not John von Neumann!

the boundary function ϕ is extended to some function $\bar{\phi}$, defined over the completion $\overline{\Omega}$ of the whole domain, i.e.,

$$\bar{\phi} \in C^2(\overline{\Omega}), \quad \bar{\phi}|_{\partial\Omega} = \phi.$$

We may now decompose the solution according to $u = v + \bar{\phi}$. For v we then obtain the Poisson equation

$$\Delta v = \Delta u - \Delta\bar{\phi} = f - \Delta\bar{\phi} =: \bar{f}$$

with corresponding *homogeneous* Dirichlet boundary condition

$$v|_{\partial\Omega} = u|_{\partial\Omega} - \bar{\phi}|_{\partial\Omega} = \phi - \phi = 0.$$

Therefore, without loss of generality, we will consider the Laplace and the Poisson equation only with *homogeneous* Dirichlet boundary conditions.

Uniqueness of the Solution.   Assume now that for a given Poisson equation at least one solution exists. (We will not deal with the existence question here; see, e.g., the textbook [217] by D. Werner.)

Theorem 1.2. *Let functions* $\phi \in C(\partial\Omega)$ *and* $f \in C(\Omega)$ *be given. Let a solution* $u \in C^2(\Omega) \cap C(\overline{\Omega})$ *exist for the Dirichlet boundary value problem*

$$\Delta u = f \quad in\ \Omega, \qquad u = \phi \quad on\ \partial\Omega.$$

Then this solution is unique.

Proof. Let $u, v \in C^2(\Omega) \cap C(\overline{\Omega})$ be two solutions of the above PDE for the prescribed inhomogeneous Dirichlet boundary condition. Then $u - v$ is the solution of a Laplace equation with homogeneous Dirichlet boundary condition

$$(u - v)|_{\partial\Omega} = 0.$$

From the energy identity (A.9), applied to the difference $u - v$, we then get

$$\int_{\Omega} |\nabla(u - v)|^2 dx = 0 \quad \Rightarrow \quad u - v = \text{const}$$

and, with the above homogeneous Dirichlet boundary condition, we finally end up with $u - v \equiv 0$ on the whole domain Ω.   □

The uniqueness proof for Robin boundary conditions is postponed until Exercise 1.4. For Neumann conditions, however, uniqueness does not hold, as shown by the following theorem.

Theorem 1.3. *Let* $g \in C(\partial\Omega)$, $f \in C(\Omega)$. *Assume there exists a solution* $u \in C^2(\Omega) \cap C(\overline{\Omega})$ *of the Neumann boundary problem*

$$\Delta u = f \quad in \ \Omega, \qquad n^T \nabla u = g \quad on \ \partial\Omega.$$

Then there exists a one-parameter family of solutions $v = u + \text{const}$ *and the boundary function* g *satisfies the compatibility condition*

$$\int_{\partial\Omega} g \, ds = \int_{\Omega} f \, dx. \tag{1.3}$$

Proof. From $n^T \nabla(u - v) = 0$ and the energy identity (A.9) follows $u - v \equiv \text{const}$. This is the stated one-parameter family of solutions. The necessary condition (1.3) is a direct consequence of the Gauss integral theorem A.3. □

Pointwise Condition.   After the uniqueness of the solution, the condition of a problem must be studied, which means that we have to find out how perturbations of the input data of a problem affect perturbations of the results. This principle was elaborated in Volume 1 (cf., for example, the derivation in Section 2.2) and in Volume 2 for ODEs; this will serve as a leitmotif here as well. Beforehand we have to decide in which way (i.e., in which mathematical space, in norm) errors should be characterized.

In Appendix A.4 we mathematically describe point perturbations of the right-hand side f of the Poisson equation (1.2) as a Dirac needle-impulse, i.e., via a Dirac δ-distribution. As derived there, this naturally leads to the *Green's function*

$$G_P(x, x_0) := \psi(\|x - x_0\|)$$

with

$$\psi(r) = \begin{cases} \dfrac{\log r}{2\pi}, & d = 2, \\[2ex] -\dfrac{1}{\omega_d(d-2)r^{d-2}}, & d > 2. \end{cases} \tag{1.4}$$

For this function we get

$$\Delta_x G_P(x, x_0) = \delta(x - x_0),$$

where we wrote Δ_x for derivatives with respect to the variable x. On the basis of the general condition concept of Volume 1, Section 2.2, the function G represents the pointwise condition as *relation of pointwise result errors with respect to pointwise input errors* of the right-hand side f in the Poisson equation.

We can be slightly more precise here. Let us insert into Green's second integral theorem (A.8) for u the solution of the Poisson equation (1.2) and for v Green's function G_P. The left-hand side of (A.8) can then be reformulated as

$$\int_{\Omega} (u\Delta G_P - G_P \Delta u) \, dx = \int_{\Omega} (u\delta(x - x_0) - G_P f) \, dx = u(x_0) - \int_{\Omega} G_P f \, dx.$$

From this we obtain

$$u(x_0) = \int_\Omega G_P(x - x_0) f(x) dx + \int_{\partial\Omega} (u n^T \nabla G_P - G_P n^T \nabla u) ds.$$

Formally speaking, the effect of pointwise perturbations is global – each point in the domain Ω is affected by the perturbation. Quantitatively, however, the effect is merely *local*, since with increasing distance from x_0 it is rapidly damped, the higher the spatial dimension is, the faster. This property of Green's function is the basis for an efficient numerical solution of PDEs of this type, especially for the construction of *adaptive* spatial grids (cf. Chapter 6).

Maximum Principle. The Laplace operator has an astonishing property (see, e.g., the textbook of F. John [132]), which plays an important role in the numerical solution process (see Section 3.2.2).

Theorem 1.4. *Let $u \in C^2(\Omega) \cap C(\overline{\Omega})$ over some bounded domain $\Omega \subset \mathbb{R}^d$. Assume that*

$$\Delta u \geq 0 \quad in\ \Omega.$$

Then

$$\max_{\overline{\Omega}} u = \max_{\partial\Omega} u.$$

Proof. The proof is done in two steps.

(a) For the time being we assume that $\Delta u > 0$ in the domain Ω and that u has a maximum point $\xi \in \Omega$. Then

$$\frac{\partial^2 u}{\partial x_k^2}(\xi) \leq 0 \quad \text{for } k = 1, \dots, d \quad \Rightarrow \quad \Delta u(\xi) \leq 0,$$

which is an obvious contradiction to the assumption.

(b) Suppose therefore now that $\Delta u \geq 0$. Let us define the auxiliary function $v := \|x\|_2^2$, for which obviously $\Delta v = 2d > 0$ in Ω holds. For all $\epsilon > 0$ we have

$$u + \epsilon v \in C^2(\Omega) \cap C(\overline{\Omega})$$

and $\Delta(u + \epsilon v) > 0$, whereby we are back at step (I). Therefore

$$\max_{\overline{\Omega}} u \leq \max_{\overline{\Omega}} (u + \epsilon v) = \max_{\partial\Omega} (u + \epsilon v) \leq \max_{\partial\Omega} u + \epsilon \max_{\partial\Omega} v,$$

and with $\epsilon \to 0$ the statement of the theorem follows finally. $\square$

Needless to say that with reversed sign $\Delta u \leq 0$ one obtains a *minimum principle*. In numerous problems of science a sign condition $\Delta u \geq 0$ or $\Delta u \leq 0$ is satisfied. Note that the above assumption $u \in C^2(\Omega) \cap C(\overline{\Omega})$ also covers domains with reentrant corners.

1.1.2   Initial Value Problem

So far, we have only studied boundary value problems for the Laplace equation. Consider now an initial value problem in $\mathbb{R}^2$ of the following form:

$$\Delta u = u_{xx} + u_{tt} = 0, \qquad u(x,0) = \phi(x), \quad u_t(x,0) = \psi(x),$$

where we assume ϕ, ψ to be given *analytic* functions. Then there exists the theorem of Cauchy–Kowalewskaya (see, e.g., [127]), which proves the existence of a *unique analytic* solution $u(x,t)$.

Example 1.5. This example dates back to the French mathematician Jacques Hadamard (1865–1963). We assume that the given functions are slightly perturbed according to either

$$\delta u(x,0) = \delta\phi = \frac{\epsilon}{n^2}\cos(nx) \quad \text{or} \quad \delta u_t(x,0) = \delta\psi = \frac{\epsilon}{n}\cos(nx).$$

Obviously the two perturbations are "small" for large n, i.e.,

$$|\delta u(x,0)| \le \frac{\epsilon}{n^2}, \quad |\delta u_t(x,0)| \le \frac{\epsilon}{n}. \tag{1.5}$$

Due to the linearity of the Laplace equation we can easily calculate the corresponding perturbations δu of the solution u and obtain (because of the uniqueness of the solution simple insertion of the results below is sufficient) either

$$\delta u(x,t) = \frac{\epsilon}{n^2}\cos(nx)\cosh(nt) \quad \text{or} \quad \delta u(x,t) = \frac{\epsilon}{n^2}\cos(nx)\sinh(nt),$$

which implies

$$|\delta u(x,t)| \sim \frac{\epsilon}{n^2}\exp(nt) \tag{1.6}$$

in both cases. A comparison of (1.5) and (1.6) shows that

$$n \to \infty: \quad |\delta u(x,0)| \to 0, \qquad |\delta u(x,t)| \to \infty \quad \text{for } t > 0.$$

Thus, the perturbation of the solution depends *discontinuously* upon the perturbation of the input data.

Hadamard called such problems *ill-posed* and demanded that a solution – additionally to its existence and uniqueness – should depend continuously on the initial data. Hereby he went one step too far: Indeed there exists a large class of "ill-posed" problems from practice that are in some sense "reasonably posed", as we can learn from some more subtle analysis.

It is clear that a perturbation with "frequency" n is amplified by a factor

$$\kappa_n(t) \sim \exp(nt).$$

Recalling Volume 1, Section 2.2, we recognize this factor as the *condition number* of the Laplace initial value problem. The amplification is the worse, the larger the frequency is; it is even unbounded, if we allow for arbitrarily high frequencies in the perturbation. At the same time we observe that such problems can well be "reasonably" treated: We only have to *cut off perturbations of higher frequencies*. Mathematically speaking, we have to *suitably restrict the space of possible solutions*. In other words: We must *regularize the problem*. This may be illustrated by the following example from medicine.

Identification of Infarct Size in the Human Heart. The rhythmic motion of the heart muscle, alternating between contraction and expansion, is generated by some interplay of elastic and electrical processes (see also Section 9.3 below). In the case of an infarct, certain "electrically dead zones" arise on the heart muscle surface, which, however, cannot be directly measured. Instead, one measures the electrical potential u on the external surface of the chest and tries to calculate backward into the interior of the body [85]. The geometrical relations are depicted in Figure 1.1.

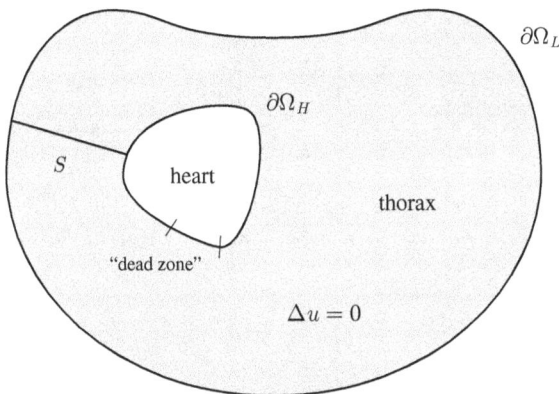

Figure 1.1. Cross section through a human chest: measurement of the potential at the surface, identification of the potential at the heart.

Let $\partial\Omega_H$ denote the surface of the heart muscle in the interior of the body and $\partial\Omega_L$ the boundary of the body to the air. In the domain in between, the Laplace equation holds (no free charges; cf. Section 2.1.1). As boundary at the air we have the Neumann condition

$$n^T \nabla u = 0 \quad \text{on } \partial\Omega_L.$$

In a "direct" problem we are given the potential at the heart, i.e.,

$$u|_{\partial\Omega_H} = V_H.$$

In the "inverse" problem, however, the potential on the chest has been measured, i.e.,

$$u|_{\partial\Omega_L} = V_L,$$

and the potential V_H on the heart muscle has to be found.

In the present case, the Laplace equation has to be solved on a doubly connected domain. That is why we cut the domain orthogonal to its boundary (see Figure 1.1): At the slit S we impose periodic boundary conditions. In Figure 1.2, left, we see that the direct problem is a Laplace boundary value problem, i.e., well-posed. On the right side, however, we perceive that the inverse problem is a Laplace initial value problem, i.e., "ill-posed". In order to be able to solve this problem nevertheless, we must "regularize" it, i.e., we must "cut off" higher frequencies in the initial data as well in the solution.

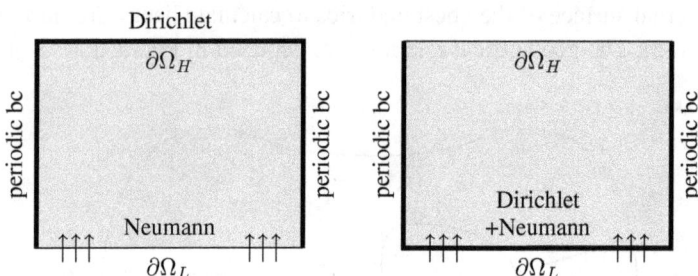

Figure 1.2. Unwind of the doubly connected domain from Figure 1.1 by means of periodic boundary conditions. *Left:* well-posed boundary value problem. *Right:* ill-posed initial value problem.

1.1.3   Eigenvalue Problem

The eigenvalue problem to the Laplace equation in $\mathbb{R}^d$ with homogeneous Dirichlet boundary conditions reads

$$-\Delta u = \lambda u \quad \text{in } \Omega \subset \mathbb{R}^d, \qquad u = 0 \quad \text{on } \partial\Omega. \tag{1.7}$$

The operator $-\Delta$ is self-adjoint and positive, its spectrum is discrete, real positive and has an accumulation point at infinity, i.e., one has

$$0 < \lambda_1 < \lambda_2 < \cdots \to \infty. \tag{1.8}$$

To each eigenvalue λ_k there exists an eigenfunction u_k, i.e.,

$$-\Delta u_k = \lambda_k u_k \quad \text{for } k = 1, 2, \ldots. \tag{1.9}$$

All eigenfunctions $\{u_k\}$, $k \in \mathbb{N}$, form an *orthogonal system*. If we introduce the inner product (L^2-product)

$$\langle u, v \rangle = \int_\Omega u(x)v(x)\,dx$$

and the Kronecker symbol

$$\delta_{jk} = \begin{cases} 1, & j = k, \\ 0, & j \neq k, \end{cases}$$

then, with the normalization

$$\langle u_j, u_k \rangle = \delta_{jk} \tag{1.10}$$

we arrive at an *orthonormal system*. Formally every solution of the Laplace equation can be represented by an expansion in terms of eigenfunctions

$$u(x) = \sum_k \alpha_k u_k(x). \tag{1.11}$$

Both the eigenvalues and the eigenfunctions depend on the boundaries and the domain Ω.

Model Problem.   Consider the unit square

$$\Omega = \,]0, 1[\, \times\,]0, 1[\, \subset \mathbb{R}^2.$$

For this domain, the eigenvalues and eigenfunctions to the Laplace equation with homogeneous Dirichlet boundary conditions are to be found. Due to the particularly simple form of the domain, we here can employ a *Fourier ansatz*

$$u(x, y) = \sum_{k,l=1}^{\infty} \alpha_{k,l} \sin(k\pi x) \sin(l\pi y), \tag{1.12}$$

which due to

$$u(0, y) = u(1, y) = u(x, 0) = u(x, 1) = 0$$

automatically satisfies the boundary conditions. We now differentiate

$$u_{xx} = \sum_{k,l=1}^{\infty} \alpha_{k,l}(-k^2\pi^2) \sin(k\pi x) \sin(l\pi y),$$

$$u_{yy} = \sum_{k,l=1}^{\infty} \alpha_{k,l}(-l^2\pi^2) \sin(k\pi x) \sin(l\pi y)$$

and insert these terms into (1.7) to get

$$\sum_{k,l=1}^{\infty} \alpha_{k,l}(k^2 + l^2)\pi^2 \sin(k\pi x) \sin(l\pi y) = \lambda \sum_{k,l=1}^{\infty} \alpha_{k,l} \sin(k\pi x) \sin(l\pi y).$$

Application of (1.10) punches the single summands as

$$(k^2 + l^2)\pi^2 \alpha_{k,l} = \lambda \alpha_{k,l}.$$

Whenever the coefficients $\alpha_{k,l}$ do not vanish, they cancel so that we obtain the eigenvalues as

$$\lambda_{k,l} = (k^2 + l^2)\pi^2, \quad k, l = 1, \ldots, \infty. \tag{1.13}$$

Obviously these special eigenvalues exactly mirror the properties (1.8). Physically speaking, the ansatz (1.12) may be interpreted as superposition of *standing plane waves*.

Rayleigh Quotient.   We now assume homogeneous Dirichlet or Neumann boundary conditions in general form as

$$\int_{\partial\Omega} u\, n^T \nabla u \, ds = 0.$$

Then application of the energy identity (A.9) supplies

$$\langle u, -\Delta u \rangle = \langle \nabla u, \nabla u \rangle > 0.$$

By insertion of an eigenfunction u_j for u and use of the definition (1.9) one obtains

$$\lambda_j = \frac{\langle \nabla u_j, \nabla u_j \rangle}{\langle u_j, u_j \rangle} > 0. \tag{1.14}$$

This formula is of restricted value, since hereby an eigenvalue can only be computed if the corresponding eigenfunction is already known – typically, however, the order is just reversed. But there is an interesting extension of this representation.

Theorem 1.6. *In the notation as introduced above we obtain for the smallest eigenvalue*

$$\lambda_1 = \min_u \frac{\langle \nabla u, \nabla u \rangle}{\langle u, u \rangle}. \tag{1.15}$$

The minimum is assumed for the corresponding eigenfunction $u = u_1$.

Proof. We start from representation (1.11) for arbitrary u. From this we conclude

$$\nabla u = \sum_k \alpha_k \nabla u_k,$$

from which by insertion we obtain the following double sum:

$$\langle \nabla u, \nabla u \rangle = \sum_{k,l} \alpha_k \alpha_l \langle \nabla u_k, \nabla u_l \rangle.$$

Application of the first integral theorem of Green (A.7) then supplies

$$\langle \nabla u, \nabla u \rangle = \sum_{k,l} \alpha_k \alpha_l \langle u_k, -\Delta u_l \rangle = \sum_{k,l} \alpha_k \alpha_l \langle u_k, \lambda_l u_l \rangle$$

and with the orthogonality relation (1.10) eventually

$$\langle \nabla u, \nabla u \rangle = \sum_k \alpha_k^2 \lambda_k, \qquad \langle u, u \rangle = \sum_k \alpha_k^2.$$

As λ_1 is the smallest eigenvalue, we may exploit $\lambda_k \geq \lambda_1$ and thus get

$$\frac{\langle \nabla u, \nabla u \rangle}{\langle u, u \rangle} = \frac{\sum_k \alpha_k^2 \lambda_k}{\sum_k \alpha_k^2} \geq \lambda_1.$$

After insertion of the first eigenfunction u_1 for u and with (1.14) the theorem is proved.   □

The quotient on the right-hand side of (1.15) is called the *Rayleigh quotient*. There are useful extensions to further lower eigenvalues $\lambda_2, \ldots, \lambda_j$.

Corollary 1.1. *Let $U_{j-1}^\perp$ denote the subspace orthogonal to the invariant subspace $U_{j-1} = \mathrm{span}\{u_1, \ldots, u_{j-1}\}$. Then*

$$\lambda_j = \min_{u \in U_{j-1}^\perp} \frac{\langle \nabla u, \nabla u \rangle}{\langle u, u \rangle}. \tag{1.16}$$

Again the minimum is assumed for the corresponding eigenfunction, here $u = u_j$.

Formula (1.15) is constructive, as it allows for an approximation of the eigenvalue via not prespecified functions u (without normalization). The same holds for formula (1.16) in a restricted sense, if one manages to get computational hold of the nested subspaces $U_1^\perp \supset \cdots \supset U_j^\perp$. Both formulas will turn out to be extremely useful in the efficient construction of *numerical* approximations of eigenvalues and eigenfunctions (see Sections 4.1.3, 5.3.4 and 7.5.2).

1.2 Diffusion Equation

The simplest time-dependent extension of the Laplace equation is the diffusion or heat equation

$$u_t(x,t) = \sigma \Delta u(x,t), \quad (x,t) \in Z, \quad \sigma > 0 \tag{1.17}$$

over the space-time cylinder $Z = \Omega \times]0, T[\subset \mathbb{R}^d \times \mathbb{R}$. In the Laplace operator only spatial derivatives occur. For the time being, let us consider only Dirichlet boundary conditions

$$u(x,t) = g(x,t), \quad x \in \partial\Omega, \quad t \in]0, T[, \tag{1.18}$$

as well as the initial conditions

$$u(x, 0) = \phi(x), \quad x \in \Omega. \tag{1.19}$$

For a unique solvability, certain additional conditions have to be satisfied. In order to be able to glue boundary and initial conditions continuously together, we need consistency conditions of the form

$$g(x, 0) = \phi(x), \quad \text{for } x \in \partial\Omega.$$

In lieu of the Dirichlet conditions (1.18) we could just as well impose Neumann or Robin boundary conditions on $\partial\Omega \times {]0, T[}$.

Fourier Expansion for Initial Boundary Value Problems.   In order to gain a first insight into the solution structure, we look at a rather simple one-dimensional example. Let $\Omega = {]0, \pi[}$ and $g(x, t) = 0$ for all $t > 0$. The the solution and the initial values can be expanded into Fourier series:

$$u(x, t) = \sum_{k \in \mathbb{Z}} A_k(t) e^{ikx}, \quad \phi(x) = \sum_{k \in \mathbb{Z}} B_k e^{ikx}. \tag{1.20}$$

The fact that the eigenfunctions of the Laplace operator on intervals are just the trigonometric functions is the deeper reason for the success of Fourier analysis in the treatment of the diffusion equation. We now insert the above solution representation into the diffusion equation (1.17) and thus obtain

$$0 = u_t - \sigma u_{xx} = \sum_{k \in \mathbb{Z}} A_k'(t) e^{ikx} - \sigma \sum_{k \in \mathbb{Z}} A_k(t)(-k^2) e^{ikx}$$

$$= \sum_{k \in \mathbb{Z}} \left(A_k'(t) + \sigma k^2 A_k(t) \right) e^{ikx}$$

for all x and t. The Fourier coefficients are unique and therefore satisfy the ordinary differential equations

$$A_k'(t) = -\sigma k^2 A_k(t).$$

Again due to the uniqueness of the Fourier coefficients the relation (1.20) for $t = 0$ implies $A_k(0) = B_k$ for the initial values. This leads to

$$A_k(t) = e^{-\sigma k^2 t} B_k.$$

Upon collecting all intermediate results we obtain

$$u(x, t) = \sum_{k \in \mathbb{Z}} B_k e^{-\sigma k^2 t} e^{ikx}.$$

This representation may be interpreted as follows. The solution u is a superposition of "standing waves" that are damped in time ($\sim k^2$), which means the more, the higher the spatial frequency k is. Each of these "damped standing waves" constitutes a characteristic spatio-temporal solution pattern denoted by the term *"mode"*. With σk^2 as *damping factor* the characteristic time $\tau = (\sigma k^2)^{-1}$ is often called the *relaxation time* of the corresponding mode. The long-term behavior $u(x, t) \to 0$ for $t \to \infty$ is called *asymptotic stability*.

Remark 1.7. Even for discontinuous initial and boundary values, a solution u exists that is arbitrarily often differentiable on each compact subset $K \subset Z$. The existence proof is done by means of the coordinate transformation (Duhamel transformation) $\xi = x/\eta, \eta = \sqrt{\sigma t}$ and a subsequent limit process for $t \to 0$.

Pointwise Condition. Just as in the (stationary) Poisson equation, we are interested in the effect of pointwise perturbations of the initial data on the solution of the (instationary) diffusion equation. Input data here will be the initial values $\phi(x)$. The corresponding Green's function for spatial dimension d is (see Appendix A.4)

$$G_D(x, t) = \frac{e^{-x^2/(4\sigma t)}}{\sqrt{2\pi}(2\sigma t)^{d/2}}.$$

For a deeper understanding we take the defining equation (A.14) from A.4:

$$(G_D)_t = \sigma \Delta G_D \quad \text{in } \mathbb{R}^d \times \mathbb{R}_+, \qquad G_D(x) = \delta(x) \quad \text{for } t = 0.$$

From this we may read off that G_D is just the "response" to initial perturbations $\delta\phi = \delta(x - x_s)$ at point x_s. As in elliptic problems the effect of the perturbation decreases fast with increasing spatial and temporal distance, but formally covers the whole domain for arbitrarily small time $t > 0$. With respect to time, the perturbations of the boundary values in point (x_s, t_s) turn out to be asymmetric: for $t < t_s$ the solution remains unchanged, while for $t > t_s$ it is affected in the whole domain.

In Figure 1.3, we illustrate the *domain of influence* as the set of all points in which the solution is influenced by a point perturbation in (x, t). Vice versa, the *domain of dependence* is that subset of points on the boundary, from which initial and boundary

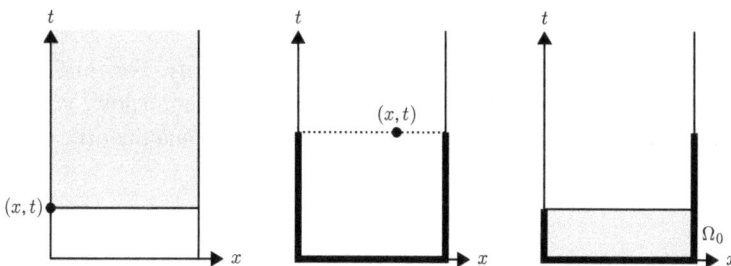

Figure 1.3. Parabolic initial boundary value problem. *Left:* domain of influence (gray area). *Center:* domain of dependence (bold line). *Right:* domain of determinacy (gray area).

data influence the solution at the point (x, t). Finally, the *domain of determinacy* comprises all points in which the solution is uniquely determined by boundary and initial data on $\Omega_0 \subset \partial Z$.

Final Boundary Value Problem.   In this problem type the diffusion equation (1.17) and the boundary values (1.18) are combined with final values instead of initial values (1.19):

$$u(x, T) = \phi(x), \quad x \in \Omega.$$

Upon reversal of the time variable by virtue of $\tau = T - t$, this problem can again be transferred into an initial boundary value problem, whereby the PDE is transformed into

$$u_\tau = -u_t = -\sigma \Delta u.$$

Thus, a *final* boundary value problem corresponds to an *initial* boundary value problem with *negative diffusion coefficient*. Again, for illustration purposes, let $\Omega = \,]0, \pi[$. We again apply the same Fourier ansatz (1.20) as above and this time obtain the solution

$$u(x, t) = \sum_{k \in \mathbb{Z}} e^{\sigma k^2 t} e^{ikx}.$$

Obviously, now the eigenmodes are not damped, but *grow* exponentially. Like the elliptic initial value problem from Section 1.1.2 the parabolic final boundary value problem is ill-posed: small highly frequent spatial perturbations at the final time increase arbitrarily fast thus making the solution discontinuously dependent on the final data. Here, too, a number of practically relevant identification problems lead to such final boundary value problems, which have to be treated by suitable *regularization*.

1.3   Wave Equation

The (linear) wave equation reads

$$u_{tt} - c^2 \Delta u = 0, \tag{1.21}$$

where the constant c denotes the *wave propagation velocity*. For this type of PDE there are various initial value problems, which we will present below. We restrict our attention to the simplest case with one spatial coordinate x and the time coordinate t:

$$u_{tt} - c^2 u_{xx} = 0. \tag{1.22}$$

At first glance, this PDE just looks like it was obtained by a mere sign change, different than the Laplace equation in $\mathbb{R}^2$. However, as we will show, this sign change implies a really fundamental change.

Initial Value Problem (Cauchy Problem). For this problem, initial values of the function and its time derivative are prescribed:

$$u(x,0) = \phi(x), \quad u_t(x,0) = \psi(x).$$

This initial value problem can be solved in closed analytic form. For this purpose, we apply the classical variable transformation dating back to the French mathematician and physicist Jean Baptiste le Rond d'Alembert (1717–1783),

$$\xi = x + ct, \quad \eta = x - ct,$$

with the reverse transformation

$$x = \frac{1}{2}(\xi + \eta), \quad t = \frac{1}{2c}(\xi - \eta).$$

This defines the transformed solution according to

$$u\big(x(\xi,\eta), t(\xi,\eta)\big) = w(\xi,\eta).$$

For simplicity we choose $c = 1$ in our intermediate calculations. The partial derivatives transform as follows:

$$u_x = w_\xi \xi_x + w_\eta \eta_x = w_\xi + w_\eta,$$
$$u_{xx} = w_{\xi\xi}\xi_x + w_{\xi\eta}\eta_x + w_{\eta\xi}\xi_x + w_{\eta\eta}\eta_x = w_{\xi\xi} + 2w_{\xi\eta} + w_{\eta\eta},$$
$$u_t = w_\xi - w_\eta,$$
$$u_{tt} = w_{\xi\xi} - 2w_{\xi\eta} + w_{\eta\eta}.$$

Under the assumption that the mixed derivatives commute, i.e., $w_{\xi\eta} = w_{\eta\xi}$, we are then led to the following PDE for w:

$$4w_{\xi\eta} = u_{xx} - u_{tt} = 0.$$

To solve it, we first integrated w.r.t. η and obtain $w_\xi = \text{const}(\xi)$, then we integrate w.r.t. ξ, which leads us to $w = \alpha(\xi) + \text{const}(\eta)$. Combining the two results, we obtain

$$w(\xi,\eta) = \alpha(\xi) + \beta(\eta) \quad \Rightarrow \quad u(x,t) = \alpha(x+t) + \beta(x-t).$$

Now, comparison with the initial values supplies

$$u(x,0) = \alpha(x) + \beta(x) = \phi(x), \quad u_t(x,0) = \alpha'(x) - \beta'(x) = \psi(x).$$

The second equation can be integrated via

$$\alpha(x) - \beta(x) = \underbrace{\int^x \psi(s)\, ds + \text{const}\,.}_{\Psi(x)} \tag{1.23}$$

From the first equation above and (1.23) one thus arrives at

$$\alpha(x) = \frac{1}{2}\big(\phi(x) + \Psi(x) + \text{const}\big) \quad \text{and} \quad \beta(x) = \frac{1}{2}\big(\phi(x) - \Psi(x) - \text{const}\big),$$

and eventually at the solution (setting again $c \neq 1$)

$$\begin{aligned}
u(x,t) &= \frac{1}{2}\big(\phi(x+ct) + \phi(x-ct)\big) + \frac{1}{2c}\big(\Psi(x+ct) - \Psi(x-ct)\big) \\
&= \frac{1}{2}\big(\phi(x+ct) + \phi(x-ct)\big) + \frac{1}{2c}\int_{x-ct}^{x+ct} \psi(s)\,ds.
\end{aligned} \tag{1.24}$$

Obviously, the family of straight lines

$$x + ct = \text{const}, \quad x - ct = \text{const}$$

plays a central role here. They are called *characteristics*, the corresponding coordinates ξ and η are the *characteristic coordinates*.

From the analytic representation above, we can directly read off the domains of influence, of dependence, and of determinacy, as shown in Figure 1.4.

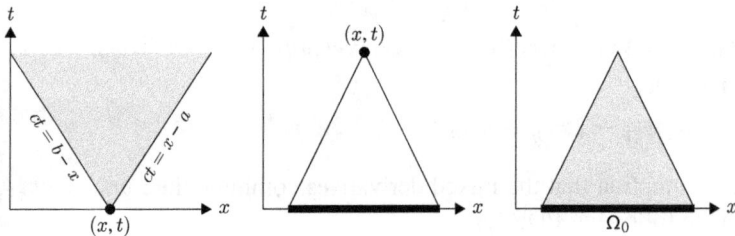

Figure 1.4. Initial value problem for the wave equation (Cauchy problem). *Left:* domain of influence (gray area). *Center:* domain of dependence (bold line). *Right:* domain of determinacy (gray area).

Condition.   The effect of initial perturbations $\delta\phi, \delta\psi$ on perturbations of the result δu can also be directly seen from (1.24). If we define the initial perturbations in the mathematical spaces

$$\delta\phi(x) \in C[a,b], \quad \delta\psi \in L_1(\mathbb{R}),$$

then we obtain

$$\delta u(x,t) \in C[a,b], \quad t > 0.$$

Clearly, this problem type is *well-posed*. Even discontinuous perturbations $\delta\phi$ and δ-distributions $\delta\psi$ would propagate along the characteristics.

Characteristic Initial Value Problem (Riemann Problem). This problem is defined by prescription of the initial values (but not time derivative)

$$u(x, 0) = \phi(x), \quad x \in [a, b],$$

and also boundary values along a characteristic (see Figure 1.5 for an illustration). This, too, leads to a well-posed initial value problem.

For a closed representation of the solution, we may again apply the d'Alembert ansatz, which, however, we have placed in Exercise 1.7. Like in the Cauchy problems a solution only exists over a restricted domain of dependence (see also Figure 1.5).

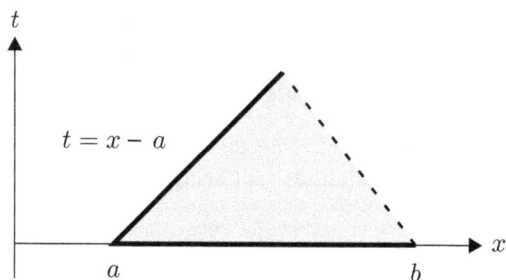

Figure 1.5. Riemann problem for the wave equation: domain of dependence ($c = 1$).

Initial Boundary Value Problem. Here, as in the Cauchy problem, initial values for both the function and its derivative are given:

$$u(x, 0) = \phi(x), \quad u_t(x, 0) = \psi(x), \qquad x \in [a, b].$$

However, additional *boundary conditions* are prescribed:

$$u(a, t) = g_a(t), \quad u(b, t) = g_b(t).$$

To assure continuity of initial and boundary values, certain *compatibility conditions* must hold:

$$\phi(a) = g_a(0), \quad \phi(b) = g_b(0).$$

In contrast to the other initial value problems (Cauchy, Riemann) the solution here is defined over an infinite stripe (domain of influence), which is depicted in Figure 1.6. The solution on this stripe can be successively constructed, starting by the Cauchy problem with triangular domain of determination over the interval $[a, b]$ via Riemann problems on the two neighboring triangular domains and so on, thus tiling the whole stripe.

Here, however, we want to derive an alternate global representation, which seems to be more appropriate for this type of problem. We proceed as in the case of the diffusion equation (Section 1.2), which is also defined over some stripe.

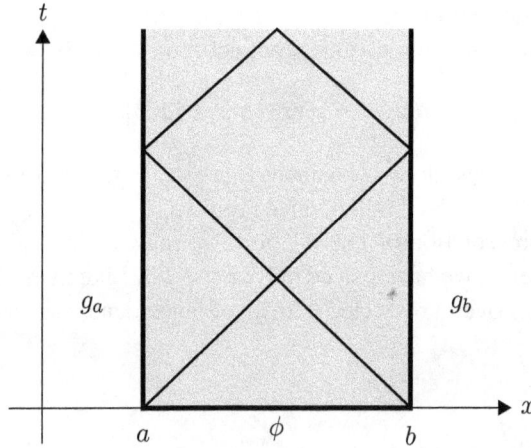

Figure 1.6. Initial boundary value problem for the wave equation: domain of dependence as infinite stripe. A possible construction leads via Cauchy and Riemann problems to a tiling of the whole stripe.

Fourier Representation.   As in Section 1.2, we start with a separation ansatz (which we will later verify on the basis of the uniqueness result originating from the above tiling construction):

$$u(x,t) = \sum_{k\in\mathbb{Z}} \alpha_k(t) u_k(x). \tag{1.25}$$

As a basis we select, in view of the general form (1.21), the eigenfunctions of the Laplace operator. For the 1D wave equation (1.22) the eigenvalues are the same as those of the 1D Laplace operator, $\lambda_k = k^2 > 0$ from (1.13), while the corresponding eigenfunctions (in complex notation) are

$$u_k = c_k e^{ikx} + d_k e^{-ikx}.$$

Insertion of the general representation (1.25) into (1.22) yields

$$u_{tt} = \sum_{k\in\mathbb{Z}} \alpha_k'' u_k = c^2 u_{xx} = -\sum_{k\in\mathbb{Z}} c^2 k^2 \alpha_k u_k \quad \Rightarrow \quad \sum_{k\in\mathbb{Z}} [\alpha_k'' + c^2 k^2 \alpha_k] u_k = 0.$$

From the orthogonality of the u_k then follows

$$\alpha_k'' = -c^2 k^2 \alpha_k \quad \Rightarrow \quad \alpha_k(t) = a_k e^{ikct} + b_k e^{-ikct}.$$

Summarizing we obtain an expansion of the form

$$u(x,t) = \sum_{k\in\mathbb{Z}} (a_k e^{ikct} + b_k e^{-ikct})(c_k e^{ikx} + d_k e^{-ikx}). \tag{1.26}$$

Here again the characteristics appear, this time as a special property of the trigonometric functions. The products

$$e^{\pm ik(x \pm ct)}$$

appearing in the representation (1.26) may be physically interpreted as *undamped
plane waves* with propagation velocity c.

1.4 Schrödinger Equation

The partial differential equation

$$i\hbar \Psi_t = H\Psi \tag{1.27}$$

with initial conditions

$$\Psi(x, 0) = \Psi_0(x),$$

named after the Viennese physicist Erwin Schrödinger (1887–1961), is the fundamental equation of physical *quantum mechanics*. Herein Ψ is the so-called *wave function*,
H the (self-adjoint) *Hamilton operator*, h the Planck constant and $\hbar = h/(2\pi)$ a usual
quantity. For a so-called *free particle* of mass m the Schrödinger equation reduces to

$$i\hbar \Psi_t = -\frac{1}{2m} \Delta \Psi.$$

Upon suitable rescaling of all constants, we obtain the model equation

$$iu_t = -\Delta u, \quad u(x, 0) = \phi. \tag{1.28}$$

for the dimensionless quantity u. It looks very much like the diffusion equation (1.17)
for $\sigma = 1$, but with the imaginary unit i as distinguishing feature. As there, we are
tempted to try a Fourier ansatz for the solution. However, the Schrödinger equation
is generally defined on an unbounded domain, mostly on the whole of $\mathbb{R}^d$. This corresponds to the fact that boundary conditions on any finite domain will not occur.
Accordingly, an expansion with eigenmodes of the form

$$u(x, t) = \sum_{k=-\infty}^{\infty} \alpha_k(t) e^{ikx}$$

would usually not work.

As an alternative mathematical tool in lieu of the Fourier analysis, the *Fourier transform* (for the analysis of a *continuous spectrum* cf. Appendix A.1) can be employed:

$$u(x, t) \to \hat{u}(k, t) = \frac{1}{\sqrt{2\pi}} \int_{\mathbb{R}^d} u(x, t) e^{-ik^T x} \, dx.$$

The corresponding synthesis step in backward direction is done by means of the *inverse Fourier transform*

$$u(x,t) = \frac{1}{\sqrt{2\pi}} \int_{\mathbb{R}^d} \hat{u}(k,t) e^{ik^T x}\, dk.$$

The associated ansatz for the Schrödinger equation then reads

$$iu_t + \Delta u = \int_{\mathbb{R}^d} e^{ik^T x}(i\hat{u}_t - |k|^2 \hat{u})\, dk = 0.$$

If we define the Fourier transform for the initial values as

$$\phi(x) \to \hat{\phi}(k) = \frac{1}{\sqrt{2\pi}} \int_{\mathbb{R}^d} \phi(x) e^{-ik^T x}\, dx,$$

then the backtransformation has the form

$$\hat{u}(k,t) = \hat{\phi}(k) e^{i|k|^2 t}.$$

From this we obtain the representation of the time dependent solution as

$$u(x,t) = \frac{1}{\sqrt{2\pi}} \int_{\mathbb{R}^d} \hat{\phi}(k) e^{i(k^T x + |k|^2 t)}\, dk.$$

This integral representation permits the following interpretation: The solution is a continuous superposition of undamped waves whose frequencies in space (wave number k) and in time ($\omega = |k|^2$) differ distinctly, as opposed to the wave equation of the previous section; the dependence on k is called *dispersion*. Note, however, that here we do not have isolated waves as "modes", since the superposition is continuous, but rather "wave packets". Moreover, in the sense of quantum mechanics, the quantities of interest are not the functions u themselves, but quantities derived from it, like "probabilities" $\|u\|^2_{L^2(\Omega)}$ to be in some subset $\Omega \subset \mathbb{R}^d$. To illustrate this, we give a simple example.

Diffluence of Wave Packets.   For the initial value we choose a Gaussian distribution in space:

$$\phi(x) = A e^{-\frac{|x|^2}{2\sigma^2}} e^{ik_0^T x}.$$

In the sense of quantum mechanics this describes the spatial probability density function at time $t = 0$ in the form of a Gaussian distribution

$$|\phi(x)|^2 = A^2 e^{-\frac{|x|^2}{\sigma^2}}.$$

By Fourier transform we are led to the solution (intermediate calculation dropped here)

$$|u(x,t)|^2 = \frac{\sigma^2}{\alpha(t)^2} A^2 \exp\left(-\frac{|x - k_0 t|^2}{\alpha(t)^2}\right) \tag{1.29}$$

with time dependent "standard variation"

$$\alpha(t) = \sqrt{\sigma^2 + t}.$$

Formula (1.29) may be interpreted as the time dependent probability density function of the wave packet. Due to $\alpha(t) \to \infty$ for $t \to \infty$ *the wave packet melts away in time.* A graphical representation is given in Figure 1.7. Any reasonable numerical approximation will have to take this effect into account.

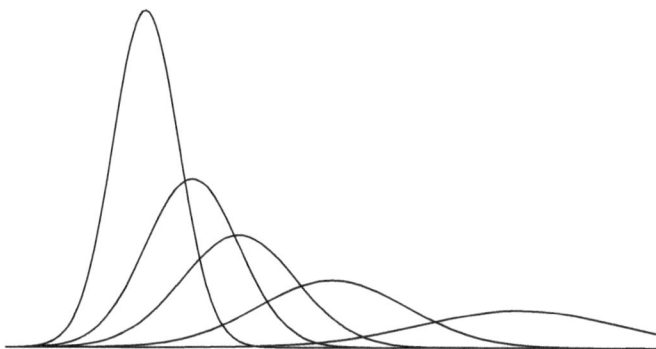

Figure 1.7. Probability density function of a "free" particle: diffluence of a wave packet for increasing time t (see (1.29)).

Uniqueness.   For the Schrödinger equation, there exists, also on unbounded domains, a formal solution

$$u(x,t) = e^{-i\Delta t}\phi(x),$$

where the notation $e^{-i\Delta t}$ stands for "semigroup", which is the resolvent of the simple model equation (1.28).

Condition.   Due to the interpretation of the wave function as probability density function we have the normalization $\|u\|^2_{L^2(\mathbb{R}^d)} = 1$ with the L^2-norm as the canonical choice. With the input ϕ and its perturbation $\delta\phi$ one has

$$\|\delta u(x,t)\|_{L^2(\mathbb{R}^d)} = \|e^{-i\Delta t}\phi - e^{-i\Delta t}(\phi + \delta\phi)\|_{L^2(\mathbb{R}^d)}$$

$$= \|e^{-i\Delta t}\delta\phi\|_{L^2(\mathbb{R}^d)} = \|\delta\phi\|_{L^2(\mathbb{R}^d)},$$

where we have used that the operator $e^{-i\Delta t}$ is unitary. The condition w.r.t the L^2-norm is therefore 1. Note that the *phase error* for the function u does not enter into this consideration. However, in quantum mechanical calculations for averages of observables, such as $\langle \psi, H\psi \rangle$ for the energy, this error is crucial.

1.5   Helmholtz Equation

This type of partial differential equation has the form

$$\Delta u + k^2 u = f, \tag{1.30}$$

in which k denotes the *wave number* and u may be complex valued. A positive sign in front of the k^2-term characterizes the *critical* case (which is solely of interest here), a negative sign the *uncritical* case. Formally, equation (1.30) looks similar to the Poisson equation (1.1). However, part of the solutions of the Helmholtz equation are completely different from those of the Poisson equation. A first insight into the different structure is given by the eigenvalue problem for the Helmholtz equation with Dirichlet boundary conditions:

$$\Delta u + k^2 u + \lambda u = 0.$$

These are obviously the same eigenfunctions as the ones for the Laplace eigenvalue problem (1.7), though with a spectrum shifted by k^2 so that

$$-k^2 < \lambda_1 < \cdots < \lambda_m \leq 0 < \lambda_{m+1} < \cdots \tag{1.31}$$

for some k^2 independent of m. The m negative eigenvalues are called *critical* eigenvalues, since they exist only in the critical case.

1.5.1   Boundary Value Problems

Let us first consider boundary value problems. As in the Poisson equation one can impose boundary conditions of Dirichlet, Neumann, or Robin type. For direct comparison with the Poisson equation, we restrict our attention to homogeneous Dirichlet boundary conditions, whereby the complex Helmholtz equation (1.30) divides into two independent real equations for the real and the imaginary part.

Existence and Uniqueness.   Suppose we select in (1.11) for both the solution of the Helmholtz equation and the source term an ansatz of the kind

$$u = \sum_{n \in \mathbb{N}} \alpha_n u_n, \quad f = \sum_{n \in \mathbb{N}} \beta_n u_n \tag{1.32}$$

in terms of eigenfunctions u_n of the Laplace operator. Then we obtain from (1.30) the equations

$$(-\lambda_n + k^2)\alpha_n = \beta_n, \quad n \in \mathbb{N}. \tag{1.33}$$

Obviously, a solution will only exist, if for each n with $\lambda_n = k^2$ also $\beta_n = 0$ holds. Similarly as with pure Neumann problems of the Poisson problem, the existence of a solution depends on the compatibility of the source term f, but additionally also on the value of the wave number k. Whenever an eigenvalue $\lambda_n = k^2$ occurs, then the solution cannot be unique, as in the pure Neumann problem for the Poisson equation.

Condition. As earlier, the pointwise dependence of the solution on local perturbations is described by the Green's function. For the Helmholtz equation this function has the form (see Appendix A.4):

$$G_H(x) = \begin{cases} \sin(k|x|)/(2k), & d = 1, \\ Y_0(k\|x\|)/4, & d = 2, \\ -\cos(k\|x\|)/(4\pi\|x\|), & d = 3. \end{cases}$$

With the exception of the case $d = 1$, this Green's function exhibits the same asymptotic decay behavior for $\|x\| \to \infty$ as the Green's function (1.4) of the Poisson equation, so the pointwise condition $|G_H(x)|$ is essentially the same. If one is not only interested in its absolute value, but also in its derivative $|\nabla G_H(x)|$, a significant difference between Poisson and Helmholtz equation appears. While for the Poisson case (1.4) the gradient of the Green's function decays quickly, this is not true for the Helmholtz case:

$$|\nabla G_P(x)| = \frac{1}{4\pi\|x\|^2}, \quad |\nabla G_H(x)| = \left|\frac{k\|x\|\sin k\|x\| - \cos k\|x\|}{4\pi\|x\|^2}\right|.$$

In particular, the derivative of the Green's function increases with wave number k.

For a better understanding of the practically relevant differences between Poisson and Helmholtz equations, we consider two further error concepts: the dependence of the solution on the *position* of a pointwise perturbation and that on the *wave number* k. If we move the perturbation $\delta(x)$ of the right-hand side by ξ to $\delta(x - \xi)$, then the perturbation of the solution is moved exactly to $G_H(x - \xi)$. In first order, the change of the perturbation is characterized by $\xi^T \nabla G_H(x)$. Here the importance of the difference in the gradients of G_P and G_H shows up. In particular, the oscillatory structure of G_H will rapidly lead to a *phase error*: A move by $\|\xi\| = \pi/k$ already leads to a complete reversal of the perturbation. A perturbation of the wave number k does not only influence the perturbation of the right-hand side, but also the solution directly. From equation (1.33) for the eigenfunction expansion we recognize that for $k^2 \approx \lambda_n$ the solution strongly depends on β_n and k. The condition w.r.t. a perturbation of the u_n-component of the right-hand side is $|-\lambda_n + k^2|^{-1}$, whereas w.r.t. k it is $2k|\beta_n|(-\lambda_n + k^2)^{-2}$.

1.5.2   Time-harmonic Differential Equations

The Helmholtz equation (1.30) arises naturally from a consideration of time-dependent PDEs in the case of *periodic excitations* either in the boundary values or in the source terms. For linear PDEs *time-harmonic* excitations, i.e., those containing only a single frequency ω, will immediately lead to time-harmonic solutions, which then can be expanded into some Fourier series. In this context we look for real valued solutions with the structure

$$\hat{u}(x,t) = \Re(e^{i\omega t} u(x)) = \frac{1}{2}(e^{i\omega t} u(x) + e^{-i\omega t}\overline{u(x)}), \quad \omega \geq 0, \tag{1.34}$$

wherein u may be complex valued.

Time-harmonic Wave Equation.   As a first example we insert (1.34) into a wave equation with periodic source term

$$\hat{u}_{tt} = c^2 \Delta \hat{u} - \hat{f} \quad \text{with} \quad \hat{f}(x,t) = \Re \left(e^{i\omega t} f(x) \right), \tag{1.35}$$

and obtain

$$\omega^2 (e^{i\omega t} u + e^{-i\omega t}\overline{u}) + c^2 \Delta (e^{i\omega t} u + e^{-i\omega t}\overline{u}) = e^{i\omega t} f + e^{-i\omega t}\overline{f}.$$

This must hold for all $t \in \mathbb{R}$ so that, after separation into real and imaginary part, we are led to the equation

$$\omega^2 u + c^2 \Delta u = f,$$

which is a Helmholtz equation (1.30) with wave number $k = \omega/c > 0$.

  Resonance. Let us recall (1.33) in connection with a Fourier expansion of the solution and the right-hand side. If at least one eigenvalue $\lambda_n = k^2$ exists and $\beta_n = 0$, then the coefficient α_n of the corresponding eigenfunction is not uniquely determined. In this case, the time-dependent original system (1.35) can oscillate permanently with arbitrary amplitude even without any excitation. This case is called *resonance*. If $\lambda_n = k^2$, but $\beta_n \neq 0$, then a periodic solution of the form (1.34) will not exist, but an unboundedly increasing linear solution of (1.35).

Time-harmonic Diffusion Equation.   The real time-harmonic ansatz (1.34) for the inhomogeneous diffusion equation $\hat{u}_t = \Delta \hat{u} - \hat{f}$ yields

$$i\omega(e^{i\omega t} u - e^{i\omega t}\overline{u}) = e^{i\omega t}(\Delta u - f) + e^{-i\omega t}(\Delta \overline{u} - \overline{f}) \quad \text{for all } t \in \mathbb{R},$$

which gives rise to a complex Helmholtz equation for u of the form

$$\Delta u - i\omega u = f. \tag{1.36}$$

Here, in contrast to the time-harmonic wave equation, real and imaginary parts are coupled through the equation itself, not only through the boundary conditions. Compared to the Laplace operator, the eigenvalues are no longer real, but instead are shifted in the direction of the imaginary axis. If we represent the homogeneous Dirichlet problem as before in (1.32), we are led to algebraic equations for the coefficients:

$$(-\lambda_n - i\omega)\alpha_n = \beta_n, \quad n \in \mathbb{N}.$$

Due to $|\lambda_n + i\omega| > 0$ for all n, existence as well as uniqueness are guaranteed, independent of $\omega \geq 0$.

Condition. The condition w.r.t. perturbations of the coefficients β_n of the right-hand side is

$$|\lambda_n + i\omega|^{-1} = \frac{1}{\sqrt{\lambda_n^2 + \omega^2}},$$

which means that both spatially (λ_n large) and temporally (ω large) high-frequency excitations are strongly damped. The complex Helmholtz equation of the type (1.36) is thus closer to the Poisson equation than to the real Helmholtz equation (1.30). Nevertheless the solutions are weakly oscillatory, as can be seen from the Green's function

$$G_H(x) = \begin{cases} \frac{1}{k}\left(\sin(k\,\|x\|) + i\cos(k\,\|x\|)\right), & d = 1, \\ \frac{1}{4}\left(Y_0(k\,\|x\|) + iJ_0(k\,\|x\|)\right), & d = 2, \\ -\frac{1}{4\pi\|x\|}\left(\cos(k\,\|x\|) - i\sin(k\,\|x\|)\right), & d = 3. \end{cases}$$

For the derivation see Appendix A.4.

Connection with the Schrödinger Equation. Different from the wave and the diffusion equation we do not expect real time-harmonic solutions for the Schrödinger equation (1.28). The complex ansatz

$$\hat{u}(x,t) = e^{-i\omega t} u(x), \quad \omega \in \mathbb{R}$$

leads to

$$\omega e^{-i\omega t} u = -e^{-i\omega t} \Delta u$$

for all t, which implies that u satisfies the equation

$$\Delta u + \omega u = 0.$$

For $\omega > 0$ we obtain the Helmholtz equation (1.30) with wave number $k = \sqrt{\omega}$, which we already know from the wave equation. For $\omega \leq 0$, however, we obtain a solution behavior similar to the one for the Poisson equation: The eigenvalues of the homogeneous Dirichlet problem are real and negative, existence and uniqueness are guaranteed independent of $\omega \leq 0$, both spatially and temporally high-frequency excitations are strongly damped. The Green's function is nonoscillatory similar to the one for the Poisson problem.

1.6 Classification

Up to now we have merely presented historical examples of PDEs of second order, i.e., of PDEs with derivatives of up to second order. Here we are going to study the general type (exemplified in $\mathbb{R}^2$ with Euclidean coordinates x, y)

$$L[u] := au_{xx} + 2bu_{xy} + cu_{yy} = f$$

in more detail. Let the coefficients a, b, c only depend on the coordinates (x, y), not on the solution u. At the start, the choice of coordinates is somewhat arbitrary, which is why we may perform an arbitrary bijective C^2-transformation of the coordinates

$$\xi = \xi(x, y), \quad \eta = \eta(x, y), \quad \xi_x \eta_y - \xi_y \eta_x \neq 0.$$

After some intermediate calculation we are led to the transformed equation

$$\Lambda[u] := \alpha u_{\xi\xi} + 2\beta u_{\xi\eta} + \gamma u_{\eta\eta} = \varphi,$$

where

$$\alpha = a\xi_x^2 + 2b\xi_x\xi_y + c\xi_y^2,$$
$$\gamma = a\eta_x^2 + 2b\eta_x\eta_y + c\eta_y^2,$$
$$\beta = a\xi_x\eta_x + b(\xi_x\eta_y + \xi_y\eta_x) + c\xi_y\eta_y.$$

Here the coefficients a, b, c depend on the coordinates $(x(\xi, \eta), y(\xi, \eta))$ as obtained from back-transformation, all other differentiation terms (of lower order) and f are included in φ.

The differential operator L transforms to Λ just as a quadratic form. The corresponding discriminants satisfy the relation

$$\beta^2 - \alpha\gamma = (b^2 - ac)(\xi_x\eta_y - \xi_y\eta_x)^2.$$

Obviously, the *sign of the discriminants* is preserved under coordinate transformation, which means it is an *invariant*. From the classification of quadratic forms one thus obtains the following *canonical forms* for partial differential equations of second order:

1. *Elliptic* class: $\Lambda[u] = \alpha \, (u_{\xi\xi} + u_{\eta\eta})$.
   Examples: *Laplace equation* or *Poisson equation*.

2. *Parabolic* class: $\Lambda[u] = \alpha \, u_{\xi\xi}$.
   Example: *diffusion equation*.

3. *Hyperbolic* class: $\Lambda[u] = \alpha \, (u_{\eta\eta} - u_{\xi\xi})$ or $\Lambda[u] = 2\beta \, u_{\xi\eta}$.
   Example: *wave equation*.

These forms refer to the terms of highest differentiation order.

The Laplace operator $-\Delta$ has infinitely many positive eigenvalues, the wave operator infinitely many positive and negative eigenvalues; this structure of the spectrum can be used just as well for classification in lieu of the above discriminant. However, some important examples fall through the grid of the above classification. For instance, we have not subsumed the *Schrödinger equation* (Section 1.4) under the parabolic class: formally it would fall in this class, but it has totally different properties than the diffusion equation (no damping of initial perturbations). Similarly, we have not subsumed the *Helmholtz equation* (Section 1.5) under the elliptic class: its critical case does not fit to the above scheme, because there, since some positive and infinitely many negative eigenvalues exist, it seems to be a kind of hybrid between the elliptic and the hyperbolic case.

Moreover, a specification of boundary conditions does not enter into the traditional classification, so that well-posedness or ill-posedness and the condition of the problem cannot be directly identified. In general parlance, the above classification is used for the respective well-posed problems.

1.7 Exercises

Exercise 1.1. In Section 1.1 we gave the Laplace operator in $\mathbb{R}^2$ in Euclidean coordinates (x, y). In science, however, *radially symmetric* problems or, in $\mathbb{R}^3$, respectively, *cylinder symmetric* problems occur. For their mathematical description *polar coordinates* (r, ϕ) defined via

$$(x, y) = (r \cos \phi, r \sin \phi)$$

are more appropriate. Verify the corresponding representation of the Laplace operator

$$\Delta u = \frac{1}{r} \frac{\partial}{\partial r} \left(r \frac{\partial u}{\partial r} \right) + \frac{1}{r^2} \frac{\partial^2 u}{\partial \phi^2}.$$

Which kind of solution representation arises naturally for a circular domain Ω?

Hint: Use a separation ansatz $u = v(r)w(\phi)$, by which the problem splits into two ordinary differential equations. For v one obtains an ODE named after Bessel, whose "interior" solutions are the Bessel functions J_k.[3]

Exercise 1.2. In Section 1.1 we gave the Laplace operator in $\mathbb{R}^3$ in Euclidean coordinates (x, y, z). For nearly *spherically symmetric* problems a formulation in *spherical coordinates* (r, θ, ϕ) defined via

$$(x, y, z) = (r \cos \phi \sin \theta, r \sin \phi \sin \theta, r \cos \theta)$$

is more appropriate. Verify the representation of the Laplace operator

$$\Delta u = \frac{1}{r^2} \frac{\partial}{\partial r} \left(r^2 \frac{\partial u}{\partial r} \right) + \frac{1}{r^2 \sin \theta} \frac{\partial}{\partial \theta} \left(\sin \theta \frac{\partial u}{\partial \theta} \right) + \frac{1}{r^2 \sin^2 \theta} \frac{\partial^2 u}{\partial \phi^2} \qquad (1.37)$$

in spherical coordinates. What kind of solution representation arises naturally?

Hint: Use a separation ansatz $u = v(r)p(\phi)q(\theta)$, by which the problem splits into three ODEs. For v one obtains, as in Exercise 1.1, a Bessel differential equation, for q an ODE named after Legendre. The complete angular part $p(\phi)q(\theta)$ leads to spherical harmonics $Y_k^l(\theta, \phi)$ with a radial l-dependent multiplicator function.[4]

Exercise 1.3. On a given domain

$$\Omega = \{rz : 0 < r < 1, z \in \Gamma\} \quad \text{with} \quad \Gamma = \{e^{i\phi} \in \mathbb{C} \simeq \mathbb{R}^2 : 0 < \phi < \alpha\}$$

solve the Poisson problem

$$\Delta u = 0 \quad \text{in } \Omega, \quad u(e^{i\phi}) = \sin(\phi\pi/\alpha) \quad \text{on } \Gamma, \quad u = 0 \quad \text{on } \partial\Omega\backslash\Gamma.$$

For which $\alpha \in]0, 2\pi[$ is $u \in C^2(\Omega) \cap C(\overline{\Omega})$?
Hint: Use the ansatz $u = z^\alpha$.

Exercise 1.4. By analogy with Theorem 1.2, show the uniqueness of a given solution $u \in C^2(\Omega) \cap C(\overline{\Omega})$ for the Robin boundary value problem

$$\Delta u = f \quad \text{in } \Omega, \quad n^T \nabla u + \alpha u = \beta \quad \text{on } \partial\Omega.$$

Which condition for α must hold?

Exercise 1.5. Let $\Omega \subset \mathbb{R}^2$ denote a bounded domain with sufficiently smooth boundary. Show that the Laplace operator Δ is only *self-adjoint* w.r.t. the inner L^2-product

$$(v, w) = \int_\Omega vw \, dx,$$

if one of the following conditions holds on the boundary

$$vn^T \nabla u = un^T \nabla v \quad \text{or} \quad n^T \nabla u + \alpha u = 0 \quad \text{with} \quad \alpha > 0.$$

Obviously, the first one is just a homogeneous Dirichlet or Neumann boundary condition, the second one a Robin boundary condition.

Exercise 1.6. Proceeding as for the diffusion equation in Section 1.2, derive an eigenmode representation for the solution of the Cauchy problem for the wave equation on the domain $\Omega =]-\pi, \pi[$. Determine the Fourier coefficients from a Fourier representation of the initial values $u(x, 0)$ and $u_t(x, 0)$. Expand the point perturbations $u(x, 0) = \delta(x - \xi)$ and $u_t(x, 0) = \delta(x - \xi)$ each into a Fourier series and calculate the thus obtained solutions. Which shapes come up for the domains of dependency?

Exercise 1.7. Consider the characteristic initial value problem for the wave equation, also known as the Riemann problem. For a prescription of boundary and initial values see Figure 1.5. Use the ansatz of d'Alembert to achieve a closed representation of the solution.

Exercise 1.8. In analogy to Figures 1.3 and 1.4, draw the domains of influence, of dependency, and of determination for elliptic boundary value problems.

Exercise 1.9. *Root cellar.* The time-harmonic heat equation (1.36) is well-suited to describe the daily and seasonal fluctuations of temperature in solid ground. How deep must a root cellar be built in order to make sure that drinks stored therein remain sufficiently cool in summer? How large are the seasonal temperature fluctuations in this cellar? Use the specific heat capacity $\kappa = 0.75\,\text{W/m/K}$ for clay.

Exercise 1.10. Calculate the (radially symmetric) solution of the 3D wave equation

$$u_{tt} = \Delta u \quad \text{in } \mathbb{R}^3 \times \mathbb{R}$$

with given initial values

$$u(x,0) = e^{-ar^2}, \quad r = |x|, \quad a > 0,$$
$$u_t(x,0) = 0.$$

Represent $u(x,t)$ as $w(r,t)$: By means of (1.37) derive a PDE for $rw(r,t)$. Which form do the characteristics have?

Exercise 1.11. Let $\hat{\Psi}_0(k)$ denote a C^∞-function with compact support. Consider

$$\Psi(x,t) = \frac{1}{\sqrt{2\pi}} \int_{-\infty}^{\infty} \hat{\Psi}_0(k) e^{i(kx - \omega(k)t)} \, dk$$

with the dispersion relation $\omega(k) = \frac{1}{2}k^2$. We want to develop a qualitative understanding of the dynamics of $\Psi(x,t)$ for large t.

1. *Principle of stationary phase.*
   By introduction of $S(x,t,k) = kx/t - \omega(k)$ for $t > 1$ one may write

   $$\Psi(x,t) = \frac{1}{\sqrt{2\pi}} \int_{-\infty}^{\infty} \hat{\Psi}_0(k) e^{itS(x,t,k)}$$

   for large t as a highly oscillatory integral. Use a plausibility argument why, for large t, this integral has essential contributions, if

   $$\frac{d}{dk} S(x,t,k_{\text{stat}}) = 0,$$

   where $k_{\text{stat}(x,t)}$ is called the stationary point. Moreover, derive that then

   $$\Psi(x,t) \sim \hat{\Psi}_0\big(k_{\text{stat}}(x,t)\big)$$

   must hold. (This statement can, of course, be made more precise, but is cumbersome to prove.)

2. Suppose that $\hat{\Psi}_0$ has an absolute maximum at some k_{max}. Where will, again for large t, the maximum of $\Psi(x,t)$ be located? With what speed will it propagate?

Chapter 2

Partial Differential Equations in Science and Technology

In the preceding chapter we discussed some elementary examples of PDEs. In the present chapter we introduce problem classes from science and engineering where these elementary examples actually arise, although in slightly modified shapes. For this purpose we selected problem classes from the application areas *electrodynamics* (Section 2.1), *fluid dynamics* (Section 2.2) and *elastomechanics* (Section 2.3). Typical for these areas is the existence of deeply ramifying hierarchies of mathematical models where different elementary PDEs dominate on different levels. Moreover, a certain knowledge of the fundamental model connections is also necessary for numerical mathematicians, in order to be able to maintain an interdisciplinary dialog.

2.1   Electrodynamics

As its name says, electro*dynamics* describes *dynamical, i.e., time-dependent* electric and magnetic phenomena. It includes (in contradiction to its name) as a subset electro*statics* for *static or stationary* phenomena.

2.1.1   Maxwell Equations

Fundamental concepts of electrodynamics are the *electric charge* as well as *electric* and *magnetic fields*. From their physical properties partial differential equations can be derived, as we will summarize here in the necessary brevity.

Conservation of Charge.   The electric charge is a conserved quantity – in the framework of this theory, charge can neither be generated nor extinguished. It is described as a continuum by some *charge density* $\rho(x) > 0$, $x \in \Omega$. From the conservation property one obtains a special differential equation, which we will now briefly derive.

Let Ω denote an arbitrary test volume. The charge contained in this volume is

$$Q = \int_\Omega \rho \, dx.$$

A change of charge in Ω is only possible via some charge flow ρu through the surface $\partial \Omega$, where u denotes a local velocity:

$$Q_t = \frac{\partial}{\partial t} \int_\Omega \rho \, dx + \int_{\partial \Omega} \rho n^T u \, ds = \int_\Omega (\rho_t + \mathrm{div}(\rho u)) \, dx = 0.$$

The quantity $j = \rho u$ is called *current density* or *charge flow density*. Now a not fully precise argument runs as follows: as Ω is arbitrary, a limit process $|\hat{\Omega}| \to 0$ enforces the above integrand to vanish. This leads to the *continuity equation* of electrodynamics:

$$\rho_t + \operatorname{div} j = 0. \tag{2.1}$$

Derivation of Maxwell Equations. In 1864 the Scottish physicist James Clerk Maxwell (1831–1879) derived certain equations (which today are named after him) and published them one year later [153]. In order to be able to mathematically model the experiments by the English physicist Michael Faraday (1791–1867), he generated the idea of a *magnetic field* $H \in \mathbb{R}^3$ and an *electric field* $E \in \mathbb{R}^3$ as space- and time-dependent physical quantities in a vacuum: He envisioned that "electric field lines" would "emanate" from electric charges.[1] From Faraday's experiments he perceived that there are no "magnetic charges" from which field lines could emanate. Instead, magnetic fields are generated by magnets or currents. The form of the Maxwell equations in the suggestive representation with differential operators div, curl, and grad (see Appendix A.2) dates back to O. Heaviside and J. W. Gibbs from 1892 [118]. We will use it in the following.

By means of the Theorem of Gauss (see also Figure A.1 for an interpretation of the div-operator), the existence of electric charges led Maxwell to the equation

$$\operatorname{div} E = \rho.$$

In a similar way, the nonexistence of magnetic charges, again via an application of the Theorem of Gauss, leads to the equation

$$\operatorname{div} H = 0.$$

Magnetic field lines are induced by some current density j, which, by means of the theorem A.7 of Stokes, boils down to the equation

$$\operatorname{curl} H = j. \tag{2.2}$$

Vice versa, the temporal change of magnetic field lines generates some current ("dynamo effect"), which can be expressed, again via the theorem of Stokes, by the equation

$$\operatorname{curl} E = -H_t.$$

Maxwell recognized (and this in his then rather unwieldy notation!), that the first and the third of these equations together with the continuity equation (2.1) give rise to a contradiction. One has, in general,

$$0 = \operatorname{div}(\operatorname{curl} H - j) = -\operatorname{div} j = \rho_t = \operatorname{div} E_t \neq 0.$$

---

[1] inspired by the experimental visualization of "magnetic fields" by virtue of iron filings in viscous fluid.

On the basis of this observation he modified, purely theoretically, equation (2.2) to obtain

$$\text{curl } H = j + E_t,$$

whereby the contradiction has been removed. *This was a magic moment for mankind*, since his new theoretical model predicted the existence of electromagnetic waves in a vacuum.[2]

A modification is obtained for materials within which electromagnetic processes take place: they are characterized by material constants like the *magnetic permeability* μ and the *dielectric constant* ϵ. From these, the Maxwell equations in materials arise in terms of the two fields *magnetic induction*

$$B = \mu H, \quad \mu = \mu_r \mu_0$$

and *electric displacement field*

$$D = \epsilon E, \quad \epsilon = \epsilon_r \epsilon_0,$$

where the quantities μ_0, ϵ_0 or $\mu_r = 1, \epsilon_r = 1$, respectively, characterize the vacuum. The quantities μ and ϵ are mostly scalar, but for *anisotropic* materials they may be *tensors*; for *ferro-magnetic* or *ferro-electric* materials they also depend nonlinearly on the corresponding fields.

In materials the system of Maxwell equations thus reads

$$\text{div } B = 0, \tag{2.3}$$
$$\text{div } D = \rho,$$
$$\text{curl } H = j + D_t,$$
$$\text{curl } E = -B_t. \tag{2.4}$$

In the above equations the current density j enters, although without any connection with the fields. The most popular model for such a connection is *Ohm's law*

$$j = \sigma E$$

in terms of an *electric conductivity* σ as material constant. This law, however, does not hold at high temperatures (e.g., in ionized plasma) or at very low temperatures (e.g., in superconducting materials).

Electrodynamic Potentials. The fact that the magnetic field is divergence free (see equation (2.3)), is due to Lemma A.1 automatically satisfied by the ansatz

$$B = \text{curl } A.$$

---

[2] Experimentally this phenomenon was not detected and proved until 1886 by the German physicist Heinrich Rudolf Hertz (1857–1894).

The quantity $A \in \mathbb{R}^3$ is called *vector potential*. The law of induction (2.4) then reads

$$\operatorname{curl}(E + A_t) = 0.$$

For any (differentiable) scalar function U the vector field ∇U lies in the nullspace of the curl-operator. Hence, again by Lemma A.1, we may set a general solution for E as

$$E = -\nabla U - A_t.$$

Here U is called the *scalar potential*. Together A and U are also called *electrodynamic potentials*. By this ansatz we have covered all possibilities for the fields E and B, which, however, we will not prove here.

Nonuniqueness. The thus defined electrodynamic potentials are not uniquely determined, but only up to a scalar function $f(x, t)$, since with

$$\tilde{A} := A + \nabla f, \quad \tilde{U} := U - f_t$$

one obtains trivially

$$\tilde{B} = \operatorname{curl} \tilde{A} = \operatorname{curl} A = B$$

as well as

$$\tilde{E} = -\nabla \tilde{U} - \tilde{A}_t = -\nabla U + \nabla f_t - A_t - \frac{\partial}{\partial t} \nabla f = E.$$

This means that the physically measurable fields derived from these nonunique potentials are nevertheless unique.

Electrostatics. In the stationary case, i.e., when all time derivatives vanish, the equations for U and A decouple. Then for the electric field only the equation

$$E = -\nabla U.$$

remains. Insertion of this relation into the Maxwell equation div $E = \rho$ then eventually supplies, with div grad $= \Delta$, the so-called *potential equation*

$$-\Delta U = \rho.$$

Due to this property, U is also called *electrostatic potential*. By comparison with Section 1.1 we recognize here the *Poisson equation*.

2.1.2 Optical Model Hierarchy

In everyday life we identify optics mostly with ray optics and think of glasses (spectacles) and lenses for telescopes and cameras. In most engineering applications optics means *wave optics*, where the wave nature of light stands in the foreground of consideration. This is described by Maxwell's equations. In particular, in wave optics we encounter nonconducting and nonmagnetic materials ($j = 0$ and $\mu_r = 1$) as well as no charges ($\rho = 0$).

Time-harmonic Maxwell Equations.   Suppose additionally that only a single frequency (or "color") ω is excited in the fields; we can then use the ansatz

$$E(x,t) = \Re(E_0(x)e^{i\omega t}).$$

This leads us to the time-harmonic Maxwell equations, often called *optical equations*, of the form

$$\begin{aligned}
\operatorname{div} H &= 0, \\
\operatorname{div}(\epsilon E) &= 0, \\
\operatorname{curl} H &= i\omega\epsilon E, \\
\operatorname{curl} E &= -i\omega\mu H
\end{aligned}$$

with $E, H \in \mathbb{C}^3$. If one applies the curl-operator to the third equation and inserts this result into the fourth one, then one obtains

$$\operatorname{curl} \frac{1}{i\epsilon\omega} \operatorname{curl} H = \operatorname{curl} E = -i\mu\omega H.$$

Suppose now that the material constants ϵ, μ depend on the position (x, y, z). Multiplication by $i\epsilon\omega$ then supplies a closed equation only for H:[3]

$$\epsilon \operatorname{curl} \frac{1}{\epsilon} \operatorname{curl} H = \epsilon\mu\omega^2 H. \tag{2.5}$$

We deliberately keep the common factor ϵ on both sides of the equation. If we reformulate (A.3) from Appendix A.2, we arrive at the fundamental equation of optics:

$$-\Delta H + \operatorname{curl} H \times \epsilon\nabla\frac{1}{\epsilon} + \underbrace{\nabla \operatorname{div} H}_{0} = \mu\epsilon\omega^2 H. \tag{2.6}$$

The Laplace operator Δ is understood to act component-wise here; the vector product (also: outer product) is written as $\times$. Thus the three components in (2.6) are coupled only via $\epsilon(x, y, z)$. Starting from this equation, successive simplifications produce a *hierarchy of mathematical models* which we will develop in the following.

Transversal Modes.   Suppose the dielectric constant depends on a preferred direction, say x, i.e.,

$$\epsilon(x, y, z) = \epsilon(x) \quad \Longrightarrow \quad \nabla\frac{1}{\epsilon} = \begin{bmatrix} -\dfrac{\epsilon_x}{\epsilon^2} \\ 0 \\ 0 \end{bmatrix},$$

---

[3] Application of curl to the fourth equation and insertion into the third one would have generated a corresponding equation only for E.

then we obtain from (2.6) and with the notation $H = (H_1, H_2, H_3)$,

$$\nabla \frac{1}{\epsilon} \times \text{curl } H = \frac{\epsilon_x}{\epsilon^2} \begin{bmatrix} 0 \\ H_{2,x} - H_{1,y} \\ H_{3,x} - H_{1,z} \end{bmatrix}.$$

As a rule, the solution of this coupled differential system is split into a direct sum of "in-plane"-solutions with $(H_1, 0, H_3)$ and "out-plane"-solutions $(0, H_2, 0)$. In the language of electrical engineering "out-plane" modes are usually called *transversal modes*. The solutions $H_2(x, y, z)$ satisfy the reduced differential equation

$$-\Delta H_2 + \frac{\epsilon_x}{\epsilon^2} H_{2,x} = \mu \epsilon \omega^2 H_2. \tag{2.7}$$

They characterize, e.g., fields in layered wave guides.

Scalar Helmholtz Equation. If one assumes in (2.6) that the dielectric constant only varies weakly, i.e., $\nabla(1/\epsilon) \approx 0$, then the second term can be dropped in all three equations. As a consequence, the fundamental optical equations decompose into three independent scalar Helmholtz equations in the critical case:

$$\Delta H + \mu \epsilon \omega^2 H = 0. \tag{2.8}$$

Solutions of interest will then arise from the choice of boundary conditions. Thus this kind of Helmholtz equation introduced in Section 1.5 as an elementary example is the simplest structural model for Maxwell's equations.

Eigenvalue Problem for Waveguides. In waveguides, e.g., in glass fiber cables, the material is in general uniformly composed along the conductor axis (let this be the z-direction). In this case the variation of the dielectric constant across the 2D cross section of the waveguide may be restricted, i.e., one has

$$\epsilon(x, y, z) = \epsilon(x, y).$$

If one additionally assumes a wave propagation that is periodic along the z-direction, i.e.,

$$H(x, y, z) = \hat{H}(x, y)e^{i\lambda z}, \tag{2.9}$$

then the scalar Helmholtz equation (2.8) reduces to a 2D eigenvalue problem for the Helmholtz equation

$$\Delta_{xy} H + \mu \epsilon \omega^2 H = \lambda^2 H.$$

In Section 6.4 below we will elaborate details for a plasmon-polariton waveguide.

Paraxial Wave Equation (Fresnel Approximation). In this model one assumes that the dependence of the magnetic field w.r.t. the z-direction can only be approximately described by a harmonic ansatz, which means that we only write

$$H(x, y, z) = \hat{H}(x, y, z)e^{i\lambda z}, \quad \hat{H}_{zz} \approx 0.$$

The second derivative w.r.t. z is then

$$H_{zz} = (\hat{H}_{zz} + 2i\lambda \hat{H}_z - \lambda^2)e^{i\lambda z}.$$

Neglecting $\hat{H}_{zz}$ then reduces (2.8) to

$$-2i\lambda \hat{H}_z = \Delta_{xy} \hat{H} + (\mu\epsilon\omega^2 - \lambda^2)\hat{H}.$$

Obviously, this is a kind of *Schrödinger equation*, where the spatial variable z takes the role of a time variable (see Section 1.4).

2.2 Fluid Dynamics

In this section we give a short introduction in those PDEs that describe the dynamics of the flow of fluids or gases. Fluids and gases together are summed up under the termf *continuous fluids*, mathematically described by the physical quantities *mass density* $\rho(x, t)$, *fluid velocity* $u(x, t)$ and pressure $p(x, t)$, all three of them dependent on the position $x \in \mathbb{R}^d$, $d = 2, 3$, and the time t. The disregarding of the microscopic particle structure of fluids is called *continuum hypothesis*.

Material Derivative. In order to derive the governing differential equations, we need some preliminary consideration. Let the motion of a particle be described by its position variable $x(t) \in \mathbb{R}^d$. Then the corresponding velocity $x_t(t) = u(x(t), t)$ and the acceleration are

$$x_{tt}(t) = \frac{du(x(t), t)}{dt} = u_x(x(t), t)x_t(t) + u_t(x(t), t) = u_x u + u_t. \quad (2.10)$$

Note that u_x is a (d, d)-matrix.[4] The nonlinear expression of the right-hand side suggests the following definition for an arbitrary given quantity f:

$$\frac{D}{Dt}f = f_t + f_x u.$$

The thus defined derivative is called *material derivative*. In this notation formula (2.10) for the acceleration can be simply written as

$$x_{tt} = \frac{Du}{Dt}.$$

---

[4] In the literature, the term $u_x u$ is also found under the formally equivalent notations $(u \cdot \nabla)u$ and $u \cdot \nabla u$, where the dot denotes the Euclidean scalar product (see also Exercise 2.3).

Insertion of $x(t)$ into an arbitrary function f with $f(t) = \hat{f}(x(t), t)$ then generally yields

$$f_t = \frac{D}{Dt} \hat{f}.$$

From this we perceive that the material derivative describes the temporal change of a quantity as seen from a moving particle. Obviously, there are two basic options for the choice of coordinates:

- *stationary* coordinates, i.e., fixed positions, also called *Euler coordinates*, where we need the concept of a material derivative;

- *moving* coordinates, also called *Lagrange coordinates*, where we can dispense of the material derivative, since it reduces to the simple time derivative.

The different choices of coordinates lead to strongly different implementations of algorithms. Sometimes mixed forms are useful.

2.2.1 Euler Equations

Starting point for the mathematical modeling is *conservation quantities*, i.e., physical quantities that are constant over time. For such quantities *conservation laws* can be derived, where we essentially follow the textbook of A. Chorin and J. Marsden [50]. In this section we first consider only *ideal isentropic* fluids – concepts which we will specify below. Throughout the text $\hat{\Omega}$ denotes an arbitrary time independent test volume in which we study the motion of the fluid.

Conservation of Mass. In this model, mass can neither be created nor destroyed. Therefore a time-dependent change of mass in $\hat{\Omega}$ can only originate from a mass flow through the boundary surface $\partial\hat{\Omega}$. This insight can be mathematically formulated as follows:

$$\frac{\partial}{\partial t} \int_{\hat{\Omega}} \rho \, dx + \int_{\partial\hat{\Omega}} \rho n^T u \, ds = 0. \tag{2.11}$$

The time derivative and the volume integral on the left-hand side commute. If we apply the Theorem of Gauss to the boundary integral (see Appendix A.3, Figure A.1), we end up with the volume integral

$$\int_{\hat{\Omega}} (\rho_t + \text{div}(\rho u)) \, dx = 0.$$

Analogously to the derivation of (2.1) we here arrive at the *continuity equation* of fluid dynamics:

$$\rho_t + \text{div}(\rho u) = 0. \tag{2.12}$$

Conservation of Momentum. A conservation of the momentum is essentially understood as the absence of any forces. Among the *forces* that act on $\hat{\Omega}$ we have to distinguish between internal and external forces:

1. *Internal forces* F_{int}. Such forces act through the surface $\partial\hat{\Omega}$, e.g., as *stresses*. The *ideal* fluids – the only ones considered here – are characterized by the fact that the internal forces act in a direction orthogonal to the surfaces, i.e., there are no tangential forces. Under this restriction there exists a scalar function, the *pressure* $p(x,t) \in \mathbb{R}$, so that $F_{int} = -pn$, where n is again the externally oriented unit normal vector. In total, the force through the surface $\partial\hat{\Omega}$ is thus

$$F_{int} = -\int_{\partial\hat{\Omega}} pn\,ds = -\int_{\hat{\Omega}} \nabla p\,dx.$$

The reformulation above is obtained by componentwise application of the Theorem of Gauss (with the componentwise divergence being the gradient).

2. *External forces* F_{ext}. Such forces act as volume forces on the whole of $\hat{\Omega}$, e.g., gravitation or electric as well as magnetic field forces. Let the thereby induced acceleration be denoted by $b(x,t) \in \mathbb{R}^d$. Then, via the volume, the external forces induce

$$F_{ext} = \int_{\hat{\Omega}} \rho b\,dx.$$

Internal and external forces together sum up to the total force

$$F = \int_{\hat{\Omega}} (-\nabla p + \rho b)\,dx.$$

With *Newton's second law* ("force equals mass times acceleration") we then obtain, as above by reduction of a volume integral to the integrand, the *momentum equation*

$$\rho\frac{Du}{Dt} = -\nabla p + \rho b. \tag{2.13}$$

For simplicity, we will drop the volume forces b in the following.

Constitutive Law.   The thermodynamics of fluids leads to a connection of mass density, pressure, and temperature in the form of a *constitutive law*

$$\rho = \rho(p,T),$$

which, together with (2.12) and (2.13) then closes the system of equations.

Let us restrict our attention to the simplest case of so-called *isentropic* fluids where the constitutive law does not depend on the temperature, which means $\rho = \rho(p)$. Here, an increase of the pressure usually induces (if at all) an increase of the mass density, hence $\rho_p \geq 0$ will hold. If $\rho_p > 0$, then it is often more convenient to use the reverse mapping $p = p(\rho)$ as constitutive law. For ideal gases one obtains from thermodynamic considerations

$$p = \frac{p_0}{\rho_0^\gamma}\rho^\gamma, \tag{2.14}$$

where the adiabatic exponent γ is a material constant (e.g., 1.4 for air) and p_0, ρ_0 are the pressure and the mass density in a reference state.

Incompressibility. Whenever $\rho_p = 0$, then the fluid is said to be *incompressible*. This property is the typical model assumption for (pure) fluids, since they (nearly) cannot be compressed under external pressure. The mass density of the *moving* fluid is constant so that $\frac{D\rho}{Dt} = 0$ holds. From the continuity equation (2.12) we then obtain

$$\operatorname{div} u = \frac{1}{\rho}\left(\frac{D\rho}{Dt} + \rho \operatorname{div} u\right) = \frac{1}{\rho}(\rho_t + \operatorname{div}(\rho u)) = 0 \qquad (2.15)$$

as the third equation. In the most frequent case of *homogeneous* fluids we additionally assume that $\rho = $ const, which implies that continuity equation (2.12) and the zero-divergence property (2.15) merge. Together with the momentum equation (2.13) we finally obtain equations that have already been derived by Leonhard Euler (1707–1783) and which are therefore today called the *incompressible Euler equations*:

$$\rho u_t + \rho u_x u = -\nabla p,$$
$$\operatorname{div} u = 0.$$

Here the pressure is determined only up to a constant. Usually it is understood as the deviation from a constant reference pressure p_0 and fixed by the normalization

$$\int_\Omega p \, dx = 0.$$

Due to the zero-divergence condition, the incompressible Euler equations have the character of a differential-algebraic system (cf. Volume 2).

Compressibility. For *compressible* fluids we have $\rho_p > 0$. This property holds for gases. In this case the constitutive law can be inserted as $p = p(\rho)$ into the momentum equation (2.13). We then obtain the *compressible barotropic Euler equations*

$$\rho u_t + \rho u_x u = -p_\rho \nabla \rho,$$
$$\rho_t + \operatorname{div}(\rho u) = 0.$$

Boundary Conditions. Apart from the initial conditions for u and possibly for ρ we have to impose boundary conditions. In general, the domain Ω will consist partly of solid walls $\partial\Omega_W$, which cannot be passed by the fluid. Hence, the normal components of the velocity will vanish there. In ideal fluids there are no tangential forces, but only normal forces, which would consequently influence the velocity at the boundary only in normal direction. Therefore, the so-called *"slip condition"*

$$n^T u = 0 \quad \text{on} \quad \partial\Omega_W \qquad (2.16)$$

determines the boundary conditions at walls completely. At inflow boundaries the mass flux must be prescribed. In many applications it is also reasonable to fix the inflow velocity normal to the boundary, but this is only one of various possible choices for boundary conditions. At outflow boundaries, the solution is already determined by the PDE and the conditions imposed on the other boundaries, such that here no further boundary condition can be specified.

Stationary Potential Flow. If one is only interested in stationary solutions of the Euler equations with $u_t = 0$, then *curl-free* flows are a rather simple class, as they may be represented as potential fields $u = \nabla\varphi$ (cf. Lemma A.1). In the *incompressible* case, the condition

$$\Delta\varphi = \operatorname{div} u = 0$$

directly leads to the Laplace equation (1.1). Then Bernoulli's law for constant mass density,

$$\frac{1}{2}|u|^2 + \frac{p}{\rho} = \text{const},$$

satisfies the momentum equation $\rho u_x u = -\nabla p$ and supplies a posteriori the density distribution.

For *compressible* fluids one obtains from the momentum equation (2.13), the property $\operatorname{curl} u = 0$, and the constitutive law (2.14) the relation

$$\frac{\nabla|u|^2}{2} + \frac{\gamma p_0}{(\gamma-1)\rho_0^\gamma}\nabla(\rho^{\gamma-1}) = 0,$$

i.e.,

$$\frac{|\nabla\varphi|^2}{2} + \frac{\gamma p_0}{(\gamma-1)\rho_0^\gamma}\rho^{\gamma-1} = \text{const} = \frac{|u_0|^2}{2} + \frac{\gamma p_0}{(\gamma-1)\rho_0},$$

where again p_0, ρ_0 are pressure and density in a reference point and u_0 the local velocity there. Upon resolution of

$$\rho(\varphi) = \rho_0\left(1 + \frac{(\gamma-1)\rho_0}{2\gamma p_0}\left(|u_0|^2 - |\nabla\varphi|^2\right)\right)^{\frac{1}{\gamma-1}}$$

and insertion into the continuity equation $\operatorname{div}(\rho\nabla\varphi) = 0$, we are finally led to the nonlinear potential equation. Its validity is restricted to the domain with positive density $\rho(\varphi) > 0$, but may be locally elliptic or even hyperbolic there.

The assumption of an irrotational flow is adequate only in certain situations,[5] even though the conservation of rotation via the Euler equation is guaranteed by the Kelvin circulation theorem, according to which the flow remains irrotational once the initial state is irrotational. Nevertheless, in some situations real fluids with frictions behave much different from ideal fluids. This aspect will be discussed in the following two sections.

2.2.2 Navier–Stokes Equations

In this section we deal with an extension of the Euler equations to *nonideal* fluids, where flowing molecules while passing each other exert forces on each other. This means that we allow for nonvanishing *internal tangential forces* (see Figure 2.1).

---

[5] R. Feynman formulated that potential flow only describes "dry water" [89]. Nevertheless, potential flows have been successfully applied to the design of airplane wings as a useful and relatively simple to compute approximation. See also Section 5.6.

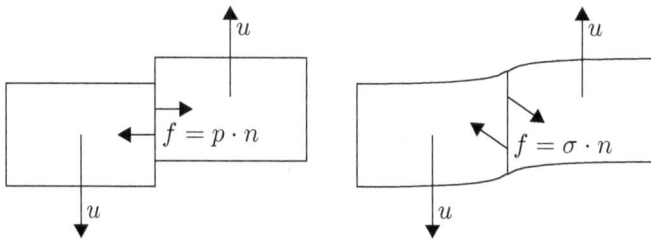

Figure 2.1. Internal tangential forces in the derivation of the Navier–Stokes equations. *Left*: ideal fluid, only normal force components. *Right*: nonideal fluid, with tangential forces involved.

In this model assumption we then write

$$F_{\text{int}} = (-pI + \sigma)n,$$

where p is again the (scalar) pressure, while σ is a matrix that represents the internal stress; that is why σ is also called *stress tensor*. This quantity is characterized by the following assumptions:

1. σ is invariant under translation and therefore depends only on u_x, i.e.,

$$\sigma = \sigma(u_x);$$

2. σ is invariant under rotation (here represented by an orthogonal matrix Q), i.e.,

$$\sigma(Qu_xQ^{-1}) = Q\sigma(u_x)Q^{-1},$$

and therefore depends only *linearly* on the components of u_x;

3. σ is symmetric:

$$\sigma = \sigma^T.$$

These three assumptions already permit a unique representation of σ as

$$\sigma = \lambda(\operatorname{div} u)I + 2\mu\epsilon$$

with the Lamé viscosity coefficients λ, μ and the *strain tensor*

$$\epsilon = \tfrac{1}{2}(u_x + u_x^T).$$

From this one gets for the internal forces

$$F_{\text{int}} = \int_{\partial\Omega} (-pI + \sigma)n \, ds = \int_{\Omega} (-\nabla p + \operatorname{div}\sigma) \, dx$$

$$= \int_{\Omega} (-\nabla p + (\lambda + \mu)\nabla \operatorname{div} u + \mu\Delta u) \, dx.$$

Analogously to (2.13) we obtain, via Newton's second law, the equation

$$\rho\frac{Du}{Dt} = -\nabla p + (\lambda + \mu)\nabla \operatorname{div} u + \mu\Delta u + \rho b. \tag{2.17}$$

Here, in contrast to (2.13), however, a higher spatial derivative arises, which means that we require more boundary conditions than for the Euler equations. Due to the internal friction between fluid and solid walls we have

$$\xi^T u = 0$$

for any tangential vector ξ. Together with $n^T u = 0$ this supplies a homogeneous Dirichlet condition, the so-called "*no-slip-condition*"

$$u|_{\partial\Omega} = 0. \tag{2.18}$$

Thus, with the equations (2.12) and (2.17) together with the constitutive law $p = p(\rho)$, the *Navier–Stokes equations* for *compressible* fluids are now complete:

$$\rho u_t + \rho u_x u = -p_\rho \nabla \rho + (\lambda + \mu)\nabla \operatorname{div} u + \mu\Delta u + \rho b, \tag{2.19a}$$

$$\rho_t + \operatorname{div}(\rho u) = 0, \tag{2.19b}$$

$$u|_{\partial\Omega} = 0. \tag{2.19c}$$

In the case of *incompressible*, homogeneous fluids in the absence of external forces we have $\operatorname{div} u = 0$ and $\rho = \rho_0 > 0$. From this one obtains the classical *Navier–Stokes equations* in the narrower sense, which means for *incompressible* fluids,

$$u_t + u_x u = -\nabla p' + \nu\Delta u,$$

$$\operatorname{div} u = 0,$$

$$u|_{\partial\Omega} = 0$$

with a scaled pressure

$$p' = \frac{p}{\rho_0}$$

and a *kinematic viscosity*

$$\nu = \frac{\mu}{\rho_0}.$$

Remark 2.1. These equations were named in honor of the French mathematician Claude Louis Marie Henri Navier (1785–1836) and the Irish mathematician and physicist Sir George Gabriel Stokes (1819–1903). Navier (1827) and, in improved form, Stokes (1845) formulated the equations independently of each other. Two years before Stokes, the French mathematician and physicist Adhémar Jean Claude Barré de Saint-Venant (1797–1886) published a correct derivation of the equations. Nevertheless the name Navier–Stokes equations prevailed.[6]

---

[6] Wikipedia.

Reynolds Number. Suppose we have certain natural scales in the problem to be solved, say U for a characteristic velocity, L for a characteristic length and T for a characteristic time (typically $T = L/U$). Then we may rescale the equations to obtain dimensionless quantities according to

$$u(x,t) \to u' = \frac{u}{U}, \quad x \to x' = \frac{x}{L}, \quad t \to t' = \frac{t}{T}.$$

This leads to the following *dimensionless* form of the incompressible Navier–Stokes equations (where we have marked the differential operators w.r.t the dimensionless variables also by an apostrophe)

$$u'_{t'} + u'_{x'}u' + \nabla'p' = \frac{1}{\text{Re}}\Delta'u',$$
$$\text{div}\, u' = 0,$$

where only a single dimensionless parameter enters, the *Reynolds number*

$$\text{Re} = \frac{LU}{\nu}.$$

By comparison with the Euler equations, one observes that, in the limit $\text{Re} \to \infty$, the incompressible Navier–Stokes equations formally migrate to the Euler equations. But the solution of the Euler equations, when interpreted as limit solution of the Navier–Stokes equations, is plain unstable. Mathematically speaking, the limiting process can simply not be performed: depending upon some not perfectly fixed critical threshold value

$$\text{Re}_{\text{krit}} \approx 1000\text{--}5000,$$

into which the boundary conditions enter, too, the solution structure splits into laminar and turbulent flow.

Laminar Flow. Below the critical threshold value Re_{krit} the equations describe a *laminar* flow without intense vortex generation. With homogeneous boundary conditions (2.18) the *similarity law* holds below the threshold: laminar flows with similar geometry and the same Reynolds number are similar. In the past, experiments for flows in the wind tunnel have been performed such that the geometry was reduced and, for compensation, the fluid was substituted with one of smaller viscosity in such a way that the Reynolds number remained the same (cf. Figure 2.2).

Structurally the Navier–Stokes equations are *parabolic* below the critical value, running asymptotically towards a stationary state. Following a suggestion by R. Glowinski and O. Pironneau (see, e.g., [211]), the mathematical problem in the stationary case can be formulated in terms of six Poisson problems, whereby again this elementary PDE comes into play.

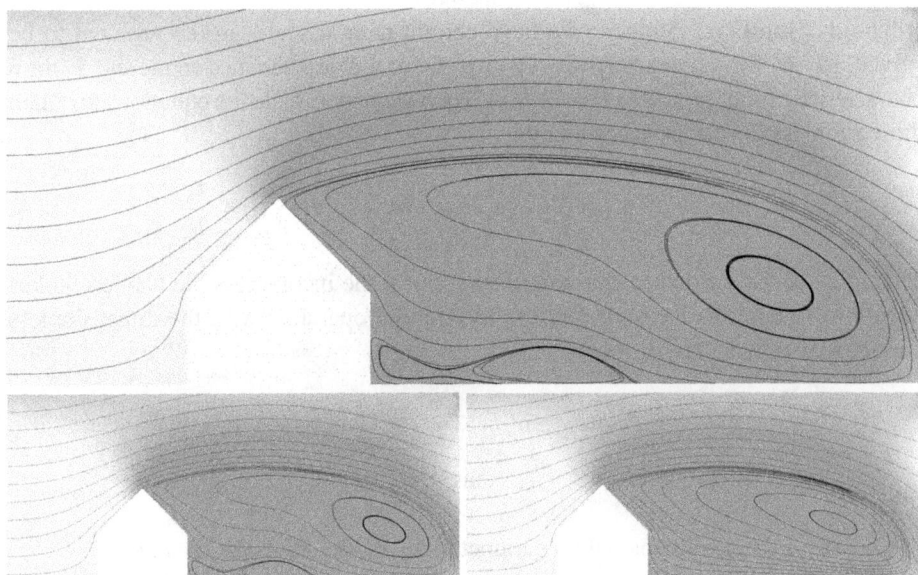

Figure 2.2. Similarity law for laminar flow. The pressure distribution is insinuated by gray coloring, the stream lines by black lines. *Top:* reference domain (Re = 200). *Bottom left:* domain reduced by a factor of two with correspondingly reduced viscosity (Re = 200); same flow profile as above. *Bottom right:* domain reduced by a factor of two, but with unchanged viscosity (Re = 100); changed flow profile.

Turbulent Flow.   Above the threshold Re_{krit} the phenomenon of turbulence occurs, a disordered and microscopically unpredictable flow pattern, which already drew the attention of Leonardo da Vinci (see Figure 2.3 and a modern visualization technique for comparison).

    The transition from laminar to turbulent flow can be characterized by the eigenvalues of the Navier–Stokes differential operator, *linearized* around the stationary laminar

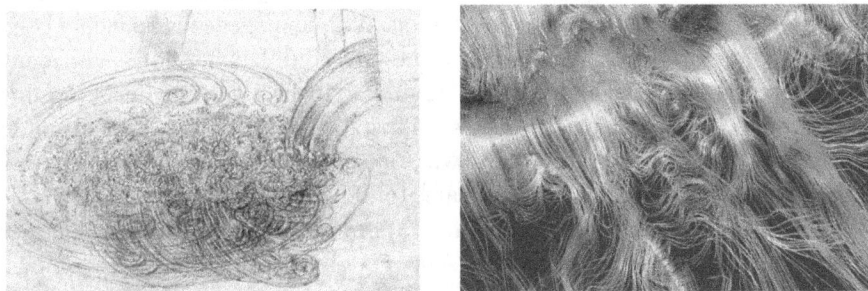

Figure 2.3. Turbulent flow. *Left*: Leonardo da Vinci (1508–1510) (see [234, p. 523]). *Right*: visualization by illuminated stream lines [233] (ZIB, Amira [190]).

flow. Below the critical threshold the eigenvalues lie in the negative complex half-plane, corresponding to the parabolic structure of the problem, as already mentioned above. With increasing Reynolds number the eigenvalues migrate into the positive half-plane, which means that the problem becomes unstable; this is the onset of turbulence. The critical Reynolds numbers for the transition to turbulence as computed from the linearization are often much larger than observed experimentally. The fact that much smaller Reynolds numbers may actually lead to transient effects in asymptotically stable laminar flows has been explained in a beautiful example by N. Trefethen et al. [202, 203]. It can be understood in elementary terms by the concept of the condition of general linear eigenvalue problems (Volume 1, Section 5.1): Laminar flows are a well-conditioned problem, whereas turbulent flows are ill-conditioned. In the turbulent case, the discretization of the linearization reveals that the left and right eigenvectors of the arising nonsymmetric matrices are "nearly orthogonal".

Remark 2.2. In the derivation of the Euler and Navier–Stokes equations we have tacitly assumed sufficient differentiability of the occurring quantities. This is not given, however, as soon as *shocks* occur, which is why in these cases numerical simulation prefers the conservation laws in their integral form (2.11). In this volume, however, we will *not* treat the numerical simulation of flows above or even close to the threshold Re_{krit} – this would be a completely different scenario. Instead, we refer to the monographs of D. Kröner [137, 138] as well as to the classic work [144] by R. LeVeque.

Remark 2.3. In 2000, the Clay Foundation, a foundation of the Boston billionaires Landon T. Clay and Lavinia D. Clay, advertised a reward of one million US dollars for the solution of each one out of ten particularly difficult problems, the so-called *millennium problems*[7]. Among these problems is also the proof of the existence and regularity of the solutions of the Navier–Stokes equations in three space dimensions for arbitrary initial data and unbounded time. This is a problem of mathematical analysis, which is why we skip it here.

2.2.3 Prandtl's Boundary Layer

For "nearly ideal" fluids (like air) with small viscosity and associated high Reynolds number the Euler equations are a good approximation for describing the dynamics of the flow. However, in 1904 the German engineer Ludwig Prandtl (1875–1953) had already found that at the boundaries of the domain Ω a *boundary layer* is generated, within which the Euler equations are *not* a good approximation [172]. This fact is modeled by a decomposition of Ω into an internal domain and a boundary layer of thickness δ. The solution process is then split into two parts: Euler equations in the

---

[7] The Russian mathematician Grigori Jakowlewitsch Perelman (born in 1966) attained fame not only for having solved one of these problems (the Poincaré conjecture), but also for rejecting afterwards both the Fields medal 2006 and the prize money of one million US dollars.

internal domain, and simplified Navier–Stokes equations in the boundary layer. At
the transition from the boundary layer to the internal domain the tangential velocity
from the solution of the Euler equations supplies the required boundary values for the
Navier–Stokes equations.

To illustrate this, we present a model calculation at a 2D-example. Let (x, y) denote position variables and (u, v) the corresponding velocity components. In these
variables, the Navier–Stokes equations (already in scaled form) read:

$$u_t + uu_x + vu_y = -p_x + \frac{1}{\mathrm{Re}}\Delta u,$$

$$v_t + uv_x + vv_y = -p_y + \frac{1}{\mathrm{Re}}\Delta v,$$

$$u_x + v_y = 0.$$

Suppose now that the boundary $\partial\Omega$ is defined by $y = 0$. Hence, the boundary layer is
generated in the variables y and v. To show this, we perform a variable transformation

$$y' = \frac{y}{\delta}, \quad v' = \frac{v}{\delta}$$

and thus obtain

$$u_t + uu_x + v'u_{y'} = -p_x + \frac{1}{\mathrm{Re}}\left(u_{xx} + \frac{1}{\delta^2}u_{y'y'}\right),$$

$$\delta v'_t + \delta uv'_x + \delta v'v'_{y'} = -\frac{1}{\delta}p_{y'} + \frac{1}{\mathrm{Re}}\left(\delta v'_{xx} + \frac{1}{\delta}v'_{y'y'}\right),$$

$$u_x + v'_{y'} = 0.$$

A connection between δ and Re, necessary in order to be able to apply some perturbation analysis, is missing. Following Prandtl we fix the thickness of the layer by the
relation

$$\delta = \frac{1}{\sqrt{\mathrm{Re}}}.$$

After multiplication of the second row by δ and insertion of Prandtl's relation into the
first two rows, one obtains

$$u_t + uu_x + v'u_{y'} = -p_x + \frac{1}{\mathrm{Re}}u_{xx} + u_{y'y'},$$

$$\frac{1}{\mathrm{Re}}(v'_t + uv'_x + v'v'_{y'}) = -p_{y'} + \frac{1}{\mathrm{Re}^2}v'_{xx} + \frac{1}{\mathrm{Re}}v'_{y'y'}.$$

If we only take into account the leading terms w.r.t Re in the limiting process $\mathrm{Re} \to \infty$,
we end up with the reduced *Prandtl's boundary layer equations*

$$u_t + uu_x + v'u_{y'} = -p_x + u_{y'y'},$$

$$0 = p_{y'},$$

$$u_x + v'_{y'} = 0.$$

A numerical implementation of the above scaling transformation obviously requires a locally orthogonal mesh at the boundary $\partial\Omega$ which has to be extremely fine with thickness $\delta = 1/\sqrt{\mathrm{Re}}$. Here long rectangles or triangles with acute angles are appropriate (cf. Figure 2.4). We will return to this aspect in Section 4.4.3. For the remaining domain, i.e., for the discretization of the Euler equations, a far coarser mesh is sufficient.

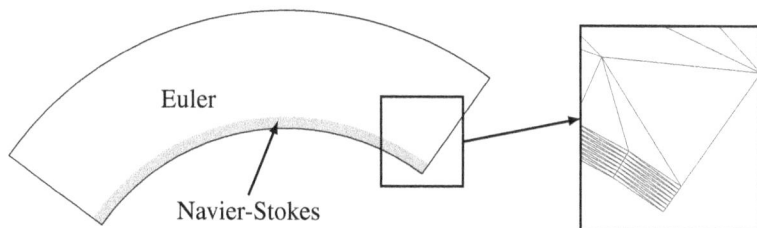

Figure 2.4. Schematic domain decomposition when combining Euler and Navier–Stokes equations. *Left*: fine mesh for Prandtl's boundary layer (gray). *Right*: zoom into the fine mesh in the boundary layer, where triangles with acute angles are appropriate for the problem (cf. Section 4.4.3).

Remark 2.4. Today numerical simulations on the basis of the Navier–Stokes equations, e.g., for the flow around even a whole airplane, are so precise that they can substitute actual wind tunnel experiments; this is why the simulations are also call *numerical wind tunnel*. They are often more useful than real experiments, because the results may be spoiled due to the spatially narrow lab situation.

2.2.4 Porous Media Equation

In this section we introduce a PDE which describes the flow of fluids through porous media. This type of equation arises, e.g., in oil reservoir simulation, in the investigation of soil pollution by leaking fluid contaminants, or the undermining of embankment dams. Characteristic of this problem class is the fact that the fluid penetrates through numerous small pores of the medium, comparable to a (even though it is solid) sponge. In some "microscopic model" the many pores would enter into the definition of a multiply-connected domain Ω, which, however, would lead to an unacceptable amount of required computational work. That is why one takes an averages of the pore structure, a mathematical limit process called *homogenization*. In such a "mesoscopic model" the medium is characterized by continuous quantities like the *absolute permeability* K and the *porosity* Φ (volume quotient of pores vs. dense material).

In the following presentation we restrict our attention to a single-phase flow, i.e., with only one (nonideal) fluid, which we describe as in the Navier–Stokes equations by the *mass density* ρ, the *viscosity* η, the *velocity* u and the *pressure* p. We start with the continuity equation for compressible flow in the 3D case

$$(\rho\Phi)_t + \mathrm{div}(\rho u) = 0.$$

In addition we need a mathematical description of the fluid flow through the porous medium. There we resort to the so-called *Darcy law*

$$u = -\frac{K}{\eta}(\nabla p - \rho g e_z),$$

where g denotes the gravitational acceleration on Earth and e_z the unit vector in the direction of the gravitational force. The French engineer Henry Philibert Gaspard Darcy (1803–1858) found this law to hold, on the basis of his own experimental investigations, when he had to design a water pump for the city Dijon. He published it in 1856 as part of a long treatise [61]. Insertion of the Darcy law into the continuity equation yields

$$(\rho\Phi)_t = \mathrm{div}\left(\frac{K}{\eta}(\rho\nabla p - \rho^2 g e_z)\right).$$

In this equation the quantities mass density ρ and pressure p occur, which in *compressible* fluids are independent. In order to connect them, we additionally define the *compressibility* by virtue of

$$c = \frac{1}{\rho}\rho_p \quad \Rightarrow \quad c\rho\nabla p = \nabla\rho.$$

Insertion of this definition finally supplies the *porous media equation*

$$(\rho\Phi)_t = \mathrm{div}\left(\frac{K}{c\eta}(\nabla\rho - c\rho^2 g e_z)\right).$$

Often the nonlinear gravitational term $c\rho^2 g e_z$ there is neglected. The equation is obviously of *parabolic type* with position dependent diffusion coefficient $K/c\eta$. A formulation of the corresponding equations for incompressible fluids is left as Exercise 2.9.

2.3   Elastomechanics

This scientific field describes the deformation of elastic bodies under the action of external forces. Dilatations of a body are denoted as *strains*, internal forces as *stresses*. Each deformation leads to a change of energy. Stable idle states are characterized by minima of the energy. Apart from the *potential energy*, which is determined by the position and shape of the body in an external force field, the *internal energy* is of particular importance. Changes of internal energy are induced by relative changes of lengths and volumes in the interior of the body.

2.3.1   Basic Concepts of Nonlinear Elastomechanics

Before we begin with a derivation of the corresponding PDEs, we will introduce some necessary concepts and notions.

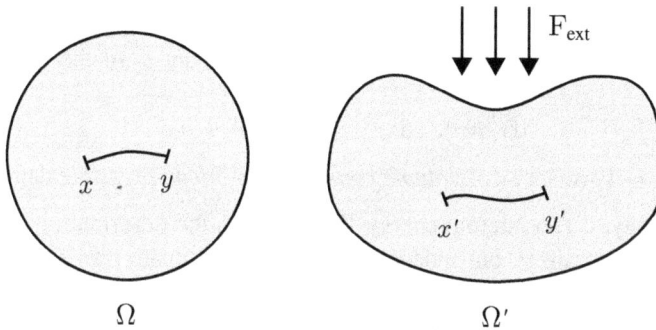

Figure 2.5. Deformation of a body under the action of external forces.

Kinematics. For a description of a state of strain we envision a locally bijective transformation $x \mapsto x'$, the *deformation*, from an undeformed body Ω to a deformed body Ω', as shown schematically in Figure 2.5. The local deviation from the identity can be written as the displacement

$$u(x) := x'(x) - x.$$

The differences between two points x, y are described by derivatives of the deviation, which may be interpreted as strains:

$$x' - y' = u(x) + x - u(y) - y = u_x(x)(x - y) + x - y + o(|x - y|).$$

This gives us the following expressions in terms of distances $|\cdot|$:

$$|x'-y'|^2 = (x'-y')^T(x'-y') = (x-y)^T(I+u_x+u_x^T+u_x^Tu_x)(x-y)+o(|x-y|^2) .$$

A linearization of this expression leads naturally to the following

Definition 2.5. *Symmetric Cauchy–Green strain tensor:*

$$C = (I + u_x)^T (I + u_x).$$

With this definition the linearization can be written in the form

$$|x' - y'|^2 = (x - y)^T C(x - y) + o(|x - y|^2).$$

The deviation from the identity is characterized by the *Green–Lagrange strain tensor*

$$E = \frac{1}{2}(C - I) = \frac{1}{2}\left(u_x + u_x^T + u_x^Tu_x\right),$$

often also called "strain", a slightly unfortunate choice of term. The quadratic term $u_x^Tu_x$ arising in C and E is called *geometric nonlinearity*.

As known from the substitution rule in multidimensional integration, the local change of volume is described by the Jacobi determinant J of the deformation according to

$$dx' = |J| \, dx, \quad J = \det(I + u_x).$$

In the case $J = 1$ we speak of *volume conserving* or *incompressible* deformations.

Internal Energy. The internal energy W^i stored in the deformation of the body is a function of the strain u, but independent of the coordinate frame of the observer, i.e., invariant under rigid body transformations of the image of u. In particular, W^i is invariant under *translation*, i.e., $W^i(u) = W^i(u + v)$ must hold for arbitrary constant vector fields v. Therefore W^i can only depend on the derivative u_x. Materials, for which the *internal energy* can be represented by some *material model* $\Psi : \mathbb{R}^{d \times d} \to \mathbb{R}$ of the form

$$W^i(u) = \int_\Omega \Psi(u_x) \, dx \tag{2.20}$$

are called *hyperelastic*. In contrast to more complex viscoelastic or plastic materials, the internal energy of which also depends on former strain states, hyperelastic materials do not have "memory" and thus represent an especially simple class of materials. Furthermore reasonable material models must satisfy the physically indispensable requirement of conservation of orientation, i.e., $J > 0$, and

$$\Psi(u_x) \to \infty \quad \text{for } J \to 0. \tag{2.21}$$

The invariance under *rotation* implies $W^i(u) = W^i(Qu)$ for all orthonormal $Q \in \mathbb{R}^{d \times d}$ with $QQ^T = I$ and $\det Q = 1$, and therefore also $\Psi(Qu_x) = \Psi(u_x)$. As u is orientation conserving and thus $J > 0$ holds, there exists a unique polar decomposition $I + u_x = QU$ with symmetric positive definite *stretch tensor* U. Hence, the internal energy can be formulated as $\Psi(U)$. Due to

$$U^T U = U^T Q^T Q U = (I + u_x)^T (I + u_x) = C$$

we also have $U = C^{1/2}$, so that the internal energy as function of the Cauchy–Green strain tensor C can be written in the form

$$W^i(u) = \int_\Omega \Psi(C) \, dx.$$

In [123] a variety of classical material models are discussed. In the special case of *isotropic* materials the internal energy does not depend on the directional orientation of the material in its initial state. This class contains, for example, glass or rubber, but not fiber materials like wood or bone. The internal energy is then invariant under rigid body transformation of the domain space of u, so that $\Psi(C) = \Psi(Q^T C Q)$ holds for all rotation matrices Q.

The tensor C can also be interpreted as an spd-matrix, so that it can be represented as $C = V^T \Lambda^2 V$ with diagonal matrix $\Lambda^2 = \mathrm{diag}(\lambda_1^2, \ldots, \lambda_d^2)$ and rotation matrix V.

If we identify $Q = V^T$, then the internal energy can be written as a function of the *principal stretches* λ_i:

$$W^i(u) = \int_\Omega \Psi(\Lambda)\, \mathrm{d}x. \tag{2.22}$$

Invariants. For $d = 3$ material models are often formulated in terms of invariants of C or E, respectively. Usually, the quantities

$$I_1 = \operatorname{tr} C = \sum_{i=1}^3 \lambda_i^2,$$

$$I_2 = \frac{1}{2}((\operatorname{tr} C)^2 - \operatorname{tr} C^2) = \sum_{i=1}^3 \prod_{j \neq i} \lambda_j^2,$$

$$I_3 = \det C = \prod_{i=1}^3 \lambda_i^2.$$

are selected, which are exactly the monomial coefficients of the characteristic polynomial for C.

Potential Energy. Any deformation of a body affects its potential energy. From the notion we expect a representation of the volume forces $f(x') = -\nabla V(x')$ as gradient field of a potential V. In reality, the arising volume forces such as gravitational or electrostatic forces are indeed frequently of this form. Under this assumption of potential forces, the potential energy then is

$$W^p(u) = \int_\Omega V(x + u)\, \mathrm{d}x.$$

Total Energy. Stable stationary states are characterized among all feasible deformations as minima of the total energy, the sum of internal and potential energy,

$$W = W^i + W^p.$$

In particular we have, for all directional derivatives $W_u(u)\delta u$ in the direction of deviations δu, the necessary condition

$$W_u(u)\delta u = \int_\Omega (\Psi_u(u) - f(x + u))\delta u \, \mathrm{d}x = 0. \tag{2.23}$$

As stationary states always exhibit an equilibrium of forces, the term $\Psi_u(u)$ can be interpreted as force density, compensated by the external force density f.

The feasible deformations are often defined by fixing the deviations on a part $\partial\Omega'_D \subset \partial\Omega'$ of the boundary. This leads to the Dirichlet boundary conditions

$$u = u_0 \quad \text{on } \partial\Omega_D.$$

Surprisingly, the total energy W is, even for *convex* potential V, in general *not convex*, which means that multiple local minima may exist. This property has consequences both for the existence theory and for the algorithms. In Section 8.3 we will present a special algorithm exemplified at a problem from craniomaxillofacial surgery.

Remark 2.6. In practical applications boundary forces are equally important, but mostly do not originate from a potential and therefore do not lead to a simple energy minimization problem. For the treatment of such forces one requires the more complex concept of the stress tensor. We do not want to go into depth here, but instead refer to the respective literature [39, 123].

Remark 2.7. Hyperelastic material models are mostly formulated as functions of the tensor C or the invariants I_1, I_2, I_3, but for a numerical treatment the equivalent formulation as a function of the tensor E is preferable. Otherwise one will obtain extinction of leading digits close to the reference configuration, i.e., for small u_x (cf. Volume 1, Section 2).

2.3.2   Linear Elastomechanics

An important simplification of equation (2.23) is its linearization around the reference configuration $u = 0$. This approximation makes sense for small strains, say with $\|u_x\| \ll 1$, which are typical for hard materials such as in engine construction or structural engineering. First, by linearization of E the geometric nonlinearity is neglected:

$$\epsilon := \tfrac{1}{2}(u_x + u_x^T) \doteq E.$$

Next, the volume forces are assumed to depend linearly on u, if at all. Consequently, a linear material law is used with quadratic internal energy $\Psi(\epsilon)$. Typically the reference configuration is the energy minimum. Thus, due to $\Psi_\epsilon(0) = 0$, there are no terms of first order in the quadratic energy Ψ. So the internal energy can be generally written as

$$\Psi^i(\epsilon) = \tfrac{1}{2}\epsilon : \mathcal{C}\epsilon = \tfrac{1}{2}\sum_{ijkl} \epsilon_{ij} \mathcal{C}_{ijkl}\epsilon_{kl} \tag{2.24}$$

with a tensor $\mathcal{C}$ of fourth order, where we use the traditional notation " $:$ " for the tensor operation "contraction". Because of the index symmetry in (2.24) and ϵ the tensor $\mathcal{C}$ can be assumed to be symmetric with $\mathcal{C}_{ijkl} = \mathcal{C}_{klij} = \mathcal{C}_{jikl}$. For isotropic materials (2.24) reduces further to the *St. Venant–Kirchhoff* model

$$\mathcal{C}_{ijkl} = \mu(\delta_{ik}\delta_{jl} + \delta_{il}\delta_{jk}) + \lambda\delta_{ij}\delta_{kl}, \quad \text{hence} \quad \mathcal{C}\epsilon = 2\mu\epsilon + \lambda(\operatorname{tr}\epsilon)I,$$

with internal energy density

$$\Psi^i(\epsilon) = \tfrac{1}{2}\lambda(\operatorname{tr}\epsilon)^2 + \mu \operatorname{tr}\epsilon^2 \tag{2.25}$$

depending on the *Lamé constants* $\lambda, \mu \geq 0$, which characterize a linear isotropic material completely. Due to $\operatorname{tr}\epsilon = \operatorname{div} u \approx \det(I + u_x) - 1$ the parameter λ characterizes the dependence of the internal energy upon the pure volume changes, while into $\operatorname{tr} \epsilon^2$ both the volume changes and the volume preserving shape changes enter.

After several reformulations, which will be presented in a more general framework later on in Chapter 4 (on the "weak" solution concept; see also Exercise 4.2) insertion of (2.25) into (2.23) leads to the classical *Lamé–Navier equations*:

$$-2\mu\Delta u - \lambda\nabla\operatorname{div} u = f \quad \text{in } \Omega, \tag{2.26}$$
$$u = u_0 \quad \text{on } \partial\Omega_D,$$
$$n^T \mathscr{C}\epsilon = 0 \quad \text{on } \partial\Omega\backslash\partial\Omega_D.$$

These equations are named after the French mathematicians Gabriel Lamé (1795–1870) and Claude Louis Marie Henri Navier (1785–1836). Here the Laplace operator is applied separately to each component of u so that we have a coupled system of three PDEs. Like in Remark 1.1 it suffices to consider homogeneous Dirichlet data $u_0 = 0$. For strongly compressible materials like cork, the value $\lambda = 0$ is a good model assumption. In this special case equation (2.26) splits into three scalar *Poisson equations* (see Section 1.1).

Missing Rotation Invariance.   Linear elastomechanics only represents a good approximation for small strains. It should be pointed out that the mathematical meaning of strain is not in agreement with the intuitive "view". As shown in Figure 2.6, rigid body motions, where the body is not really "strained", look like large relative strains u_x. In particular, for strong compression, linear elastomechanics may even lead to orientation-changing strains with $\det(I + u_x) < 0$, which obviously does not cor-

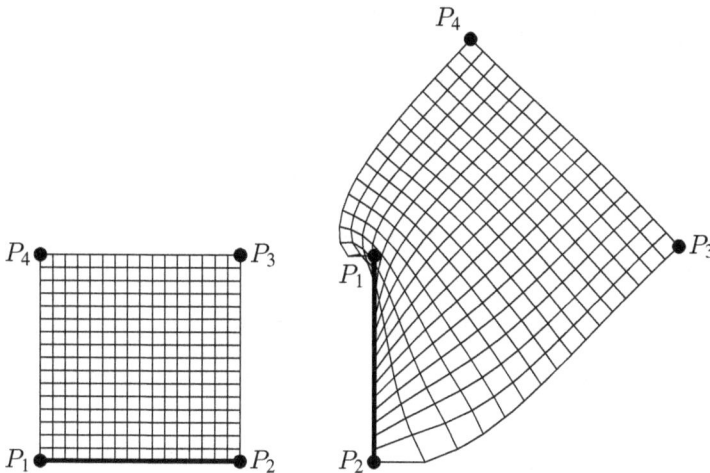

Figure 2.6. Missing rotation invariance of linear elastomechanics. *Left:* reference configuration. *Right:* deformed configuration via rotation of the marked edge by 90°.

rectly describe physical reality (see the neighborhood of P_1 in Figure 2.6). In such cases the geometric nonlinearity should be taken into account by all means.

Theorem 2.8. *The Green–Lagrange strain tensor E is invariant under rotation. This does not hold for its linearization ϵ.*

Proof. Let us represent the rotation by an orthogonal matrix Q. Let $\phi(x) = x + u(x)$ denote a deformation and

$$\hat{\phi}(x) = Q\phi(x) = x + (Qu(x) + (Q - I)x).$$

The corresponding deviations are $u(x)$ and $\hat{u}(x) = Qu(x) + (Q - I)x$. Now we have

$$2\hat{\epsilon} = \hat{u}_x + \hat{u}_x^T = Qu_x + Q - I + u_x^T Q^T + Q^T - I.$$

For $u_x = 0$ we get $2\hat{\epsilon} = Q + Q^T - 2I \neq 0$, so that ϵ cannot be rotation invariant, whereas for the full Green–Lagrange tensor we calculate

$$
\begin{aligned}
2\hat{E} &= 2\hat{\epsilon} + (Qu_x + Q - I)^T (Qu_x + Q - I) \\
&= Qu_x + Q + u_x^T Q^T + Q^T - 2I + u_x^T Q^T Qu_x + Q^T Qu_x - Qu_x \\
&\quad + u_x^T Q^T Q + Q^T Q - Q - u_x^T Q^T - Q^T + I \\
&= Qu_x + Q + u_x^T Q^T + Q^T - 2I \\
&\quad + u_x^T u_x + u_x - Qu_x + u_x^T + I - Q - u_x^T Q^T - Q^T + I \\
&= u_x + u_x^T + u_x^T u_x \\
&= 2E.
\end{aligned}
$$

□

Linear Elastodynamics.   Time-dependent motion of elastic bodies is described by Newton's law. On one hand, the force acting on a test volume is given as the product of mass and acceleration, and on the other hand, it is given by equation (2.23). This is why we get

$$\rho u_{tt} = f + \Psi_u. \tag{2.27}$$

This is, obviously, in agreement with physical experience, since it implies momentum and energy conservation (see Exercise 2.11). Upon assuming small strains again and therefore applying linear elastomechanics, we obtain

$$\rho u_{tt} = f + 2\mu \Delta u + \lambda \nabla \operatorname{div} u, \tag{2.28}$$

which, for strongly compressible materials with $\lambda = 0$, leads to three decoupled *wave equations* (see Section 1.3).

A special case of (2.27) of practical importance is periodic oscillations with small deviation; this is part of *elastoacoustics*, the science which describes the typical vibration behavior of buildings, engines, and vehicles. Due to the normally small oscillation amplitudes in the acoustic frequency regime and consequently small strains, linear elastomechanics is fully sufficient. In general, temporally constant potentials may be neglected there.

The excitation acts via temporally periodic boundary deviations. Due to the linearity of the Lamé–Navier equations (2.26), the resulting oscillation can be computed for each frequency separately. As in Section 1.5.2 we insert the time-harmonic ansatz

$$u(t) = \Re(\bar{u}e^{i\omega t}), \quad f(t) = \Re(\bar{f}e^{i\omega t})$$

into equation (2.28), which leads to

$$-\omega^2\rho\bar{u} = \bar{f} + 2\mu\Delta\bar{u} + \lambda\nabla\,\mathrm{div}\,\bar{u}.$$

For $\lambda = 0$ the system again splits into three decoupled scalar PDEs, this time of *Helmholtz type* (see Section 1.5).

2.4 Exercises

Exercise 2.1. Given are two electric point charges q_1 at position $x_1 \in \mathbb{R}^3$ and q_2 at position x_2 at a distance $|x_1 - x_2| = 1$.

1. Calculate the electric field by superposition of the single potentials.

2. Produce a schematic drawing of the field in the neighborhood of the charges for the two special cases $q_1 = q_2$ and $q_1 = -q_2$. Discuss in particular the decay behavior of the field for $|x| \to \infty$.

Exercise 2.2. Consider an infinitely long, infinitely thin electric conductor. Let a current with density $j = 1$ flow through the conductor. Calculate the resulting magnetic field.
Hint: Exploit the symmetry and Theorem A.7 of Stokes.

Exercise 2.3. Show that by formal interpretation of the gradient operator as vector $\nabla = (\frac{\partial}{\partial x_1}, \ldots, \frac{\partial}{\partial x_d})^T$ the following statements hold:

$$\nabla^T u = \mathrm{div}\,u, \quad (u^T\nabla)u = u^T\nabla u = u_x u \quad \text{und} \quad \nabla \times u = \mathrm{rot}\,u.$$

Exercise 2.4. *Helmholtz decomposition.* For an unknown vector field $q : \mathbb{R}^3 \to \mathbb{R}^3$ let the following equations be satisfied:

$$\mathrm{div}\,q = \rho, \quad \mathrm{curl}\,q = j.$$

Represent q as a sum of a curl-free vector field α and a divergence-free vector field β (see Lemma A.1). Derive the differential equations for the two vector fields. Is the decomposition unique?

Exercise 2.5. Derive the form of the Navier–Stokes equations in a rotating plane coordinate system

$$x(t) = r(t)(\cos(\omega t) + \sin(\omega t)), \quad y(t) = r(t)(-\sin(\omega t) + \cos(\omega t)).$$

Which of the terms describes the centrifugal force, which the Coriolis force?

Exercise 2.6. Consider an incompressible viscous flow in a domain $\Omega \subset \mathbb{R}^d$, $d = 2, 3$ with "no-slip" boundary conditions $u|_{\partial\Omega} = 0$. Show that

$$\int_\Omega u^T \nabla p \, dx = 0$$

and

$$\frac{d}{dt} \int_\Omega \rho |u|^2 \, dx \leq 0 \quad \Rightarrow \quad \nu \geq 0$$

hold, which means that a "negative viscosity" cannot occur.

Exercise 2.7. Consider the stationary flow of a viscous incompressible fluid in $\mathbb{R}^2$ in the direction parallel to the x-axis between two fixed infinite walls $y = 0$ and $y = 1$. Find a solution for u and p under the following assumptions:

$$u = u(y), \quad p = p(x), \quad p(0) = 0, \quad p(1) > 0.$$

Exercise 2.8. *Shallow water equation.* Water is a frictionless incompressible fluid. Let $x = (x_1, x_2, z)$ denote the spatial coordinates. Consider the motion of water in a gravitational field g in the direction of the negative z-axis. For constant mass density ρ the velocity $u = (u_1, u_2, v)$ and the pressure p are determined by the Euler equations with an additional gravitational term

$$u_t + u_x u = -\frac{1}{\rho} \nabla p - g e_z,$$

$$\text{div } u = 0.$$

From this, a simplified model of the wave motion in shallow water is to be derived. Let the following assumptions hold:

- the water depth is small compared to the wave length of sound in water;
- vertical accelerations v_t can be neglected, i.e., in z-direction, each volume element is in a hydrostatic equilibrium;
- the horizontal velocity components are independent of the depth, which means $u_1 = u_1(x_1, x_2), u_2 = u_2(x_1, x_2)$.

Derive the shallow water equation under these assumptions.

Remark: This equation, with an additional quadratic term for ground friction, describes the propagation of a *tsunami*.

Exercise 2.9. In Section 2.2.4 we derived the porous medium equation for compressible fluids. Modify this equation for *incompressible* fluids such as water.

Exercise 2.10. For a vector field q let only $\operatorname{curl} q = j$ and $\operatorname{div} q = \rho$ be known. Give a general representation of the vector field.
  Hint: Apply a separation ansatz $q = \alpha + \beta$ with

$$\operatorname{div} \alpha = \rho, \quad \operatorname{curl} \alpha = 0,$$

$$\operatorname{div} \beta = 0, \quad \operatorname{curl} \beta = j.$$

Exercise 2.11. From equation (2.27) conclude the conservation of the mechanical total energy in the form

$$W = \int_{\Omega} \left(\Psi(u) + V(x+u) + \tfrac{1}{2}\rho|u_t|^2 \right) dx.$$

Chapter 3

Finite Difference Methods for Poisson Problems

In this chapter we turn to the *numerical solution* of PDEs. We start with the simplest class of methods, the difference methods; an alternative class will be introduced in Chapter 4. For reasons of presentation, we restrict ourselves to Poisson problems, in most cases with homogeneous Dirichlet boundary conditions,

$$-\Delta u = f \quad \text{in } \Omega \subset \mathbb{R}^d, \quad u|_{\partial\Omega} = 0, \tag{3.1}$$

where we choose space dimension $d = 2$ throughout.

3.1   Discretization of Standard Problem

As standard problems we select Poisson problems over regular domains, in the simplest case over the *unit square* $\Omega =]0, 1[^2$. In its discretization we turn to a *mesh* Ω_h with *mesh size h*. For the unit square we obtain an *equidistant* mesh (cf. Figure 3.1)

$$\Omega_h := \{(x_i, y_j) \mid 0 \le i, j \le M\}, \quad (x_i, y_j) = (ih, jh)$$

for $Mh = 1$. In analogy to its continuous counterpart, we define the discrete boundary $\partial\Omega_h := \Omega_h \cap \partial\Omega$. Over this mesh, the continuous function u is replaced by a *discrete function* $u_h : \Omega_h \to \mathbb{R}$ with $u_{i,j} := u_h(x_i, y_j)$.

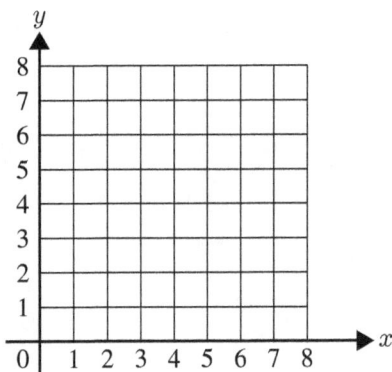

Figure 3.1. Equidistant grid over the unit square: $M = 8$, which means $N = 81$ nodes.

Discretization of the Laplace Operator.   The Laplace operator Δ is approximated by a discrete Laplace operator Δ_h. For a smooth solution this is realized in the interior of Ω_h by the symmetric choice

$$u_{xx}(x_i, y_j) = \frac{1}{h^2}(u_{i-1,j} - 2u_{i,j} + u_{i+1,j}) + \mathcal{O}(h^2),$$

$$u_{yy}(x_i, y_j) = \frac{1}{h^2}(u_{i,j-1} - 2u_{i,j} + u_{i,j+1}) + \mathcal{O}(h^2).$$

Upon dropping $\mathcal{O}(h^2)$ terms, we are led to the definition of the discrete Laplace operator Δ_h via

$$-h^2 \Delta_h u_h(x_i, y_j) = (4u_{i,j} - u_{i-1,j} - u_{i+1,j} - u_{i,j-1} - u_{i,j+1}), \quad 1 \le i, j \le M-1,$$

or schematically also as the two-dimensional 5-*point star*, which is composed of two one-dimensional 3-*point stars* as follows:

$$
\begin{bmatrix}
\cdot & \frac{-1}{h^2} & \cdot \\
\frac{-1}{h^2} & \frac{4}{h^2} & \frac{-1}{h^2} \\
\cdot & \frac{-1}{h^2} & \cdot
\end{bmatrix}
=
\begin{bmatrix}
\frac{-1}{h^2} & \frac{2}{h^2} & \frac{-1}{h^2}
\end{bmatrix}
+
\begin{bmatrix}
\frac{-1}{h^2} \\
\frac{2}{h^2} \\
\frac{-1}{h^2}
\end{bmatrix}.
\tag{3.2}
$$

With the discrete right-hand side $(f_h)_{ij} = f(x_i, y_j)$ one obtains as discretization of (3.1)

$$-\Delta_h u_h = f_h. \tag{3.3}$$

3.1.1   Discrete Boundary Value Problems

In the following we present the discretization for Dirichlet boundary conditions and for homogeneous Neumann boundary conditions in detail. The case of *Robin boundary conditions* for $d = 2$ is left as Exercise 3.1.

Dirichlet Boundary Conditions

In view of Remark 1.1 we consider only the *homogeneous* case. For the mesh Ω_h given above this leads to the conditions

$$u_{0,j} = u_{M,j} = u_{i,0} = u_{i,M} = 0, \quad i = 0, \ldots, M, \quad j = 0, \ldots, M.$$

If we drop the anyway fixed boundary data, we obtain from (3.3) exactly N linear equations for $N = (M-1)^2$ unknowns $u_{1,1}, \ldots, u_{M-1,M-1}$. Upon ordering the unknowns $u_{i,j}$ not in a two-dimensional grid, but as a vector $u \in \mathbb{R}^N$, then $-h^2 \Delta_h$ may be interpreted as a matrix A_h. Thus (3.3) comes out to be a system of linear equations

$$A_h u_h = h^2 f_h. \tag{3.4}$$

For historical reasons the matrix A_h is called *stiffness matrix*.

Numberings.   The actual shape of the vector u_h and the matrix A_h depends on the numbering, i.e., the linear ordering of the doubly-indexed nodes. A change of the numbering acts directly as a simultaneous permutation of rows and columns of A_h. For illustration, we now give a few examples with $M = 4$.

Natural ordering. This name is often given to the row-wise numbering

$$(i, j) \mapsto (M - 1)(i - 1) + j.$$

We thereby obtain a mesh as represented in Figure 3.2, left, together with its numbering.

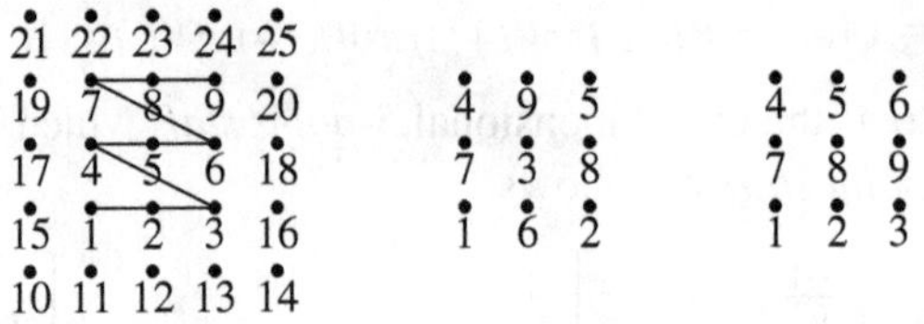

Figure 3.2. Numberings in comparison ($M = 4, N = 9$). *Left:* natural ordering. *Center:* checkerboard pointwise ordering. *Right:* checkerboard row-wise ordering.

For $N = 9$ unknowns, the stiffness matrix is obtained as

$$A_h = \begin{bmatrix}
4 & -1 & & \cdot & -1 & & & \cdot & \\
-1 & 4 & -1 & \cdot & & -1 & & & \cdot \\
& -1 & 4 & \cdot & & & -1 & \cdot & \\
& \cdot & \cdot & \cdot & \cdot & \cdot & \cdot & \cdot & \cdot & \cdot \\
-1 & & & \cdot & 4 & -1 & & \cdot & -1 \\
& -1 & & \cdot & -1 & 4 & -1 & \cdot & & -1 \\
& & -1 & \cdot & & -1 & 4 & \cdot & & & -1 \\
& & \cdot & \cdot & \cdot & & & \cdot & \cdot & \cdot \\
& & & \cdot & -1 & & & \cdot & 4 & -1 \\
& & & & \cdot & -1 & & \cdot & -1 & 4 & -1 \\
& & & & & & -1 & \cdot & & -1 & 4 \\
\end{bmatrix}. \qquad (3.5)$$

Upon row partitioning we obtain the *block tridiagonal* shape ($M = 4, N = 9$)

$$A_h = \begin{bmatrix}
T_3 & -I_3 & \\
-I_3 & T_3 & -I_3 \\
& -I_3 & T_3
\end{bmatrix}, \quad T_3 = \begin{bmatrix}
4 & -1 & \\
-1 & 4 & -1 \\
& -1 & 4
\end{bmatrix}, \quad -I_3 = \begin{bmatrix}
-1 & & \\
& -1 & \\
& & -1
\end{bmatrix}.$$

Checkerboard pointwise or red-black point ordering. This numbering is given in Figure 3.2, middle, and leads to the matrix

$$
A_h = \begin{bmatrix} 4I_5 & H \\ H^T & 4I_4 \end{bmatrix}, \quad H = \begin{bmatrix} -1 & -1 & & & \\ -1 & & -1 & & \\ -1 & -1 & -1 & -1 \\ & -1 & & -1 & \\ & & -1 & -1 \end{bmatrix}. \tag{3.6}
$$

This shape is a special case of the "property $\mathcal{A}$" as introduced by D. Young [225] for the analysis of linear iterative methods. Such a structure occurs essentially only for matrices originating from difference methods over uniform meshes.

Definition 3.1. *Property $\mathcal{A}$.* This property is defined by the existence of a permutation matrix P such that

$$
PAP^T = \begin{bmatrix} D_r & L^T \\ L & D_b \end{bmatrix} \tag{3.7}
$$

with positive diagonal matrices D_r, D_b (r/b for "red/black") and a rectangular coupling matrix L.

Red black line or zebra ordering. In this case, one has the numbering 3.2, right, thus obtaining a matrix of the shape

$$
A = \begin{bmatrix} T_3 & 0 & -I_3 \\ 0 & T_3 & -I_3 \\ -I_3 & -I_3 & T_3 \end{bmatrix}.
$$

Remark 3.2. In principle, all numberings are equivalent. However, since the shape of the stiffness matrix depends strongly on the numbering, any solution algorithms, should be, if at all possible, invariant under row or column permutation, respectively (see Section 5.3.1). If this is not the case, then the question of an optimal numbering comes up for the corresponding algorithms (example: Gauss-Seidel method; see Section 5.2.2).

  The rather simple matrix structures shown here arise only for purely rectangular domains, not for L-shaped domains (see Exercise 3.2) or even for domains with curvilinear boundaries (see Section 3.3.2).

Neumann Boundary Conditions

We start from homogeneous Neumann boundary conditions

$$n^T \nabla u = 0.$$

In this case the continuous problem is determined only up to an additive constant (cf. Theorem 1.3). This basic structure will also occur in the discrete problem, i.e., the matrices originating from the discretization will be expected to be *singular*. Beyond that and in view of the subsequent Section 3.1.2, the originating matrix will be symmetric, thus in total *symmetric positive semidefinite*, in order to inherit the eigenvalue structure of the operator $-\Delta$ correctly into the discrete.

For the discretization of Neumann boundary conditions there are essentially two variants, which we will now briefly present. In both cases, unknowns $u_{-1,j}$ beyond the boundary are defined (see Figure 3.3).

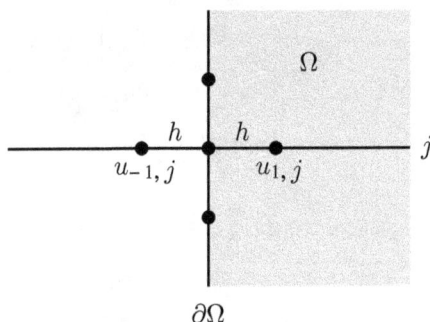

Figure 3.3. External boundary point with Neumann boundary conditions.

Symmetric Discretization. Here we discretize the boundary derivative symmetrically according to

$$0 = n^T \nabla u = \frac{1}{2h}(u_{-1,j} - u_{1,j}) + \mathcal{O}(h^2).$$

Obviously, this discretization of the normal derivative possesses the same approximation order $\mathcal{O}(h^2)$ as the discretization of the Laplace operator. Thus we set at the boundary

$$u_{-1,j} = u_{1,j}.$$

This leads to the 4-*point star*

$$4u_{0,j} - u_{0,j+1} - u_{0,j-1} - 2u_{1,j} = h^2 f_{0,j}.$$

Insertion of these boundary terms into the stiffness matrix then yields

$$
A_h =
\begin{bmatrix}
4 & -2 & & \cdot & -2 & & & \cdot & \\
-1 & 4 & -1 & \cdot & & -2 & & \cdot & \\
& -2 & 4 & \cdot & & & -2 & \cdot & \\
& \cdot & \cdot & \cdot & \cdot & \cdot & \cdot & \cdot & \cdot \\
-1 & & & \cdot & 4 & -2 & & \cdot & -1 \\
& -1 & & \cdot & -1 & 4 & -1 & \cdot & & -1 \\
& & -1 & \cdot & & -2 & 4 & \cdot & & & -1 \\
& \cdot & \cdot & \cdot & \cdot & \cdot & \cdot & \cdot & \cdot & \cdot \\
& & & \cdot & -2 & & & \cdot & 4 & -2 \\
& & & & \cdot & -2 & & \cdot & -1 & 4 & -1 \\
& & & & & \cdot & -2 & \cdot & & -2 & 4
\end{bmatrix},
$$

where $N = (M+1)^2$, in our example therefore $M = 2$, $N = 9$. Due to the property

$$
A_h e = 0, \quad e^T = (1, \ldots, 1) \in \mathbb{R}^N \tag{3.8}
$$

this matrix is obviously *singular*, as required. However, it is *nonsymmetric*, which means that it can have complex eigenvalues – thus differing from the Laplace operator (cf. Section 3.1.2). A small comfort is given in Exercise 3.3.

Unsymmetric Discretization.   Here we discretize the boundary derivative unsymmetrically according to

$$
0 = n^T \nabla u = \frac{1}{h}(u_{-1,j} - u_{0,j}) + \mathcal{O}(h). \tag{3.9}
$$

The approximation order $\mathcal{O}(h)$ for the boundary derivative is reduced by one compared with the discretization of the Laplace operator. However, as we will show in Section 3.2 below, this does not perturb the convergence order in total. At the boundary we set

$$
u_{-1,j} = u_{0,j}.
$$

As stiffness matrix we obtain an (N, N)-matrix of the form

$$
A_h =
\begin{bmatrix}
2 & -1 & & \cdot & -1 & & & \cdot & & \\
-1 & 3 & -1 & \cdot & & -1 & & \cdot & & \\
 & -1 & 2 & \cdot & & & -1 & \cdot & & \\
\cdot & \cdot & \cdot & \cdot & \cdot & \cdot & \cdot & \cdot & \cdot & \cdot \\
-1 & & & \cdot & 3 & -1 & & \cdot & -1 & \\
 & -1 & & \cdot & -1 & 4 & -1 & \cdot & & -1 \\
 & & -1 & \cdot & & -1 & 3 & \cdot & & & -1 \\
\cdot & \cdot & \cdot & \cdot & \cdot & \cdot & \cdot & \cdot & \cdot & \cdot \\
 & & & \cdot & -1 & & & \cdot & 2 & -1 \\
 & & & & & -1 & & \cdot & -1 & 3 & -1 \\
 & & & & & & -1 & \cdot & & -1 & 2 \\
\end{bmatrix},
$$

where again $N = (M + 1)^2$, hence in our example $M = 2$, $N = 9$. The relation (3.8) holds as well; the matrix is therefore again *singular*. This time, however, it is also *symmetric*, in total then *symmetric positive semidefinite*.

Summary. Symmetric discretization at the boundary leads to a nonsymmetric matrix, unsymmetric discretization at the boundary to a symmetric matrix, which is preferable. Due to the singularity of the stiffness matrix, we have to dispose about the undefined degree of freedom to make the problem uniquely solvable. This is usually done via fixing one of the values u_h at the boundary, sometimes also by discretization of the compatibility condition (1.3).

3.1.2   Discrete Eigenvalue Problem

In this section we will deal with the eigenvalue problem that arises after discretization of the Poisson model problem. We again assume an equidistant mesh

$$\Omega_h := \{(x_i, y_j) | 0 \leq i, j \leq M\}, \quad (x_i, y_j) = (ih, jh)$$

over the unit square. Instead of the continuous Poisson eigenvalue problem (1.7), here one obtains the discrete Poisson eigenvalue problem

$$-\Delta_h u_h(x_i, y_j) = \lambda_h u_h(x_i, y_j).$$

If we collect all nodes, then an eigenvalue problem from linear algebra (see, e.g., Volume 1, Chapter 5) emerges:

$$A_h u_h = \lambda_h u_h.$$

As derived in Section 3.1.1, one obtains, for *Dirichlet boundary conditions* and natural ordering of the unknowns, a *symmetric positive definite* (N, N)-matrix of the structure (let $m = M - 1$)

$$
A_h = \begin{bmatrix} T_m & -I_m & & & \\ -I_m & T_m & \ddots & & \\ & \ddots & \ddots & \ddots & \\ & & \ddots & \ddots & -I_m \\ & & & -I_m & T_m \end{bmatrix}, \quad T_m = \begin{bmatrix} 4 & -1 & & & \\ -1 & 4 & \ddots & & \\ & \ddots & \ddots & \ddots & \\ & & \ddots & \ddots & -1 \\ & & & -1 & 4 \end{bmatrix},
$$

$$
-I_m = \begin{bmatrix} -1 & & \\ & \ddots & \\ & & -1 \end{bmatrix}.
$$

Like the operator $-\Delta$, this matrix also has a real positive spectrum.

Because of the extreme simplicity of our mesh we may, just as in the continuous case, make an ansatz in terms of a Fourier expansion:

$$
u_h(x_i, y_j) = \sum_{k,l=1}^{M-1} \alpha_{k,l} \sin(k\pi x_i) \sin(l\pi y_j),
$$

whereby the Dirichlet boundary conditions are automatically satisfied. In view of (3.2), we may decompose the discrete Laplace operator in $\mathbb{R}^2$ into two one-dimensional Laplace operators according to

$$
\Delta_h = \Delta_{h,x} + \Delta_{h,y}.
$$

Thus we obtain

$$
\Delta_h u_h(x_i, y_j) = \sum_{k,l=1}^{M-1} \alpha_{k,l} \Big(\big(\Delta_{h,x} \sin(k\pi x_i)\big) \sin(l\pi y_j) \\ + \sin(k\pi x_i) \big(\Delta_{h,y} \sin(l\pi y_j)\big) \Big).
$$

It is therefore sufficient to determine the eigenvalues $\{\lambda_{h,k}\}$ in our simple model problem from the one-dimensional discrete eigenvalue problem

$$
\Delta_{h,x} \sin(k\pi x_i) = \lambda_{h,k} \sin(k\pi x_i) .
$$

For ease of writing, we define $s_i = \sin(k\pi i h)$. We can go back to the trigonometric three-term recurrence (see Volume 1, Section 2.3.2, Example 2.27):

$$
s_{i+1} = 2\cos(k\pi h)s_i - s_{i-1}.
$$

Insertion into our eigenvalue problem then supplies

$$-\Delta_{h,x}s_i = \frac{1}{h^2}(2s_i - (s_{i+1} + s_{i-1})) = 2\frac{1 - \cos(k\pi h)}{h^2}s_i = 4\frac{\sin^2\left(\frac{1}{2}k\pi h\right)}{h^2}s_i.$$

In this way we obtain the desired eigenvalues in the one-dimensional case

$$\lambda_{h,k} = 4\frac{\sin^2\left(\frac{1}{2}k\pi h\right)}{h^2}. \tag{3.10}$$

Upon collecting pieces we get

$$-\Delta_h u_h(x_i, y_j) = \sum_{k,l=1}^{M-1} \alpha_{k,l}(\lambda_{h,k} + \lambda_{h,l}) \sin(k\pi x_i)\sin(l\pi y_j).$$

From this intermediate result we may directly read off the eigenfunctions

$$\varphi_{h,k,l}(x_i, y_j) = \sin(k\pi x_i)\sin(l\pi y_j), \quad k,l = 1,\ldots, M-1 \tag{3.11}$$

and the associated eigenvalues

$$\lambda_{h,k,l} = 4\frac{\sin^2\left(\frac{1}{2}k\pi h\right) + \sin^2\left(\frac{1}{2}l\pi h\right)}{h^2}, \quad k,l = 1,\ldots, M-1.$$

The discrete eigenvalues exactly mimic the distribution (1.13) of the continuous spectrum as can be seen from

$$\lim_{h\to 0} \lambda_{h,k,l} = (k^2 + l^2)\pi^2, \quad k,\, l = 1,\ldots,\infty,$$

where we exploited that $h \to 0$ implies $M \to \infty$. The eigenvalues of the matrix A_h differ from those of the discrete Laplace operator $-\Delta_h$ by a scaling factor h^2, (cf. (3.3) and (3.4)). Thus we get the minimum and the maximum eigenvalue of the matrix as

$$\lambda_{\min}(A_h) = 8\sin^2\left(\tfrac{1}{2}\pi h\right), \quad \lambda_{\max}(A_h) = 8\cos^2\left(\tfrac{1}{2}\pi h\right) = 8 - \lambda_{\min}(A_h). \tag{3.12}$$

Homogeneous *Neumann boundary conditions* (with $N = (M+1)^2$) are satisfied by the corresponding ansatz

$$u_h(x_i, y_j) = \sum_{k,l=0}^{M} \alpha_{k,l} \cos(k\pi x_i)\cos(l\pi y_j).$$

The calculation runs just as before. As eigenfunctions we again obtain the trigonometric functions above, as eigenvalues become accordingly

$$\lambda_{h,k,l} = 4\frac{\sin^2\left(\frac{1}{2}k\pi h\right) + \sin^2\left(\frac{1}{2}l\pi h\right)}{h^2}, \quad k,l = 0,\ldots, M.$$

This yields, with $Mh = 1$, in particular

$$\lambda_{\min}(A_h) = 0, \quad \lambda_2(A_h) = 4\sin^2\left(\tfrac{1}{2}\pi h\right), \quad \lambda_{\max}(A_h) = 8. \tag{3.13}$$

In the following, we will now and then make detailed use of the matrix eigenvalues for Dirichlet and Neumann boundary conditions.

3.2 Approximation Theory on Uniform Grids

In the preceding section we introduced elementary finite difference methods for the example of our Poisson model problems. In the present section, we want to investigate, how the discrete solutions u_h thus defined converge towards the continuous solution u. For simplicity, we start from a Poisson problem over the unit square:

$$-\Delta u = f, \quad x \in \Omega =]0, 1[^2, \quad u|_{\partial\Omega} = 0.$$

The corresponding discretization then reads

$$-\Delta_h u_h = f_h, \quad u_h|_{\partial\Omega_h} = 0.$$

The discrete Laplace operator Δ_h represents the discretization via the 5-point star introduced above. With homogeneous Dirichlet boundary conditions we obtain a linear system of equations

$$A_h u_h = h^2 f_h \tag{3.14}$$

with $N = (M - 1)^2$ unknown components of u_h, where, due to the unit square, the relation $Mh = 1$ holds. So the dimension N of the equation system is coupled to the mesh size h according to

$$N \approx \frac{|\Omega|}{h^2},$$

where we have $|\Omega| = 1$ for the unit square.

Discretization Error. For the *global* discretization error over the complete mesh Ω_h we define the mesh function

$$\epsilon_h := u_h - u|_{\Omega_h}.$$

Of interest here is the dependence of the error ϵ_h on h. In order to quantify the errors, we will arrange the values $\epsilon_h(x_i, y_j)$ as components in a long vector $\epsilon_h \in \mathbb{R}^N$ and recur on vector norms.

Norms. Having solved the linear systems (3.14), we can measure the algebraic errors for fixed dimension N in some *vector norm*

$$|\epsilon_h|_p \quad \text{for } 1 \leq p \leq \infty.$$

However, for an evaluation of the global discretization errors we need to identify suitable norms $\|g_h\|$, for discretizations $g_h = g|_{\Omega_h}$ of fixed functions g, that "survive" the limit process $h \to 0$, i.e., that lead to norms in the respective function spaces.

Maximum Norm ($p = \infty$). We begin with the case

$$|g_h|_\infty = \max_{k=1,\dots,N} |g_{h,k}|.$$

For $h \to 0$ we get $N \to \infty$ as well as, for continuous g:

$$|g_h|_\infty \to \max_{(x,y) \in \Omega} |g(x,y)| = \|g\|_\infty.$$

This means that for every pair (h, N) we can define

$$\|g_h\|_\infty := |g_h|_\infty = \max_{k=1,\ldots,N} |g_{h,k}|.$$

In summary, the maximum norm is a suitable norm for a convergence theory in the function space $L^\infty(\Omega)$.

Euclidean norm ($p = 2$).   Next we investigate

$$|g_h|_2^2 = \sum_{k=1}^{N} g_{h,k}^2.$$

For $h \to 0$ this expression is *unbounded* and therefore not useful for a convergence theory. We require a suitable scaling that compensates for the growing number of terms in the sum. For this reason, we define

$$\|g_h\|_2^2 := \frac{1}{N} |g_h|_2^2 = \frac{1}{N} \sum_{k=1}^{N} g_{h,k}^2.$$

For $h \to 0$ and continuous g we can pick the two-dimensional trapezoidal sum as approximation of the two-dimensional integral (for the one-dimensional case cf., for example, Volume 1, Chapter 9):

$$h^2 |g_h|_2^2 = h^2 \sum_{\Omega_h} g_h(x,y)^2 \to \int_\Omega g(x,y)^2 \, dx \, dy = \|g\|_2^2.$$

Obviously, this scaling leads to the function space $L^2(\Omega)$.

In $\mathbb{R}^N$, the norms $|\cdot|_\infty$ and $|\cdot|_2$ are of course equivalent. This equivalence, however, breaks down in the limit case $N \to \infty$ or $h \to 0$, respectively. Instead we have

$$h \|g_h\|_\infty \le \|g_h\|_2 \le \|g_h\|_\infty, \tag{3.15}$$

which is obtained from the elementary estimation chain

$$h^2 \|g_h\|_\infty^2 = \frac{1}{N} \max_k g_{h,k}^2 \le \frac{1}{N} \sum_{k=1}^{N} g_{h,k}^2 = \|g_h\|_2^2 \le \frac{1}{N} N \max_k g_{h,k}^2 = \|g_h\|_\infty^2.$$

Energy norm.   The finite dimensional energy norm for fixed mesh size h reads

$$|g_h|_A^2 = g_h^T A_h g_h. \tag{3.16}$$

Compared with the Euclidean norm it has the advantage that it is invariant under componentwise scaling of the vector g_h, i.e., under pointwise scaling of the grid function g_h. For the limit process $h \to 0$, however, we get

$$g_h^T A_h g_h = -h^2 \sum_{\Omega_h} g(x, y)\Delta_h g(x, y) \to -h^2 N \int_\Omega g(x, y)\Delta g(x, y)\, dx\, dy,$$

so that the expression (3.16) is reasonable only for $d = 2$. With the special scaling

$$\|g_h\|_A^2 \sim h^{d-2}|g_h|_A^2$$

the limit process can also be performed for other space dimensions. For homogeneous Dirichlet boundary conditions we thus obtain a norm equivalent to the norm in the Sobolev space $H_0^1(\Omega)$ (see Appendix A.5). For homogeneous Neumann boundary conditions, however, one only obtains some seminorm in $H^1(\Omega)$ (cf. Theorem 1.3).

3.2.1 Discretization Error in L^2

We begin with the theory in $L^2(\Omega)$, since this is relatively easy to derive. In order to first achieve *local consistency*, we apply the 5-point star to the exact solution u and obtain for $(x, y) \in \Omega_h \backslash \partial\Omega$ and $u \in C^4(\overline{\Omega})$

$$-\Delta_h u(x, y) = \frac{1}{h^2}\big(2u(x, y) - u(x + h, y) - u(x - h, y)$$
$$+ 2u(x, y) - u(x, y + h) - u(x, y - h)\big)$$
$$= -\frac{1}{h^2}\Big(h^2 u_{xx}(x, y) + \frac{h^4}{12}u_{xxxx}(\xi, y)$$
$$+ h^2 u_{yy}(x, y) + \frac{h^4}{12}u_{yyyy}(x, \eta)\Big)$$
$$= -\Delta u(x, y) + \frac{h^2}{12}\big(u_{xxxx}(\xi, y) + u_{yyyy}(x, \eta)\big),$$

where we have applied the mean value theorem in $\mathbb{R}^1$ along each coordinate direction separately. Due to the linearity of the operators Δ, Δ_h we have

$$-\Delta_h(u_h - u|_{\Omega_h}) = \frac{1}{12}h^2\big(u_{xxxx}(\xi, y) + u_{yyyy}(x, \eta)\big).$$

Note that in this difference the right-hand side $f|_{\Omega_h} = f_h$ has dropped out. Thus we obtain an error equation of the kind

$$-\Delta_h \epsilon_h = h^2 R_h, \tag{3.17}$$

where we have

$$\|R_h\|_\infty \le \tfrac{1}{6}\|u^{(4)}\|_\infty \tag{3.18}$$

for the right-hand side. The corresponding linear equation system then reads

$$A_h \epsilon = h^4 R \tag{3.19}$$

with the notation $R = (R_1, \ldots, R_N)^T$. Beyond that we have to fix the boundary conditions. Thus we arrive at the following convergence theorems.

Theorem 3.3. *Consider the Poisson model problem (3.1) for $u \in C^4(\overline{\Omega})$ with Dirichlet boundary conditions on the unit square Ω_h for mesh size $h \leq 1$. Then the global discretization error comes out as*

$$\|\epsilon_h\|_2 \leq \frac{\|u^{(4)}\|_\infty}{6\lambda_{\min}} h^4 \leq \frac{\|u^{(4)}\|_\infty}{48} h^2. \tag{3.20}$$

Proof. For Dirichlet conditions the matrix A_h is *symmetric positive definite*, from which

$$0 < \lambda_{\min} |x|_2^2 \leq x^T A_h x \leq \lambda_{\max} |x|_2^2$$

can be concluded. Application of this relation to (3.19) yields

$$\lambda_{\min} |\epsilon|_2^2 \leq \epsilon^T A_h \epsilon = h^4 \epsilon^T R \leq h^4 |\epsilon|_2 |R|_2. \tag{3.21}$$

If we cancel the common factor $|\epsilon|_2$ on both sides and introduce the scaling factor h^2, then we can pass on to the L^2-norm according to

$$\lambda_{\min} \|\epsilon_h\|_2 \leq h^4 \|R_h\|_2 \leq h^4 \|R_h\|_\infty \leq \tfrac{1}{6} h^4 \|u^{(4)}\|_\infty.$$

From (3.12) we can take the expression for the minimal eigenvalue at Dirichlet boundary conditions and estimate a lower bound as

$$\lambda_{\min} = 8 \sin^2 \left(\tfrac{1}{2}\pi h\right) \geq 8h^2 \quad \text{for } h \leq 1.$$

This finally supplies the desired error estimation.   □

In the case of homogeneous Neumann boundary conditions the above estimation technique is not directly applicable, since then $\lambda_{\min} = 0$. Therefore we have to dig a bit deeper.

Theorem 3.4. *Consider the Poisson model problem for $u \in C^4(\overline{\Omega})$ with homogeneous Neumann boundary conditions on the unit square Ω_h for mesh size h. Assume we have made provisions for the degree of freedom undetermined according to 1.3. Then we have for the global discretization error*

$$\|\epsilon_h\|_2 \leq \frac{\|u^{(4)}\|_\infty}{6\lambda_2} h^4 \leq \frac{\|u^{(4)}\|_\infty}{24} h^2, \quad h \leq 1. \tag{3.22}$$

Proof. In this case the matrix A_h is singular, i.e., *positive semidefinite*. Therefore we start from (3.8), which means that the nullspace of A_h is spanned by $e^T = (1, \ldots, 1)$. Due to $e^T e = N$ the projectors

$$P = I - \frac{1}{N} e e^T, \quad P^\perp = I - P$$

are orthogonal, i.e., $P^2 = P$, $P^T = P$. With these projectors we have $A_h P = A_h$, $P A_h = A_h$ as well as the complementary result $A_h P^\perp = 0$, $P^\perp A_h = 0$. Application of $P^\perp$ to the linear equation system (3.19) supplies

$$0 = P^\perp A_h \epsilon = h^4 P^\perp R.$$

Application of P yields the reduced linear equation system

$$(P A_h P)(P \epsilon) = h^4 P R,$$

from which the unknowns $P \epsilon$ can be determined. The degree of freedom $P^\perp \epsilon$ still missing can be fixed in a simple way: as given at the end of Section 3.1.1, we merely choose a discrete value such that $P^\perp u_h = P^\perp u$, from which $P^\perp \epsilon = 0$ immediately follows. Hence, $|\epsilon|_2 = |P \epsilon|_2$. In order to be able to modify the remaining steps of the above proof, we need to have a closer look into the eigenvalue problem for $P A_h P$. Let

$$A_h \eta_i = \lambda_i \eta_i, \quad i = 1, \ldots, N,$$

where $\eta_1 \sim e$ corresponds to the eigenvalue $\lambda_1 = 0$. As the eigenvector system is orthogonal, we have $e^T \eta_i = 0$, $i = 2, \ldots, N$. Application of the projector P then generates the reduced eigenvalue problem

$$(P A_h P)(P \eta_i) = \lambda_i (P \eta_i).$$

Due to

$$P \eta_i = \eta_i - \frac{1}{N} (e^T \eta_i) e = \eta_i, \quad i = 2, \ldots, N$$

the nonvanishing eigenvalues and the corresponding eigenvectors are preserved. However, the eigenvalue $\lambda_{\min} = 0$ falls out of the spectrum due to $Pe = 0$, which means that we have

$$0 < \lambda_2 < \cdots < \lambda_{\max}.$$

Therefore, in order to estimate $|P \epsilon|_2$, we need only, compared to the Dirichlet case, substitute $\lambda_{\min}$ by λ_2 from (3.13) and come up with a lower bound in the usual way. Thus we finally obtain (3.22).   □

Energy Norm.   The convergence behavior in the energy norm is obtained in passing from the two previous theorems. For this purpose, we return to the intermediate step (3.21) in the proof of Theorem 3.3, which is also valid for Theorem 3.4:

$$\epsilon^T A_h \epsilon = h^4 \epsilon^T R \le h^4 \, |\epsilon|_2 \, |R|_2 \, .$$

For $d = 2$, we obtain with either Theorem 3.3 or 3.4

$$\|\epsilon_h\|_{A_h}^2 \le h^2 \|\epsilon_h\|_2 \, \|R_h\|_2 \le \text{const } h^4,$$

from which we immediately get the bound

$$\|\epsilon_h\|_{A_h} \le \text{const } h^2.$$

3.2.2   Discretization Error in L^∞

At first glance it seems that we merely need to apply the left inequality in (3.15) to derive an L^∞-result from the L^2-results (3.20) or (3.22), respectively. Upon following this procedure, we would, however, lose one power of h, which is why we turn here to a more subtle estimation technique. For this purpose we use a discrete variant of the continuous maximum principle, which dates back to [115, 132] and was provided in Section 1.1.1, Theorem 1.4.

Discrete Maximum Principle.   Let Δ_h be the discretization of the Laplace operator defined by the 5-point star.

Theorem 3.5. *Let u_h denote the discrete solution of the Poisson equation over a d-dimensional equidistant mesh Ω_h. Assume that $\Delta_h u_h = f_h > 0$. Then*

$$\max_{\Omega_h} u_h = \max_{\partial\Omega_h} u_h.$$

Proof. Let $(x_h, y_h) \in \Omega_h$ be an arbitrary *internal* node. Let $B := \{(\xi_h, \eta_h) \in \Omega_h : |(\xi_h, \eta_h) - (x_h, y_h)| = h\}$ denote the set of neighboring nodes. The 5-point star has a factor -4 in its center and factors 1 in all four neighboring nodes of B. From the relation $\Delta_h u_h > 0$ we then conclude that

$$u_h(x_h, y_h) < \tfrac{1}{4} \sum_{(\xi_h, \eta_h) \in B} u_h(\xi_h, \eta_h) \le \max_{(\xi_h, \eta_h) \in B} u_h(\xi_h, \eta_h).$$

Hence, the point (x_h, y_h) is no maximum. This means that the maximum must be realized in the boundary points of the mesh Ω_h.   □

Compared with the proof of Theorem 1.4 this proof is a bit simpler: as we have assumed strict positivity, we only needed the first step. The second step comes into play, when we want to gain an estimation for the modulus of u_h, and this for arbitrary f_h. This leads us first to the following auxiliary result.

Lemma 3.6. *Let u_h denote the discrete solution of the Poisson equation $\Delta_h u_h = f_h$ on the d-dimensional equidistant mesh Ω_h. In contrast to Theorem 3.5 the values of f_h may now have an arbitrary sign. Then*

$$\max_{\Omega_h} |u_h| \leq \max_{\partial \Omega_h} |u_h| + C \|f_h\|_\infty \qquad (3.23)$$

with a constant C depending only on the domain Ω.

Proof. Suppose the maximum of u_h is achieved at the boundary, then statement (3.23) is already satisfied independent of the second term. Assume therefore that the maximum is achieved at an *interior* node ξ. As in the second part of the proof of Theorem 1.4, we define $v(x) := \|\xi - x\|_2^2$. Obviously, we get $v(\xi) = 0$, $\|v(x)\|_\infty \leq 2dC$ with a constant which only depends on the domain Ω as well as $\Delta_h v = \Delta v = 2d > 0$. In order to exploit the former Theorem 3.5, we set

$$w_h = u_h + \gamma v, \quad \gamma = \frac{1}{2d}\left(\epsilon + \max_{\Omega_h}(-f_h)\right).$$

For this auxiliary function we have

$$\Delta_h w_h = f_h + 2d\gamma = f_h + \epsilon + \max_{\Omega_h}(-f_h) \geq \epsilon > 0,$$

which, by application of Theorem 3.5, immediately supplies

$$\max_{\Omega_h} w_h = \max_{\partial \Omega_h} w_h .$$

We collect all intermediate results and obtain

$$\max_{\Omega_h} u_h = u_h(\xi) = w_h(\xi) \leq \max_{\Omega_h} w_h = \max_{\partial \Omega_h} w_h$$

$$\leq \max_{\partial \Omega_h} u_h + \gamma \|v\|_\infty = \max_{\partial \Omega_h} u_h + C(\epsilon + \max_{\Omega_h}(-f_h)).$$

With $\epsilon \to 0$ this yields

$$\max_{\Omega_h} u_h \leq \max_{\partial \Omega_h} u_h + \frac{\|v\|_\infty}{2d} \max_{\Omega_h} |f_h| \leq \max_{\partial \Omega_h} u_h + C \|f_h\|_\infty.$$

In perfect analogy this relation also holds for $-u_h$, i.e.,

$$\max_{\Omega_h}(-u_h) \leq \max_{\partial \Omega_h}(-u_h) + C \|f_h\|_\infty.$$

We can therefore use either of the two estimates for the positive and the negative components of u_h. After collecting these results we finally arrive at statement (3.23). □

After these preparations, we are now ready to prove an L^∞-estimate of the discretization error.

Theorem 3.7. *Consider the Poisson model problem for $u \in C^4(\overline{\Omega})$ on the unit square Ω_h for mesh size h. For Dirichlet boundary conditions, the global discretization error can be bounded by*

$$\|\epsilon_h\|_\infty \le \tfrac{1}{6}\|u^{(4)}\|_\infty C \, h^2. \tag{3.24}$$

For homogeneous Neumann boundary conditions we get (with unsymmetric discretization)

$$\|\epsilon_h\|_\infty \le \left(\tfrac{1}{2}\|u^{(2)}\|_\infty + \tfrac{1}{6}\|u^{(4)}\|_\infty C\right) h^2. \tag{3.25}$$

Proof. We apply Lemma 3.6 to the error equation (3.17) and for now obtain, without taking special boundary conditions into account,

$$\max_{\Omega_h} |\epsilon_h| \le \max_{\partial\Omega_h} |\epsilon_h| + C h^2 \|R_h\|_\infty.$$

For Dirichlet boundary conditions we have that $\epsilon_h|_{\partial\Omega_h} = 0$, wherefrom, with (3.18), we immediately get (3.24).

For homogeneous Neumann boundary conditions with unsymmetric discretization (see (3.9)), one obtains

$$\max_{\partial\Omega_h} |\epsilon_h| \le \tfrac{1}{2}h^2 \|u^{(2)}\|_\infty,$$

i.e., also a term $\mathcal{O}(h^2)$. Hence, equation (3.25) is shown, too.   □

Remark 3.8. In the presence of *reentrant corners* with internal angle $\alpha\pi > \pi$, we in general have $u \notin C^4(\overline{\Omega})$. Therefore, in all estimates containing a factor $\|u^{(4)}\|_\infty$, the convergence order drops to $\mathcal{O}(h^{1+\frac{1}{\alpha}})$ (here without proof).

3.3   Discretization on Nonuniform Grids

Poisson problems appear as subtasks in many problems from science and engineering, e.g., in fluid dynamics (see Chapter 2.2) – there, however, mostly in connection with rather complex boundary geometries. Moreover, the diffusion coefficients $\sigma(x)$ or the right sides $f(x)$ often exhibit strongly *localized* effects for $x \in \Omega$, which one would like to resolve on fine meshes to gain better accuracy, while coarser meshes with less nodes would do in other parts of the domain. In both cases one will realize nonequidistant meshes with a problem adapted number of unknowns, thus achieving tolerable computing times. In what follows we will investigate how finite difference methods can cope with this kind of challenge.

3.3.1   One-dimensional Special Case

The essential effect of nonuniform discretizations in finite difference methods can already be seen in the one-dimensional case, which we will therefore elaborate first.

Let $\Omega_h = \{x_1 < x_2 < \cdots < x_N\}$ with intermediate nodes $x_{i+1/2} := \frac{1}{2}(x_i + x_{i+1})$. In real-life applications, the Laplace operator usually must be extended via some spatially dependent (material dependent) *diffusion coefficient* $\sigma(x) > 0$ according to

$$\Delta u \to \operatorname{div}(\sigma(x)\nabla u).$$

The coefficient σ often exhibits discontinuities (jumps) (see Exercise 3.5). A symmetric discretization of this differential operators dates back to C. E. Pearson [168]:

$$\operatorname{div}(\sigma\nabla u)|_{x_i} \approx \frac{\sigma(x_{i+1/2})u'(x_{i+1/2}) - \sigma(x_{i-1/2})u'(x_{i-1/2})}{x_{i+1/2} - x_{i-1/2}},$$

$$u'(x_{i\pm1/2}) \approx \frac{u_{i\pm1} - u_i}{x_{i\pm1} - x_i}.$$

This is a discretization in terms of a nonuniform 3-*point star*

$$\operatorname{div}(\sigma\nabla u)|_{x_i} \approx \alpha_i u_{i+1} - \gamma_i u_i + \beta_i u_{i-1}$$

with

$$\alpha_i = \frac{2\sigma(x_{i+1/2})}{(x_{i+1} - x_{i-1})(x_{i+1} - x_i)} > 0,$$

$$\beta_i = \frac{2\sigma(x_{i-1/2})}{(x_{i+1} - x_{i-1})(x_i - x_{i-1})} > 0,$$

$$\gamma_i = \alpha_i + \beta_i > 0.$$

For homogeneous Dirichlet boundary conditions ($u_0 = u_{N+1} = 0$) the resulting matrix A_h has a tridiagonal shape

$$\begin{bmatrix} \gamma_1 & -\alpha_1 & & & \\ -\beta_2 & \gamma_2 & \ddots & & \\ & \ddots & \ddots & -\alpha_{N-1} & \\ & & -\beta_N & \gamma_N \end{bmatrix}. \tag{3.26}$$

A reasonable discretization should, also for nonuniform meshes, inherit the symmetric structure of the differential operator to the matrix A_h. From the requirement $A_h^T = A_h$ we immediately get $\alpha_i = \beta_{i+1}$, which for the Pearson discretization means

$$\frac{2\sigma(x_{i+1/2})}{(x_{i+1} - x_{i-1})(x_{i+1} - x_i)} = \frac{2\sigma(x_{i+1/2})}{(x_{i+2} - x_i)(x_{i+1} - x_i)}.$$

This implies the condition

$$x_{i+2} - x_{i+1} = x_i - x_{i-1},$$

which characterizes two telescoped equidistant meshes with in general different mesh sizes. If we repeat this condition with an index reduced by one and add, we obtain

$$x_{i+2} - x_i = x_i - x_{i-2}.$$

Thus we are back to an *equidistant* grid (over each second node) – and this already in $\mathbb{R}^1$ on the basis of the simple requirement of symmetry of the arising stiffness matrix!

Tensor Product Grids. An extension from space dimension $d = 1$ to $d > 1$ is usually done via tensor products. Thereby nonuniform discretizations give rise to the problem that locally refined meshes in the interior of the domain continue up to the boundaries (see Figure 3.4). As a consequence, the number of nodes increases beyond a value necessary for a sufficiently accurate representation of the solution. In addition, the symmetry violation as worked out above (in one space dimension) throws us back to essentially *equidistant* grids.

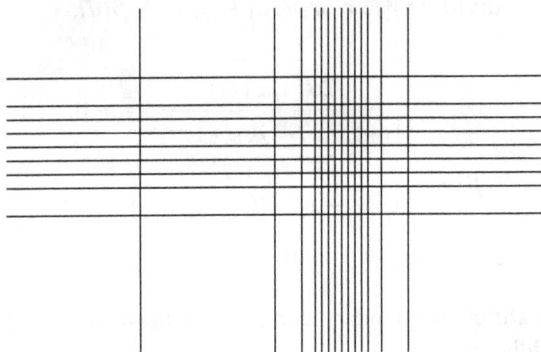

Figure 3.4. Local mesh refinement in tensor product grids continues far beyond the desirable regime.

3.3.2   Curved Boundaries

For finite difference methods, the representation of curved boundaries is a technical challenge which deserves careful consideration. We start from a situation (in 2D) as depicted in Figure 3.5.

Dirichlet Boundary Conditions. Suppose a 5-point star centered at u_0 crosses the boundary $\partial\Omega$ between u_0 and u_1 in $u_{1'}$ at a distance of ξh from u_0. Then, depending on ξ, three different modifications of the 5-point star arise naturally.

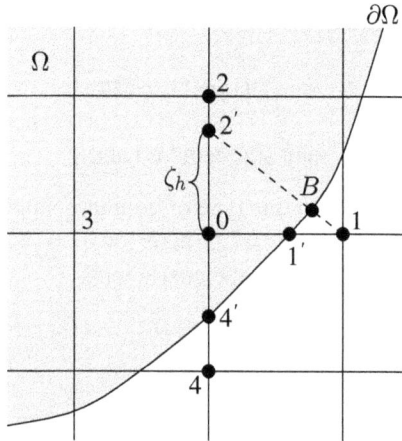

Figure 3.5. Definition of auxiliary points for finite difference methods at curved boundaries.

Linear interpolation of an interior point u_0 ($\xi < \frac{1}{2}$). The situation is schematically depicted in Figure 3.6, left. With

$$u_0 = u_3 + \frac{1}{1+\xi}(u_{1'} - u_3) = \frac{\xi}{1+\xi}u_3 + \frac{1}{1+\xi}u_{1'}$$

the value u_0 can be expressed by u_3 and the known boundary value $u_{1'}$ up to $\mathcal{O}(h^2)$ and inserted into the 5-point star centered around u_3. Analogously, one proceeds for the star around u_2 via expressing u_0 by u_2 and $u_{4'}$.

Linear interpolation at a boundary point $u_{1'}$ ($\xi > \frac{1}{2}$). The situation is shown in Figure 3.6, right. Here, up to $\mathcal{O}(h^2)$ we obtain

$$u_{1'} = u_0 + \xi(u_1 - u_0),$$

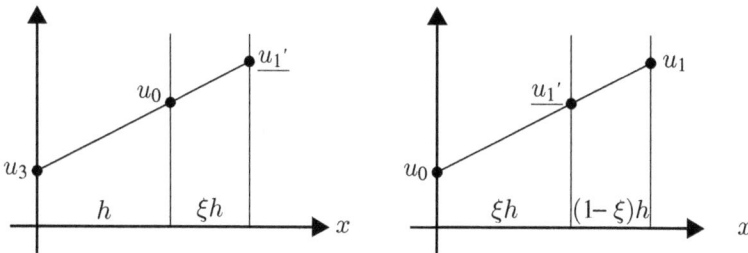

Figure 3.6. Linear interpolation at interior points (*left*) or boundary points (*right*) for an adaptation of the discretized Laplace operator to curved boundaries.

which yields the representation

$$u_1 = \frac{1}{\xi}\left(u_{1'} - (1 - \xi)u_0\right).$$

This can be inserted into the 5-point star centered at u_0.

Shortley–Weller approximation. In this type of boundary approximation we recall the Pearson discretization in the case $d = 1$ (here for $\sigma = 1$): In row i of (3.26) we defined the nonuniform 3-point star by the coefficients

$$[-\beta_i, \gamma_i, -\alpha_i],$$

where

$$\alpha_i = \frac{2}{h_E(h_W + h_E)}, \quad \beta_i = \frac{2}{h_W(h_W + h_E)}, \quad \gamma_i = \alpha_i + \beta_i = \frac{2}{h_W h_E},$$

with the notation W/E for West/East and

$$h_W = x_i - x_{i-1}, \quad h_E = x_{i+1} - x_i.$$

For the extension to $d = 2$ we merely add the two 3-point stars in the x- and y-direction in the style of (3.2) and introduce correspondingly h_N, h_S with N/S for North/South. In this way we obtain a nonuniform 5-*point star* of the form

$$\begin{bmatrix} & \cdot & \frac{-2}{h_N(h_N+h_S)} & \cdot & \\ \frac{-2}{h_W(h_W+h_E)} & \left(\frac{2}{h_W h_E} + \frac{2}{h_N h_S}\right) & \frac{-2}{h_E(h_W+h_E)} & \cdot \\ & \cdot & \frac{-2}{h_S(h_N+h_S)} & \cdot & \end{bmatrix}$$

If one identifies the different h-values, then one obtains the 5-point star in the uniform case (see (3.2)). The transfer of this method to the case $d > 2$ should be clear enough, even though one runs out of geographic directions beyond north, west, south, east.

Neumann Boundary Conditions.   By means of linear interpolation

$$u_{2'} = u_0 + \zeta(u_2 - u_0)$$

and Euler discretization of the normal derivative according to

$$0 = n^T \nabla u = \frac{u_1 - u_{2'}}{\sqrt{1 + \zeta^2}\, h} + \mathcal{O}(h)$$

the value u_1 may be expressed in terms of u_0 and u_2:

$$u_1 = u_1(u_0, u_2) + \mathcal{O}(h^2).$$

Along this line the 5-point star around u_0 may be modified in a similar way as worked out above for Dirichlet boundary conditions (see [90]). On one hand, this approximation is easier to implement, on the other hand, however, it is relatively inaccurate, since neither the position of the boundary point B nor the unknown boundary value $u(B)$ (see Figure 3.5) enter into the formula.

Patching.   The analysis given so far in this section offers a series of clear indications that finite difference methods are only well-suited for the discretization of elliptic PDEs, if

1. equidistant meshes are appropriate for the problem at hand, and

2. the computational domain Ω has a sufficiently simple shape.

As a compromise between complex geometry and uniformity of meshes special, *domain decomposition methods* have gained in popularity, where individual *patches* with locally uniform meshes, but with different mesh sizes, are introduced. An appropriate combination of these different meshes in the process of the solution of the arising structured equation systems requires careful consideration (see, e.g., Section 7.1.4).

Boundary-fitted Grids.   For problems of fluid dynamics in particular, the computation of boundary fitted grids has been algorithmically worked out (see Section 2.2.3). We do not want to go into details, but only give an idea of the method in Figure 3.7.

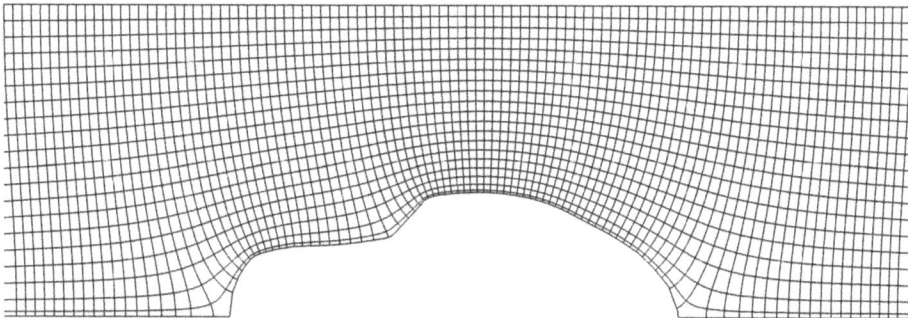

Figure 3.7. Boundary fitted grid around a car. (In two space dimensions it is cheaper.)

3.4   Exercises

Exercise 3.1. In Section 3.1.1 we gave the discretization of Neumann boundary conditions by finite difference methods. Carry the derivation of the discretization over to the one for Robin boundary conditions and discuss the discretization error as well as the structure of the arising matrix. (Space dimension $d = 2$ is enough.)

Exercise 3.2. Study the discretization of Dirichlet boundary conditions for L-shaped domains with reentrant rectangular corner in analogy to the representation in Section 3.1.1 with respect to the structure of the arising matrix. (Space dimension $d = 2$ is sufficient.)

Exercise 3.3. Let A_h denote the matrix arising from the symmetric boundary discretization of Neumann boundary conditions (see Section 3.1.1). Prove that for the eigenvalues λ of this matrix one has at least

$$\Re\lambda \geq 0.$$

Hint: Theorem of Gerschgorin.

Exercise 3.4. Consider the Helmholtz equation

$$-\Delta u + cu = f \quad \text{in }]0, 1[^2$$

with homogeneous Dirichlet boundary conditions. Show that, in the noncritical case $c > 0$, for uniform finite difference discretization the approximation order $\|\epsilon_h\|_\infty = \mathcal{O}(h^2)$ is achieved. Does this hold for the critical case $c < 0$ as well? Study in particular the situation that one of the eigenvalues of the problem is "close to" 0.

Exercise 3.5. *Dew point.* When insulating the external wall of a house the position of the so-called dew point must be taken into account by all means. This is the exact point at which water vapor precipitates. In order to avoid damage to load-bearing walls by wetness, the dew point should lie outside the wall. In Figure 3.8 the situation is depicted schematically. The question is: How *thick* should the insulating wall be to assure that the wall remains dry?

As a mathematical model the one-dimensional stationary heat equation with prescribed negative outside temperature and positive inside temperature may be

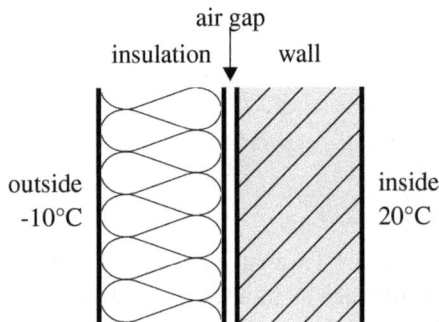

Figure 3.8. Outside insulation of house walls.

picked. For a 20 cm thick brick wall, set a specific heat conductivity of 1 W/K/m, for the 3 mm thick air gap set 0.026 W/K/m. Assume expanded polystyrene as insulating material with a specific heat conductivity of 0.04 W/K/m. For the heat transfer number, set 2 W/K/m^2 inside and 25 W/K/m^2 outside. At a relative room air humidity of 50% precipitation occurs at 9.3°C. Do not take changes of the water vapor pressure into account.

Chapter 4

Galerkin Methods

As in the preceding chapter, the present one also deals with the *numerical* solution of
PDEs. However, in contrast to the finite difference methods presented in Chapter 3,
which can be studied and understood with rather elementary means, the Galerkin methods to be treated here require more mathematically challenging tools, in particular the
functional analysis of Sobolev spaces. A corresponding introductory basis is laid in
Appendix A.5.

Galerkin methods arise in two typical forms, as *global* Galerkin methods (spectral
methods; see Section 4.2) and as *local* Galerkin methods (finite element methods; see
Sections 4.3 and 4.4). In the first section we define the concept of "weak" solutions
which lies at the root of all Galerkin methods.

4.1 General Scheme

For the purpose of simplifying the presentation we again start from the simple Poisson
equation

$$-\Delta u = f, \tag{4.1}$$

where still we leave the boundary condition open. The solution u must be contained in
the function space $C^2(\Omega) \cap C(\overline{\Omega})$. Such a solution is also called *strong* solution, since
the regularity requirements are so strong that in many situations such a solution does
not even exist. In what follows we therefore want to generalize the solution concept
to less smooth functions, which leads us to *weak* solutions.

4.1.1 Weak Solutions

From Appendix A.3 we pick (A.7) from Green's Theorem A.5 for scalar functions
$u, v \in C^2(\Omega)$. Upon exchanging the left- and the right-hand side there, we arrive at

$$\int_\Omega \nabla u^T \nabla v \, dx - \int_{\partial\Omega} v n^T \nabla u \, ds = \int_\Omega (-\Delta u) v \, dx.$$

If u is a solution of the above Poisson equation, then we have

$$\int_\Omega \nabla u^T \nabla v \, dx - \int_{\partial\Omega} v n^T \nabla u \, ds = \int_\Omega f v \, dx.$$

If, in addition, we assume that both u and v satisfy homogeneous Dirichlet boundary
conditions on a boundary subset $\partial\Omega_D$ and homogeneous Neumann boundary condi-
tions on the remaining $\partial\Omega_N = \partial\Omega \setminus \partial\Omega_D$, then the second integral on the left-hand

side will vanish. Under this assumption we get:

$$\int_\Omega \nabla u^T \nabla v \, dx = \int_\Omega f v \, dx. \tag{4.2}$$

This relation is surprising: while in the formulation (4.1) obviously only functions $u \in C^2(\Omega) \cap C(\overline{\Omega})$ could be permitted, the presentation (4.2) permits the much weaker condition $u, v \in H^1(\Omega)$ in terms of the Sobolev space (see Appendix A.5)

$$H^1(\Omega) = \{v \in L^2(\Omega) \, : \, \nabla v \in L^2(\Omega)\}.$$

If we adjoin the Dirichlet part of the boundary conditions, then $u, v \in H_0^1(\Omega)$ holds with

$$H_0^1(\Omega) = \{v \in H^1(\Omega) : v|_{\partial\Omega_D} = 0\}.$$

The relations thus derived are usually written in some short-hand notation. For this purpose we define a symmetric bilinear form, the *energy product*:

$$a(u, v) = \int_\Omega \nabla u^T \nabla v \, dx, \quad u, v \in H_0^1(\Omega). \tag{4.3}$$

The name stems from the fact that the functional $a(u, u)$ can be physically interpreted as energy, which justifies the notion *energy norm* for

$$\|u\|_a = a(u, u)^{\frac{1}{2}}, \quad u \in H_0^1(\Omega) .$$

Furthermore we interpret, for $f \in L^2(\Omega)$, the expression

$$\int_\Omega f v \, dx$$

as an application of the linear functional $f \in H_0^1(\Omega)^*$ to $v \in H_0^1(\Omega)$ and thus use the notation $\langle f, v \rangle$ for the dual pairing. With these definitions, equation (4.2) can now be written in the form

$$a(u, v) = \langle f, v \rangle \quad \text{for all } v \in H_0^1(\Omega). \tag{4.4}$$

This equation, or (4.2), respectively, is denoted as the weak formulation of the PDE (4.1), a solution $u \in H_0^1(\Omega)$ of (4.4) correspondingly as a *weak* solution.

The bilinear form (4.3) induces a symmetric positive definite operator $A : H_0^1(\Omega) \to H_0^1(\Omega)^*$ by

$$\langle Au, v \rangle = a(u, v) \quad \text{for all } u, v \in H_0^1(\Omega) ,$$

so that (4.4) can be equivalently written as an equation in $H_0^1(\Omega)^*$,

$$Au = f. \tag{4.5}$$

Depending on the purpose, either the notation (4.4) or (4.5) will be more appropriate; we will use both of them in different contexts.

The basic idea of Galerkin methods now consists of replacing the infinite dimensional space $H^1(\Omega)$ with a finite dimensional subspace, the *ansatz space* V_h, which means approximating the continuous solution u by a discrete solution u_h, i.e.,

$$u \in H^1(\Omega) \;\rightarrow\; u_h \in V_h \subset H^1(\Omega). \tag{4.6}$$

Here the index h denotes some "grain size" of the discretization, characterized by a mesh size h or a number N of degrees of freedom. The idea (4.6) is realized replacing the weak formulation (4.4) by a *discrete weak formulation*

$$a(u_h, v_h) = \langle f, v_h \rangle, \quad \text{for all } v_h \in V_h \subset H_0^1(\Omega). \tag{4.7}$$

In order to solve this system, we may assume that $V_h = \text{span}\{\varphi_1, \ldots, \varphi_N\}$. Therefore u_h can be expanded as follows:

$$u_h(x) = \sum_{i=1}^{N} a_i \, \varphi_i(x) = a_h^T \varphi(x), \quad x \in \Omega,$$

where $\varphi^T = (\varphi_1, \ldots, \varphi_N)$ denotes the selected basis functions or *ansatz functions* and $a_h^T = (a_1, \ldots, a_N)$ the unknown coefficients. If we now set $v_h = \varphi_i$ successively for $i = 1, \ldots, N$ in (4.7), we then obtain, due to the bilinearity of $a(\cdot, \cdot)$, for the time being

$$\sum_{i=1}^{N} a_i \, a(\varphi_i, \varphi_j) = \langle f, \varphi_j \rangle, \quad j = 1, \ldots, N.$$

If we additionally define the vector $f_h \in \mathbb{R}^N$ of the right-hand side and the matrix $A_h = (A_{ij})$ via

$$f_h^T = (\langle f, \varphi_1 \rangle, \ldots, \langle f, \varphi_N \rangle), \quad A_h = (A_{ij}) = \big(a(\varphi_i, \varphi_j)\big), \tag{4.8}$$

then the above relation can be written in short-hand notation as a *system of linear equations*

$$A_h a_h = f_h. \tag{4.9}$$

The computation of A_h, f_h according to (4.8) is often called *assembly*. It is frequently more time-consuming than the solution of the equation system (4.9).

In comparison with finite difference methods (see Section 3) we generally observe that the matrix A_h defined above is, by construction, *symmetric* and, depending on the boundary conditions, either *positive definite* or *semidefinite* – and this independent of the choice of basis.

4.1.2   Ritz Minimization for Boundary Value Problems

A formulation alternative to (4.4) employs the *variational functional*

$$\Phi(v) := \tfrac{1}{2}a(v, v) - \langle f, v\rangle.$$

For this functional an important *minimization property* holds.

Theorem 4.1. *Let u be a solution of the Poisson equation (4.1). Then*

$$\Phi(u) \le \Phi(v) \quad \textit{for all } v \in H^1(\Omega).\tag{4.10}$$

Proof. Assume u is a classical solution of (4.1) for homogeneous Dirichlet or Neumann boundary conditions. By means of the classical Lagrange embedding $v = u + \epsilon\delta u$ with $\epsilon > 0$ and $\delta u \in H^1(\Omega)$ we then get

$$\Phi(u + \epsilon\delta u) = \frac{1}{2}a(u + \epsilon\delta u, u + \epsilon\delta u) - \langle f, u + \epsilon\delta u\rangle$$

$$= \frac{1}{2}a(u, u) + \epsilon a(\delta u, u) + \tfrac{1}{2}\epsilon^2 a(\delta u, \delta u) - \langle f, u\rangle - \epsilon\langle f, \delta u\rangle$$

$$= \Phi(u) + \epsilon\underbrace{\langle -\Delta u - f, \delta u\rangle}_{=0} + \frac{1}{2}\epsilon^2\underbrace{\langle -\Delta\delta u, \delta u\rangle}_{\ge 0}$$

$$\ge \Phi(u).$$

Here we have taken into account that the operator $-\Delta$ is positive symmetric.   □

The minimization principle presented in Theorem 4.1, often referred to as the *variational principle*, additionally supplies so-called "natural" boundary conditions.

Theorem 4.2. *Let $u \in C^2(\Omega) \cap C(\overline{\Omega})$ be a solution of the weak formulation*

$$a(u, v) + \int_{\partial\Omega} \alpha uv \, ds = \langle f, v\rangle + \int_{\partial\Omega} \beta v \, ds \quad \textit{for all } v \in H^1(\Omega).\tag{4.11}$$

Then it is also solution of the Poisson equation

$$-\Delta u = f$$

with Robin boundary conditions

$$n^T \nabla u + \alpha u = \beta.$$

Proof. As a first step, we again apply Theorem (A.7) of Green to (4.11) and obtain

$$0 = \int_\Omega (\nabla v^T \nabla u - fv) \, dx + \int_{\partial\Omega} (\alpha u - \beta)v \, ds$$

$$= -\int_\Omega (\Delta u + f)v \, dx + \int_{\partial\Omega} (n^T \nabla u + \alpha u - \beta)v \, ds.\tag{4.12}$$

It must be shown that the integrand is zero almost everywhere if each of the above integrals vanishes. As a first step, we choose an interior point, $\xi \in \Omega$, a "nonvanishing" test function v with support in a ball $B(\xi; \epsilon)$. In the limit $\epsilon \to 0$ the boundary integrals from (4.12) diminish more quickly than the domain integrals so that we, due to continuity of u, eventually obtain $-(\Delta u + f) = 0$ in Ω.

As a second step, we choose a boundary point $\xi \in \partial\Omega$ and a test function v with support of length ϵ along the boundary $\partial\Omega$ and a thickness ϵ^2 extending into the interior of the domain Ω; this time the domain integrals in (4.12) tend faster to 0 than the boundary integrals. In the limit $\epsilon \to 0$ we therefore obtain the Robin boundary condition

$$n^T \nabla u + \alpha u - \beta = 0.$$ □

Special Boundary Conditions. If $\alpha = \beta = 0$ in (4.12), then only the *homogeneous Neumann boundary conditions*

$$n^T \nabla u = \frac{\partial u}{\partial n}\bigg|_{\partial\Omega} = 0$$

remain, which are often called *natural boundary conditions*. Both the Neumann boundary condition and the more general Robin boundary condition necessarily involve the space $H^1(\Omega)$.

For $\alpha \to \infty$ one obtains *homogeneous Dirichlet boundary conditions*, which in the calculus of variations are often called *essential boundary conditions*, in contrast to the natural boundary conditions. In order for the limit process to make sense, the space must be restricted to $H_0^1(\Omega)$. For space dimensions $d > 1$, however, arbitrary H^1-functions need not be continuous (see Sobolev Lemma, Appendix A.5), so that the realization of the boundary conditions must be considered with care. The meaning of $v|_{\partial\Omega} = 0$ can be made precise also for discontinuous H^1-functions (in the sense of the trace theorem; again, see Appendix A.5).

Jump Conditions at Interior Edges. In many practical problems material jumps occur. They lead, e.g., to a modified Poisson problem with discontinuous diffusion coefficient σ. In this case the discontinuous edges induce a "natural interface condition" which must be taken into account.

Corollary 4.1. *Consider a Poisson problem over the union of two domains, $\Omega = \text{int}(\overline{\Omega}_1 \cup \overline{\Omega}_2)$, with common boundary $\Gamma = \overline{\Omega}_1 \cap \overline{\Omega}_2$. Let the diffusion coefficient be discontinuous according to*

$$\sigma(x) = \sigma_1, \quad x \in \Omega_1, \qquad \sigma(x) = \sigma_2, \quad x \in \Omega_2.$$

Let u denote the solution of the weak formulation

$$a(u, v) + \int_{\partial\Omega} \alpha u v \, ds = \langle f, v \rangle + \int_{\partial\Omega} \beta v \, ds \quad \text{for all } v \in H^1(\Omega)$$

in terms of the energy product

$$a(u, v) = \int_{\Omega_1} \nabla u^T \sigma_1 \nabla v \, dx + \int_{\Omega_2} \nabla u^T \sigma_2 \nabla v \, dx$$

and the source term $f \in L^2(\Omega)$. Then it is also the solution of the Poisson equation

$$- \operatorname{div}(\sigma \nabla u) = f \quad in \ \Omega$$

for the Robin boundary conditions

$$n^T \sigma \nabla u + \alpha u = \beta \quad on \ \partial\Omega$$

and the flow condition

$$[\![n^T \sigma \nabla u]\!]_\Gamma = 0 \tag{4.13}$$

with the notation

$$[\![g]\!]_\Gamma (x) = \lim_{\epsilon \to 0} n^T \big(g(x + \epsilon n) - g(x - \epsilon n) \big).$$

Note that the definition of the *jump* $[\![g]\!]_\Gamma$ is independent of the orientation of the normal vector n on Γ.

The proof follows the lines of the proof of Theorem 4.2 and is therefore left to Exercise 4.9. The term $\sigma \nabla u$ is called "flow"; the condition (4.13) determines the *continuity of flows in normal direction*, hence the name. As for an arbitrary domain Ω, the decomposition into Ω_1 and Ω_2 can be chosen such that the normal vector of the common boundary Γ points into an arbitrary direction, the flow must be continuous.

Discrete Minimization Problem.   The key idea $H^1(\Omega) \to V_h \subset H^1(\Omega)$ can be realized in this framework substituting the infinite dimensional minimization problem (4.10) by a finite dimensional minimization problem, i.e.,

$$\Phi(u) = \min_{v \in H^1(\Omega)} \Phi(v) \quad \to \quad \Phi(u_h) = \min_{v_h \in V_h} \Phi(v_h).$$

With the above notation for the vectors a_h, f_h and the matrix A_h we may then write

$$\Phi(u_h) = \tfrac{1}{2} a_h^T A_h a_h - f_h^T a_h = \min.$$

Since the matrix A_h is symmetric positive definite or semidefinite, respectively, the condition for the minimum appears as equation (4.9), i.e., the system of linear equations we are already acquainted with

$$A_h a_h = f_h.$$

Approximation Properties. The relation between the *Galerkin approximation* u_h and the weak solution u for boundary value problems is summarized in the following theorem.

Theorem 4.3. *Let u_h be the discrete solution of (4.7) and u the weak solution of (4.4). Then the following holds:*

1. *orthogonality:*
$$a(u - u_h, v_h) = 0 \quad \text{for all } v_h \in V_h; \tag{4.14}$$

2. *minimization property:*
$$a(u - u_h, u - u_h) = \min_{v_h \in V_h} a(u - v_h, u - v_h). \tag{4.15}$$

Proof. For the proof of the orthogonality relation we show
$$a(u - u_h, v_h) = a(u, v_h) - a(u_h, v_h)$$
$$= \langle f, v_h \rangle - \langle f, v_h \rangle = 0.$$

The minimization property is then a direct consequence:

$a(u - v_h, u - v_h)$
$$= a(u - u_h + u_h - v_h, u - u_h + u_h - v_h)$$
$$= a(u - u_h, u - u_h) - 2a(u - u_h, u_h - v_h) + a(u_h - v_h, u_h - v_h)$$
$$= a(u - u_h, u - u_h) + a(u_h - v_h, u_h - v_h)$$
$$\geq a(u - u_h, u - u_h). \qquad \square$$

In Figure 4.1, a graphical interpretation of this elementary and central theorem is given. If one replaces there the infinite dimensional space $H^1(\Omega)$ by some finite dimensional "finer" and therefore larger space $V_{h'} \supset V_h$, then orthogonality and minimization property hold in a similar way. This leads us to an auxiliary result that will be important for the development of adaptive algorithms.

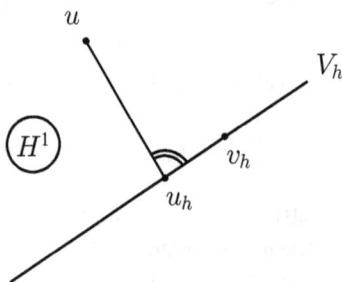

Figure 4.1. Geometric interpretation of Theorem 4.3: orthogonality relation of the corresponding $u_h, v_h \in V_h \subset H^1(\Omega)$ and $u \in H^1(\Omega)$ with respect to the energy product.

Corollary 4.2. *Let* $V_{h'} \supset V_h$ *and* $u_{h'}, u_h$ *the corresponding Galerkin approxima-tions. Then*

$$a(u_{h'} - u, u_{h'} - u) = a(u_h - u, u_h - u) - a(u_{h'} - u_h, u_{h'} - u_h). \qquad (4.16)$$

The proof follows in detail that of Theorem 4.3.

Remark 4.4. For good reasons we want to draw attention to a distinguishing feature of *Hilbert spaces* compared to *Banach spaces*: the reduction of suitably chosen error norms when extending the ansatz space, here exemplified at the energy norm accord-ing to (4.16), only holds in Hilbert spaces. It is not generally true in Banach spaces, but holds there, at best, asymptotically; however, it is unclear at which index the asymp-totics actually sets on.

Weak Formulation vs. Ritz Minimization.   Obviously, for the above simple elliptic boundary value problem the two discretization approaches – via the weak formulation (Section 4.1.1) or via the Ritz minimization (Section 4.1.2) – are equivalent. There are, however, problems where one of the approaches is better suited than the other. As an example, the Helmholtz equation in the critical case does not correspond to a minimization problem, due to the indefiniteness of its spectrum, since $\Phi''(u)$ is then also indefinite. Nevertheless, its weak formulation can still be interpreted as a prob-lem for the determination of a *stationary point*, since the linear system still represents the condition $\Phi'(u) = 0$. For other problems, however, the minimization method is preferable, e.g., for obstacle problems, which can be interpreted as nondifferen-tiable minimization problems (for further reading we recommend the paper [106] by C. Gräser and R. Kornhuber).

Different Galerkin discretizations differ only in the choice of the basis $\{\varphi_k\}$. For global bases, one arrives at *spectral methods*, which will be presented in Section 4.2. For a choice of local bases, *finite element methods* come up as the most prominent realization, which we will deal with in Sections 4.3 and 4.4. Before that, however, we want to have a deeper look into weak solutions in connection with eigenvalue problems.

4.1.3   Rayleigh–Ritz Minimization for Eigenvalue Problems

In this section we consider *selfadjoint* eigenvalue problems only, for simplicity subject to Dirichlet boundary conditions

$$Au = \lambda u, \quad u \in H_0^1(\Omega), \quad A^* = A.$$

For such problems, the eigenspectrum is real and discrete, multiple eigenvalues are possible. Let the eigenvalues be numbered according to $\lambda_1 \leq \lambda_2 \leq \cdots$. For *Poisson problems* (cf. (1.8)) we have in particular

$$0 < \lambda_1^P < \lambda_2^P < \cdots,$$

for *Helmholtz problems in the critical case* (cf. (1.31)) we have

$$-k^2 < \lambda_1^H < \cdots < \lambda_m^H \le 0 < \lambda_{m+1}^H < \cdots$$

for an m depending on k^2, where $\lambda_n^H = \lambda_n^P - k^2$. In this simple case, the computation of the eigenvalues λ_n^H can be reduced to the computation of the eigenvalues λ_n^P. The associated (common) eigenvectors form an orthogonal system w.r.t. the L^2-product. If only positive eigenvalues arise, as in the Poisson problem, then the eigenvectors form an orthogonal system also w.r.t. the energy product. In the indefinite or the semidefinite case, however, the energy product is no longer a scalar product, so that an orthogonality can no longer be defined along with it.

In view of weak solutions $u_h \in V_h \subset H_0^1$, we obtain finite dimensional discrete eigenvalue problems of the form

$$a(u_h, v_h) = \lambda_h \langle u_h, v_h \rangle \quad \text{for all } v_h \in V_h \in H_0^1(\Omega),$$

where $a(\cdot, \cdot)$ denotes the energy product induced by A. If we insert a basis here, we arrive at an algebraic eigenvalue problem in finite dimension $N = \dim(V_h)$ of the form

$$A_h u_h = \lambda_h M_h u_h, \quad A_h^* = A_h, \tag{4.17}$$

where M_h denotes the positive definite mass matrix. The discrete eigenvectors form an orthogonal system w.r.t. the discrete energy product as well as the scalar product $\langle M_h \cdot, \cdot \rangle_h$, where $\langle \cdot, \cdot \rangle_h$ represents the finite dimensional Euclidean product.

For the numerical solution of such problems there are two principal approaches. The first one treats the linear problem (4.17) as an underdetermined system of linear equations for u_h with a parameter λ_h, to be suitably determined (see also Section 7.5.1). This approach, however, has turned out to be not robust enough in difficult applications, due to the necessary computation of the kernel of $A_h - \lambda_h M_h$; this is why we will essentially skip this approach here.

Rayleigh Quotient Minimization.   The second approach goes back to Theorem 1.6. For simplicity, we again set Dirichlet boundary conditions: then, following (1.15), we have

$$\lambda_1 = \min_{u \in H_0^1} \frac{a(u, u)}{\langle u, u \rangle} = \frac{a(u_1, u_1)}{\langle u_1, u_1 \rangle}.$$

Note that this also holds in the indefinite case, even though then the energy product is not a scalar product.

If we turn to the Galerkin discretization, we then have

$$\lambda_{1,h} = \min_{u_h \in V_h} \frac{a(u_h, u_h)}{\langle u_h, u_h \rangle} = \frac{a(u_{1,h}, u_{1,h})}{\langle u_{1,h}, u_{1,h} \rangle}, \tag{4.18}$$

or, respectively, the algebraic form

$$\lambda_{1,h} = \min_{u_h \in \mathbb{R}^N} \frac{\langle A_h u_h, u_h \rangle_h}{\langle M_h u_h, u_h \rangle_h} = \frac{\langle A_h u_{1,h}, u_{1,h} \rangle_h}{\langle M_h u_{1,h}, u_{1,h} \rangle_h}.$$

The function $u_{1,h} \in V_h$ will be naturally interpreted as Galerkin approximation of the eigenfunction u_1. Due to $V_h \subset H_0^1$ and the minimization property we immediately get

$$\lambda_1 \le \lambda_{1,h}. \tag{4.19}$$

Computation of Eigenvalue Clusters. For a *sequential* computation of the m smallest eigenvalues $\lambda_1, \ldots, \lambda_m$ one is tempted to apply formula (1.16) in weak formulation, which means

$$\lambda_{m,h} = \min_{v_h \in V_{m-1,h}^\perp} \frac{\langle A v_h, v_h \rangle}{\langle v_h, v_h \rangle}. \tag{4.20}$$

This form is also useful as a starting point for a *simultaneous* approximation of the m eigenvalues as well as the corresponding invariant subspace of the eigenfunctions, an idea suggested independently in 1980 by D. E. Longsine and S. F. McCormick [146] and, in a somewhat refined version, in 1982 by B. Döhler [78]. Let

$$U_{m,h} = (u_{1,h}, \ldots, u_{m,h}) \quad \text{with} \quad U_{m,h}^T M_h U_{m,h} = I_m$$

denote the column orthogonal (N, m)-matrix of the discrete eigenfunctions to be computed and $\Lambda_h = \mathrm{diag}(\lambda_{1,h}, \ldots, \lambda_{m,h})$ the diagonal matrix of the unknown discrete eigenvalues. Then the corresponding algebraic eigenproblem in the invariant subspace can be written as

$$A_h U_{m,h} = M_h U_{m,h} \Lambda_h. \tag{4.21}$$

In order to compute the eigenfunctions, we consider $U_{m,h}$ as unknowns and minimize the trace of the (m, m)-matrix

$$\min_{U_{m,h}^T M_h U_{m,h} = I_m} \mathrm{tr}(U_{m,h}^T A_h U_{m,h}) = \mathrm{tr}\, \Lambda_h. \tag{4.22}$$

Thus for selfadjoint eigenvalue problems the m smallest eigenvalues are filtered out. Note that for clusters of "close-by" eigenvalues, the problem of calculating single eigenvalues and corresponding eigenvectors is ill-conditioned, whereas the calculation of the invariant subspace is well-conditioned (cf., for example, Volume 1, Section 5.1 or the classical textbook for numerical linear algebra by G. H. Golub and C. F. van Loan [104]).

4.2 Spectral Methods

This class of methods is widely used in theoretical physics and chemistry as well as in the engineering sciences [47, 48]. Instead of the notation V_h for the general ansatz space we here introduce the notation

$$S_N = \text{span}\{\varphi_1, \ldots, \varphi_N\},$$

in a natural way showing the number N of ansatz functions as an index. As global bases various orthogonal systems are usually taken. For details of their efficient numerical computation we refer to Volume 1, Section 6.

Tensor product ansatz. Nearly without exception, multidimensional spectral methods are realized via tensor product ansatz spaces

$$S_N = \prod_{j=1}^{d} S^j,$$

where each S^j is only defined for space dimension $d_j = 1$. This goes with a separation ansatz of the form

$$S^j = \text{span}\{\phi_1^j, \ldots, \phi_{N_j}^j\}, \quad S_N = \text{span}\{\phi(x) = \prod_{j=1}^{d} \phi_{k_j}^j(x_j) : 1 \le k_j \le N_j\}.$$

$$\tag{4.23}$$

Then we obviously have

$$N = N_1 \cdots N_d.$$

For general rectangular domains in $\mathbb{R}^d$, products of orthogonal polynomials will be chosen as bases. For circular domains in $\mathbb{R}^2$, products of trigonometric polynomials and cylinder functions are suited, for spherical domains in $\mathbb{R}^3$ spherical harmonics (i.e., products of trigonometric functions and Legendre polynomials; cf. Volume 1, Section 6).

4.2.1   Realization by Orthogonal Systems

For the following presentation we extend the above Poisson model equation to a slightly more general (linear) elliptic differential equation. For this purpose we consider an operator $A : H \to H^*$ on some Hilbert space H; assumed to be *symmetric positive*, A is clearly a generalization of the operator $-\Delta$.

Expansion in Terms of Eigenfunctions.   Let $\{\varphi_k\}$ denote the eigenfunctions of A. With a suitable normalization they form an *orthonormal* basis. Under the present assumption all eigenvalues λ_n are positive.

First we consider the *stationary* case

$$Au = f.$$

Due to the completeness of H the right-hand side f can be represented as

$$f = \sum_{k=1}^{\infty} f_k \varphi_k \quad \text{with } f_k = \langle f, \varphi_k \rangle .$$

Herein $\langle \cdot, \cdot \rangle$ again denotes the dual pairing for H^*, H. The continuous solution can also be expanded according to

$$u = \sum_{k=1}^{\infty} a_k \varphi_k, \quad \text{where} \quad \langle \varphi_k, \varphi_n \rangle = \delta_{kn} . \tag{4.24}$$

As the Galerkin ansatz we therefore choose

$$u_N = \sum_{k=1}^{N} a_k \varphi_k .$$

Insertion of this expansion into the PDE supplies

$$Au = \sum_{k=1}^{N} a_k A\varphi_k = \sum_{k=1}^{N} a_k \lambda_k \varphi_k .$$

Due to the orthogonality of the φ_k we have

$$0 = \langle Au - f, \varphi_n \rangle = \left\langle \sum_{k=1}^{N} \langle a_k \lambda_k - f_k \rangle \varphi_k, \varphi_n \right\rangle = a_n \lambda_n - f_n$$

for all n. Given the first N eigenvalues, one obtains the first N coefficients

$$a_n = \frac{f_n}{\lambda_n}, \quad n = 1, \dots, N .$$

Obviously, these coefficients do not depend upon the truncation index N.

In the *time-dependent* case we start from the *parabolic* PDE

$$u_t = -Au .$$

The usual separation ansatz

$$u(x, t) = \sum_{k=1}^{\infty} a_k(t) \varphi_k(x)$$

then leads, by insertion of the expansion, to

$$\sum_{k=1}^{\infty} a_k' \varphi_k = u_t = -Au = -\sum_{k=1}^{\infty} a_k A\varphi_k = -\sum_{k=1}^{\infty} a_k \lambda_k \varphi_k .$$

Due to the orthogonality of the φ_k one obtains infinitely many *decoupled* ordinary differential equations:

$$a_k' = -\lambda_k a_k.$$

The expansion of the initial values $u(x,0)$ in terms of eigenfunctions supplies the required initial values

$$a_k(0) = \langle u(x,0), \varphi_k(x) \rangle.$$

These values also do not depend upon the truncation index N. Thus we obtain as Galerkin approximation the discrete solution

$$u_N(x,t) = \sum_{k=1}^{N} a_k(0) e^{-\lambda_k t} \varphi_k(x).$$

For realistic problems in the sciences, the numerical solution of the eigenvalue problem to A is, as a rule, more time-consuming than the solution of the boundary value problem. An exception is the treatment of perturbed operators $A_1 + \varepsilon A_2$ (see Exercise 4.5). In the standard case the unknown solution is expanded in terms of more general orthogonal bases $\{\varphi_k\}$, a case that we will investigate next.

Expansion in Terms of General Orthogonal Bases.   Let us start from expansions with respect to general *orthonormal* bases such as trigonometric functions, cylinder functions or spherical harmonics.

Again we first consider the *stationary*, which means the elliptic case $Au = f$. For the right-hand side f there exists an expansion with coefficients

$$f_k = \langle f, \varphi_k \rangle.$$

The finite dimensional ansatz

$$u_N = \sum_{k=1}^{N} a_k \varphi_k$$

then leads to

$$\langle Au, \varphi_n \rangle = \sum_{k=1}^{N} a_k \langle A\varphi_k, \varphi_n \rangle, \quad n = 1, \ldots, N.$$

As in Section 4.1 we define a matrix A_h with elements $A_{ij} = \langle \varphi_i, A\varphi_j \rangle$ as well as vectors $f_h^T = (f_1, \ldots, f_N)$, $a_h^T = (a_1, \ldots, a_N)$. In perfect analogy with (4.9), we obtain an (N, N)-system of linear equations

$$A_h a_h = f_h. \tag{4.25}$$

The matrix A_h is symmetric positive definite or semidefinite, respectively, as expected from the introductory Section 4.1. However, in the case of interest here the coefficients

a_h usually depend upon the truncation index N, so that we should more precisely write $a_k^{(N)}$ whenever this difference plays a role.

In the *time-dependent*, i.e., parabolic case $u_t = -Au$, we again try the usual separation ansatz

$$u(x,t) = \sum_k a_k(t)\varphi_k(x).$$

From this, with the above notation, we obtain the *coupled* linear differential equation system

$$a_h' = -A_h a_h. \tag{4.26}$$

Note that this system is *explicit* in terms of the derivatives, in contrast to the case of finite element methods, which we will treat in Section 4.3 below. The differential equation system above is in general "stiff" (cf. our short introduction in Section 9.1 in the context of parabolic PDEs or the more elaborate treatment for ordinary differential equations in Volume 2). Summarizing, for given timestep size τ, linear equations of the form

$$(I + \tau A_h) a_h = \tau g_h$$

with a right-hand side g_h not further specified here are to be solved numerically. Note that the above matrix $I + \tau A_h$ for $\tau > 0$ is symmetric positive definite.

Approximation of Integrals.   The bulk of computational cost for the solution of systems (4.25) or (4.26) is incurred in the actual computation of the integrals $\langle \varphi_i, A\varphi_j \rangle$ and $\langle f, \varphi_i \rangle$. For general nonspecific choice of basis the matrix A_h is *dense*. However, for the popular orthonormal bases with $\langle \varphi_i, \varphi_k \rangle = \delta_{ik}$ one additionally has relations of the kind

$$\varphi_i' \in \operatorname{span}\{\varphi_{i-2}, \ldots, \varphi_{i+2}\},$$

so that certain partial blocks of A_h will vanish. In view of such properties, careful consideration will identify numerous zero elements in advance, so that a coupled block structure arises. A sufficiently accurate approximation of the nonzero elements in the matrix and the right-hand side leads us to the topic of "numerical quadrature", as elaborated in Volume 1, Section 9 – there, however, only for the case $d = 1$. This is usually sufficient, since spectral methods are mostly realized over tensor product spaces (see (4.23) above). For a clever choice of orthogonal basis, Gauss–Christoffel quadrature can be exploited (see Volume 1, Section 9.3).

Solution of Linear Equation Systems.   Once the matrix A_h and the right-hand side f_h have been assembled, an (N, N)-system of linear equations with symmetric positive definite matrix must be solved, both in the stationary and in the time dependent case. In the special case of Fourier spectral methods the generated equation system may be solved using fast Fourier transform (FFT) methods at a computational amount of $\mathcal{O}(N \log N)$ (see, e.g., Volume 1, Section 7.2). For different bases, Cholesky decomposition of dense or block-structured matrices is as a rule applied, as we presented in Volume 1, Section 1.4, and will therefore skip here. If, however, spectral methods for a large number N of degrees of freedom arise, then Cholesky solvers consume too many operations per elimination , e.g., $\mathcal{O}(N^3)$ in the dense case. *This property is an explicit disadvantage of spectral methods.* Therefore, for very large N a *preconditioned conjugate gradient method* will be applied (see, e.g., Volume 1, Sections 8.3 and 8.4, or Section 5.3.2 below); the choice of an efficient preconditioner plays a central role.

Truncation Index.   An economic choice of the truncation index N is of particular importance, both in view of the number of elements in A_h, f_h to be computed and with respect to the size of the equation systems to be solved. That is why we will carefully study this question in Section 4.2.3 below; for this purpose, the following approximation theory will be needed.

4.2.2   Approximation Theory

In this section we study the question of how the discretization error $\|u - u_N\|$ depends on the truncation index N in a suitable norm. As a model for our presentation we choose the following minor modification of the Poisson equation

$$-\operatorname{div}(\sigma(x)\nabla u) = f, \quad x \in \Omega, \tag{4.27}$$

with appropriate Robin boundary conditions

$$\sigma(x)n^T\nabla u + \alpha(x)u = \beta, \quad x \in \partial\Omega, \tag{4.28}$$

where $\sigma(x) \geq \sigma_{\min} > 0$ and $\alpha(x) \geq 0$ are assumed; the coefficient σ can be physically interpreted as *diffusion coefficient*. For this PDE one obtains, instead of (4.3), the energy product

$$a(u, v) = \int_\Omega \sigma(x)\nabla u^T\nabla v \, dx, \quad u, v \in H^1(\Omega).$$

This product induces the compatible energy norm $\| \cdot \|_a$. With these definitions Theorem 4.2 holds unchanged.

In view of the fact that spectral methods are usually realized via tensor product spaces (see (4.23)), we restrict our attention to the treatment of the one-dimensional case, i.e., we study only approximation errors of $u_h^j(x_j)$ for the separated variables $x_j \in \mathbb{R}^1$ and drop the index j.

Fourier–Galerkin Methods.   For ease of presentation we choose a simple domain, the interval $\Omega = \,]{-1}, 1[$ in (4.27), with homogeneous Dirichlet boundary conditions $u(-1) = u(1) = 0$ instead of Robin boundary conditions (4.28). Let $u \in C^s(\overline{\Omega})$ be a periodic function with (complex) Fourier representation

$$u = \sum_{k=-\infty}^{\infty} \gamma_k \, e^{ik\pi x}, \quad \gamma_{-k} = \overline{\gamma}_k \in \mathbb{C}.$$

This suggests a spectral method in terms of Fourier series with the ansatz space

$$S_N = \left\{ v = \sum_{k=-N}^{k=N} \gamma_k e^{ik\pi x} : \gamma_{-k} = \overline{\gamma}_k \in \mathbb{C} \right\}.$$

For $\sigma \equiv$ const we would be done, as in this case the Fourier expansion is already exactly the eigenfunction expansion. For the case $\sigma = \sigma(x) \neq$ const the approximation error can be estimated as follows.

Theorem 4.5. *Let $u \in C^s(\overline{\Omega})$ periodic solution of equation (4.27) for homogeneous Dirichlet boundary conditions. Let*

$$u_N = \sum_{k=-N}^{N} c_k^{(N)} e^{ik\pi x}, \quad c_{-k}^{(N)} = \overline{c}_k^{(N)} \in \mathbb{C}$$

be the corresponding Fourier–Galerkin approximation. Then, in the energy norm,

$$\|u - u_N\|_a \leq \frac{C_s}{N^{s-3/2}}. \tag{4.29}$$

Proof. We start from the minimal property (4.15) in the energy norm. As a comparison function we take the truncated Fourier series of the solution, written as

$$v_N = \sum_{k=-N}^{N} \gamma_k \, e^{ik\pi x}, \quad \gamma_{-k} = \overline{\gamma}_k \in \mathbb{C}.$$

Then

$$\|u - u_N\|_a^2 \leq \|u - v_N\|_a^2 = \left\| \sum_{|k|>N} \gamma_k \, e^{ik\pi x} \right\|_a^2 = a(u - v_N, u - v_N)$$

$$= \int_{-1}^{1} \sum_{|k|>N} ((ik\pi)\gamma_k \, e^{ik\pi x})^2 \mathrm{d}x$$

$$\leq \pi^2 \int_{-1}^{1} \sum_{|k|>N} k^2 |\gamma_k|^2 \mathrm{d}x.$$

From Appendix A.1 we take the property

$$|\gamma_k| \leq \frac{M_s}{k^{s+1}}, \quad M_s = \max_{x \in \Omega} |u^{(s)}|$$

and thus obtain the estimate (for monotone decreasing integrand the rectangular sum is the sub sum associated with the integral)

$$\|u - u_N\|_a^2 \leq 2\pi^2 M_s^2 \sum_{k=N+1}^{\infty} \frac{1}{k^{2s-2}} \leq 2\pi^2 M_s^2 \int_{k=N}^{\infty} \frac{1}{k^{2s-2}} \mathrm{d}k = \frac{C_s^2}{N^{2s-3}}$$

where

$$C_s = \pi M_s \sqrt{\frac{2}{2s-3}}.$$

This is just (4.29).   □

Remark 4.6. For *analytical* functions (i.e., $u \in C^\omega(\overline{\Omega})$) the following estimate can be derived:

$$\|u - u_N\|_{L^2(\Omega)} \leq \text{const } \rho^N, \quad \rho = e^{-\sigma} < 1. \tag{4.30}$$

The proof requires complex analytical means that we do not want to assume here. Interested readers should refer to P. Henrici [119].

Theorem 4.7. *Let* $u \in C^N(\overline{\Omega})$ *with* $\Omega =]-1, 1[$ *denote the solution of the equation* (4.27) *for Robin boundary conditions* (4.28) *and* $u_N \in S_N$ *the corresponding Galerkin approximation. Then*

$$\|u - u_N\|_a \leq \sqrt{2(1+\alpha)} \frac{\|u^{(N)}\|_{L^\infty(\Omega)}}{N!} \left(\frac{1}{2}\right)^{N-1}. \tag{4.31}$$

Proof. We may start from the energy norm for Robin boundary conditions

$$\|u - v\|_a^2 = \int_{-1}^{1} (u' - v')^2 dx + \alpha (u(-1) - v(-1))^2 + \alpha (u(1) - v(1))^2.$$

Here we choose a function v, for which a good estimate of the interpolation error can be found. In view of the integral let $v' \in P_{N-1}$ be a function interpolating u' at N Chebyshev nodes (cf. Volume 1, Section 7.1, Theorem 7.19). The remaining open degree of freedom for $v \in P_N$ may be fixed by the condition

$$u(\xi) = v(\xi), \quad \xi \in [-1, 1],$$

where we leave the choice of ξ still open. Then

$$\|u' - v'\|_{L^\infty(\Omega)} \leq \frac{\|u^{(N)}\|_{L^\infty(\Omega)}}{N!} \left(\frac{1}{2}\right)^{N-1},$$

$$|u(1) - v(1)| = \left|\int_{\xi}^{1} (u' - v') dx\right| \leq (1 - \xi) \frac{\|u^{(N)}\|_{L^\infty(\Omega)}}{N!} \left(\frac{1}{2}\right)^{N-1}$$

as well as

$$|u(-1) - v(-1)| = \left|-\int_{\xi}^{-1} (u' - v') dx\right| \leq (1 + \xi) \frac{\|u^{(N)}\|_{L^\infty(\Omega)}}{N!} \left(\frac{1}{2}\right)^{N-1}.$$

Insertion supplies

$$\|u - v\|_a^2 \leq (2 + 2\alpha(1 + \xi^2)) \left(\frac{\|u^{(N)}\|_{L^\infty(\Omega)}}{N!} \left(\frac{1}{2}\right)^{N-1}\right)^2.$$

Obviously, the best estimate is obtained for the choice $\xi = 0$, which confirms the above result. $\square$

In principle, we now might restrict our attention to the special Gevrey function class $\mathcal{G}_1$, for which

$$\frac{\|u^{(N)}\|_{L^\infty(\Omega)}}{N!} \le \text{const}, \quad u \in \mathcal{G}_1$$

holds. Then, instead of (4.31), we would come up with a formula similar to (4.30) with the special value $\rho = 0.5$. Unfortunately, the class $\mathcal{G}_1$ is *not* characterized by the condition

$$\frac{\|u^{(N)}\|_{L^\infty(\Omega)}}{N!} \approx \text{const},$$

which drastically reduces its usefulness within adaptive numerical algorithms.

Spectral Methods for Other Orthogonal Systems.   The *Chebyshev–Galerkin method* can be derived from the Fourier–Galerkin method by the transformation $x = \cos y$. As optimal interpolation points one will naturally select Chebyshev nodes (see Volume 1, Section 7.1.4), defined via

$$x_i = \cos\left(\frac{2i+1}{2N+2}\pi\right) \quad \text{for } i = 0, \dots, N.$$

Consequently, one obtains an estimate analog to (4.29). The application of *Legendre polynomials* leads to comparable estimates (see, e.g., the survey article [105] by D. Gottlieb, M. Y. Hussaini, and S. Orszag).

Assessment.   Global Galerkin methods, or spectral methods, respectively, are quite popular, since they are easy to implement, compared with the local methods developed in Section 4.3 below. Depending upon the domain various *tensor product approaches* are suitable. At the same time, this feature is their dominant disadvantage: Their efficiency is restricted to relatively simple geometries and right-hand sides without strongly localized phenomena, i.e., they are of limited use for problems from science and engineering where several temporal and spatial scales are simultaneously present. Moreover, in practical applications the proven (asymptotic) exponential convergence is often not observable (see also the form of the estimate (4.31)), which in sufficiently complex, but practically relevant problems exhibits an unclear dependence on the truncation index N.

4.2.3   Adaptive Spectral Methods

Essential for the efficiency of spectral methods is the appropriate choice of a basis that is well-suited for the problem in combination with the *truncation index N*, the number of terms in the expansion. Concerning the choice of this index, a certain carelessness has crept in for quite a number of complex problems in science and engineering. Thus, often "solutions" are arrived at which appear to "look good", but which are far too inaccurate, or even simply wrong. Indeed, on the one hand the truncation index N should be small enough to keep the amount of work tolerable (this aspect is usually observed), but on the other hand large enough to achieve a desired approximation quality (this aspect is far too often overlooked). That is why we present a possibility here of how to realize a "reasonable" compromise.

Let N^* denote a desired "optimal" truncation index, at which the approximation u_{N^*} roughly meets a prescribed accuracy, i.e.,

$$\epsilon_{N^*} = \|u_{N^*} - u\| \approx \text{TOL} \qquad (4.32)$$

for a prescribed error tolerance TOL and a *problem adapted* norm $\| \cdot \|$. For an estimation of N^* the following algorithmic components are necessary.

1. *Nested ansatz spaces:* instead of a single ansatz space with fixed dimension N, a set of nested ansatz spaces for *variable dimension*

$$N < N' < N'' \quad \Rightarrow \quad S_N \subset S_{N'} \subset S_{N''}$$

are to be realized for the computation of successively more accurate approximations

$$u_N, \ u_{N'}, \ u_{N''} .$$

2. *A posteriori error estimator:* a numerically easily accessible estimate $[\epsilon_N]$ of the truncation error ϵ_N is needed. This can usually be gained from the projection property of the Galerkin method. Let

$$\|\epsilon\|_a^2 = a(\epsilon, \epsilon)$$

denote the *energy norm*, then, following (4.16), the relation

$$\|u_N - u\|_a^2 = \|u_N - u_{N'}\|_a^2 + \|u_{N'} - u\|_a^2$$

holds. Under the assumption (to be algorithmically verified afterwards)

$$\|u_{N'} - u\|_a \ll \|u_N - u\|_a$$

we have

$$[\epsilon_N] = \|u_N - u_{N'}\|_a \approx \|u_N - u\|_a$$

as a useful error estimator. Thus we replace (4.32) algorithmically by the condition

$$[\epsilon_N] \approx \text{TOL} . \qquad (4.33)$$

3. *Approximation theory:* The theory of the truncation error supplies presented
   above (cf. (4.30) or, in a restricted sense also (4.31))

$$\epsilon_N \approx C\rho^N \tag{4.34}$$

with a problem dependent factor $\rho < 1$, if u is sufficiently smooth. Obviously,
the two unknown quantities C, ρ have to be estimated from available data of the
problem and exploited to derive a suggestion for N^*.

With these components we now construct an algorithm for the *adaptive determination
of the truncation index N*. From the three computed approximations for the successive
indices $N < N' < N''$ we gain the two error estimates

$$[\epsilon_N] = \|u_N - u_{N''}\|_a \approx \|u_N - u\|_a \approx C\rho^N \tag{4.35}$$

and

$$[\epsilon_{N'}] = \|u_{N'} - u_{N''}\|_a \approx \|u_{N'} - u\|_a \approx C\rho^{N'}$$

and, from their quotient, a numerical estimate for ρ:

$$\frac{[\epsilon_{N'}]}{[\epsilon_N]} \approx \rho^{N'-N} \Rightarrow \rho \approx [\rho] := \left(\frac{[\epsilon_{N'}]}{[\epsilon_N]}\right)^{\frac{1}{N'-N}}.$$

Under the generic assumption $[\epsilon_{N'}] < [\epsilon_N]$ the property $[\rho] < 1$ is assured. Taking
logarithms on both sides supplies

$$\log[\rho] = \frac{1}{N' - N} \log \frac{[\epsilon_{N'}]}{[\epsilon_N]}.$$

Upon collecting (4.33) and (4.35) one obtains

$$\frac{\text{TOL}}{[\epsilon_N]} \approx \rho^{N^*-N},$$

from which an estimate for N^* results as

$$N^* - N = \log \frac{\text{TOL}}{[\epsilon_N]} / \log[\rho].$$

Putting the pieces together, we arrive at the following suggestion for a computationally
available "optimal" truncation index

$$N^* = N + (N' - N) \log \frac{\text{TOL}}{[\epsilon_N]} / \log \frac{[\epsilon_{N'}]}{[\epsilon_N]}.$$

From this formula, the following strategy is obtained:

- If $N^* \leq N''$, then the best already computed approximation $u_{N''}$ may be taken as the solution. In order to gain a corresponding error estimate, we will merely apply (4.35) again and thus obtain

$$[\epsilon_{N''}] \approx [\epsilon_{N'}][\rho]^{N''-N'} = [\epsilon_{N'}]\left(\frac{[\epsilon_{N'}]}{[\epsilon_N]}\right)^{\frac{N''-N'}{N'-N}}.$$

- If $N^* > N''$, then the ansatz space must be extended replacing the indices $\{N, N', N''\}$ by $\{N', N'', N^*\}$, assembling the equation system (4.25) for dimension N^* and solving it.

- If the solution process is repeated in the context of bulky computations, then the above derivation allows for an "economization" of the expansion, which means a reduction to as few terms as possible.

Instead of the *energy norm* in the above adaptive algorithm, a comparable algorithm for an arbitrary norm $\| \cdot \|$ can be derived (see Exercise 4.3; cf. also (4.30)).

Example 4.8. *RC-generator.* The adaptive spectral method described here was elaborated in 1994 in [94] for the one-dimensional case of periodic orbits, and slightly improved in [68, Chapter 8]. An example from this paper is the RC-generator in Figure 4.2, left. Here U_1, U_2, U_3 denote electric voltages, I_1, I_2, I_3 electric currents, R_1, R_2 Ohm resistors, C_1, C_2 capacities and V an amplifier whose characteristic curve is given by a function $f : U_1 \rightarrow U_2$ depending on a parameter λ. Figure 4.2, right, shows the three voltages over one period vs. $R_1 = R_2 = C_3 = 1, C_1 = 3, \lambda = 9.7946$, computed to required relative precision $\text{TOL} = 10^{-5}$. In Figure 4.3 we show part of an adaptive pathfollowing strategy [68, Section 5] with respect to the parameter λ, wherein the adaptively determined number N varies with the parameter values.

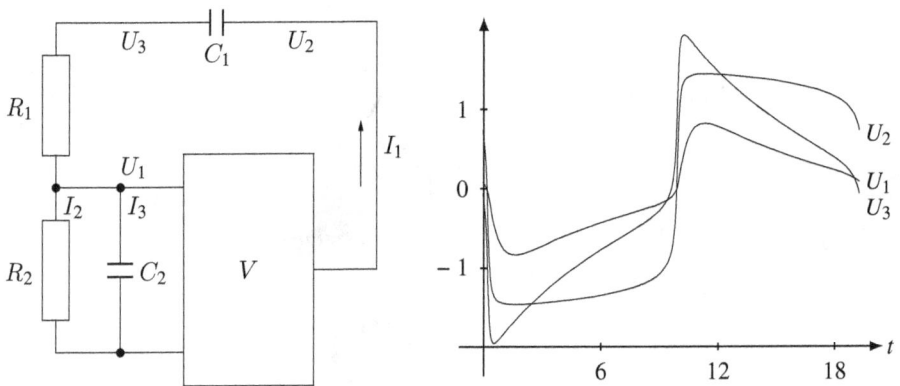

Figure 4.2. RC-generator. *Left:* circuit diagram. *Right:* solution for special parameter set.

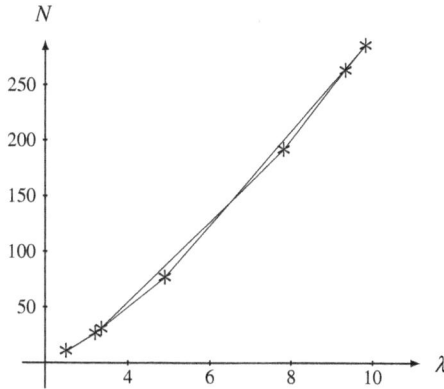

Figure 4.3. RC-generator. *Adaptively* determined number N of Fourier coefficients in dependence on parameter λ. The upper right corner point $*$ is the value for Figure 4.2, right.

Remark 4.9. In [152], C. Mavriplis suggested a heuristic adaptive spectral method for the special case of expansions in terms of Legendre polynomials, where she has chosen a slightly different scaling than in (4.24). Using the notation introduced here, the basic idea can be represented as follows. She starts out from an ansatz similar to (4.34)

$$a_n \doteq C e^{-\sigma n} = C \rho^n \quad \text{with} \quad \rho = e^{-\sigma}, \ \sigma > 0.$$

From this, she tries to determine the two problem dependent parameters C, σ by solving the linear least squares problem (see, e.g., Volume 1, Section 3.1)

$$\sum_{n=N-m}^{N} (\ln a_n)^2 = \sum_{n=N-m}^{N} (\ln C - \sigma\, n)^2 = \min,$$

where $m \in \{3, \dots, 6\}$, following certain rules. Needless to say, this approach may require several trials with N, m, until, asymptotically, some $\sigma > 0$ or $\rho < 1$, respectively, is achieved. In the successful case, she arrives at the following estimate for the truncation error:

$$\epsilon_N^2 = \sum_{n=N+1}^{\infty} a_n^2 \doteq C^2 \frac{\rho^{2N+2}}{1 - \rho^2} =: [\epsilon_N]^2.$$

This estimator is implemented within a *spectral element* method, where spectral expansions (of different lengths N) are used within tensor product elements (i.e., Cartesian products of intervals). If a computed $\sigma > 0$ in one of the coordinates of the tensor products comes "too close" to 0, the element in this coordinate is reduced (see [124]). In this way she obtains an elementary mesh refinement technique for spectral element methods.

4.3   Finite Element Methods

We again start from the weak formulation of the Poisson equation,

$$a(u, v) = \langle f, v \rangle \quad \text{for all } v \in H_0^1(\Omega),$$

and the corresponding discrete weak formulation

$$a(u_h, v_h) = \langle f, v_h \rangle \quad \text{for all } v_h \in V_h \subset H_0^1(\Omega),$$

where, for simplicity, we assume polygonal or polyhedral domains. In the preceding Section 4.2 we had discussed *global* bases for the representation of the Galerkin approximation u_h. Now, in this section, we turn to representations by *local* bases. Local ansatz functions lead to extremely sparse matrices A_h, which is an essential advantage compared to global ansatz functions. The most prominent method of this kind are the finite element methods (often abbreviated as FEM), which, due to their geometric variability, are well established in engineering and biosciences.

Remark 4.10. This kind of methods was already suggested in 1941 in a seminar talk by R. Courant, there especially for the simplest case of linear finite elements in space dimension $d = 2$ in the framework of the calculus of variations. The suggestion was published in early 1943 (see [55]). This paper also contains an appendix, inserted later, where an example was calculated: There the accuracy could be increased with a moderate increase of the number of nodes.[1] The paper, however, became widely forgotten. It was not until 1956 the the US engineers M. J. Turner et al. [205] reinvented the method, obviously without any knowledge of the article by Courant. In their paper they introduced "triangular *elements*" as substructures of larger elastic structures and showed that "rectangular *elements*" can be decomposed into triangular elements. In the 1960's, independent of research in the West, Chinese mathematicians associated with Feng Kang developed an associated method, published in Chinese by 1965 in [88].[2]

4.3.1   Meshes and Finite Element Spaces

In Volume 1 we already learned about local representations in the examples of Bézier techniques (Section 7.3) and cubic splines (Section 7.4), there, however, only for $d = 1$; in this case, finite intervals are subdivided into subintervals, on which local polynomials are defined. For $d > 1$, domains $\Omega \subset \mathbb{R}^d$ are also decomposed into (simple) subdomains for which local polynomials are defined; in contrast to the one-dimensional case, however, it turns out that there is a close connection between

---

[1] Interestingly enough, Courant already indicated that a weakness of his method was its missing error control.

[2] These mathematicians had access to Western mathematical literature only after the end of the 'cultural revolution', i.e., after 1978.

the geometric subdivision elements and the local representations defined there. These decompositions are called *grids*, or *meshes*. Simplicial decompositions, i.e., decompositions consisting only of triangles or tetrahedra, are called *triangulations*.

In two-space dimensions, triangles and quadrilaterals are used almost exclusively. The earlier ones are preferred for the triangulation of complex domains, whereas the latter ones are better suited for Cartesian structured grids (Figure 3.7). Correspondingly, in three dimensions tetrahedra and hexahedra are applied. Moreover, thin boundary layers are often discretized by prisms. For the sometimes desired coupling of tetrahedral and hexahedral grids, pyramids are useful (Figure 4.4).

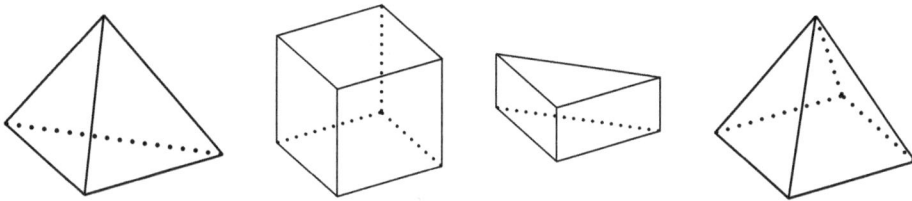

Figure 4.4. Simple element geometries for $d = 3$: (i) tetrahedra, (ii) hexahedra, (iii) prisms, (iv) pyramids

The construction of Galerkin discretizations and the keeping track of their degrees of freedom is simplest when the faces and edges of the elements fit perfectly to one another. In this case one speaks of *conformal*, otherwise of *nonconformal* meshes (see Figure 4.5). This description, rather vague so far, will now be made precise, thus fixing the frame for the subsequent sections.

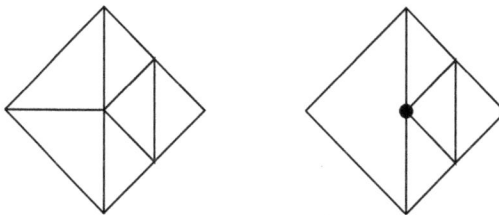

Figure 4.5. Triangulations. *Left:* conformal. *Right:* nonconformal. The nonconformal nodes (marked by •) are often called "hanging" nodes.

Definition 4.11. *Grid or mesh.* Let $\Omega \subset \mathbb{R}^d$, $d \in \{2, 3\}$, denote a domain and Σ a finite system of closed connected sets of subdomains of $\overline{\Omega}$. A subset $M \subset \Sigma$ is called *conformal*, if for all $s_1, s_2 \in M$ with $s_1 \cap s_2 \in M$ also $s_1 = s_2$ holds.

Let $\mathcal{T} = \{T \in \Sigma : \text{int } T \neq \emptyset\}$ denote the set of *elements*. Σ is called *grid* or *mesh*,[3] whenever the following properties hold:

1. $\mathcal{T}$ covers Ω: $\overline{\Omega} = \bigcup_{T \in \mathcal{T}} T$;

2. $\mathcal{T}$ is conformal;

3. $\Sigma \cup \partial\Omega$ is closed under intersection of sets: $s_1, s_2 \in \Sigma \cup \partial\Omega \Rightarrow s_1 \cap s_2 \in \Sigma$.

If for some $F \in \Sigma$ there exist exactly two $T_1, T_2 \in \mathcal{T}$ with $F = T_1 \cap T_2$, then F is called *boundary face*. The boundary faces form a set $\mathcal{F}$. For $d = 2$ this set is identical with the set $\mathcal{E}$ of the *edges*. For $d = 3$, an element $E \in \Sigma$ is called edge if for some $T \in \mathcal{T}$ exactly two $F_1, F_2 \in \mathcal{F}$, $F_1, F_2 \subset T$ exist, so that $E = F_1 \cap F_2$. Finally, $\mathcal{N} = \{E_1 \cap E_2 : E_1, E_2 \in \mathcal{E}, E_1 \neq E_2\}$ is the set of *vertices*. A grid is called *conformal* if $\mathcal{E}$ and $\mathcal{F}$ are conformal.

From the above list of the various popular element geometries it becomes clear that grids where the elements $T \in \mathcal{T}$ are polytopes have a special meaning. For such grids we can assume that all faces are contained in $\mathcal{F}$, and that also the edges are in $\mathcal{E}$ and the nodes in $\mathcal{N}$.

For quite a number of statements about FEM the size of the elements is essential, which we will fix in the following.

Definition 4.12. *Mesh size.* For $T \in \mathcal{T}$ let $h_T = \max_{x,y \in T} |x - y|$ denote the *diameter*. The *mesh size* $h \in L^\infty(\Omega)$ is defined by $h(x) = h_T$ for $x \in \text{int } T$ as a piecewise constant function on Ω. A family $(\Sigma_i)_{i \in I}$ of meshes is called *uniform* if the quotient $\max h_i / \min h_i$ is bounded independently of $i \in I$.[4] Analogously to h_T the diameters h_F of faces and h_E of edges are defined. For $x \in \mathcal{N}$ the notation $\omega_x = \bigcup_{T \in \mathcal{T}: x \in T} T$ means a *patch* around x, and h_x its diameter. In a similar way, patches ω_E are defined around edges and ω_F around boundary faces.

Of course, the diameters h_F, h_E and h_x cannot be understood as functions over Ω, due to the overlapping of patches ω_F, ω_E and ω_x. Apart from the size of elements, also the *shape* of elements is of interest.

Definition 4.13. *Shape regularity.* For $T \in \mathcal{T}$ let $\rho_T = \max_{B(x,r) \subset T} r$ denote the *incircle radius*, which we may understand as piecewise constant function ρ on Ω. A family $(\Sigma_i)_{i \in I}$ of meshes is called *shape regular* or *quasi-uniform* if the functions h_i / ρ_i are bounded independently of i.

The generation of meshes for given domains in two- and especially three-space dimensions is a complex task, in particular if the meshes are conformal and the triangles satisfy conditions on the shape of elements which will be developed in Section 4.4.3. We do not want to go into depth here, but instead merely refer to Section 6.2.2 on mesh *refinements* as well as on the comprehensive literature on the subject (see, e.g., [93]).

Given a mesh $\mathcal{T}$, we are now ready to specify an ansatz space $V_h \subset H^1(\Omega)$ as $V_h = S_h^p(\mathcal{T})$ by requiring that the ansatz functions should be piecewise polynomial.

Definition 4.14. *Finite element space.* For a mesh Σ over $\Omega \subset \mathbb{R}^d$ let

$$S_h^p(\mathcal{T}) = \{u \in H^1 : \text{for all } T \in \mathcal{T} : u|_T \in \mathbb{P}_p\}$$

denote the conformal *finite element space* of order p. In most cases, the mesh is obtained from the context, which is why in general we only write S_h^p. Whenever p remains unspecified, we simply write S_h.

With this definition of an ansatz space, the solution $u_h \in S_h$ is already fixed, together with its approximation error; this will be dealt with later in Section 4.4. Beyond that a crucial aspect is the choice of a *basis* of S_h. This is not only important for the implementation treated below in Section 4.3.3, but also defines, via (4.8), the shape of the matrix A_h and thus decides the efficiency of the solution of system (4.9) (see Chapter 5).

4.3.2 Elementary Finite Elements

In the following we explicate the basic ideas at the example of the Helmholtz equation in the uncritical case (see Section 1.5)

$$-\Delta u + u = f \quad \text{with} \quad a(u, v) = \int_\Omega (\nabla u^T \nabla v + uv)\, dx$$

first in $\mathbb{R}^1$, extending it afterwards into $\mathbb{R}^2$ and $\mathbb{R}^3$. Readers interested in more details about the realization of finite element methods may want to take a look at the textbook of H. R. Schwarz [185].

Finite elements in $\mathbb{R}^1$

On a bounded domain $\Omega =]a, b[$, a possibly nonequidistant subdivision yields a nonuniform grid with nodes

$$\mathcal{N} = \{a = x_1, x_2, \ldots, x_N = b\}$$

and local grid sizes $h_i = x_{i+1} - x_i$. On this grid we now choose *local ansatz functions* φ_i with the aim of substituting, after (4.8), the PDE (4.4) with the linear equation system (4.9). Entries $(A_h)_{ij}$ in the matrix A_h vanish if the supports of φ_i and φ_j do not overlap. In order to obtain a matrix which is as sparse as possible, we will choose ansatz functions with small support. As we do not want to restrict ourselves to ansatz spaces that vanish at the nodes x_i, we will need ansatz functions with a support of at least two neighboring subintervals.

Linear Elements over Intervals.   The simplest ansatz choice consists of piece-wise linear functions φ_i associated with nodes x_i that cover exactly two neighboring subintervals $[x_{i-1}, x_i]$ and $[x_i, x_{i+1}]$ and have the value 1 in the grid nodes x_i (see Figure 4.6, right). At the boundaries one will naturally have only one subinterval. Due to the Lagrange property $\varphi_i(x_j) = \delta_{ij}$ the representation of the solution $u_h = \sum_{i=1}^{N} a_i \varphi_i$ is particularly intuitive, since $u_h(x_i) = a_i$ holds. Thus, by solving the equation system (4.9) not only the coefficients of the solution representation, but also the values at the nodes are computed.

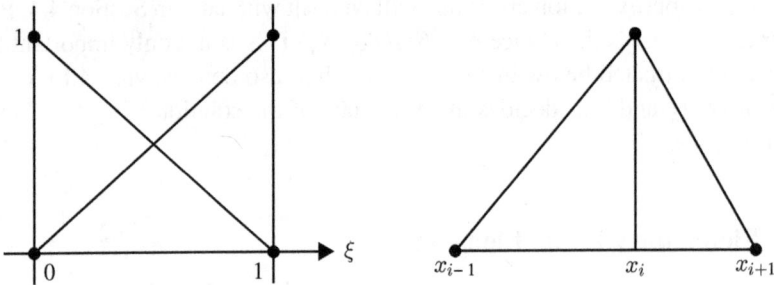

Figure 4.6. Linear finite elements in $\mathbb{R}^1$. *Left:* shape functions ϕ_j over one subinterval. *Right:* ansatz function φ_i over neighboring subintervals.

In practice, assembling the system (4.9) has proven to be better implemented via considering the individual subintervals than via considering the ansatz functions. In this way the solution u_h on each subinterval can be represented by *shape functions* ϕ_j. As the ansatz functions φ_i are composited by shape functions, we deliberately chose a similar notation – nevertheless, shape functions and ansatz functions have to be carefully distinguished.

We now pick an arbitrary subinterval $I_i = [x_i, x_{i+1}]$. In order to obtain a unified local representation, we introduce a scaled variable $\xi \in [0, 1]$ and a bijective affine *transformation*

$$x_{I_i} : \overline{I} \to I_i, \quad \xi \mapsto x_{I_i}(\xi) = x_i + h_i \xi,$$

whereby we obtain all subintervals I_i as images of a common *reference interval* $\overline{I} = [0, 1]$. Vice versa, $\xi_{I_i}(x) = (x - x_i)/h_i$ is a mapping from I_i onto $\overline{I}$. Whenever the interval I_i is clear from the context, we simply write $x(\xi)$ and $\xi(x)$.

On the reference interval we make the linear ansatz

$$u_h(\xi) = \alpha_1 + \alpha_2 \xi.$$

The two local parameters α_1, α_2 can be converted into the unknowns on interval I_i via

$$u_i = u_h(0) = \alpha_1, \quad u_{i+1} = u_h(1) = \alpha_1 + \alpha_2.$$

Thus we obtain the standardized local representation

$$u_h(\xi) = (1 - \xi)u_i + \xi u_{i+1} = u_i \phi_1(\xi) + u_{i+1}\phi_2(\xi), \qquad (4.36)$$

expressed in terms of shape functions

$$\phi_1 = 1 - \xi, \quad \phi_2 = \xi. \qquad (4.37)$$

Note that the shape functions realize a "partition of unity" and therefore the property

$$\phi_1(\xi) + \phi_2(\xi) = 1$$

holds in every point $\xi \in [0, 1]$. In Figure 4.6, left, we show such shape functions, which can be seen to be quite different from the corresponding *ansatz function* over the *nonequidistant* grid $\mathcal{T}$ shown in Fig, 4.6, right. The ansatz functions are obtained by composition of the shape functions:

$$\varphi_i(x) = \begin{cases} \phi_2(\xi(x)), & x \in I_{i-1}, \\ \phi_1(\xi(x)), & x \in I_i, \\ 0, & \text{elsewhere.} \end{cases} \qquad (4.38)$$

Let us now turn to the assembling of the equation system (4.9) according to (4.8). Similar to the solution representation (4.36) we again decompose the matrix A_h into contributions from the various intervals:

$$(A_h)_{ij} = \sum_{k=1}^{N} \int_{I_k} (\nabla\varphi_i(x)^T \nabla\varphi_j(x) + \varphi_i(x)\varphi_j(x))\, dx.$$

Due to (4.38) we can replace the ansatz functions φ_i and φ_j for $k-1 \le i, j \le k+1$ by the shape functions $\phi_n(\xi(x))$ and $\phi_m(\xi(x))$, $n, m \in \{1, 2\}$ – for other values of i or j, the integral vanishes anyway. Let us define the *element matrix* A_{I_k} by

$$(A_{I_k})_{nm} = \int_{I_k} \left(\nabla_x\phi_n(\xi(x))^T \nabla_x\phi_m(\xi(x)) + \phi_n(\xi(x))\phi_m(\xi(x)) \right) dx$$

$$= \int_{\tilde{I}} \left(\nabla_\xi\phi_n(\xi)^T \xi'(x(\xi))\xi'(x(\xi))\nabla_\xi\phi_m(\xi) + \phi_n(\xi)\phi_m(\xi))\det(x'(\xi)) \right) d\xi$$

$$= \int_{\tilde{I}} (h_k^{-2}\nabla\phi_n^T \nabla\phi_m + \phi_n\phi_m)h_k\, d\xi. \qquad (4.39)$$

Then the total matrix A_h can be obtained by adding up the element matrices. The entries of the $(2, 2)$-element matrices A_{I_k} need only be positioned correctly in the global matrix A_h (see Figure 4.7). Formally speaking, this can be written in terms of

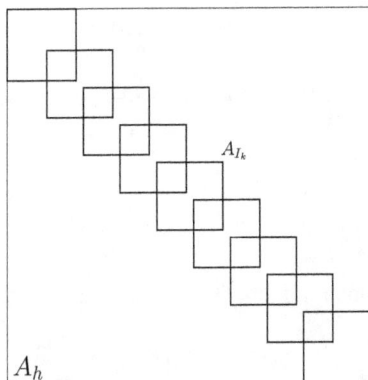

Figure 4.7. Composition of global matrix A_h from local element matrices A_{I_k} for $d = 1$.

an extremely sparse distribution matrix P_k which only defines the right allocation of indices:

$$A_h = \sum_{k=1}^{N} P_k A_{I_k} P_k^T, \quad P_k \in \mathbb{R}^{(N+1)\times 2} \quad \text{with} \quad (P_k)_{in} = \delta_{ik}\delta_{n1} + \delta_{(i+1)k}\delta_{n2}.$$
(4.40)

The form (4.39) suggests a simplified representation of the element matrices, since the integrals over the reference interval can be calculated independently of the target interval I_k

$$A_{I_k} = \frac{1}{h_k}S + h_k M \quad \text{with} \quad S = \begin{bmatrix} 1 & -1 \\ -1 & 1 \end{bmatrix}, \quad M = \frac{1}{6}\begin{bmatrix} 2 & 1 \\ 1 & 2 \end{bmatrix}.$$
(4.41)

In close analogy to the concepts of elastomechanics (see Section 2.3) the matrix S above is called the local *stiffness matrix* and M the local *mass matrix*. S is symmetric positive semidefinite, whereas M is symmetric positive definite.

Quadratic Elements over Intervals.   Upon making a quadratic ansatz

$$u_h(\xi) = \alpha_1 + \alpha_2\xi + \alpha_3\xi^2$$

over the reference interval $\overline{I} = [0, 1]$ we have to choose a further nodal point in order to be able to convert the three local parameters $\alpha_1, \alpha_2, \alpha_3$ into function values. As a rule one introduces $u_{i+\frac{1}{2}} = u_h((x_i + x_{i+1})/2),$[5] i.e., the value in the center $\xi = \frac{1}{2}$. A short intermediate calculation then supplies the local representation

$$u_h(\xi) = u_i\phi_1(\xi) + u_{i+1}\phi_2(\xi) + u_{i+\frac{1}{2}}\phi_3(\xi)$$
(4.42)

---

[5] For special problems an unsymmetric interior point can be chosen.

with

$$\phi_1(\xi) = 2\left(\xi - \tfrac{1}{2}\right)(\xi - 1), \quad \phi_2(\xi) = 2\xi\left(\xi - \tfrac{1}{2}\right), \quad \phi_3(\xi) = 4\xi(1 - \xi). \quad (4.43)$$

Such shape functions are depicted in Figure 4.8, left; obviously they again realize a "partition of unity". For later purposes we also show an alternative, but equivalent set of shape functions in Figure 4.8, right, in which the boundary oriented elements are replaced by linear elements (see Exercise 4.10); these are no longer a partition of unity.

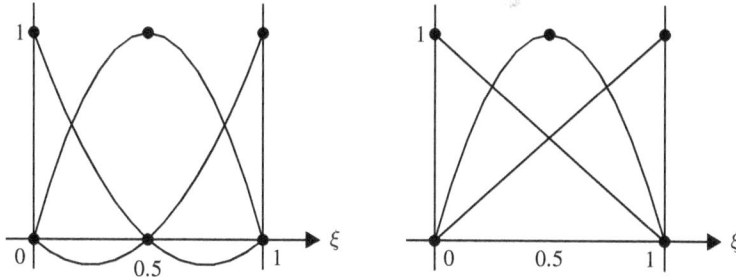

Figure 4.8. Quadratic finite elements in $\mathbb{R}^1$. *Left:* Lagrangian shape functions. *Right:* equivalent hierarchical shape functions.

The computation of the matrix A_h is performed just as before via the element matrices A_{I_k}, only the local stiffness and mass matrices look slightly different (see Exercise 4.11).

Finite elements in $\mathbb{R}^2$

We now deal with meshes over $\Omega \subset \mathbb{R}^2$, on whose elements $T \in \mathcal{T}$ local representations u_h are defined. Due to their convenient subdivision property, we first treat simplicial meshes, i.e., the in general *nonuniform* decomposition of Ω into triangles. For ease of presentation, we restrict ourselves to polygonal domains.

Transformation onto Reference Triangles.   In analogy to our procedure in $\mathbb{R}^1$ we transform each triangle T by an affine transformation onto a common reference triangle $\overline{T}$, where we introduce scaled coordinates (ξ, η); this is depicted in Figure 4.9.
   The reference triangle $\overline{T} = \overline{P_1 P_2 P_3}$ with vertices $\{(0, 0), (1, 0), (0, 1)\}$ in a (ξ, η)-plane is mapped by the *affine* transformation

$$x = x_1 + (x_2 - x_1)\xi + (x_3 - x_1)\eta$$
$$y = y_1 + (y_2 - y_1)\xi + (y_3 - y_1)\eta$$

to the triangle $T = P_1 P_2 P_3$ with vertices $\{(x_1, y_1), (x_2, y_2), (x_3, y_3)\}$. The partial derivatives of a function $u(x, y)$ on the triangle T transform according to

$$u_x = u_\xi \xi_x + u_\eta \eta_x$$
$$u_y = u_\xi \xi_y + u_\eta \eta_y$$

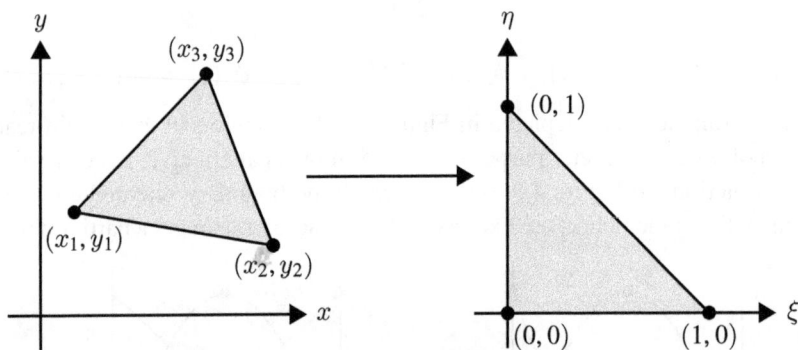

Figure 4.9. Local coordinate transformation of triangles. *Left:* arbitrary triangle T from triangulation $\mathcal{T}$. *Right:* standardized reference triangle $\overline{T}$.

with the partial derivatives of the inverse transformation as

$$\xi_x = \frac{y_3 - y_1}{J}, \quad \eta_x = \frac{y_2 - y_1}{J},$$
$$\xi_y = \frac{x_3 - x_1}{J}, \quad \eta_y = \frac{x_2 - x_1}{J}.$$

Here

$$J = x_\xi y_\eta - y_\xi x_\eta = (x_2 - x_1)(y_3 - y_1) - (x_3 - x_1)(y_2 - y_1)$$

is the corresponding functional determinant. It describes the local change of the differential volume element

$$\mathrm{d}x\,\mathrm{d}y = J\,\mathrm{d}\xi\,\mathrm{d}\eta.$$

We have $J = \text{const}$ over each T. Since the reference triangle $\overline{T}$ has the area $\frac{1}{2}$, the original triangle T has the area $\frac{1}{2}J$. For simplicity, we require $J > 0$, thus equipping all triangles with the same orientation.

Linear Elements over Triangles.   We start from the representation

$$u_h(\xi, \eta) = \alpha_1 + \alpha_2 \xi + \alpha_3 \eta$$

in scaled coordinates over $\overline{T}$. It contains three parameters $\alpha_1, \alpha_2, \alpha_3$, which can be related via

$$u_j := u_h(\overline{P}_j), \quad j = 1, 2, 3$$

to the three vertices of the reference triangle; obviously, linear finite elements "require" triangles as a geometric basis. Upon proceeding, as in the case $d = 1$, we come up with the local representation

$$u_h(\xi, \eta) = u_1 \phi_1(\xi, \eta) + u_2 \phi_2(\xi, \eta) + u_3 \phi_3(\xi, \eta)$$

in terms of the *shape functions*

$$\phi_1 = 1 - \xi - \eta, \quad \phi_2 = \xi, \quad \phi_3 = \eta. \tag{4.44}$$

Obviously, the essential properties of these shape functions are

$$\phi_i(\overline{P}_j) = \delta_{ij}, \quad i, j = 1, 2, 3,$$

as well as the partition of unity

$$\sum_{i=1}^{3} \phi_i(\xi, \eta) = 1, \quad \xi, \eta \in \overline{T}.$$

Behavior at the edges. As shown in Figure 4.10, we consider two neighboring triangles $T_1 = P_1 P_2 P_3$ and $T_2 = P_2 P_4 P_3$ in a triangulation. Let $u_h^{(1)}, u_h^{(2)}$ denote the local linear representation on each of the two triangles. On the common edge $E = P_2 P_3$ the two points P_2 and P_3 define exactly one linear function in $\mathbb{R}^1$, which is why we have

$$u_h^{(1)}|_E = u_h^{(2)}|_E.$$

In other words: $u_h \in C(\overline{\Omega})$. However, for linear finite elements we have $u_h \notin C^1(\Omega)$, since there exist kinks at the edges. As the tangential component of $\nabla u_h(x)$ at the edges is continuous, only *jumps of the normal components* may occur at the edges, which we denote by

$$\left[\!\left[\frac{\partial u_h}{\partial n} \right]\!\right]_E = [\![n^T \nabla u_h]\!]_E.$$

Of course, the finite jumps of the piecewise constant derivatives do not spoil the basic fact that $u_h \in H^1(\Omega)$. Therefore the notion of *conformal* finite elements has been established.

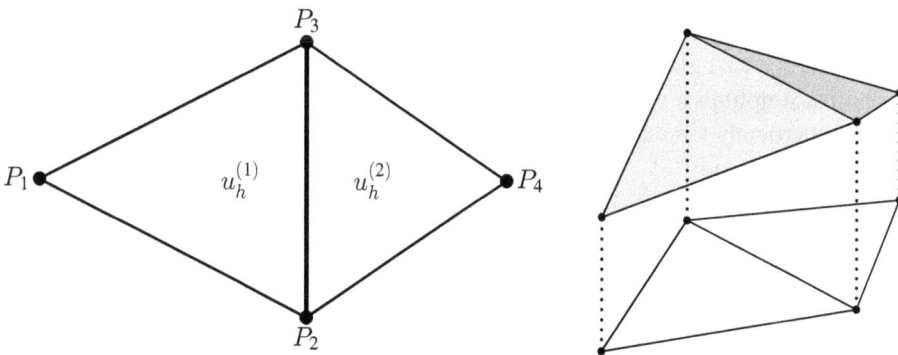

Figure 4.10. Linear finite elements: continuity and kinks at the edges.

Bilinear Elements over Parallelograms.   This type of finite elements realizes a *tensor product ansatz* composed from one-dimensional linear elements

$$u_h(\xi, \eta) = (\beta_1 + \beta_2\xi)(\gamma_1 + \gamma_2\eta)$$
$$= \alpha_1 + \alpha_2\xi + \alpha_3\eta + \alpha_4\xi\eta.$$

We have four parameters to be represented via four reference points of a *reference square* (cf. Figure 4.11, right). Obviously, this is a direct extension of the representation for triangles: therefore, the transformation remains the same and the coordinate transformations based on it merely need to be adapted; back-transformation of the reference quadrangle thus leads to local parallelograms (cf. Figure 4.11, left). The calculation runs just like in the case of linear finite elements: as shape functions we obtain products of one-dimensional shape functions, which is why we will dispense with any further elaboration here.

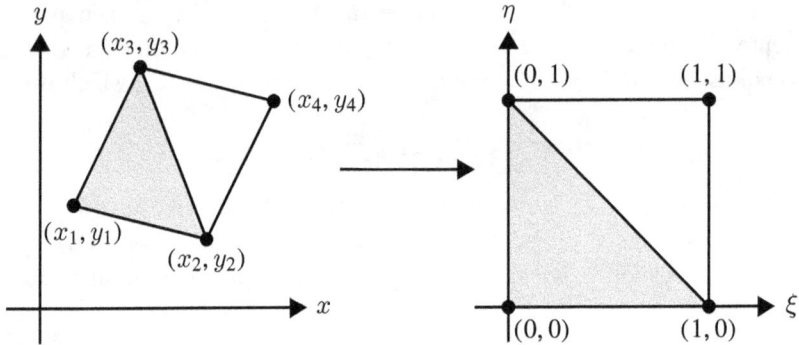

Figure 4.11. Local coordinate transformation for bilinear finite elements (cf. Figure 4.9).

Quadratic Elements over Triangles.   The complete quadratic ansatz

$$u_h(\xi, \eta) = \alpha_1 + \alpha_2\xi + \alpha_3\eta + \alpha_4\xi^2 + \alpha_5\xi\eta + \alpha_6\eta^2$$

contains six parameters. Therefore u_h can again be represented by six selected points on a reference triangle. The selection of the three vertices and three additional edge midpoints simplifies a possible transfer in the course of a successive subdivision into four geometrically similar triangles (see Figure 4.12). The corresponding shape functions explicitly read

$$\phi_1 = (1 - \xi - \eta)(1 - 2\xi - 2\eta)$$
$$\phi_2 = \xi(2\xi - 1)$$
$$\phi_3 = \eta(2\eta - 1)$$
$$\phi_4 = 4\xi(1 - \xi - \eta)$$
$$\phi_5 = 4\xi\eta$$
$$\phi_6 = 4\eta(1 - \xi - \eta).$$

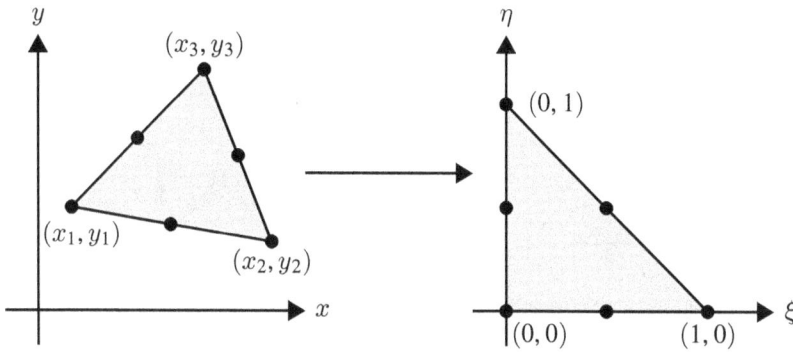

Figure 4.12. Local coordinate transformation for quadratic finite elements.

Just as for linear finite elements, here, too, local stiffness matrices S and mass matrices M can be calculated from which a large symmetric positive definite matrix A_h for an equation system of the type (4.9) can be assembled (see Section 4.3.3).

Isoparametric Triangular Elements.   This kind of finite elements aims at representing curved boundaries on rather coarse meshes by few unknowns. Instead of a linear transformation onto the reference triangle, *curvilinear elements* apply a general bijective transformation

$$(x, y) = F(\xi, \eta)$$

from the reference triangle $\overline{T}$ to the triangle $T \in \mathcal{T}$. In the simplest case F is represented directly by the shape functions ϕ_i on $\overline{T}$:

$$x = \sum_i \gamma_i \phi_i(\xi, \eta), \quad y = \sum_i \delta_i \phi_i(\xi, \eta).$$

This ansatz is called *isoparametric* due to the equivalent parametrization of ansatz space and transformation. An example is isoparametric quadratic finite elements, in which the coefficients γ_i and δ_i are given by the coordinates of the nodal points (x_i, y_i) of T (see Figure 4.13).

The functional determinant $J(\xi, \eta) = \det F'(\xi, \eta)$ now depends upon the point (ξ, η) in the reference triangle. Again we require that $J(\xi, \eta) \geq c > 0$ and thus obtain that F is invertible. In order to assure this condition, however, the possible positions of the points (x_i, y_i) have to be restricted in some nontrivial way (see Exercise 4.7).

Consequently, the ansatz functions that can be represented on T,

$$v(x, y) = \sum_i v(x_i, y_i)\phi_i\big(\xi(x, y), \eta(x, y)\big), \tag{4.45}$$

are no longer polynomials, but rational functions. The continuity beyond element boundaries remains valid (see Exercise 4.6), the error estimates get significantly more intricate [51] compared to the approximation theory for linear elements in Section 4.4.

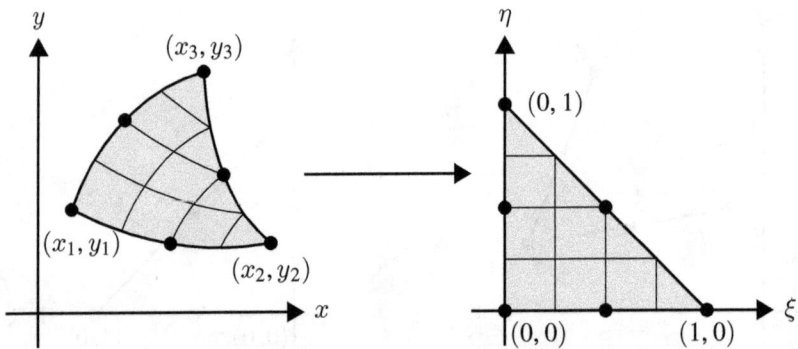

Figure 4.13. Local coordinate transformation for isoparametric finite elements.

Finite Elements in $\mathbb{R}^3$

Similar to the 2D case, we decompose bounded domains $\Omega \in \mathbb{R}^3$ into subdomains. Simplicial decomposition of polyhedral domains leads us especially to *tetrahedra*, over which local representations u_h are defined. Proceeding like in $\mathbb{R}^2$, we transform all tetrahedra $T \in \mathcal{T}$ by an affine transformation onto a common reference tetrahedron $\overline{T}$, where we again introduce scaled coordinates (ξ, η, ζ).

Linear Elements over Tetrahedra.   In Euclidean coordinates the general linear representation reads

$$u_h(\xi, \eta, \zeta) = \alpha_1 + \alpha_2 \xi + \alpha_3 \eta + \alpha_4 \zeta.$$

It contains four parameters $\alpha_1, \ldots, \alpha_4$, which we can relate via

$$u_j := u_h(\overline{P}_j), \quad j = 1, \ldots, 4$$

to the four vertices of the reference tetrahedron (see the original nodes in Figure 4.14, left). Obviously, for linear finite elements in $\mathbb{R}^3$ we require tetrahedra as geometrical basis.

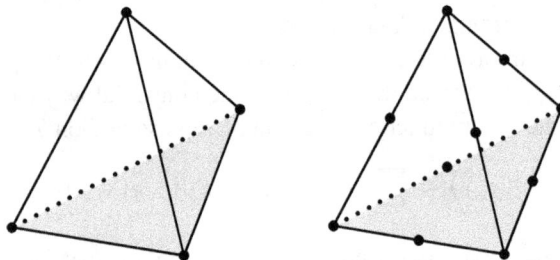

Figure 4.14. Reference nodes in tetrahedra. *Left:* linear finite elements. *Right:* quadratic finite elements.

From this, just as in the case $d = 2$, we obtain the local representation

$$u_h(\xi, \eta, \zeta) = u_1\phi_1 + u_2\phi_2 + u_3\phi_3 + u_4\phi_4.$$

in terms of *shape functions* defined by the property

$$\phi_i(\overline{P}_j) = \delta_{ij}, \quad i, j = 1, \ldots, 4$$

and the partition of unity.

At the faces $F \subset \partial T$ of the tetrahedra $T \in \mathcal{T}$ *kinks* occur, comparable to the kinks at the edges in the 2D case (see Figure 4.10). Correspondingly, this leads to jumps in the normal derivatives

$$[\![n^T \nabla u_h]\!]_{F \in \mathcal{F}} \neq 0.$$

Quadratic Elements over Tetrahedra.   In this case the general ansatz in coordinates ξ, η, ζ, i.e., in the reference tetrahedron $\overline{T}$, contains exactly ten parameters, which leads us to ten points in the reference tetrahedron: This can be conveniently achieved by the addition of the edge midpoints compared to linear elements (see Figure 4.14, right). The rest of the procedure runs as before, which is why we will not go into more detail here.

4.3.3   Realization of Finite Elements

After the examples given in the preceding subsections we are now ready to present a slightly more abstract frame for arbitrary ansatz orders. We will restrict our attention to simplicial decompositions of polytopes Ω. On quadrilaterally or hexahedrically structured meshes, finite elements of arbitrary ansatz order can be constructed by tensor product ansatzes – this intriguing simplicity, however, is widely lost in adaptive mesh refinement (see Chapter 7).

Shape Functions over the Reference Simplex

Given the reference simplex

$$\overline{T} = \{\xi \in \mathbb{R}^d : \xi \geq 0, \ |\xi|_1 \leq 1\},$$

let us span the local polynomial ansatz space $\mathbb{P}_{p,d}$ of order p and dimension

$$K_{p,d} = \binom{d + p}{d}$$

again by the shape functions $\phi_k \in \Phi_{p,d}$, $k = 1, \ldots, K_{p,d}$. There, essentially two approaches are in common use: *Lagrangian* and *hierarchical* shape functions.

Lagrangian Shape Functions.   These shape functions are given by the interpolation conditions

$$\phi_k(\xi_j) = \delta_{kj}, \quad k, j = 1, \ldots, K_{p,d} \tag{4.46}$$

at $K_{p,d}$ interpolation nodes ξ_j. There the choice of the ξ_j is not fully free: on the one hand, the shape functions must be uniquely defined via (4.46), and on the other hand, one aims at keeping the intriguing simplicity of the direct identification of the interpolation nodes by global ansatz functions. In order to ensure continuity beyond element boundaries, the nonvanishing shape functions on a boundary must be uniquely determined by the ξ_j on that boundary, so that exactly $K_{p,d-1}$ nodes must lie on each boundary. They must be the same for the two neighboring elements, which requires some symmetry. For edges and vertices associated conditions must hold, in particular the vertices of the reference simplex must belong to the interpolation nodes.

The simplest choice of interpolation nodes is certainly the *equidistant tensor product mesh*

$$\xi_j \in \mathbb{N}^d / p \cap \overline{T} = \{\xi \in \overline{T} : p\xi \in \mathbb{N}^d\}.$$

For $p \in \{1, 2\}$ we get them by the shape functions presented in the preceding sections (see Figures 4.6, 4.8 left, 4.9, 4.12, and 4.14). For larger p (see Figure 4.15) the Lagrangian finite element methods gradually move to spectral methods; these hybrids are therefore also called spectral element methods. However, polynomial interpolation of higher order on equidistant meshes is ill-conditioned (see Volume 1) and should not be applied for $p > 8$ roughly. On the basis of different criteria, especially the associated Lebesgue constant for the interpolation, various nodal sets have been suggested [198, 210] which allow for Lagrangian shape functions of significantly higher order – a unified theory, however, does not yet seem to exist. A general drawback of Lagrangian shape functions is, however, that an increase of the ansatz order goes with a change of the shape functions.

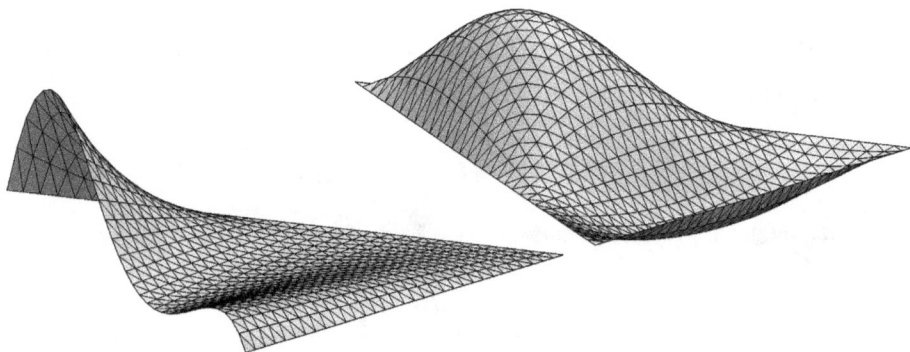

Figure 4.15. Quartic Lagrangian shape functions. *Left:* edge. *Right:* triangle.

Hierarchical Shape Functions. In this case shape functions are constructed such that an increase of the ansatz order merely requires an extension of the basis. This corresponds to a recursive decomposition of the ansatz space $\mathbb{P}_{p+1,d} = \mathbb{P}_{p,d} \oplus \boldsymbol{P}_{p+1,d}$ according to the polynomial order, a feature that is particularly required for hierarchical error estimators (see Section 6.1.4). The simplest example of such a hierarchical extension was already given in Figure 4.8, right.

The simplest possible assignment of ansatz functions to shape functions is obtained, if the hierarchical extensions are distributed to edges, faces, and elements, i.e., $\boldsymbol{P}_{p,d} = \boldsymbol{P}_{\mathcal{E},p,d} \oplus \boldsymbol{P}_{\mathcal{F},p,d} \oplus \boldsymbol{P}_{\mathcal{T},p,d}$. Thereby the edge functions $\phi_{E,p}$ associated with some edge E vanish on vertices and on all other edges, the face functions $\phi_{F,p}$ (only for $d = 3$, of course) vanish on all vertices and edges and on all other faces. The local element functions $\phi_{T,p}$ vanish even on the whole boundaries $\partial \overline{T}$ (see Figure 4.16).

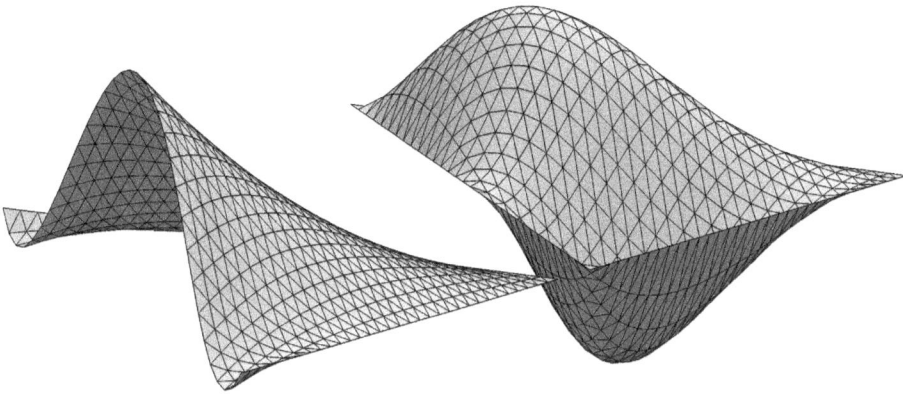

Figure 4.16. Quartic hierarchical shape functions. *Left:* edge. *Right:* triangle.

Shape functions can be expressed in *barycentric coordinates* in a convenient way. Each point ξ in the reference simplex $\overline{T}$ can be uniquely written as a convex combination of the $d + 1$ vertices ξ_i (see Exercise 4.12):

$$\xi = \sum_{i=1}^{d+1} \lambda_i(\xi)\xi_i, \quad \lambda_i(\xi) \in [0,1], \quad \sum_{i=1}^{d+1} \lambda_i(\xi) = 1.$$

The coefficients $\lambda_i(\xi)$ are the barycentric coordinates of ξ. Then the $d + 1$ linear shape functions (4.37) and (4.44) from the preceding section are just the barycentric coordinates of the corresponding vertices: $\phi_i = \lambda_i$. On an interval we obtain, e.g., the quadratic shape function $\phi_{12} = 4\lambda_1\lambda_2$. So we can define the edge function on some edge E between vertices ξ_i and ξ_j as

$$\phi_{E,p} = \lambda_i\lambda_j q_{p-2}(\lambda_i - \lambda_j),$$

where $q_k(x)$ is a polynomial of degree k on $[-1, 1]$. Because of the difference $\lambda_i - \lambda_j$ this definition of edge functions necessarily leads to a dependence upon the *local orientation* of edge E. In order to be able to glue the edge functions of neighboring elements directly together to global ansatz functions, we require that q_k be either an even function ($q_k(x) = q_k(-x)$), in which case the orientation does not play a role, or an odd function ($q_k(x) = -q_k(-x)$), in which case any orientation change corresponds to a mere sign change. Since q_k is a polynomial, the first case will occur for even k, the second case for odd k. For the q_k, various systems of orthogonal polynomials may be selected, such as Chebyshev or Legendre polynomials (see Volume 1).

For $d > 1$ we define element functions by

$$\phi_{T,p,k} = \prod_{i=1}^{d+1} \lambda_i \prod_{i=1}^{d} q_{k_i}(\lambda_i - \lambda_{i+1}),$$

where the d-dimensional multiindex k with $|k| = p - d - 1$ specifies the element function. Here the difference $\lambda_{d+1} - \lambda_1$ does not appear, so that the thus defined basis of $P_{T,p,d}$ need not be symmetric w.r.t symmetries of $\overline{T}$. This will not bother us, since the element functions are completely local and therefore need not be glued together with neighboring elements to get global ansatz functions.

Slightly more difficult is the situation with faces in three-space dimensions. Consider a face F spanned by the vertices $\xi_{j_1}, \xi_{j_2}, \xi_{j_3}$. The associated face functions can be defined in a similar way as the element functions in 2D by

$$\phi_{F,p,k} = \prod_{i=1}^{3} \lambda_{j_i} \prod_{i=1}^{2} q_{k_i}(\lambda_{j_i} - \lambda_{j_{i+1}}). \tag{4.47}$$

Because of the lack of symmetry, a direct assignment of shape functions to ansatz functions is no longer possible. Instead the ansatz functions have to be built from linear combinations of shape functions. For details see [189].

An alternative is the recursive definition

$$\phi_{F,p,k} = \psi_k(\lambda_{j_1}, \lambda_{j_2}, \lambda_{j_3}) \prod_{i=1}^{3} \lambda_{j_i}, \quad \psi_k \in \Phi_{p-3,d-1}, \tag{4.48}$$

$$\phi_{T,p,k} = \psi_k(\lambda_1, \ldots, \lambda_d) \prod_{i=1}^{d+1} \lambda_i, \quad \psi_k \in \Phi_{p-d-1,d}, \tag{4.49}$$

which starts from $\Phi_{0,d} = \{\lambda_1, \ldots, \lambda_{d+1}\}$. This version leads to completely symmetric shape function sets and thus makes a direct assignment of shape functions to ansatz functions possible [235]. A disadvantage of this approach is that the thus defined spaces

$$S_p^{\text{sym}} = \sum_{k=0}^{p} P_{k,d} \supset \mathbb{P}_{p,d}$$

are slightly larger than the complete polynomial spaces and contain shape functions of degree $p + 1$.

Assembly

As in the preceding section, the equation system (4.9) is generated in two steps: First, for each element $T \in \mathcal{T}$, the local matrices A_T and the right-hand sides f_T are computed, which then, second, are scattered to get the global matrix A_h and the right-hand side f_h. The whole process is called *assembly*. Here we only consider the generation of the matrix A_h; the computation of the right-hand side is done in a similar, but a bit simpler way.

Let us consider the slightly more general bilinear form

$$a(u, v) = \int_{\Omega} (\nabla u^T \sigma \nabla v + \gamma u v)\, dx$$

with space dependent coefficients $\sigma \in \mathbb{R}^{d \times d}$ and $\gamma \in \mathbb{R}$. We explicitly mention that this also covers nonlinear problems (see Chapter 8) and time-dependent problems (see Chapter 9), since there the previous iterate or the previous timestep enter into the coefficients σ and γ.

Computation of Element Matrices. The computation of the entries A_{ij} according to (4.8) is decomposed as in Section 4.3.2 into the entries $(A_T)_{ij} = a_T(\phi_i, \phi_j)$ of the local elements, which are integrated either over the element T or, after substitution by the transformation $x = x(\xi)$, over the reference element $\overline{T}$:

$$(A_T)_{ij} = \int_T \left(\nabla_x \left(\phi_i\left(\xi(x)\right) \right)^T \sigma(x) \nabla_x \left(\phi_j\left(\xi(x)\right) \right) + \gamma(x)\phi_i\left(\xi(x)\right)\phi_j\left(\xi(x)\right) \right) dx$$

$$= \int_{\overline{T}} \left(\nabla\phi_i^T B(\xi)\nabla\phi_j + \gamma\left(x(\xi)\right)\phi_i\phi_j \right) \det\left(x'(\xi)\right) d\xi.$$

Here the values $B(\xi) = x'(\xi)^{-1}\sigma(x(\xi))x'(\xi)^{-T}$ and $\phi_i, \phi_j \in \Phi_{p,d}$ run through all shape functions.

The conceptually simplest possibility to compute the integrals is their direct numerical quadrature (see Volume 1, Section 9.3):

$$(A_T)_{ij} \approx \sum_{l=1}^{n} w_l \left(\nabla\phi_i(\xi_l)^T B(\xi_l)\nabla\phi_j(\xi_l) + \gamma\left(x(\xi_l)\right)\phi_i(\xi_l)\phi_j(\xi_l) \right) \det\left(x'(\xi_l)\right).$$

$$(4.50)$$

Efficient Gaussian quadrature formulas over parallelograms or parallelepipeds are easily obtained from tensor products of the one-dimensional quadrature formulas. For triangles and tetrahedra, however, analytical expressions for the computation of nodal points and weights for arbitrary order are not known – they must determined in each separate case [54, 189]. The number of integration nodes increases thereby with the order of the quadrature formula. Assuming that in nonlinear and time-dependent problems σ as well as ϕ_i are polynomials of degree p, and that we want to integrate (4.50) exactly, we will need to apply a quadrature formula of order $3p - 2$. Thus the computational amount for the element matrices grows quite rapidly with the ansatz order (see Table 4.1).

Table 4.1. Required number of integration nodes for numerical quadrature over simplices due to [189] and [149].

order	1	2	3	4	5	6	7	8	9	10	15	19
triangle	1	3	4	6	7	10	12	15	19	25	48	73
tetrahedron	1	4	5	11	14	24	31	43	53		330	715

For piecewise-constant coefficients and affine transformations the decomposition of the element matrices A_T into local stiffness and mass matrices analogous to (4.41) is an alternative to direct numerical quadrature. Let $\{B_m\}, m = 1, \ldots, M = d(d+1)/2$ denote a basis of the symmetric (d, d)-matrices and assume that the entries

$$(S_m)_{i,j} = \int_{\overline{T}} \nabla \phi_i^T B_m \nabla \phi_j \, d\xi \quad \text{for } m = 1, \ldots, M$$

have been computed beforehand once. Then we can represent the spatially constant matrix B as

$$B = \sum_{m=1}^{M} \beta_m B_m.$$

In a similar way we define the mass element matrix

$$M_{ij} = \int_{\overline{T}} \phi_i \phi_j \, d\xi.$$

Then

$$A_T = \left(\sum_{m=1}^{M} \beta_m S_m + \gamma M \right) \det(x'). \tag{4.51}$$

This kind of computation of the element matrices can be performed with much less effort than numerical quadrature in certain situations (see Exercise 4.17).

Assignment of Global Degrees of Freedom.   As in (4.40) the global matrix A_h must collect all element matrices A_T via

$$A_h = \sum_{T \in \mathcal{T}} P_T A_T P_T^T. \tag{4.52}$$

The distribution matrices P_T come by definition of the global ansatz functions

$$\varphi_i|T(x) = \sum_{k=1}^{K_{p,d}} (P_T)_{i,k}\phi_k\big(\xi(x)\big)$$

as linear combinations of the shape functions, since

$$(A_h)_{ij} = a(\varphi_i,\varphi_j) = \sum_{T\in\mathcal{T}} a_T(\varphi_i,\varphi_j)$$

$$= \sum_{T\in\mathcal{T}} \sum_{k,l=1}^{K_{p,d}} (P_T)_{i,k}(P_T)_{j,l}\underbrace{a_T(\phi_i\circ\xi,\phi_j\circ\xi)}_{=(A_T)_{kl}} = \Big(\sum_{T\in\mathcal{T}} P_T A_T P_T^T\Big)_{ij}.$$

For Lagrangian shape functions a direct identification of shape functions and ansatz functions via the interpolation nodes is appropriate. For $T \subset \operatorname{supp}\varphi_i$ there exists exactly one $k \in \{1,\ldots,K_{p,d}\}$ such that $(P_T)_{il} = \delta_{lk}$; for elements $\hat{T}$ outside of the support of φ_i, however, the row i of $P_{\hat{T}}$ has only zero entries. The Lagrange basis thus defined is, especially for linear elements, also called *nodal basis*, since each node x of the mesh is associated with exactly one ansatz function with support ω_x:

$$\Phi_h^1 = \{\varphi_x \in S_h^1 : \varphi_x(v) = \delta_{xv} \text{ for all } v \in \mathcal{N}\}. \tag{4.53}$$

Therefore, for the computation of (4.52) only the global index i associated with each element T and each shape function ϕ_k will be needed.

Just as the hierarchical shape functions, the hierarchical ansatz functions are subdivided into element functions $\varphi_{T,k}$ with $\operatorname{supp}\varphi_{T,k} = T$, face functions $\varphi_{F,k}$ with $\operatorname{supp}\varphi_{F,k} = \omega_F$ (only for $d = 3$, of course), edge functions $\varphi_{E,k}$ with $\operatorname{supp}\varphi_{E,k} = \omega_E$ (for $d > 1$) and nodal functions φ_x with $\operatorname{supp}\varphi_x = \omega_x$. Strictly speaking, this subdivision serves the sole purpose to be able to define ansatz functions in a comparably simple manner.

For element functions, the direct assignment

$$\varphi_{T,k} = \phi_{T,k} \quad \text{for } k = 1,\ldots,K_{p-2,d}$$

can be made as for the Lagrangian shape functions; the same holds for nodal elements. For edge and face functions, however, the assignment cannot be so simple, since at least the sign of $(P_T)_{ik}$ depends upon the orientation of the the edge or face in the corresponding element. Therefore, in addition to the global index i of the corresponding ansatz function, a sign must be stored. This is enough for the definition of hierarchical shape functions (see (4.48)–(4.49)). In the nonsymmetric variant (4.47), however, some face ansatz functions can only be written as linear combinations of shape functions, so that individual rows of P_T are densely populated.

Remark 4.15. In contrast to meshes with rectangular or hexahedral elements, finite elements over simplicial triangulations have two advantages. On the one hand, arbitrarily complex domains are much simpler and with acceptable approximation properties to decompose into triangles or tetrahedra. On the other hand, a refinement of such elements may generate an even finer conformal mesh, a property important for the adaptive solution of the arising equations (see Chapter 7). Adaptive rectangular or hexahedral meshes lead to nonconformal meshes, which, in principle, can also occur for simplicial triangulations. The desired continuity is then algebraically ensured by the introduction of so-called "hanging nodes" (see Figure 4.5), which means nodes which have no genuine degree of freedom assigned but which use the interpolated values from the ansatz functions of the neighboring nodes.

From this occurrence, nonconformal finite elements, where the ansatz space V_h is not a subspace of V, have to be carefully distinguished. This unnatural ansatz, also called "variational crime", requires an extended convergence theory, where, in particular, the best-approximation property is lost. Nevertheless, such approaches are sometimes advantageous, since they are either much simpler to implement, e.g., at plate or shell models in elastomechanics (see [39]), or, with the freedom thusly gained, certain desirable properties can be realized, e.g., divergence-free elements of lower order for fluid dynamics problems.

4.4 Approximation Theory for Finite Elements

In this section we treat the discretization error in finite element methods. We will focus our interest on the energy norm $\| \cdot \|_a$ and the L^2-norm $\| \cdot \|_{L^2(\Omega)}$. For simplicity, we restrict our consideration to the maximum mesh size $h = \max_{T \in \mathcal{T}} h_T$, which is appropriate for *uniform* families of meshes and, in particular, for equidistant grids. We first treat boundary value problems, afterwards eigenvalue problems.

4.4.1 Boundary Value Problems

In the following we perform the estimation of the approximation error of FE methods over equidistant grids first for the energy norm, then for the L^2-norm.

Estimation of the Energy Error

As a starting point for the approximation theory we slightly reformulate the projection property (4.15) for Galerkin methods:

$$\|u - u_h\|_a = \inf_{v_h \in S_h} \|u - v_h\|_a. \tag{4.54}$$

The first idea now is to insert the *interpolation error* into the above right-hand side, thus obtaining an upper bound for the discretization error. Let I_h denote the interpolation operator. Then the interpolation error can be written in the form $u - I_h u$. However, some caution should be taken: the interpolation operator I_h is only well-defined for *continuous* solutions u. But, so far, we have only considered $u \in H_0^1(\Omega)$, which in general does not imply continuity. For a formulation of error estimates we thus need Sobolev spaces of higher *regularity*. The spaces $H^m(\Omega)$ for $m \geq 1$ defined in the Appendix A.5 are characterized by the existence of weak m−th order derivatives in $L^2(\Omega)$. The connection between the Sobolev spaces H^m and the continuity class C is given by the Sobolev embedding lemma, which we may pick from Appendix A.5 to use it in our case here:

Lemma 4.16. *Let $\Omega \subset \mathbb{R}^d$. Then*

$$H^m(\Omega) \subset C(\bar{\Omega}) \quad \Leftrightarrow \quad m > \frac{d}{2}.$$

Obviously, the regularity of the solution is correlated with the spatial dimension. If we start from $u \in H^1$, i.e., $m = 1$, then we end up with the disappointing result $d = 1$. Therefore, in order to cover the practically relevant case $d \leq 3$, at least $m = 2$ is required. This regularity, however, is only ensured under restrictions, as demonstrated by the following lemma.

Lemma 4.17. *Let Ω be convex or have a smooth boundary and let $f \in L^2(\Omega)$. Then the solution u of (4.4) lies in $H^2(\Omega)$ and the estimate*

$$\|u\|_{H^2(\Omega)} \leq c\|f\|_{L^2(\Omega)}$$

holds.

A proof can be found, e.g., in [101].

A problem, whose solution lies in $H^2(\Omega)$ for all $f \in L^2(\Omega)$ is called H^2-*regular*. If we assume this in the following, then we are on the safe side only if the domain Ω is convex or has a smooth boundary. This excludes, e.g., domains with *reentrant corners* (see Remarks 4.20 and 4.22 below).

Lemma 4.18. *Let Ω be a convex domain in $\mathbb{R}^d$ for $d \leq 3$. Let $u \in H^2(\Omega)$ be a solution of the Poisson equation*

$$a(u, v) = \langle f, v \rangle, \quad v \in H_0^1(\Omega).$$

Assume it is interpolated by linear finite elements continuous over the whole Ω. Then the interpolation error is bounded by

$$\|u - I_h u\|_{H^1(\Omega)} \leq ch\|u\|_{H^2(\Omega)},$$

where the constant c depends on the domain Ω and in particular on the shape regularity of the mesh.

A proof of this statement can be skipped; this kind of estimate was derived in Volume 1, Section 7.1.3, although only for $d = 1$. With Lemma 4.18 we are now ready for an estimation of the discretization error in the energy norm.

Theorem 4.19. *Let $\Omega \subset \mathbb{R}^d$ for $d \leq 3$ be convex and $u \in H^2(\Omega)$ weak solution of the Poisson equation. Let $u_h \in S_h$ denote the Galerkin solution of*

$$a(u_h, v_h) = \langle f, v_h \rangle, \quad v_h \in S_h \subset H_0^1(\Omega)$$

with linear, conformal finite elements. Then for the discretization error we have the estimate

$$\|u - u_h\|_a \leq ch\|f\|_{L^2(\Omega)}. \tag{4.55}$$

Proof. We insert $v_h = I_h u$ into the right-hand side of (4.54) and apply Lemma 4.18 and 4.17:

$$\|u - u_h\|_a \leq \|u - I_h u\|_a \leq ch\|u\|_{H^2(\Omega)} \leq ch\|f\|_{L^2(\Omega)}. \qquad \square$$

Remark 4.20. The above derivation excludes domains with *reentrant corners*. In order to get a corresponding estimate, one may employ Sobolev spaces with noninteger regularity index m which lie in some suitable sense "between" the Sobolev spaces with integer indices. For domains $\Omega \subset \mathbb{R}^2$ we have $u \in H^{1-\epsilon+1/\alpha}(\Omega)$ for every $\epsilon > 0$, depending upon the interior angle $\alpha\pi > \pi$ at the corner [108]. As a consequence, one may obtain an error estimate of the form

$$\|u - u_h\|_a \leq ch^{1/\alpha-\epsilon}.$$

For one-sided Neumann problems the angle α in the above exponent must be exchanged by 2α (see the illustrating Example 6.23 in Section 6.3.2).

Estimation of the L^2-error

In the L^2-norm we have one differentiation order less than in the energy norm. That is why we expect, for H^2-regular problems, a better error result of the kind

$$\|u - u_h\|_{L^2(\Omega)} \leq ch^2\|f\|_{L^2(\Omega)}.$$

However, such a result is harder to prove as, e.g., (4.55) in Theorem 4.19 above. For this purpose we require some method often named in the literature as *duality trick* or also *Aubin–Nitsche trick*; indeed J. P. Aubin [8] in 1967 and J. A. Nitsche [160] in 1968 invented this "trick" independently of each other.

Theorem 4.21. *In addition to the assumptions of Theorem 4.19 let the problem be H^2-regular. Then the following estimates hold:*

$$\|u - u_h\|_{L^2(\Omega)} \leq ch\|u - u_h\|_a, \tag{4.56}$$

$$\|u - u_h\|_{L^2(\Omega)} \leq ch^2\|f\|_{L^2(\Omega)}. \tag{4.57}$$

Proof. Let $e_h = u - u_h$ denote the approximation error. The "trick" (which we will encounter again within a more general framework in Section 6.1.5) now is that we introduce some $z \in H_0^1(\Omega)$ as weak solution of the Poisson equation with right-hand side e_h

$$a(z, v) = \langle e_h, v \rangle, \quad v \in H_0^1(\Omega).$$

Let $z_h \in S_h$ denote the Galerkin solution corresponding to z. If we choose $v = e_h$ in the above equation, then we first obtain the reformulation

$$\|e_h\|_{L^2(\Omega)}^2 = a(z, e_h) = a(z, e_h) - \underbrace{a(z_h, e_h)}_{=0} = a(z - z_h, e_h).$$

Application of the Cauchy–Schwarz inequality for the energy product and of the result (4.55) for $z - z_h$ then leads to

$$\|e_h\|_{L^2(\Omega)}^2 \leq \|z - z_h\|_a \|e_h\|_a \leq ch \|e_h\|_{L^2(\Omega)} \|e_h\|_a.$$

Upon dividing the common factor $\|e_h\|_{L^2(\Omega)}$ on both sides one obtains statement (4.56). Finally, application of (4.55) for e_h directly supplies (4.57). □

Remark 4.22. As for the energy error estimate we here also have an estimation result differing from (4.57) for Dirichlet problems over domains with *reentrant corners*: for interior angles $\alpha\pi$ with $\alpha > 1$, we get

$$\|u - u_h\|_{L^2(\Omega)} \leq ch^{2/\alpha - \epsilon} \|f\|_{L^2(\Omega)}.$$

Again for one-sided Neumann problems α should be exchanged by 2α (illustrated in Example 6.23 in Section 6.3.2).

Remark 4.23. We have always assumed that Ω can be exactly represented by the mesh and thus have restricted ourselves to polytopes $\Omega = \operatorname{int} \bigcup_{T \in \mathcal{T}} T$. However, if Ω has arbitrary curvilinear boundaries, then an exact representation of the domain by some triangulation is impossible in general, which introduces an additional approximation error (see, e.g., the textbooks by G. Dziuk [82] or D. Braess [39]).

4.4.2 Eigenvalue Problems

In this section we return to self-adjoint eigenvalue problems (see Section 4.1.3 above and the notation introduced there). In the following we want to find out to which extent the finite element approximations $\lambda_{k,h}, u_{k,h}$ differ from the continuous variables λ_k, u_k. As in the preceding sections, we restrict our attention to linear finite elements, if not stated otherwise.

FE-approximation of Eigenvalues.   For the time being, let us consider the simplest case $k = 1$, in physics often denoted as the "ground state". The following theorem dates back to G. M. Vainikko [206] in 1964 (published first in Russian, here quoted from the classical textbook by G. Strang and G. J. Fix [194]).

Theorem 4.24. *Let $\lambda_{1,h}$ denote the linear finite element approximation of the lowest eigenvalue $\lambda_1 > 0$ and $u_1 \in H_0^2(\Omega)$ the corresponding eigenfunction. Then, for sufficiently small mesh sizes $h \leq \bar{h}$, one has*

$$0 \leq \lambda_{1,h} - \lambda_1 \leq ch^2 \lambda_1 \|u_1\|_{H^2(\Omega)} \tag{4.58}$$

with a generic constant c.

Proof. The here presented proof is a slight simplification of the proof in [194]. The left-hand side of (4.58) follows from (4.19). Without loss of generality we may assume that the eigenfunction is normalized as $\langle u_1, u_1 \rangle = 1$, the Rayleigh-quotient is anyway invariant under scaling.

The trick of the proof consists in introducing the Ritz projection $Pu_1 \in V_h$ of the exact eigenfunction u_1, for which, by definition, one has $a(u_1 - Pu_1, Pu_1) = 0$ (see Figure 4.1). Assume that V_h is "sufficiently fine", so that $Pu_1 \neq 0$. We now insert this projection as comparative element into (4.18) and obtain

$$\lambda_{1,h} = \min_{u_h \in V_h} \frac{a(u_h, u_h)}{\langle u_h, u_h \rangle} \leq \frac{a(Pu_1, Pu_1)}{\langle Pu_1, Pu_1 \rangle}.$$

For the numerator we directly obtain from the definition of Pu_1

$$a(Pu_1, Pu_1) \leq a(u_1, u_1) = \lambda_1.$$

For the denominator one gets

$$\langle Pu_1, Pu_1 \rangle = \langle u_1, u_1 \rangle - 2\langle u_1, u_1 - Pu_1 \rangle + \langle u_1 - Pu_1, u_1 - Pu_1 \rangle \geq 1 - \sigma(h),$$

where we have defined some quantity $\sigma(h)$ as

$$\sigma(h) = |2\langle u_1, u_1 - Pu_1 \rangle - \langle u_1 - Pu_1, u_1 - Pu_1 \rangle|.$$

We now assume (and assure later) that $\sigma(h) \leq 1/2$ can be chosen; thus we have confirmed our earlier assumption $Pu_1 \neq 0$, since in this case we would get $\sigma(h) = 1$. This permits the estimate

$$\lambda_{1,h} \leq \frac{\lambda_1}{1 - \sigma(h)} \leq \lambda_1(1 + 2\sigma(h)) \quad \Leftrightarrow \quad \lambda_{1,h} - \lambda_1 \leq 2\sigma(h)\lambda_1. \tag{4.59}$$

Application of the triangle inequality and the Cauchy–Schwarz inequality to $\sigma(h)$ supplies

$$\sigma(h) \leq 2\|u_1\|_{L^2(\Omega)} \|u_1 - Pu_1\|_{L^2(\Omega)} + \|u_1 - Pu_1\|_{L^2(\Omega)}^2$$

$$= \|u_1 - Pu_1\|_{L^2(\Omega)} \left(2 + \|u_1 - Pu_1\|_{L^2(\Omega)}\right).$$

Upon employing the L^2-error estimate (4.57) we obtain

$$\|u_1 - Pu_1\|_{L^2(\Omega)} \leq ch^2 \|u_1\|_{H^2(\Omega)}. \tag{4.60}$$

Let us now choose an upper bound $\bar{h}$ of the mesh size h small enough such that $\sigma(h) \leq 1/2$ can be assured, i.e.,

$$\|u_1 - Pu_1\|_{L^2(\Omega)} \leq c\bar{h}^2 \|u_1\|_{H^2(\Omega)} \leq \frac{1}{2 + \sqrt{6}} \quad \Rightarrow \quad \sigma(h) \leq \sigma(\bar{h}) \leq 1/2.$$

Then the bracket $(2 + \|u_1 - Pu_1\|_{L^2(\Omega)})$ can be bounded by a constant. Finally, insertion into (4.59) directly leads to statement (4.58) of the theorem with a (new) generic constant c. $\square$

In principle, Theorem 4.24 can be transferred to higher eigenvalues (i.e., for $k > 1$) in an elementary way, based on the representation (1.16) from Corollary 1.1 in both the continuous case or its weak formulation (4.20), respectively. Both cases, however, require the assumption that an invariant subspace $U_{j-1} = \text{span}(u_1, \ldots, u_{j-1})$ or its weak counterpart $V_{j-1,h} = \text{span}(u_{1,h}, \ldots, u_{j-1,h})$, respectively, is known to sufficient accuracy, so that in its orthogonal complement $U_{j-1}^{\perp}$ or $V_{j-1,h}^{\perp}$, resp., the Rayleigh-quotient minimization can be performed. That is why we already indicated in Section 4.1.3 that for higher eigenvalues the approach of B. Döhler [78] is recommended, where eigenvalues and eigenvectors are computed *simultaneously*. Here we do not go into more detail on the approximation quality of the higher eigenvalues, but instead postpone this question until Section 7.5.2, where we introduce an extension of the idea of Döhler to multigrid methods.

FE-approximation of Eigenfunctions.   Before plunging into details, let us focus on an elementary, but important connection of the approximations for eigenvalues and eigenvectors.

Lemma 4.25. *For $k = 1, \ldots, N$ the following identity holds*

$$a(u_k - u_{k,h}, u_k - u_{k,h}) = \lambda_k \|u_k - u_{k,h}\|_{L^2(\Omega)}^2 + \lambda_{k,h} - \lambda_k. \tag{4.61}$$

Proof. For the eigenfunctions u_k and $u_{k,h}$ we choose the special normalization $\|u_k\|_{L^2(\Omega)} = \|u_{k,h}\|_{L^2(\Omega)} = 1$. Hence, from the Rayleigh quotients we immediately see that $a(u_k, u_k) = \lambda_k$ and $a(u_{k,h}, u_{k,h}) = \lambda_{k,h}$. Then we conclude

$$a(u_k - u_{k,h}, u_k - u_{k,h}) = \lambda_k - 2\lambda_k \langle u_k, u_{k,h} \rangle + \lambda_{k,h}$$
$$= \lambda_k(2 - 2\langle u_k, u_{k,h}\rangle) + \lambda_{k,h} - \lambda_k.$$

By reformulation of the right bracket towards a quadratic form with the special normalization for u_k and $u_{k,h}$ we confirm (4.61). $\square$

This identity is extremely useful in the context of adaptive algorithms. As a consequence, the order of the approximation error of the eigenfunctions in the energy norm
is only half the order of the approximation error of the eigenvalues. This quite obviously holds independent of the order of the finite elements. The main remaining is to
derive an estimate of the error in the L^2-norm, which we will perform in the following
approximation theorem. There again we restrict out interest to the case $k = 1$ where
the essential structure is already visible, as far as it is important for adaptive meshes.

Theorem 4.26. *Let $\lambda_1 > 0$ and $\lambda_1 \leq \lambda_{1,h} < \lambda_{2,h} < \cdots$ for the linear finite element approximations of the eigenvalues. Then, for sufficiently small mesh sizes h, the
estimates*

$$\|u_1 - u_{1,h}\|_{L^2(\Omega)} \leq Ch^2 \|u_1\|_{H^2(\Omega)} \tag{4.62}$$

in the L^2-norm and

$$\|u_1 - u_{1,h}\|_a \leq Ch\sqrt{\lambda_1}\|u_1\|_{H^2(\Omega)}$$

in the energy norm hold, each with a generic constant C.

Proof. As an orientation we again take the textbook [194], here for the special case
of linear finite elements. Just as in the proof of Theorem 4.24 we define Pu_1 as the
Ritz-projection. As normalization we select $\|u_1\|_{L^2(\Omega)} = \|u_{1,h}\|_{L^2(\Omega)} = 1$, which
directly implies $a(u_1, u_1) = \lambda_1$ and $a(u_{1,h}, u_{1,h}) = \lambda_{1,h}$.

In a first step, we prove the L^2-estimate. As $Pu_1 \in V_h$, it can be expanded in terms
of the orthogonal system of the discrete eigenfunctions ($N = \dim(V_h)$):

$$Pu_1 = \sum_{j=1}^{N} \alpha_j u_{j,h} \quad \text{with} \quad \alpha_j = \langle Pu_1, u_{j,h} \rangle.$$

The trick of this proof is now to isolate the first term of this expansion. For this purpose,
let $\alpha_1 = \langle Pu_1, u_{1,h} \rangle > 0$, which can be assured by a choice of the sign in $u_{1,h}$. By
the usual orthogonal decomposition we then obtain

$$\|Pu_1 - \alpha_1 u_{1,h}\|_{L^2(\Omega)}^2 = \sum_{j=2}^{N} \langle Pu_1, u_{j,h} \rangle^2.$$

Each of the above terms in the sum can be converted by virtue of the auxiliary result

$$(\lambda_{j,h} - \lambda_1)\langle Pu_1, u_{j,h} \rangle = \lambda_1 \langle u_1 - Pu_1, u_{j,h} \rangle.$$

The different left- and right-hand terms above can be verified to be equal by

$$\lambda_{j,h}\langle Pu_1, u_{j,h} \rangle = a(Pu_1, u_{j,h}) + a(u_1 - Pu_1, u_{j,h}) = a(u_1, u_{j,h}) = \lambda_1 \langle u_1, u_{j,h} \rangle .$$

Insertion then yields

$$\|Pu_1 - \alpha_1 u_{1,h}\|^2_{L^2(\Omega)} = \sum_{j=2}^{N}\left(\frac{\lambda_1}{\lambda_{j,h} - \lambda_1}\right)^2 \langle u_1 - Pu_1, u_{j,h}\rangle^2.$$

As the first eigenvalue is isolated, one can define some constant ρ, for which holds

$$\frac{\lambda_1}{\lambda_{j,h} - \lambda_1} \le \rho \quad \text{for } j > 1.$$

Inserting this relation into the course of the calculation we obtain

$$\|Pu_1 - \alpha_1 u_{1,h}\|^2_{L^2(\Omega)} \le \rho^2 \sum_{j=2}^{N}\langle u_1 - Pu_1, u_{j,h}\rangle^2 \le \rho^2 \sum_{j=1}^{N}\langle u_1 - Pu_1, u_{j,h}\rangle^2,$$

which, in summary, supplies the intermediate result

$$\|Pu_1 - \alpha_1 u_{1,h}\|_{L^2(\Omega)} \le \rho\|u_1 - Pu_1\|_{L^2(\Omega)}.$$

This result, in turn, may be inserted into the triangle inequality

$$\|u_1 - \alpha_1 u_{1,h}\|_{L^2(\Omega)} \le \|u_1 - Pu_1\|_{L^2(\Omega)} + \|Pu_1 - \alpha_1 u_{1,h}\|_{L^2(\Omega)}$$
$$\le (1 + \rho)\|u_1 - Pu_1\|_{L^2(\Omega)}.$$

At this point we are close to the finish:

$$\|u_1 - u_{1,h}\|_{L^2(\Omega)} \le \|u_1 - \alpha_1 u_{1,h}\|_{L^2(\Omega)} + \|(\alpha_1 - 1)u_{1,h}\|_{L^2(\Omega)}$$
$$= \|u_1 - \alpha_1 u_{1,h}\|_{L^2(\Omega)} + |\alpha_1 - 1|.$$

Using the normalization of the eigenfunctions, the second term can be estimated by means of a further triangle inequality according to

$$1 - \|u_1 - \alpha_1 u_{1,h}\|_{L^2(\Omega)} \le \|\alpha_1 u_{1,h}\|_{L^2(\Omega)} = \alpha_1 \le 1 + \|u_1 - \alpha_1 u_{1,h}\|_{L^2(\Omega)},$$

which immediately yields

$$-\|u_1 - \alpha_1 u_{1,h}\|_{L^2(\Omega)} \le \alpha_1 - 1 \le \|u_1 - \alpha_1 u_{1,h}\|_{L^2(\Omega)},$$

which, in turn, is equivalent to $|\alpha_1 - 1| \le \|u_1 - \alpha_1 u_{1,h}\|_{L^2(\Omega)}$. Hence, we obtain above

$$\|u_1 - u_{1,h}\|_{L^2(\Omega)} \le 2\|u_1 - \alpha_1 u_{1,h}\|_{L^2(\Omega)} \le 2(1 + \rho)\|u_1 - Pu_1\|_{L^2(\Omega)}.$$

Now we are finally able to employ (4.60), which leads us to

$$\|u_1 - u_{1,h}\|_{L^2(\Omega)} \leq C \|u_1 - Pu_1\|_{L^2(\Omega)} \leq C h^2 \|u_1\|_{H^2(\Omega)}.$$

This is statement (4.62) of the theorem.

The proof for the *energy norm* is significantly shorter. It is simply based on the identity (4.61) for $k = 1$: The desired estimate is obtained by inserting (4.62) into the first term on the right and (4.58) into the difference of the second and the third term. The leading orders in the mesh size h are obtained by a restriction to "sufficiently small" mesh sizes $h \leq \bar{h}$ similar to the procedure in the proof of the preceding Theorem 4.24. □

As expected from Lemma 4.25, the approximation error of the eigenfunction $u_{1,h}$ in the L^2-norm possesses the same order as the error of the eigenvalue $\lambda_{1,h}$, whereas in the energy norm only the square root of the order is realizable. A theory covering the general case $k \geq 1$, even for *multiple* eigenvalues as well as higher order finite elements can be found in the voluminous paper [12] by I. Babuška and J. E. Osborn – there, however, only for uniform meshes.

4.4.3   Angle Condition for Nonuniform Meshes

In the previous sections we have only considered the case of uniform or even equidistant meshes, mainly for ease of presentation. For the solution of practically important problems, however, *problem adapted* meshes are essential, which in general are *nonuniform*; the actual construction of such meshes will be treated in Section 6.2. For the extension of the above estimates to general nonuniform meshes not only the local mesh size h_T, the longest sides of triangle T, but also the *shape* of the elements T play an important role. In their meanwhile classical analysis P. G. Ciarlet and P.-A. Raviart [52] characterized the shape of triangles by means of the inscribed circle radius ρ_T: for *shape regular* triangles they demand the condition

$$\frac{h_T}{\rho_T} \leq C_T, \quad T \in \mathcal{T},$$

where the constant $C_T \geq 1$ should be "not too large"; for the geometric situation see Figure 4.17. For a long time, based on this analysis, the ruling opinion had been to avoid "acute" angles, which means "large" constants C_T, in finite elements. The two limiting cases are illustrated in Figure 4.18.

A more subtle analysis, however, shows that only "obtuse" interior angles should be avoided. For the special case $d = 2$ with linear finite elements there is an elementary geometric proof by J. L. Synge [197] (dating back to 1957), on which the following presentation is based.

We start by studying only the *interpolation error*, which via (4.54) enters into the estimate of the *discretization error*.

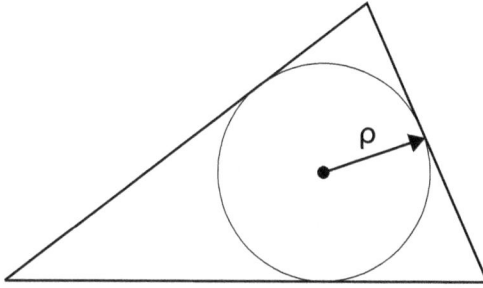

Figure 4.17. Inscribed circle for the characterization of the shape of a triangle.

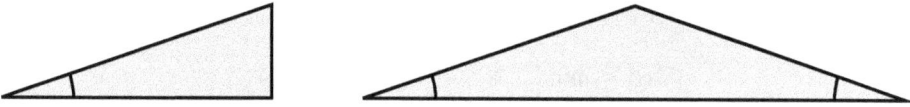

Figure 4.18. Triangles with large constant C_T. *Left:* acute interior angle ($C_T \approx 7.2$). *Right:* two acute and one obtuse interior angle ($C_T \approx 12.3$).

Theorem 4.27. *Consider a triangle $T = P_1 P_2 P_3$ with maximum interior angle γ. Let I_h be the linear interpolation operator over the vertices P_i. Then the following pointwise estimate holds:*

$$|\nabla(u - I_h u)| \leq h \left(1 + \frac{2}{\sin \gamma} \right) \|u\|_{C^{1.1}(T)}.$$

Proof. Let $h_1 e_1 = P_2 - P_1$ and $h_2 e_2 = P_3 - P_1$ with $\|e_i\| = 1$, and assume an interior angle β located at P_1. We consider the interpolation error $w := u - I_h u$ with $w(P_i) = 0$. Due to the theorem of Rolle there exist points x_1 and x_2 on these edges such that

$$w_x(x_i) e_i = \frac{\partial w}{\partial e_i}(x_i) = 0, \quad i = 1, 2.$$

For $x \in T$, the pointwise estimate

$$|w_x(x) e_i| \leq h \|w\|_{C^{1.1}(T)} = h \|u\|_{C^{1.1}(T)} \tag{4.63}$$

holds due to the Lipschitz continuity of w_x and $\|x_i - x\| \leq h$. From the bound for the directional derivatives we now obtain a bound for the whole gradient. We start with

$$|\nabla w| = \max_{\|v\|=1} w_x v.$$

Representing v as a linear combination of the edge vectors e_i, say $v = \alpha_1 e_1 + \alpha_2 e_2$, we obtain from (4.63)

$$|\nabla w| = \max_{\|v\|=1} w_x(\alpha_1 e_1 + \alpha_2 e_2) \leq \max_{\|v\|=1} (|\alpha_1| + |\alpha_2|) h \|u\|_{C^{1.1}(T)}. \tag{4.64}$$

W.l.o.g. we may orient the coordinate frame such that $e_1 = (1, 0)^T$, which implies $e_2 = (\cos \beta, \sin \beta)^T$. With $v = [v_1, v_2]^T$ we then have

$$
\begin{bmatrix} 1 & \cos \beta \\ & \sin \beta \end{bmatrix} \begin{bmatrix} \alpha_1 \\ \alpha_2 \end{bmatrix} = \begin{bmatrix} v_1 \\ v_2 \end{bmatrix}.
$$

Due to $|v_i| \leq 1$ we obtain the bound

$$
|\alpha_1| + |\alpha_2| \leq 1 + \frac{2}{\sin \beta}.
$$

Insertion into (4.64) supplies an estimate from each of the angles β_i, $i = 1, 2, 3$ at one of the three corner points P_1, P_2, P_3, which gives rise to

$$
|\nabla w| \leq \min_i \left(1 + \frac{2}{\sin \beta_i} \right) h \|u\|_{C^{1,1}(T)}.
$$

Obviously, right angles will provide the smallest bounds, while acute and obtuse angles will deliver large bounds. For a better understanding let us return to Figure 4.18: on the left, the above sin-factor will be small for the acute angle, but close to 1 for the other two nearly right angles; on the right, the above denominator will be small for all three angles, so that a blow-up of the gradient error bound is unavoidable. This confirms the statement of the theorem.   □

Remark 4.28. The proof is based on a *stability argument*: Within the representation $v = \alpha_1 e_1 + \alpha_2 e_2$, the terms in the sum are each required to be "not too much" larger than the sum. This argument translates directly to tetrahedra with $d = 3$, only that the interpretation of the angles is a bit more complex. In Section 7.1, we will again encounter this kind of stability condition, there in a different context.

The pointwise assertion of Theorem 4.27 can be extended to the whole domain in the form

$$
\|u - I_h u\|_{C^1(\Omega)} \leq \left(1 + \frac{2}{\sin \gamma} \right) h \, \|u\|_{C^{1,1}(\Omega)}.
$$

Here the assumption $u \in C^{1,1}(\Omega)$ is disturbing. An analysis by P. Jamet [130] from 1976, here given in the version of I. Babuška and A. K. Aziz [10] from the same year (see also F. A. Bornemann [34]), actually supplies the improved (and more natural) result

$$
\|u - I_h u\|_{H^1(\Omega)} \leq \frac{ch}{\cos(\gamma/2)} \|u\|_{H^2(\Omega)}.
\tag{4.65}
$$

The result carries over to the case $d = 3$ (see M. Křížek [135]). In the one or the other form the clear message reads:

In triangulations obtuse *interior angles (see Figure 4.18, right) should be avoided, whereas* acute *interior angles (see Figure 4.18, left) are well acceptable in view of the discretization errors.*

Remark 4.29. This theoretical insight has important consequences in a number of application problems; e.g., for strong *anisotropies* slim triangles, i.e., individual acute angles as in Figure 4.18, left, are indispensable for theoretical reasons. They may be caused either by anisotropic coefficients (material constants) or by the geometry of the domain, e.g., in a Prandtl boundary layer (see Figure 2.4). Furthermore the scaled anisotropic Poisson equation

$$-\epsilon u_{xx} - u_{yy} = f$$

or also the scaled convection dominated equation

$$-\epsilon \Delta u + v^T \nabla u = f(u)$$

reflect such a problem situation.

4.5 Exercises

Exercise 4.1. Verify the subsequent statements for the symmetric bilinear form

$$a(u, v) = \int_0^1 x^2 \nabla u^T \nabla v \, dx, \quad u, v \in H_0^1(]0, 1[).$$

1. a is positive definite, but not $H_0^1(]0, 1[)$-elliptic.
2. The problem
$$J(u) := \frac{1}{2} a(u, u) - \int_0^1 u \, dx = \min$$
   does not possess a solution in $H_0^1(]0, 1[)$.

Identify the corresponding differential equation.

Exercise 4.2. Derive the Lamé-Navier equation (2.26) from the vanishing of the directional derivatives (2.23) using the St. Venant–Kirchhoff material law (2.25). For this purpose assume a linear potential V. What would change for nonlinear V?

Hint: As in the transition from the weak formulation to the strong formulation in Section 4.1.2, use Theorem A.3 of Gauss.

Exercise 4.3. In Section 4.2.3 an algorithm for the adaptive determination of the truncation index N in spectral methods has been derived on the basis of the *energy norm*. As an alternative, let $\| \cdot \|$ denote a norm induced by a scalar product $(\cdot, \cdot)$, with respect to which the applied basis is orthonormal. Which partial steps and results of the derivation in Section 4.2.3 change? In what sense does the numerical evaluation simplify?

Exercise 4.4. In Section 4.2.3 an algorithm for the adaptive choice of the truncation index N in spectral methods has been derived that is based on the theoretical
approximation model (4.30) for analytical functions $u \in C^{\omega}(\Omega)$. Derive an alternative algorithm that instead employs the approximation model

$$\epsilon_N \approx C \, N^{-s},$$

whereby only $u \in C^s(\Omega)$ is assumed to hold. Compare this algorithm with the
one worked out in Section 4.2.3.

Exercise 4.5. Let L_1 and L_2 be two self-adjoint differential operators for functions over a bounded domain Ω. Let $\{\varphi_i\}$ denote an orthonormal system corresponding to the operator L_1. The problem stated is to solve the differential
equation

$$(L_1 + \epsilon L_2)u = f.$$

Consider the transition to the weak formulation. Let u_N denote the Galerkin approximation w.r.t. $\{\varphi_i\}$. Derive a linear equation system of the form

$$(I - \epsilon A)u_N = b.$$

Construct some fixed point iteration and study the conditions for its convergence.

Exercise 4.6. Show that the isoparametric quadratic finite elements defined
by (4.45) generate continuous ansatz functions.

Exercise 4.7. For isoparametric quadratic elements, let the "edge midpoint"
(x_m, y_m) between two vertices (x_a, y_a) and (x_b, y_b) of a triangle T be given by

$$\frac{1}{2} \begin{bmatrix} x_a + x_b \\ y_a + y_b \end{bmatrix} + s \begin{bmatrix} y_b - y_a \\ x_a - x_b \end{bmatrix}$$

where $|s| \le S$ (see Figure 4.19). Determine a maximal S so that for the determi-
nant J the condition $J > 1/2$ remains assured.

Exercise 4.8. Let Ω_h be the equidistant finite difference mesh on $\Omega = \,]0, 1[^2$
with mesh size $h = M^{-1}$. By uniform subdivision of the local quadrilaterals
along their diagonal one obtains a finite element triangulation. Consider the Poisson model problem

$$-\Delta u = f \quad \text{in } \Omega.$$

1. Compare the global finite difference matrix generated by the 5-point star
   with the stiffness matrix generated by linear finite elements for the case of
   homogeneous Dirichlet boundary conditions.

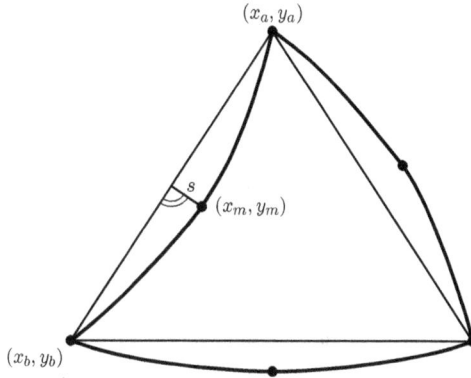

Figure 4.19. Exercise 4.7: Positioning of edge midpoints in isoparametric quadratic elements.

2. Which FD-discretization of the boundaries corresponds to the FE-discretization for the case of Neumann boundaries?

Exercise 4.9. Work out a proof of Corollary 4.1 following step by step the proof of Theorem 4.2.

Exercise 4.10. Prove that the shape functions in Figure 4.8, left and right, span the same space.

Exercise 4.11. Calculate the local stiffness and mass matrices for the quadratic shape functions (4.42) and (4.43) for $d = 1$.

Exercise 4.12. Show that barycentric coordinates on simplices are unique. Determine the barycentric coordinates of the reference triangle and the reference tetrahedron as functions of the Euclidean coordinates.

Exercise 4.13. Explain in some detail, which meaning the Synge Lemma has for the choice of a triangulation. Discuss the qualitative difference between slim triangles with very large and very small interior angles at the example of interpolation for $u(x, y) = x(1 - x)$ on the meshes depicted in Figure 4.20 over $]0, 1[^2$ for $\epsilon \to 0$.

Exercise 4.14. Given the nodal basis Φ_h^1 on simplicial meshes, show the following properties:

1. There exists a constant $c > 0$ that is independent of the elliptic operator A and the mesh size h such that

$$c^{-1}\alpha_a h^{d-2} \leq \langle A\varphi, \varphi \rangle \leq c C_a h^{d-2}$$

holds for all $\varphi \in \Phi_h^1$.

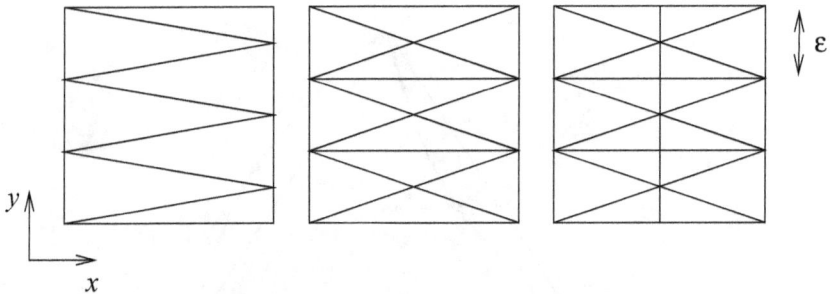

Figure 4.20. Exercise 4.13: triangulations with acute interior angles.

2. The condition of the mass matrix is bounded: There exists a constant $c > 0$ independent of h such that

$$c^{-1}\|v\|^2_{L^2(\Omega)} \le \sum_{\xi \in \mathcal{N}} v(\xi)^2 \|\varphi_\xi\|^2_{L^2(\Omega)} \le c\|v\|^2_{L^2(\Omega)}$$

holds for all $v \in S^1_h$.

Exercise 4.15. Let $\Omega \subset \mathbb{R}^d$ denote a domain with $0 \in \Omega$. Let $u(x) := |x|^\alpha$. Under what condition on d and $\alpha \in \mathbb{R}$ is $u \in H^1(\Omega)$ true?

Exercise 4.16. Determine the dimension of the hierarchical extension spaces $P_{p,d}$.

Exercise 4.17. We want to study ansatz functions of order p in space dimension d. For the time being, let the coefficients in the PDE be constant. By estimating the number of local element matrices to be summed find out for which p and d the computation of the element matrices according to (4.50) is more economic than according to (4.51). Extend the representation (4.51) for nonconstant coefficients and find out again, for which p and d this extension is more efficient than direct numerical quadrature.

Chapter 5

Numerical Solution of Linear Elliptic Grid Equations

In the two preceding chapters we studied various methods of how to discretize elliptic PDEs: finite difference methods in Chapter 3, spectral methods in Section 4.2, and finite element methods in Sections 4.3 and 4.4. For linear PDEs, any of these discretization methods gives rise to systems of linear equations. We have thus substituted a problem in the *infinite dimensional* space $H_0^1(\Omega)$ by a problem in the *finite dimensional* space $\mathbb{R}^N$, though with possibly large N. The *ellipticity* of the underlying PDE is reflected in the fact that the generated matrices are *symmetric positive definite or semidefinite*.

In spectral methods (Section 4.2) the solution of the arising linear equation systems can be dealt with using the methods of Volume I, (see the Remarks at the end of Section 4.2.1). This is why we focus here exclusively on such linear systems that come up by *finite difference* or *finite element* discretizations; they are called elliptic systems of *grid equations*. The generated spd-matrices are *sparse*. For sufficiently complex problems from engineering applications *mesh generators* are applied, the construction of which demands a lot of mathematics and computer science themselves, which, however, is omitted here.

For $d = 1$ tridiagonal matrices arise. For $d = 2$ and simple domains Ω *block tridiagonal matrices* of the principal shape

$$
A = \begin{bmatrix}
A_{11} & A_{12} & & & \\
A_{21} & A_{22} & A_{23} & & \\
& \ddots & \ddots & \ddots & \\
& & A_{q-1,q-2} & A_{q-1,q-1} & A_{q-1,q} \\
& & & A_{q,q-1} & A_{qq}
\end{bmatrix}
$$

are obtained where A_{ij} are sparse block matrices. For real life geometries of Ω, however, the matrices may exhibit some rather complex sparsity pattern. This holds all the more for $d = 3$. Due to the global domains of dependency of elliptic boundary value problems all discrete variables are coupled, which structurally means that the arising equation systems cannot be decoupled into smaller mutually independent subsystems by a reordering of the unknowns.

For large systems of linear grid equations, three classes of algorithms may be useful, which will be discussed subsequently:

1. direct elimination methods as treated in Section 5.1
   (for a preparation see, e.g., Volume 1, Section 1);

2. iterative solvers as treated in Section 5.2 and 5.3
   (for a preparation see, e.g., Volume 1, Section 8);

3.] hierarchical iterative solvers, where additionally the hierarchical mesh structure
   is exploited, as treated in Section 5.5.

The convergence analysis for methods of type 2 leads to the phenomenon of "smoothing" of the iterative errors, which we will investigate in Section 5.4. It lays the basis for the construction of methods of type 3, which we will treat in Section 5.5 and later, in more depth, in Chapter 7.

5.1   Direct Elimination Methods

In this section we deal with the direct solution of systems of linear grid equations whose matrices A are symmetric positive definite and sparse. The *sparsity* $s(A)$ is defined as the portion of nonzero entries in A. In the case of finite differences for Poisson problems in space dimension d we encounter, ignoring any missing neighboring nodes on the boundaries, just $2d + 1$ nonzeros per row so that

$$\text{NZ}(A) \le (2d + 1)N, \quad s(A) = \frac{\text{NZ}(A)}{N^2} \le \frac{2d + 1}{N} \ll 1. \tag{5.1}$$

For spd-matrices A Cholesky factorization is a general tool. In Volume 1, Section 1.4 we introduced two slightly different variants from which, in our context, mainly the *rational* Cholesky factorization is usual. Here the spd-matrix is factorized according to

$$A = LDL^T \tag{5.2}$$

with lower triangular matrix L, whose diagonal elements are all 1, and a diagonal matrix D, whose elements are positive, whenever A is positive definite. This factorization can also be generalized to *symmetric indefinite* matrices, as arise, e.g., from the discretization of the Helmholtz equation in the critical case (where in the diagonal also $(2, 2)$-blocks are permitted to allow for a numerically stable solution). The implementation for *dense* systems requires a storage of $\mathcal{O}(N^2)$ and a computing time of $\mathcal{O}(N^3)$. Direct elimination methods, however, may gain efficiency by exploiting the sparse structure of the arising matrices, which is why they are also called *direct sparse solvers* (see, e.g., the beautiful survey article by I. Duff [5] and the relatively recent monograph by T. Davis [63]). Direct sparse solvers consist of two important algorithmic pieces, a symbolic factorization and a coupling of matrix assembly with the Cholesky factorization, the principles of which we will present subsequently.

5.1.1   Symbolic Factorization

In the course of factorization additional nonzero elements are generated, so-called *fill-in elements*, often also just called *fill elements*. The less such elements are generated, the more efficient the sparse solver is. The number $NZ(L)$ of nonzeros in the factorization matrix L depends on the order of unknowns. In order to characterize an arbitrary numbering of unknowns, we therefore introduce a permutation matrix P, for which $PP^T = P^T P = I$ is known to hold. Then the equation system $Au = b$ can be written in the form

$$(PAP^T)(Pu) = Pb. \tag{5.3}$$

In particular for large sparse matrices a symbolic *pre*factorization is implemented preceding the actual numerical factorization.

Graph Representation.   Behind all partial steps lies the representation of the sparse matrix A by virtue of the corresponding *graph* $\mathcal{G}(A)$ (cf. Volume 1, Section 5.5). *Nodes* of the graph are the indices $i = 1, \ldots, N$ of the unknowns u, *edges* between index i and j exist where $a_{ij} \neq 0$; due to the symmetry of the matrix A the graph is *undirected*. Note that a reordering of the discrete unknowns does not change the graph, i.e., each undirected graph is invariant under the choice of the permutation matrix P.

Symbolic Elimination.   The aim is to find a simultaneous permutation of rows and columns in (5.3) such that as little fill-in as possible is generated. For this purpose, the elimination process is run symbolically before the numerical computation is performed. Along this route, starting from $\mathcal{G}(A)$ an extended $\mathcal{G}(L)$ is generated. In order to present the basic idea, we recall Gaussian elimination from Volume 1, Section 1.2: Suppose the elimination has already proceeded to some $(N - k + 1, N - k + 1)$ remainder matrix A'_k. On this matrix, we may perform the k-th elimination step

$$l_{ik} := a_{ik}^{(k)}/a_{kk}^{(k)} \qquad \text{for } i = k + 1, \ldots, N,$$

$$a_{ij}^{(k+1)} := a_{ij}^{(k)} - l_{ik}a_{kj}^{(k)} \quad \text{for } i, j = k + 1, \ldots, N$$

if the *pivot element* $a_{kk}^{(k)}$ does not vanish, which in our case of spd-matrices is guaranteed. We now consider this algorithmic piece under the aspect of the generation of fill-in. The basic pattern is a linear combination of the actual row i with the pivot row $k < i$, i. e.

$$\text{new row } i = \text{old row } i - l_{ik} \text{ pivot row } k.$$

We start from the fact that the nonzero structure of the given matrix A has been stored symbolically. Then the above numerical algorithm yields the following step k of a formal algorithm:

- the nonzero pattern of the subdiagonal column k of L is just the one of column k of the remainder matrix A'_k to be worked on;

- for $i = k + 1, \ldots, N$ rhe nonzero pattern of the new row i is the union of the nonzero patterns of the old row i with the one of the pivot row k.

Having run through all the symbolic steps for $k = 1, \ldots, N$ we have thus preallocated the storage space necessary for Gaussian elimination and stored the detailed information in the graph $\mathcal{G}(L)$.

Optimal Ordering.   For our Poisson model problem on the unit square there exists a theoretical solution of the ordering problems, an *optimal* ordering, the so-called *nested dissection*. In Figure 5.1, left, we graphically represent such an ordering strategy for a $(10, 10)$-mesh; in Figure 5.1, right, we show the structure of the corresponding matrix PAP^T. Clearly, one will not at first glance see that this distribution of the elements of A will produce the minimal fill-in in Cholesky factorization. The proof of optimality for the model problem was already given by A. George in 1973. For irregular finite element meshes, however, only suboptimal heuristics are left, which was published in 1978 by A. George and J. W.-H. Liu (see their monograph [100]). Today such strategies can be found in all direct sparse solvers, in two as well as in three space dimensions.

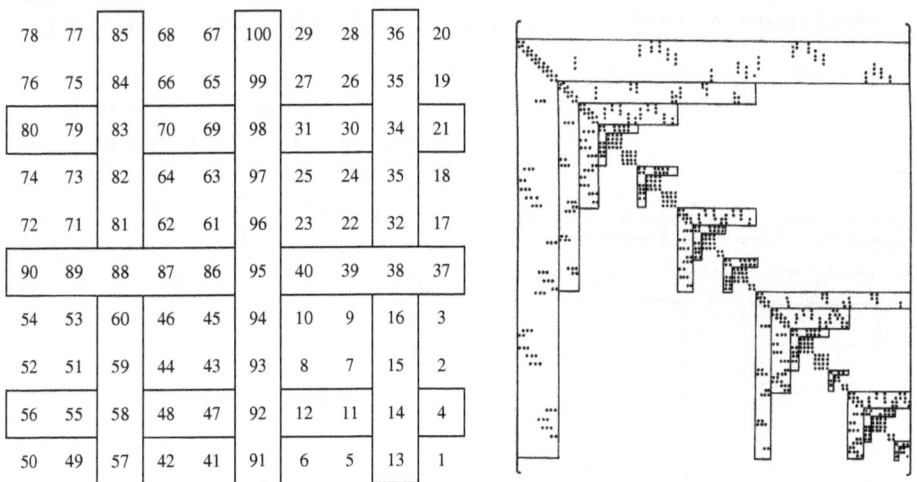

78	77	85	68	67	100	29	28	36	20
76	75	84	66	65	99	27	26	35	19
80	79	83	70	69	98	31	30	34	21
74	73	82	64	63	97	25	24	35	18
72	71	81	62	61	96	23	22	32	17
90	89	88	87	86	95	40	39	38	37
54	53	60	46	45	94	10	9	16	3
52	51	59	44	43	93	8	7	15	2
56	55	58	48	47	92	12	11	14	4
50	49	57	42	41	91	6	5	13	1

Figure 5.1. Poisson model problem [100]. *Left:* "nested dissection" ordering over a $(10, 10)$-mesh. *Right:* corresponding matrix structure.

A finer analysis of the problem of optimal ordering shows that is is in general NP-hard (quoted due to [63]). As a consequence, "good" heuristics, which means "suboptimal" orderings, are acceptable as sufficiently efficient. Aside from the method of *nested dissection* a *minimum degree* algorithm has evolved, oriented at the graph $\mathcal{G}(A)$. The methods of *Cuthill–McKee* and *reversed Cuthill–McKee*, which are also popular, merely reduce the bandwidth of each row in the arising matrix L, which is why they are often also called *envelope* methods; in this concept the fast *Level-3 BLAS* routines can be applied efficiently. This and further heuristics are elaborated in the book [63] by T. Davis.

Numerical Solution.   After symbolic factorization the permutation P is determined in the form of an *elimination tree*, so that the explicit numerical factorization

$$\overline{A} = PAP^T = LDL^T$$

can be performed on the allocated storage space. Then the following partial steps remain to be performed:

1. ordering of the components of the right-hand side $\overline{b} = Pb$;

2. forward substitution $Lz = \overline{b}$;

3. backward substitution $DL^T\overline{u} = z$;

4. backordering of the computed unknowns $u = P^T\overline{u}$.

Numerical Stability.   In general, the symbolic minimization of the fill-in does not guarantee numerical stability. In [63] one or two steps of *iterative refinement* of the results on the basis of the residual

$$\overline{r}(\overline{u}) = \overline{b} - \overline{A}\overline{u}$$

at the same mantissa length are suggested (cf. also Volume 1, Section 1.3). Let $\overline{u}^{(0)}$ denote the approximate solution from the Cholesky method, then the iterative refinement reads, in short-hand notation,

$$LL^T\delta\overline{u}_i = \overline{r}(\overline{u}_i), \quad \overline{u}_{i+1} = \overline{u}_i + \delta\overline{u}_i, \quad i = 0, 1.$$

However, experience seems to show that this refinement reduces the residuals $\overline{r}(\overline{u}_i)$, but not the errors $\overline{u} - \overline{u}_i$.

Software.   On the basis described here, a number of codes exist, from which we only want to mention as examples SuperLU [64] and PARDISO [178]; see also the software list at the end of the book.

5.1.2   Frontal Solvers

This kind of algorithm by I. Duff et al. [81] is particularly suited in the context of finite element methods and has achieved top acceptance in a series of application areas. The idea here is to couple the assembly of the matrix A directly with the subsequent rational Cholesky factorization, which means to perform a concurrent factorization of the matrix in the course of the computation of the individual finite elements.

Upon recalling Section 4.3.3 we start from the decomposition of the global matrix into its element matrices, which anyway come up in the assembly process:

$$A = \sum_{T \in \mathcal{T}} A(T).$$

The factorization (5.2) is to be computed. Let $D = \operatorname{diag}(d_1, \ldots, d_N)$ and let $l_1, \ldots, l_N$ denote the subdiagonal columns of L, then the factorization may also be written elementwise as

$$A = \sum_{k=1}^{N} d_k l_k l_k^T.$$

The column vector l_k cannot be computed before the column of A corresponding to the variable k has been assembled (see, e.g., Volume 1, Section 1.4). In view of that fact, the nodes (variables) are divided into (i) already assembled ones, which have been treated in previous steps of the Cholesky factorization, (ii) newly to be assembled ones, which are to be factorized next, and (iii) ones not yet finally assembled, which therefore do not yet contribute to the factorization. In Figure 5.2 we give a simple example due to [5].

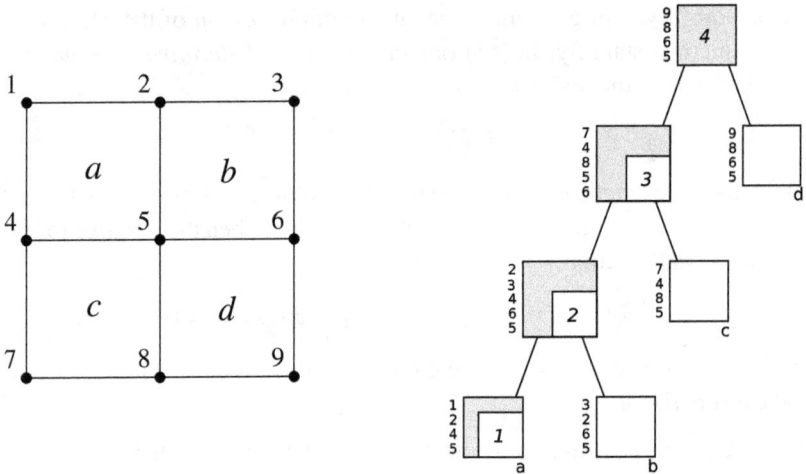

Figure 5.2. Illustrative example for frontal method (due to [5]). *Left*: order of assembly steps on the FE-mesh. *Right*: corresponding elimination steps.

The simple mesh in Figure 5.2, left, has four finite elements, whose element matrices are computed (assembled) in the order a, b, c, d. After the assembly of a only the node 1 is finished, which implies that the corresponding column 1 of L (equivalent to row 1 of L^T) can be computed. After the assembly of b the same holds for the new nodes $2, 3$, after c then for the nodes $7, 4$ and, finally, after d for the nodes $9, 8, 6, 5$. In Figure 5.2, right, the associated submatrices with their column- and row-wise factorization parts (gray) for the new nodes can be found. A substep of the matrix factorization is represented in Figure 5.3. Along these lines dense "small" factor matrices (gray areas in Figures 5.2 and 5.3) arise which can be efficiently worked down by *Level*-3 *BLAS* routines.

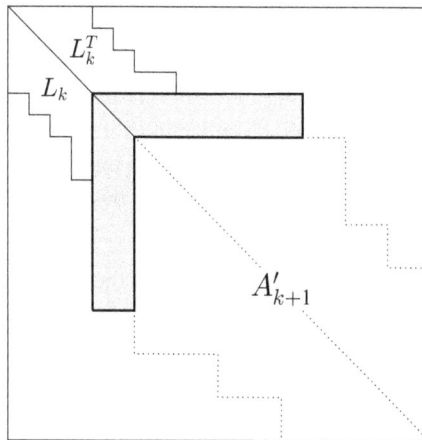

Figure 5.3. Illustrative example for frontal method (due to [5]): Structure of submatrix from some intermediate step of the assembly. (The normalized upper triangular matrix L_k^T is not stored, of course.)

In the frontal method, too, there is a variant for *indefinite* systems in which diagonal $(2, 2)$-blocks are introduced to improve the numerical stability. After running through the frontal method, only the above four algorithmic partial steps have to be performed.

Numerical Stability. An extension of the frontal method to unsymmetric matrices implements a *threshold pivoting*

$$\frac{|a_{ij}|}{\max_k |a_{kj}|} \geq \text{relpiv}, \quad 0 < \text{relpiv} \leq 1,$$

where the parameter relpiv is to be selected by the user. This strategy comprises a compromise between minimization of fill-in and numerical stability. An alternative is the *static* pivot strategy, where "too small" pivot elements are replaced by some user-prescribed minimal quantity. The perturbed system is then solved in a numerically stable way and its solution afterwards improved, e.,g., by iterative refinement.

Multifrontal Method. This is a special variant of the frontal method on parallel computers. There the factorization process starts at different finite elements that are mutually decoupled. By clever distribution an additional efficiency gain can be achieved; for more details, again see the survey article by I. Duff et al. [5].

Software. A rather mature code for the multifrontal method is MUMPS (acronym for **MU**ltifrontal **M**assively **P**arallel **S**olver) (see [6] and the software list at the end of the book).

5.2   Matrix Decomposition Methods

In Sections 5.2 and 5.3 we present iterative solvers for large linear grid equations, in particular under the aspect of discretized elliptic PDEs. For this problem class an iterative solution is anyway reasonable, since we are only interested in a solution up to discretization error accuracy. Any additional effort beyond that would waste computing time.

Linear Iterative Methods.   The common idea of the subsequent Sections 5.2.1 and 5.2.2 is: as the original system of equations is "too costly" to solve, we substitute the matrix A by a nonsingular matrix B such that systems of the form $Bz = w$ with given right-hand side w are "easy" to solve. Formally, we are led to the iteration

$$Bu_{k+1} = Bu_k + (b - Au_k)$$

or, slightly reformulated,

$$u_{k+1} - u = u_k - u - B^{-1}(Au_k - b) = G(u_k - u) \tag{5.4}$$

with $G = I - B^{-1}A$ as *iteration matrix*.

Convergence of Symmetrizable Iterative Methods.   Sufficient for the convergence of the linear iteration (5.4) is that the spectral radius satisfies $\rho(G) < 1$. To start with, we derive a basic result, which will be helpful in convergence proofs for various iterative methods.

Theorem 5.1.   *Given the iteration matrix $G = I - B^{-1}A$. Let B also be symmetric positive definite. Then, under the assumption*

$$\langle v, Av \rangle < 2 \langle v, Bv \rangle, \quad v \in \mathbb{R}^N, \tag{5.5}$$

we get the following result:

$$\rho(G) = |\lambda_{\max}(G)| < 1. \tag{5.6}$$

Proof. First we have to observe that with B and A symmetric, the iteration matrix G will not be symmetric, in general. However, we have

$$(BG)^T = (B(I - B^{-1}A))^T = (I - AB^{-1})B = B(I - B^{-1}A) = BG.$$

This means that G is symmetric w.r.t the discrete "energy" product $\langle \cdot, B \cdot \rangle$. As a consequence, the eigenvalues λ of the eigenvalue problem are real and the eigenvectors are B-orthogonal due to

$$Gv = \lambda v \quad \Leftrightarrow \quad BGv = \lambda Bv \quad \Leftrightarrow \quad (B^{1/2}GB^{-1/2})B^{1/2}v = \lambda B^{1/2}v, \quad \lambda \in \mathbb{R}.$$

From this the definition of the spectral radius via the B-norm immediately follows. Insertion then yields

$$\rho(G) = \sup_{v \neq 0} \left| \frac{\langle v, B(I - B^{-1}A)v \rangle}{\langle v, Bv \rangle} \right| = \sup_{v \neq 0} \left| 1 - \frac{\langle v, Av \rangle}{\langle v, Bv \rangle} \right|.$$

Because of the positivity of A and B together with assumption (5.5) we have

$$0 < \frac{\langle v, Av \rangle}{\langle v, Bv \rangle} < 2 \quad \text{for all } v \neq 0,$$

from which we finally obtain $\rho(G) < 1$.   □

Matrix Decomposition.   We begin with the classical iterative methods for large linear equation systems with *symmetric positive definite* matrices. Their construction principle is a *decomposition* of the matrix A according to

$$A = L + D + L^T,$$

with $D = \text{diag}(a_{11}, \ldots, a_{NN})$ *positive* and

$$L = \begin{bmatrix} 0 & \cdots & & \cdots & 0 \\ a_{21} & \ddots & & & \vdots \\ \vdots & \ddots & \ddots & & \vdots \\ a_{N,1} & \cdots & & a_{N,N-1} & 0 \end{bmatrix}$$

a *sparse* singular lower triangular matrix.

Diagonal Dominance.   The decomposition above is based on the important property of diagonal dominance of matrices that originate from the discretization of elliptic PDEs. *Strong diagonal dominance* means that

$$a_{ii} > \sum_{j \neq i} |a_{ij}|.$$

Expressed in terms of the matrix decomposition $A = L + D + L^T$ this means that

$$\|D^{-1}(L + L^T)\|_\infty < 1, \quad \|D^{-1}A\|_\infty < 2, \quad \|D^{-1}L\|_\infty < 1. \tag{5.7}$$

This property holds, e.g., in the noncritical case of the Helmholtz equation (see Section 1.5), but not in the Poisson equation: just recall our model problem in $\mathbb{R}^2$, where diagonal elements (4) are the negative sum of the nondiagonal elements (-1) in the same row, e.g., due to the 5-point star. For the Poisson equation only *weak diagonal dominance*

$$a_{ii} \geq \sum_{j \neq i} |a_{ij}|,$$

holds, equivalent to

$$\|D^{-1}(L + L^T)\|_\infty \leq 1, \quad \|D^{-1}A\|_\infty \leq 2, \quad \|D^{-1}L\|_\infty \leq 1. \tag{5.8}$$

5.2.1 Jacobi Method

In this method one substitutes the matrix A by $B_J = D$, which yields the classical Jacobi iteration

$$Du_{k+1} = -(L + L^T)u_k + b, \quad G_J = I - D^{-1}A, \tag{5.9}$$

written componentwise as

$$a_{ii}u_{i,k+1} = -\sum_{j \neq i} a_{ij}u_{j,k} + b_i.$$

The method is obviously invariant under reordering of the nodes. Note that all components of u_{k+1} can be computed independently from each other, i.e., in *parallel*. In fact, the Jacobi iteration is the prototype of more general *parallel subspace correction methods*, which we will treat in Section 7.1.3; they supply the theoretical key to the class of *additive* multigrid methods (see Section 7.3.1).

Convergence. In Volume 1, Theorem 8.6, we already showed convergence of the Jacobi method under the assumption of strong diagonal dominance (5.7). Here we repeat the proof in the unified frame of Theorem 5.1.

Theorem 5.2. *Under the assumption of strong diagonal dominance (5.7) the Jacobi iteration converges for all spd-matrices.*

Proof. By construction, $B_J = D$ is symmetric positive definite and we have

$$\|D^{-1}A\|_2 \leq \|D^{-1}A\|_\infty < 2.$$

With this, we can directly verify assumption (5.5) in Theorem 5.1

$$\langle v, Av \rangle \leq \|D^{-1}A\|_2 \langle v, Dv \rangle < 2\langle v, Dv \rangle,$$

which then readily yields $\rho < 1$.   □

For a more detailed analysis according to (5.6) we have to search the real spectrum of G_J for the eigenvalue largest in modulus.

As an example we take the matrix $A = A_h$ in (3.5) for our Poisson model problem in $\mathbb{R}^2$ with Dirichlet boundary conditions (presented there for $N = 9$). For this matrix we get $D^{-1} = \frac{1}{4}I$, so that

$$G_J = I - \frac{1}{4}A.$$

Upon recalling (3.12) we obtain

$$\lambda_{\min}(A_h) = 8\sin^2\left(\frac{1}{2}\pi h\right) = \mathcal{O}(h^2), \quad \lambda_{\max}(A_h) = 8\cos^2\left(\frac{1}{2}\pi h\right) = 8\big(1 - \mathcal{O}(h^2)\big),$$

which immediately yields

$$\rho(G_J) = \max \left\{ \left| 1 - 2\sin^2\left(\frac{1}{2}\pi h\right) \right|, \left| 1 - 2\cos^2\left(\frac{1}{2}\pi h\right) \right| \right\} = 1 - 2\sin^2\left(\frac{1}{2}\pi h\right).$$

For the reduction of the error after k iterations by the factor $\epsilon \approx \rho(G_J)^k$ we thus will need, in our Poisson model problem,

$$k_J \approx \frac{|\log \epsilon|}{|\log(1 - \frac{1}{2}\pi^2 h^2)|} \approx 2\frac{|\log \epsilon|}{\pi^2 h^2} \tag{5.10}$$

Jacobi iterations. Not only for this model problem, but generally we get:

$$\rho(G_J) = 1 - \mathcal{O}(h^2) < 1 \quad \Rightarrow \quad k_J \sim \frac{1}{h^2}. \tag{5.11}$$

The proof is postponed to Section 7.1.3, where it can be performed simpler in a more general theoretical framework.

Obviously, for $h \to 0$, the convergence my slow down arbitrarily. In d space dimensions, due to (5.1), the computational amount per iteration is $\sim N$ and with (5.11) the total amount is

$$W_J \sim |\log \epsilon| N^{1+2/d}. \tag{5.12}$$

In contrast to direct elimination methods the solution of elliptic grid equations by the Jacobi method gets simpler, the larger the space dimension is. This stems from the fact that the number of unknowns grows faster than the discrete pathlength of the information transport in the grid.

Damped Jacobi Method.   This is a variant of the classical Jacobi method of the form

$$Du_{k+1} = -\omega(L + L^T)u_k + (1 - \omega)Du_k + b, \quad G_J(\omega) = I - \omega D^{-1}A.$$

If we restrict the parameter ω to $0 < \omega \leq 1$, then these iterates may be interpreted as convex combination of "old" and "new" iterates. In view of Theorem 5.2 we obtain the following convergence result.

Theorem 5.3. *Assume weak diagonal dominance (5.8) and $0 < \omega < 1$. Then the damped Jacobi iteration converges for all spd-matrices.*

Proof. By construction, we have $B_J = \frac{1}{\omega}D$, which is symmetric positive definite. We again apply Theorem 5.1. So we need to check assumption (5.5). We have

$$\|\omega D^{-1}A\|_2 \leq \omega \|D^{-1}A\|_\infty \leq 2\omega < 2, \tag{5.13}$$

from which, similar as in the proof of Theorem 5.2, eventually $\rho < 1$ follows.   □

Upon performing again the eigenvalue analysis for our model problem, then we get $G_J(\omega) = I - \frac{\omega}{4}A$ and thus

$$\rho(G_J(\omega)) = \max\left\{\left|1 - 2\omega\sin^2\left(\tfrac{1}{2}\pi h\right)\right|, \left|1 - 2\omega\cos^2\left(\tfrac{1}{2}\pi h\right)\right|\right\}$$
$$= 1 - 2\omega\sin^2\left(\tfrac{1}{2}\pi h\right).$$

In our model problem this iteration converges for $\omega < 1$, even more slowl than the undamped classical Jacobi iteration. For $\omega = 1/2$, however, it exhibits an important property, which we will further investigate in Section 5.4.

5.2.2 Gauss–Seidel Method

In the Gauss–Seidel method (GS) one chooses the matrix $B_{GS} = D + L$, a nonsingular lower triangular matrix. This leads to the *forward substitution*

$$(D + L)u_{k+1} = -L^T u_k + b,$$

to be written componentwise as the classical Gauss–Seidel iteration

$$a_{ii}u_{i,k+1} + \sum_{j<i} a_{ij}u_{j,k+1} = -\sum_{j>i} a_{ij}u_{j,k} + b_i.$$

In contrast to the Jacobi method, the GS-method is not invariant under reordering of nodes. The components of u_{k+1} cannot be computed independently from each other, but rather *sequentially* with the advantage that they can be directly overwritten in the algorithm so that memory space is saved. In fact, the GS-method is the proto-type of more general *sequential subspace correction methods*, which we will treat in Section 7.1.2; they supply the theoretical key to the class of *multiplicative* multigrid methods (see Section 7.3.2).

The corresponding iteration matrix is

$$G_{GS} = -(D + L)^{-1}L^T = I - (D + L)^{-1}A.$$

In contrast to the Jacobi iteration, the GS-iteration is not symmetrizable, which is why Theorem 5.1 cannot be applied. For the proof of convergence one therefore chooses some detour via a symmetrizable method.

Symmetric Gauss–Seidel Method (SGS). Upon reversing the order of running through the indices in the GS-method, one is led to the *backward substitution*

$$(D + L^T)u_{k+1} = -Lu_k + b.$$

This corresponds to the iteration matrix adjoint w.r.t. the matrix A

$$G_{GS}^* = I - (D + L)^{-T}A = -(D + L)^{-T}L,$$

for which

$$AG_{GS}^* = (AG_{GS})^T$$

holds. If one combines the forward and the backward substitution, one arrives at the SGS-method:

$$G_{SGS} = G_{GS}^* G_{GS} = (D+L)^{-T} L (D+L)^{-1} L^T = I - B_{SGS}^{-1} A,$$

where some intermediate calculations supply the form

$$B_{SGS} = (D+L) D^{-1} (D+L)^T = A + L D^{-1} L^T. \tag{5.14}$$

This method still is dependent upon the order of the nodes. In contrast to the (damped and undamped) Jacobi as well as the ordinary GS-method, it requires *two* matrix-vector multiplies per step.

Convergence of SGS and GS.   The convergence rates of the SGS- and GS-methods satisfy an elementary relation.

Theorem 5.4. *Let $G_{SGS} = G_{GS}^* G_{GS}$ denote the connection between the iteration functions of the SGS- and GS-method. Then*

$$\rho(G_{SGS}) = \|G_{SGS}\|_A = \|G_{GS}\|_A^2.$$

Proof. We abbreviate the notation to $G_s = G_{SGS}, G = G_{GS}$. As G_s is symmetric w.r.t. the scalar product $\langle \cdot, \cdot \rangle_A$, we get

$$\rho(G_s) = \|G_s\|_A = \sup_{v \neq 0} \frac{\langle v, AG^*Gv \rangle}{\langle v, Av \rangle}$$

$$= \sup_{v \neq 0} \frac{\langle AGv, Gv \rangle}{\langle Av, v \rangle} = \sup_{v \neq 0} \frac{\|Gv\|_A^2}{\|v\|_A^2} = \|G\|_A^2. \qquad \square$$

Whenever one of the methods converges, then both of them converge.

Theorem 5.5. *The classical and the symmetric Gauss–Seidel iteration converge for all spd-matrices.*

Proof. Since the matrix B_{SGS} is symmetric positive definite, we do the proof first for the SGS-method. In order to be able to apply Theorem 5.1, we merely need to verify (5.5): Due to (5.14) one even obtains

$$\langle v, B_{SGS} v \rangle = \langle v, Av \rangle + \langle L^T v, D^{-1} L^T v \rangle > \langle v, Av \rangle. \tag{5.15}$$

With Theorem 5.1 we therefore have $\|G_{SGS}\|_A = \rho(G_{SGS}) < 1$. Finally, application of Theorem 5.4 assures the convergence of the GS-iteration, too. $\qquad \square$

A more subtle analysis of the GS-iteration exploits the fact that the arising matrices A frequently satisfy the *property* $\mathcal{A}$ (see Definition 3.1). (This property, however, only holds for equation systems from finite difference methods on uniform meshes or in rather restricted cases of finite element meshes.) Under this assumption one is able to prove that

$$\|G_{GS}\|_A = \rho(G_J)^2.$$

As a consequence, we obtain, with the help of (5.10),

$$k_{GS} = \frac{1}{2}k_J \approx \frac{|\log \epsilon|}{\pi^2 h^2}.$$

The total amount

$$W_{GS} = \mathcal{O}(|\log \epsilon| N^{1+2/d})$$

asymptotically behaves just as for the Jacobi method. A more detailed convergence analysis of all of the iteration methods presented here is postponed to Section 5.4 below.

5.3 Conjugate Gradient Method

The conjugate gradient method (in short: *CG-method*) uses the fact that the arising matrix A is symmetric positive definite in some different manner perfectly. It was independently invented in 1952 by M. R. Hestenes and E. Stiefel [120]. An elementary derivation can be found, e.g., in Volume 1, Section 8.3.

5.3.1 CG-Method as Galerkin Method

The classical derivation of the CG-method started from the idea to construct a method for the solution of a system of N linear equations in N iterations. The modern derivation considers the method as a Galerkin method in $\mathbb{R}^N$. This approach nicely agrees with Galerkin methods for the discretization of elliptic PDEs (see Section 4 above).

Galerkin Approach.   The basic idea of Galerkin methods is transferred here to finite dimension: instead of the computationally inaccessible solution $u \in \mathbb{R}^N$, one computes an approximation $u_k \in U_k \subset \mathbb{R}^N$ with dim $U_k = k \ll N$. It is defined as the solution of the minimization problem

$$\|u_k - u\|_A = \min_{v \in U_k} \|v - u\|_A, \tag{5.16}$$

where again

$$\| \cdot \|_A^2 = \langle \cdot, A \cdot \rangle$$

denotes the *discrete energy norm* and $\langle \cdot, \cdot \rangle$ the Euclidean inner product. In view of Theorem 4.3 we then have

$$\langle u - u_k, Av \rangle = 0 \quad \text{for all} \quad v \in U_k.$$

Let the *residuals* be defined as

$$r_k = b - Au_k = A(u - u_k).$$

Then

$$\langle r_k, v \rangle = 0 \quad \text{for all} \quad v \in U_k.$$

Let us introduce a traditional way of putting it: vectors v, w are called *A-orthogonal* (in the original paper *A-conjugate*, hence the name of the method), whenever

$$\langle v, Aw \rangle = 0.$$

Now let $p_1, \ldots, p_k$ be an A-orthogonal basis of U_k, i.e.,

$$\langle p_k, Ap_j \rangle = \delta_{kj} \langle p_k, Ap_k \rangle.$$

In terms of this basis the A-orthogonal projection $P_k : \mathbb{R}^n \to U_k$ can be written as

$$u_k = P_k u = \sum_{j=1}^{k} \frac{\langle p_j, A(u - u_0) \rangle}{\langle p_j, Ap_j \rangle} p_j$$

$$= \sum_{j=1}^{k} \underbrace{\frac{\langle p_j, r_0 \rangle}{\langle p_j, Ap_j \rangle}}_{=:\alpha_j} p_j.$$

Note that the choice of the discrete energy norm led to the fact that the coefficients α_j can be evaluated. Starting from some given initial value u_0 (often $u_0 = 0$) and the corresponding residual $r_0 = b - Au_0$ we have the recursions

$$u_k = u_{k-1} + \alpha_k p_k \quad \text{and} \quad r_k = r_{k-1} - \alpha_k Ap_k.$$

As approximation spaces U_k with $U_0 = \{0\}$ we choose the *Krylov spaces*

$$U_k = \text{span}\{r_0, Ar_0, \ldots, A^{k-1}r_0\} \quad \text{for } k = 1, \ldots, N.$$

With this choice, the following result is helpful for the computation of an A-orthogonal basis.

Lemma 5.6. *Let $r_k \neq 0$. Then:*

1. $\langle r_i, r_j \rangle = \delta_{ij} \langle r_i, r_i \rangle$ *for* $i, j = 0, \ldots, k$,
2. $U_{k+1} = \text{span}\{r_0, \ldots, r_k\}$.

For the proof of this lemma we refer to Volume 1, Section 8.3, Lemma 8.15.

With the help of this lemma, we construct the A-orthogonal basis vectors p_k as follows: if $r_0 \neq 0$ (otherwise u_0 would be the solution), we set $p_1 = r_0$. Lemma 5.6

now states for $k > 1$ that either r_k vanishes, or the vectors $p_1, \ldots, p_{k-1}$ and r_k are linearly independent and span U_{k+1}. In the first case, we get $u = u_k$ and are done. In the second case, we obtain by virtue of the choice

$$p_{k+1} = r_k - \sum_{j=1}^{k} \frac{\langle r_k, Ap_j \rangle}{\langle p_j, Ap_j \rangle} p_j = r_k - \underbrace{\frac{\langle r_k, Ap_k \rangle}{\langle p_k, Ap_k \rangle}}_{=: \beta_k} p_k, \qquad (5.17)$$

an orthogonal basis of U_{k+1}. Upon inserting (5.17), the evaluation of α_k and β_k can be further simplified. Since

$$\langle u - u_0, Ap_k \rangle = \langle u - u_{k-1}, Ap_k \rangle = \langle r_{k-1}, p_k \rangle = \langle r_{k-1}, r_{k-1} \rangle,$$

we get

$$\alpha_k = \frac{\langle r_{k-1}, r_{k-1} \rangle}{\langle p_k, Ap_k \rangle},$$

and, due to

$$-\alpha_k \langle r_k, Ap_k \rangle = \langle -\alpha_k Ap_k, r_k \rangle = \langle r_k - r_{k-1}, r_k \rangle = \langle r_k, r_k \rangle$$

obtain

$$\beta_k = \frac{\langle r_k, r_k \rangle}{\langle r_{k-1}, r_{k-1} \rangle}.$$

At this point, we have collected all pieces of the algorithm.

Algorithm 5.7. *CG-Algorithm.*
$u_k := \mathrm{CG}(A, b, u_0)$

$\quad p_1 = r_0 = b - Au_0$
$\quad$ **for** $k = 1$ **to** $k_{\max}$ **do**
$\qquad \alpha_k \quad = \frac{\langle r_{k-1}, r_{k-1} \rangle}{\langle p_k, Ap_k \rangle}$
$\qquad u_k \quad = u_{k-1} + \alpha_k p_k$
$\qquad$ **if** accurate **then exit;**
$\qquad r_k \quad = r_{k-1} - \alpha_k Ap_k$
$\qquad \beta_k \quad = \frac{\langle r_k, r_k \rangle}{\langle r_{k-1}, r_{k-1} \rangle}$
$\qquad p_{k+1} = r_k + \beta_k p_k$
$\quad$ **end for**

Per iterative step the computational amount is one matrix-vector multiply Ap and the evaluation of some scalar products, i.e., $\mathcal{O}(N)$ operations. The required memory is four vectors, (u, r, p, Ap). In contrast to matrix decomposition methods here no individual entries A need to be touched – it suffices to have a routine for matrix-vector multiplies.

It is worth noting that the CG-method is *invariant under reordering of the nodes*, as for the determination of the coefficients only scalar products need to be evaluated, which possess this invariance.

Convergence.   On the basis of the above derivation the convergence of the CG-method will reasonably be studied in terms of the energy norm.

Theorem 5.8. *Let $\kappa_2(A)$ denote the condition number of the matrix A w.r.t. the Euclidean norm. Then the approximation error of the CG-method can be estimated as*

$$\|u - u_k\|_A \leq 2 \left(\frac{\sqrt{\kappa_2(A)} - 1}{\sqrt{\kappa_2(A)} + 1} \right)^k \|u - u_0\|_A . \tag{5.18}$$

For the proof we refer to Volume 1, Section 8.3, Theorem 8.17.

Corollary 5.1. *In order to reduce the error in the energy norm by a factor ϵ, i.e.,*

$$\|u - u_k\|_A \leq \epsilon \|u - u_0\|_A,$$

one requires roughly

$$k \sim \frac{1}{2} \sqrt{\kappa_2(A)} \ln(2/\epsilon)$$

CG iterations.

For the proof see Volume 1, Corollary 8.18.

For our Poisson model problem we have (see Section 3.1.2),

$$\kappa_2(A) = \frac{\lambda_{\max}}{\lambda_{\min}} \sim 1/h^2,$$

from which we may conclude

$$k \sim \frac{|\log \epsilon|}{h} \tag{5.19}$$

and, finally, for space dimension d obtain the computational amount

$$W_{\mathrm{CG}} \sim |\log \epsilon| N^{1+1/d} .$$

Comparison with (5.12) reveals a clear efficiency gain – nevertheless the CG-method, just like the Jacobi and the GS-method, also converges disappointingly slowly for large problems.

5.3.2   Preconditioning

As can be seen from (5.19) the ordinary CG-method is far too slow for general discretized elliptic PDEs, since there the number N of equations is usually large. The method can be sped up if one manages to transform the linear system in such a way that a significantly smaller *effective* condition number is generated, which, in turn, reduces the number of necessary iterations.

Basic Idea.   Instead of the linear equation system $Au = b$ with a symmetric positive definite (N, N)-matrix A, we may also formulate an equivalent problem, for each symmetric positive definite matrix B, of the form[1]

$$\overline{A}\overline{u} = b \quad \text{with} \quad \overline{A} = AB^{-1} \quad \text{and} \quad \overline{u} = Bu.$$

An *efficient* preconditioner B should essentially exhibit the following properties:

1. *cheap*: the evaluation of $B^{-1}r_k$ should preferably require only $\mathcal{O}(N)$ operations, or at least $\mathcal{O}(N \log N)$;

2. *optimal* (or *nearly optimal*, respectively): the *effective* condition number $\kappa(AB^{-1})$ should be sufficiently small, preferably $\mathcal{O}(1)$ independent of N (or $\mathcal{O}(\log N)$, respectively);

3. *permutation invariant*: B should be invariant under reordering of the nodes.

If B is also symmetric positive definite, then the transformed matrix $\overline{A}$ is no longer self-adjoint w.r.t. the Euclidean scalar product $\langle \cdot, \cdot \rangle$, but still w.r.t. the induced inner product $\langle \cdot, B^{-1} \cdot \rangle$:

$$\langle AB^{-1}v, B^{-1}w \rangle = \langle B^{-1}v, AB^{-1}w \rangle.$$

Formally speaking, the CG-method can again be applied, replacing the inner products according to

$$\langle v, w \rangle \to \langle v, B^{-1}w \rangle, \quad \langle v, Aw \rangle \to \langle v, \overline{A}w \rangle = \langle v, AB^{-1}w \rangle.$$

From this we immediately get the following transformed iteration:

$$p_1 = r_0 = b - AB^{-1}\overline{u}_0$$

for $k = 1$ **to** $k_{\max}$ **do**

$$\alpha_k \quad = \frac{\langle r_{k-1}, B^{-1}r_{k-1} \rangle}{\langle AB^{-1}p_k, B^{-1}p_k \rangle}$$

$$\overline{u}_k \quad = \overline{u}_{k-1} + \alpha_k p_k$$

if accurate **then exit**;

$$r_k \quad = r_{k-1} - \alpha_k AB^{-1}p_k$$

$$\beta_k \quad = \frac{\langle r_k, B^{-1}r_k \rangle}{\langle r_{k-1}, B^{-1}r_{k-1} \rangle}$$

$$p_{k+1} = r_k + \beta_k p_k$$

end for

Of course, we are interested in an iteration for the original solution $u = B^{-1}\overline{u}$ and therefore replace the line for the iterates $\overline{u}_k$ by

$$u_k = u_{k-1} + \alpha_k B^{-1} p_k.$$

---

[1] Note the difference in notation compared to Volume 1, Section 8.4, where we used B instead of B^{-1}.

Now the p_k only appear in the last line without the prefactor B^{-1}. Let us therefore introduce the (A-orthogonal) vectors $q_k = B^{-1} p_k$. Thus we arrive at the following economic version of the method, the *preconditioned conjugate gradient method*, often briefly called *PCG-method*.

Algorithm 5.9. PCG-*Algorithm.*
$u_k := \mathrm{PCG}(A, b, u_0, B)$

$$r_0 = b - Au_0, q_1 = \bar{r}_0 = B^{-1} r_0, \sigma_0 = \langle r_0, \bar{r}_0 \rangle$$

for $k = 1$ **to** $k_{\max}$ **do**

$$\alpha_k \quad = \frac{\sigma_{k-1}}{\langle q_k, Aq_k \rangle}$$

$$u_k \quad = u_{k-1} + \alpha_k q_k$$

$$[\gamma_k^2 \quad = \sigma_{k-1}\alpha_k = \|u_{k+1} - u_k\|_A^2]$$

if accurate **then exit**;

$$r_k \quad = r_{k-1} - \alpha_k Aq_k$$

$$\bar{r}_k \quad = B^{-1} r_k$$

$$\sigma_k \quad = \langle r_k, \bar{r}_k \rangle$$

$$\beta_k \quad = \frac{\sigma_k}{\sigma_{k-1}}$$

$$q_{k+1} = \bar{r}_k + \beta_k q_k$$

end for

Compared with the CG-method, the PCG-method only requires one additional application of the symmetric *preconditioner* B per iterative step. Attentive readers have probably observed that we added a line $[\cdots]$ compared with Algorithm 5.7. Obviously, the relation

$$\|u_{k+1} - u_k\|_A^2 = \alpha_k^2 \|q_k\|_A^2 = \sigma_{k-1}\alpha_k = \gamma_k^2 \tag{5.20}$$

holds and is independent of the preconditioner B. This result will turn out to be useful in Section 5.3.3 in the context of the adaptive determination of an optimal truncation index k^*.

Convergence. The PCG-method may be interpreted as a CG-method for the system $AB^{-1}y = b$ with respect to the scalar products $\langle \cdot, B^{-1} \cdot \rangle$ and $\langle \cdot, B^{-1}AB^{-1} \cdot \rangle$. Therefore Theorem 5.8 immediately supplies the following convergence result.

Theorem 5.10. *The approximation error for the PCG-method preconditioned with B can be estimated by*

$$\|u - u_k\|_A \leq 2 \left(\frac{\sqrt{\kappa} - 1}{\sqrt{\kappa} + 1} \right)^k \|u - u_0\|_A ,$$

where

$$\kappa = \kappa_2(AB^{-1}) = \kappa_2(B^{-1}A)$$

is the matrix condition number w.r.t. the Euclidean norm.

Note that the condition is independent of the underlying scalar product. The transition from AB^{-1} to $B^{-1}A$ is done via a similarity transformation.

Choice of Preconditioner.   In what follows we first present preconditioners that are constructed from the view of pure linear algebra, i.e., they merely exploit the fact that A is symmetric positive definite and sparse. Below, in Sections 5.5.2 and 7.2.2, we will introduce much more efficient preconditioners, which take the theoretical background of elliptic PDEs into account.

Jacobi preconditioner. The simplest choice in the decomposition $A = D + L + L^T$ is

$$B = D.$$

This realization exactly corresponds to one Jacobi iteration (5.9). This choice, however, has an impact only for nonuniform meshes, whereas for uniform meshes one merely has $B \approx 2d \ I$ (ignoring the boundary effects), which (essentially) does not change the condition number κ.

Symmetric Gauss–Seidel (SGS) preconditioner. A better convergence rate is achieved with the SGS-preconditioner, where B is again symmetric (see (5.14)). In contrast to the Jacobi preconditioner the SGS-preconditioner depends upon the order of the nodes, whereby the permutation invariance of the CG-method is destroyed.

Incomplete Cholesky factorization (IC-factorization). In the course of the sparse Cholesky factorization only "small" *fill-in elements* appear in general outside the sparse structure of A. The idea of the incomplete Cholesky factorization is now to just drop these fill-in elements. Let

$$P(A) := \{(i, j) : a_{ij} \neq 0\}$$

denote the index set of the nonvanishing elements of the matrix A. Then we construct, in lieu of L, a matrix $\tilde{L}$ with

$$P(\tilde{L}) \subset P(A),$$

by proceeding just as in the Cholesky factorization, but setting $\tilde{l}_{ij} := 0$ for all $(i, j) \notin P(A)$. The computational amount for this factorization is $\mathcal{O}(N)$. The expectation behind this idea is that

$$A \approx \tilde{L}\tilde{L}^T =: B.$$

As the preconditioner B is available in factorized form, an efficient modification of Algorithm 5.9 can be implemented by defining the quantities $q_k = \tilde{L}^{-T} p_k$ and $\bar{r}_k = \tilde{L}^{-1} r_k$.

In [155], a proof of the existence and numerical stability of such a method was given for the case when A is a so-called *M-matrix*, i.e.,

$$a_{ii} > 0, \quad a_{ij} \le 0 \quad \text{for} \quad i \neq j \quad \text{and} \quad A^{-1} \ge 0 \text{ (elementwise)}.$$

Such matrices actually occur in the discretization of simple PDEs (see [208]), including our model problems. In many problems from applications, a drastic speed-up of the CG-method is observed, far beyond the class of M-matrices, which are accessible to proofs. The method is particularly popular in industry, since it is robust and does not require any deeper insight into the underlying special structure of the problem (such as multigrid methods). However, the decomposition depends on the order of the nodes, which again spoils the invariance of the CG-method.

Remark 5.11. For nonsymmetric matrices there exists a popular algorithmic extension of the IC-preconditioner, the ILU-preconditioner. The basic idea is similar to the IC-factorization, whereas the theoretical basis is even smaller.

5.3.3 Adaptive PCG-method

For the construction of an efficient PCG-algorithm we are left with the question of when to truncate the iteration, i.e., to decide when an iterate u_k is accepted as "sufficiently close" to the solution u. In Algorithm 5.9 this is marked by the line

$$\text{if accurate then exit.}$$

The following strategies for an error criterion have been considered:

(a) *Residual error.* The most popular idea consists of the prescription of a tolerance for the residual r_k in the form

$$\|r_k\|_2 \le \text{TOL} .$$

However, due to

$$\lambda_{\min}(A) \|u - u_k\|_A^2 \le \|r_k\|_2^2 \le \lambda_{\max}(A) \|u - u_k\|_A^2 ,$$

this is not very well suited for ill-conditioned A, i.e., the convergence of the residual does not say too much about the convergence of the iterates.

(b) *Preconditioned residual error.* As the preconditioned residual $\bar{r}_k = B^{-1} r_k$ is anyway computed, a criterion of the kind

$$\|\bar{r}_k\|_{B^{-1}} \le \overline{\text{TOL}}$$

is easily implemented. For efficient preconditioners, i.e., with small condition number $\kappa(AB^{-1})$, this is a gain compared to the residual error, since

$$\lambda_{\min}(AB^{-1}) \|u - u_k\|_A^2 \leq \|\bar{r}_k\|_2^2 \leq \lambda_{\max}(AB^{-1}) \|u - u_k\|_A^2 .$$

Nevertheless the discrepancy may be still too large, even for "good" preconditioners (with h-independent bounded effective condition number).

(c) *Discrete energy error* [66]. For a direct estimation of

$$\epsilon_k = \|u - u_k\|_A$$

we may use the Galerkin orthogonality

$$\epsilon_0^2 = \|u - u_0\|_A^2 = \|u - u_k\|_A^2 + \|u_k - u_0\|_A^2.$$

By means of (5.20) we conveniently calculate the second term in the course of the PCG iteration

$$\|u_k - u_0\|_A^2 = \sum_{j=0}^{k-1} \|u_{j+1} - u_j\|_A^2 = \sum_{j=0}^{k-1} \gamma_j^2.$$

This yields

$$\epsilon_k^2 = \epsilon_0^2 - \sum_{j=0}^{k-1} \gamma_j^2.$$

However, the term ϵ_0 is unknown. Upon ignoring any rounding errors we have $\epsilon_N = 0$, from which we directly get with the above results

$$\epsilon_k^2 = \sum_{j=k}^{N-1} \gamma_j^2.$$

Instead of the here required $N - k$ additional iterations we realize an error estimator by only $n - k$ additional iteration ($n \ll N$) according to

$$[\epsilon_k]_n^2 = \sum_{j=k}^{n-1} \gamma_j^2 \leq [\epsilon_k]_N^2 = \epsilon_k^2. \tag{5.21}$$

This line of thoughts suggests the truncation criterion

$$[\epsilon_k]_n \leq \text{TOL}_A, \tag{5.22}$$

where $n - k$ need not be chosen "too large". Note that this truncation criterion is independent of the preconditioner.

In the asymptotic phase of the PCG-iteration linear convergence

$$\epsilon_{k+1} \leq \Theta \epsilon_k, \quad \Theta < 1, \tag{5.23}$$

arises, which can be exploited in the form

$$\epsilon_k^2 = \epsilon_{k+1}^2 + \gamma_k^2 \leq (\Theta \epsilon_k)^2 + \gamma_k^2 \quad \Rightarrow \quad \epsilon_k \leq \frac{\gamma_k}{\sqrt{1-\Theta}} =: \bar{\epsilon}_k.$$

Correspondingly, one is led to a simplified truncation criterion

$$\bar{\epsilon}_k \leq \text{TOL}_A,$$

which just demands one additional PCG iteration. Note that the onset of the asymptotic phase depends strongly on the choice of preconditioner.

The estimator (5.21) for the exact energy error inherits its iterative monotonicity, since

$$[\epsilon_{k+1}]_n \leq [\epsilon_k]_n, \quad \epsilon_{k+1} \leq \epsilon_k.$$

As relative energy error we define the expression

$$\delta_k = \frac{\sqrt{[\epsilon_k]}}{\|u_k\|_A} \approx \frac{\|u - u_k\|_A}{\|u_k\|_A}.$$

If we specify the initial value as $u_0 = 0$, which is a quite popular option in all iterative solvers, then we get

$$\|u_{k+1}\|_A^2 = (1 + \delta_k^2)\|u_k\|_A^2,$$

as well as the monotonicity

$$\|u_{k+1}\|_A \geq \|u_k\|_A \quad \Leftrightarrow \quad \delta_{k+1} \leq \delta_k. \tag{5.24}$$

The truncation criterion (5.22) in terms of the energy error is quintessential for the adaptive solution of discretized linear as well as nonlinear elliptic PDEs and will come up again in a similar form in Sections 5.5, 7.4, 8.1.2, and 8.2.2.

5.3.4 A CG-variant for Eigenvalue Problems

Here we want to derive a variant of the CG-method that was suggested in 1982 especially for the minimization of the Rayleigh quotient (1.15) by B. Döhler [78]. This variant also supplies the theoretical basis for the construction of an adaptive multigrid method in Section 7.5.2.

Suppose we have to compute the minimal eigenvalue λ, assumed to be isolated. In Section 5.3.1 we showed that the classical CG-method is based on the minimization of the *quadratic* objective function

$$f(u) = \tfrac{1}{2}\|b - Au\|_2^2.$$

The approximate solutions u_k, the residuals r_k and the corrections p_k are obtained recursively according to

$$p_{k+1} = r_k + \beta_k p_k, \quad u_{k+1} = u_k + \alpha_{k+1} p_{k+1}, \quad r_{k+1} = r_k - \alpha_{k+1} A p_{k+1}.$$

In the recursion the orthogonality

$$p_k^T r_k = 0 \tag{5.25}$$

holds. The aim is now to modify this algorithm for the minimization of a special *nonlinear* objective function

$$R(u) = \frac{u^T A u}{u^T M u}$$

in some suitable way. As will turn out, the recursion for the residuals drops out totally, whereas the computation of the iterates u_k as well as the corrections p_k is obtained from some two-dimensional eigenvalue problem.

To begin with, we observe that in the classical CG-method the residual r_k can be represented as

$$r_k = -\nabla f(u_k).$$

This inspires the idea to replace the residual by the gradient

$$g(u_k) := \nabla R(u_k).$$

If we set the normalization $u_k^T M u_k = 1$, then we obtain for the *gradient* the expression

$$g(u_k) = 2 (A - R(u_k)M) u_k,$$

which shows that the gradient is identical to the residual of the eigenvalue problem (4.17) up to a prefactor of 2. In addition, we have the property

$$u_k^T g(u_k) = 0. \tag{5.26}$$

Due to the homogeneity $R(\sigma u) = R(u)$ for arbitrary $\sigma \neq 0$ we extend the above CG-ansatz to

$$u_{k+1} = \sigma_{k+1}(u_k + \alpha_{k+1} p_{k+1}). \tag{5.27}$$

In the sense of repeated induction we first assume that the correction p_{k+1} is given. Then we get an optimal coefficient α_{k+1} from the condition

$$\lambda_{k+1} = \min_{\alpha} R(u_k + \alpha p_{k+1}), \tag{5.28}$$

from which we immediately conclude the orthogonality relation

$$0 = \frac{\partial}{\partial \alpha} R(u_k + \alpha p_{k+1})\Big|_{\alpha_{k+1}} = g(u_{k+1})^T p_{k+1}. \tag{5.29}$$

In principle, one could use this relation to obtain the coefficient by solving a quadratic equation in α (cf. Exercise 5.10).

However, here we want to follow a more elegant idea due to B. Döhler [78]. There a two-dimensional subspace

$$\tilde{U}_{k+1} := \operatorname{span}\{u_k, p_{k+1}\} \subset \mathbb{R}^N$$

is introduced first and the minimization problem projected in the form

$$\lambda_{k+1} = \min_{u \in \tilde{U}_{k+1}} R(u) \quad \text{with} \quad u_{k+1} = \arg\min_{u \in \tilde{U}_{k+1}} R(u). \tag{5.30}$$

Let

$$U_{k+1} = [u_k, p_{k+1}]$$

denote the corresponding $(N, 2)$-matrix for the given basis, then we may write the new iterate in the form

$$u_{k+1} = U_{k+1}\xi \quad \text{with} \quad \xi^T = \sigma(1, \alpha)$$

in terms of coefficients σ, α to be determined. Along this line the symmetric $(2, 2)$-matrices

$$\overline{A}_{k+1} := U_{k+1}^T A U_{k+1}, \quad \overline{M}_{k+1} := U_{k+1}^T M U_{k+1} \tag{5.31}$$

may be defined. Then the projected Rayleigh quotient can be written as

$$\overline{R}_{k+1}(\xi) := \frac{\xi^T \overline{A}_{k+1}\xi}{\xi^T \overline{M}_{k+1}\xi}.$$

The two eigenvalues

$$\lambda_{k+1} = \min_{\xi} \overline{R}_{k+1}(\xi), \quad \mu_{k+1} = \max_{\xi} \overline{R}_{k+1}(\xi)$$

are associated with the two-dimensional eigenvalue problem

$$(\overline{A}_{k+1} - \lambda \overline{M}_{k+1})\xi = 0, \tag{5.32}$$

which can be solved in an elementary way. In the nondegenerate case we obtain the two values $\mu_{k+1} > \lambda_{k+1}$ for these eigenvalues, and the projected eigenvectors

$$\xi_{k+1}^T = \sigma_{k+1}(1, \alpha_{k+1}), \quad \bar{\xi}_{k+1}^T = \bar{\sigma}_{k+1}(1, \bar{\alpha}_{k+1})$$

as well as the original eigenvector approximations

$$u_{k+1} = \sigma_{k+1}(u_k + \alpha_{k+1}p_{k+1}), \quad \bar{u}_{k+1} = \bar{\sigma}_{k+1}(u_k + \bar{\alpha}_{k+1}p_{k+1}).$$

The four coefficients $\alpha_{k+1}, \bar{\alpha}_{k+1}, \sigma_{k+1}, \bar{\sigma}_{k+1}$ are supplied by the solution of the eigenvalue problem

$$(u_{k+1}, \bar{u}_{k+1})^T A(u_{k+1}, \bar{u}_{k+1}) = \begin{bmatrix} \lambda_{k+1} & 0 \\ 0 & \mu_{k+1} \end{bmatrix}, \tag{5.33}$$

in connection with the normalization

$$(u_{k+1}, \bar{u}_{k+1})^T M(u_{k+1}, \bar{u}_{k+1}) = \begin{bmatrix} 1 & 0 \\ 0 & 1 \end{bmatrix}. \tag{5.34}$$

This decomposition is obviously sufficient for the orthogonality relation

$$\bar{u}_{k+1}^T g(u_{k+1}) = 0. \tag{5.35}$$

A comparison with the orthogonality relations (5.29) and (5.35) with (5.25) leads us, one index lower, to the identification $p_k = \bar{u}_k$. Instead of the CG-recursion for the correction we thus set

$$p_{k+1} = -g(u_k) + \beta_k \bar{u}_k, \tag{5.36}$$

where the coefficients β_k are still to be determined. As it turns out, for this purpose we have to replace the nonlinear function $R(u)$ by its locally *quadratic* model to obtain useful expressions.

During the intermediate calculation to follow we drop all indices. As the *Hessian matrix* we obtain

$$H(u) = 2\big(A - R(u)M - g(u)(Mu)^T - Mu\,g(u)^T\big). \tag{5.37}$$

Taylor expansion and neglection of terms from the second order on then supplies

$$R(u + \alpha p) \doteq R(u) + \alpha p^T g(u) + \tfrac{1}{2}\alpha^2 p^T H(u)p =: J(u, p). \tag{5.38}$$

Here we now insert the explicit form (5.36) and obtain, again dropping all indices,

$$J(u, -g + \beta\bar{u}) = R(u) + \alpha(-g + \beta\bar{u})^T g + \tfrac{1}{2}\alpha^2(-g + \beta\bar{u})^T H(u)(-g + \beta\bar{u}).$$

Now the coefficient β must be chosen such that J attains its minimum. From this requirement we get

$$0 = \frac{\partial}{\partial\beta} J(u, -g + \beta\bar{u}) = \alpha\bar{u}^T g + \alpha^2(-g + \beta\bar{u})^T H(u)\bar{u}.$$

The term linear in α vanishes due to (5.35) by the induction hypothesis for index k. From (5.37) and the orthogonality relation (5.35) we conclude in the nondegenerate case that $\bar{u}^T H(u)\bar{u} > 0$, which then yields the condition

$$\beta = \frac{g(u)^T H(u)\bar{u}}{\bar{u}^T H(u)\bar{u}}$$

for a unique minimum of J. Upon application of (5.33) and (5.34), after introducing the indices again, we finally obtain the conveniently evaluable expression

$$\beta_k = \frac{g(u_k)^T (A - \lambda_k M)\bar{u}_k}{\mu_k - \lambda_k}.$$

Having completed this preparation, we have now collected nearly all the pieces of the nonlinear CG for the minimization of the Rayleigh quotient. Only the initial values are still missing. For this purpose, we choose an initial approximation u_0 for the unknown eigenvector u and some linearly independent initial value $\bar{u}_0$, most conveniently $\bar{u}_0 = g(u_0)$ because of (5.26).

Algorithm 5.12. RQCG-*Algorithm* in $\mathbb{R}^N$.
$(\tilde{\lambda}, \tilde{u}) = \mathrm{RQCG}(A, M, u_0, k_{\max})$

$\quad p_1 := -\bar{u}_0 := -g(u_0)$
$\quad \textbf{for } k = 1 \textbf{ to } k_{\max} \textbf{ do}$
$\qquad \overline{A}_k := [u_{k-1}, p_k]^T A [u_{k-1}, p_k]$
$\qquad \overline{M}_k := [u_{k-1}, p_k]^T M [u_{k-1}, p_k]$
$\qquad$ Solve the eigenvalue problem $(\overline{A}_k - \lambda \overline{M}_k)\xi = 0.$
$\qquad\quad$ Result: eigenvalues λ_k, μ_k, eigenvectors $u_k, \bar{u}_k$.
$\qquad \textbf{if } k < k_{\max} \textbf{ then}$
$\qquad\quad \beta_k := \frac{g(u_k)^T (A - \lambda_k M)\bar{u}_k}{\mu_k - \lambda_k}$
$\qquad\quad p_{k+1} := -g(u_k) + \beta_k \bar{u}_k$
$\qquad \textbf{end if}$
$\quad \textbf{end for}$
$\quad (\tilde{\lambda}, \tilde{u}) = (\lambda_k, u_k)$

Convergence. The question of whether the u_k, p_{k+1} are linearly independent is checked within the above algorithm while solving the $(2,2)$-eigenvalue problem (5.32). Under this assumption and due to the homogeneity of R we then have

$$\lambda_{k+1} = R(u_{k+1}) = \min_\alpha R(u_k + \alpha p_{k+1}) < R(u_k) = \lambda_k.$$

From this we obtain by induction the one-sided *monotonicity*

$$\lambda \le \lambda_j < \lambda_{j-1} < \cdots < \lambda_0. \tag{5.39}$$

This property is extremely useful for accuracy monitoring within the algorithm, unless alternatively the accuracy of the eigenvector approximations is checked. We want to mention explicitly that this monotony is independent of the quadratic substitute model (5.38), which is merely needed to determine the coefficients β.

A more detailed analysis of the convergence of the method was given in [224], although for a slightly modified variant. Asymptotically, the Rayleigh quotient $R(u)$ approaches a quadratic function with Hesse matrix $H(u)$. For an analysis of the asymptotic convergence of the RQCG-method the theory of the CG-method may serve as a good approximation. According to Theorem 5.8 we require the spectral condition number $\kappa_2(H)$. Let the eigenvalues of the generalized eigenvalue problems be ordered according to (indices here characterize the different eigenvalues, no longer the different approximations of the lowest eigenvalue)

$$\lambda_1 < \lambda_2 \le \cdots \le \lambda_N,$$

and assume there exists a gap between λ_1 and the remaining spectrum. If, for simplicity, we insert the exact eigenfunction u, then we get

$$\kappa_2(H) = \frac{\lambda_N - \lambda_1}{\lambda_2 - \lambda_1}.$$

With this result, the asymptotic convergence rate is given by

$$\rho = \frac{\sqrt{\kappa_2(H)} - 1}{\sqrt{\kappa_2(H)} + 1}.$$

Preconditioning.　In the process of successive mesh refinement we asymptotically get $\lambda_N \to \infty$ and thus $\rho \to 1$. Hence, in order to realize the algorithm for large N in an efficient way, some preconditioning will be necessary. This can be achieved by some *shift* according to

$$g(u) \to C^{-1} g(u) \quad \text{with} \quad C = A + \mu M, \ \mu > 0.$$

In Section 7.5.2 we will introduce a multigrid method for Rayleigh quotient minimization which works efficiently without such a spectral shift.

Extension to Eigenvalue Clusters.　The variant presented here can be extended to the case when multiple eigenvalues $\lambda_1, \ldots, \lambda_q$ are sought simultaneously. In this case the computation of a single eigenvector is replaced by the computation of an orthogonal basis of the corresponding invariant subspace (see (4.21)). This is realized by minimization of the trace (4.22) in lieu of the single Rayleigh quotient. An extension to nonselfadjoint eigenvalue problems is also possible, which, for reasons of numerical stability, is based on the Schur canonical form.

5.4　Smoothing Property of Iterative Solvers

As shown in the preceding sections, the asymptotic convergence behavior of the classical iterative methods (Jacobi and Gauss–Seidel methods) as well as the (not preconditioned) CG-method is not particularly good. In this section we want to study this convergence behavior in some more detail. In Section 5.4.1 we compare the iterative methods treated so far in a simple illustrative example. The smoothing behavior observed in it will be analyzed for the special case of the Jacobi method in Section 5.4.2 and for more general iterative methods in Section 5.4.3.

5.4.1　Illustration for the Poisson Model Problem

For the purpose of illustration we again choose our notorious model problem.

Example 5.13. Given a two-dimensional Poisson problem with homogeneous Dirichlet boundary conditions,

$$-\Delta u = 0 \quad \text{in } \Omega, \qquad u|_{\partial\Omega} = 0,$$

here with right-hand side $f = 0$, which implies $u = 0$. Suppose the problem has been discretized by the 5-point star over a $(33, 33)$-mesh. In passing, let us mention that the thus generated matrix A has the "property $\mathcal{A}$", see Definition 3.1. For the initial value u_0 of the iteration we apply uniformly $[0, 1]$-distributed random numbers.

Figure 5.4 demonstrates a comparison of the convergence behavior in the energy norm of the iterative methods treated so far; note, however, that the SGS-method requires two A-evaluations per iteration, as opposed to only one in all other methods.

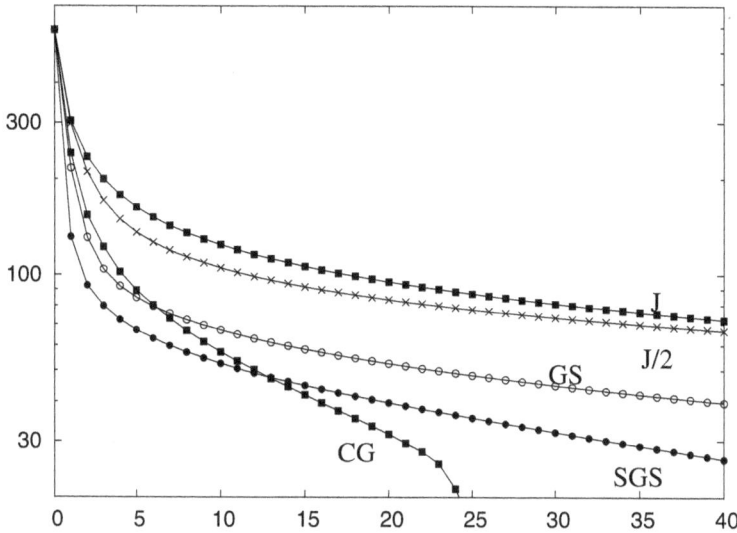

Figure 5.4. Poisson model problem 5.13: Comparison of iterative energy errors $\|u - u_k\|_A$ versus number of iterations: classical Jacobi method (J), damped Jacobi method with $\omega = 1/2$ (J/2), Gauss–Seidel method (GS), symmetric Gauss–Seidel method (SGS), and CG-method (CG).

In Figure 5.5 the *total* behavior of the iterative error for the classical (undamped) and the damped Jacobi method over the chosen mesh is represented. In this detailed representation the error is seen to rapidly "smooth" in the beginning, but remains nearly unaltered afterwards. While in the classical (undamped) Jacobi iteration a high-frequency error part is still visible, the error of the damped Jacobi method is significantly "smoothed". In Figure 5.6, left, the behavior of the Gauss–Seidel method (GS) is presented for comparison. For the symmetric Gauss–Seidel method (SGS) the iterations 5,10, 20 would look just like the GS iterations 10, 20, 40 given here, which nicely illustrates Theorem 5.4. Obviously, the GS-method "smoothes" more efficiently than the Jacobi method. For comparison we show in Figure 5.6, right, the behavior of the (not preconditioned) CG-method, which seems to be the clear winner.

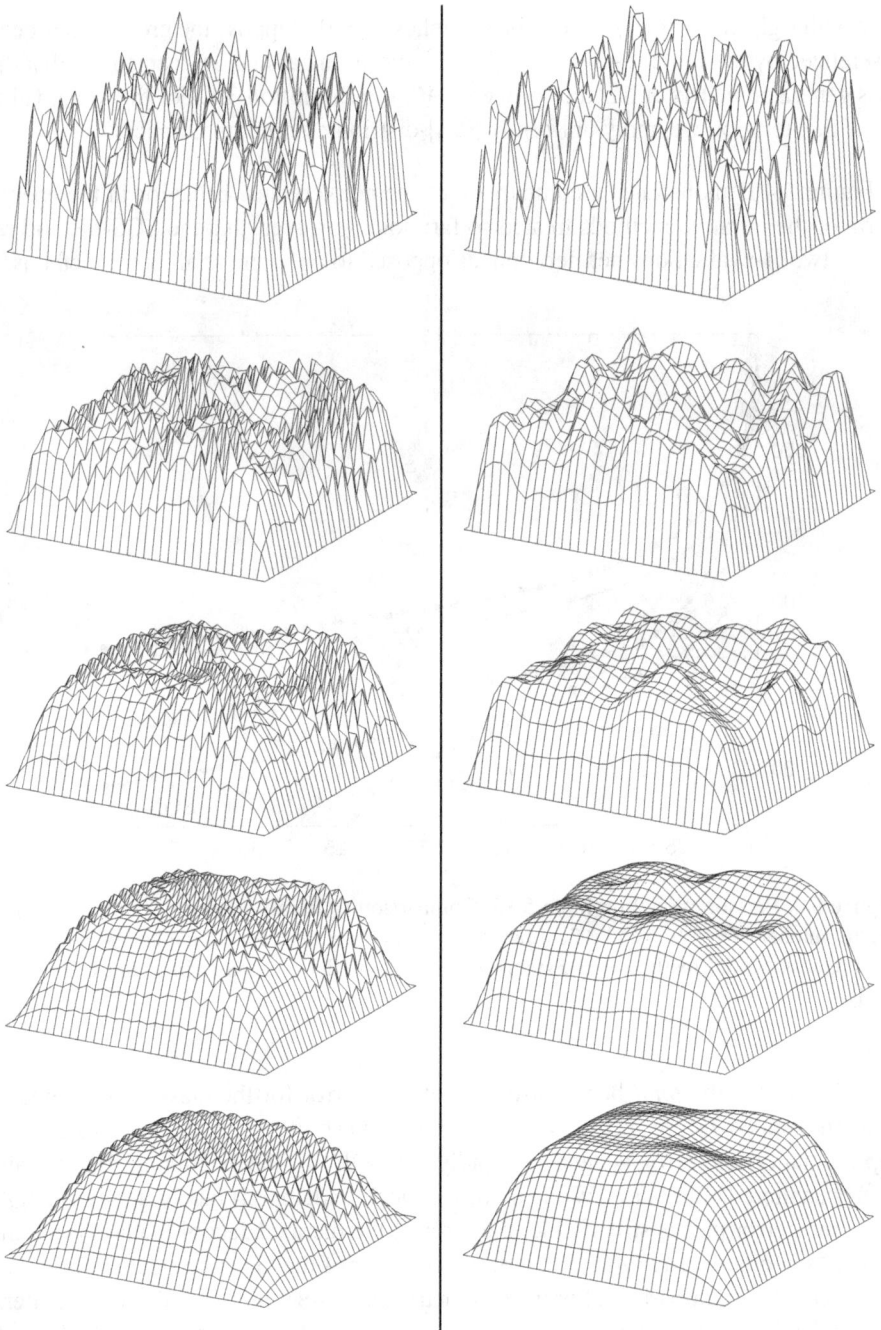

Figure 5.5. Poisson model problem 5.13: iterative Errors $u_k - u$ after 1, 5, 10, 20, and 40 iterations. *Left:* classical Jacobi method ($\omega = 1$). *Right:* damped Jacobi method ($\omega = 0.5$).

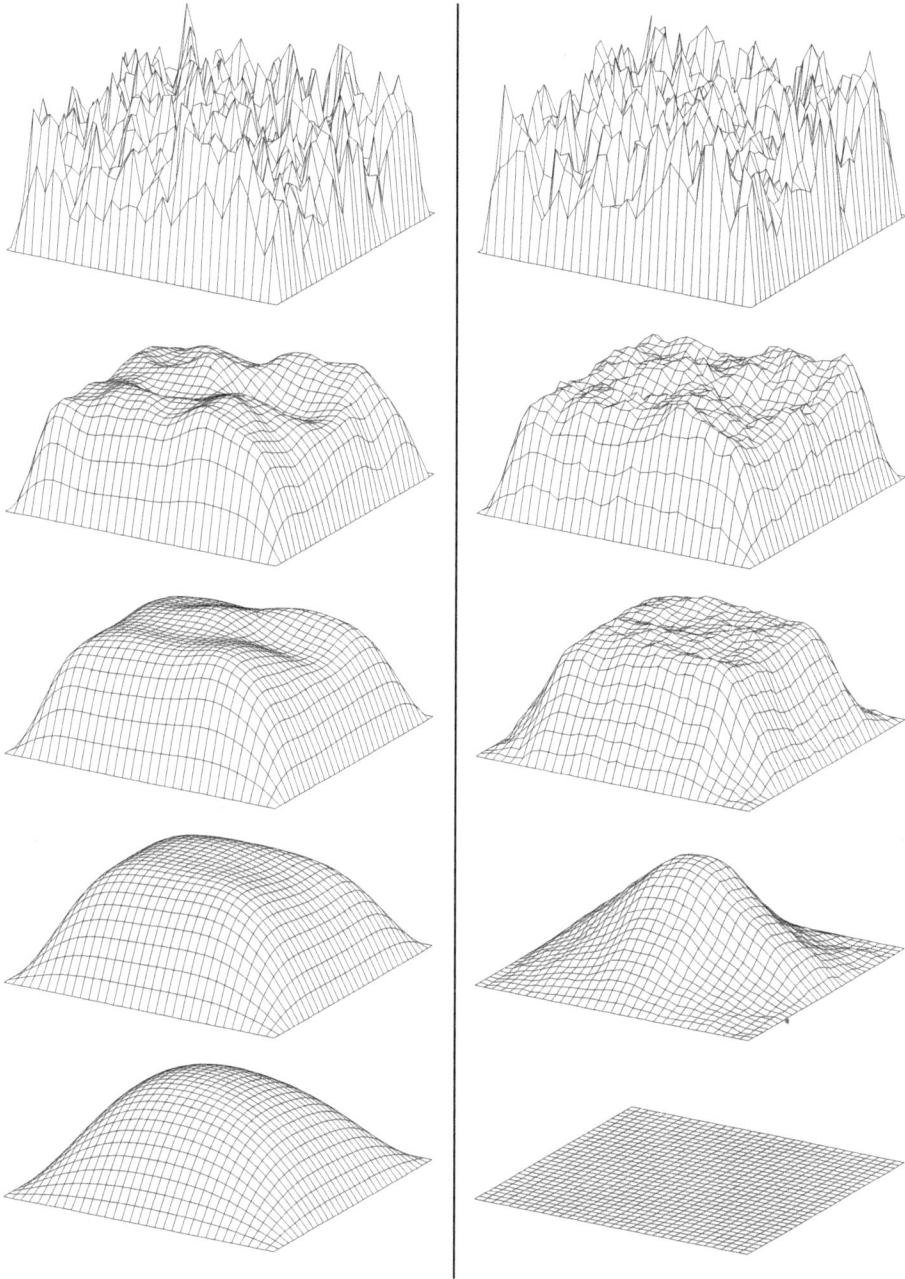

Figure 5.6. Poisson model problem 5.13: iterative errors $u_k - u$ after 1, 5, 10, 20, and 40 iterations. *Left:* Gauss–Seidel method. *Right:* method of conjugate gradients.

Summarizing, all iterative methods converge *in the starting phase much faster than in the asymptotic phase*. In what follows we want to understand this typical convergence profile.

5.4.2 Spectral Analysis for Jacobi Method

We now want to study in more detail the behavior shown in Figure 5.5, first for the *classical* and then for the *damped Jacobi method*. Mathematically speaking, the observed "smoothing" means a suppression of higher frequencies in the iterative error. As worked out in Section 3.1.2, the eigenfunctions in our model problem are just trigonometric functions, to be interpreted as "standing" waves with low and high spatial frequencies. We may therefore decompose the iterative error into these very eigenfunctions. An essential insight can already be gained in the 1D case, as has also been seen in Section 3.1.2.

For the Poisson model problem in $\mathbb{R}^1$ with Dirichlet boundary conditions we get the eigenvalues from (3.10), taking the scale factor of h^2 into account, as

$$\lambda_{h,k} = 4\sin^2\left(\tfrac{1}{2}k\pi h\right),$$

and the corresponding eigenfunctions from (3.11) as

$$\varphi_{h,k}(x_i) = \sin(k\pi x_i), \quad k = 1, \ldots, M-1,$$

where $x_i = ih$, $Mh = 1$. Obviously, these eigenfunctions represent standing waves on the grid Ω_h with spatial frequencies $k\pi$. Expansion of the iterative error w.r.t. this orthogonal basis supplies

$$u_{k+1} - u = \sum_{l=1}^{M-1} \alpha_l^{k+1}\varphi_{h,l}, \quad u_k - u = \sum_{l=1}^{M-1} \alpha_l^k\varphi_{h,l}.$$

We thus have

$$u_{k+1} - u = G_{\mathrm{J}}(\omega)(u_k - u),$$

where the iteration matrix in 1D due to $D^{-1} = \tfrac{1}{2}I$ has the shape

$$G_{\mathrm{J}}(\omega) = I - \tfrac{1}{2}\omega A.$$

From this, by virtue of the orthogonality of the basis, one obtains the componentwise recursion

$$\alpha_l^{k+1} = \sigma_l(\omega)\alpha_l^k$$

with the factor

$$\sigma_l(\omega) = 1 - 2\omega\sin^2(l\pi h/2), \tag{5.40}$$

which represents the damping of the error component with frequency $l\pi$. In Figure 5.7 this factor is depicted graphically, for both the classical case $\omega = 1$ and the damped

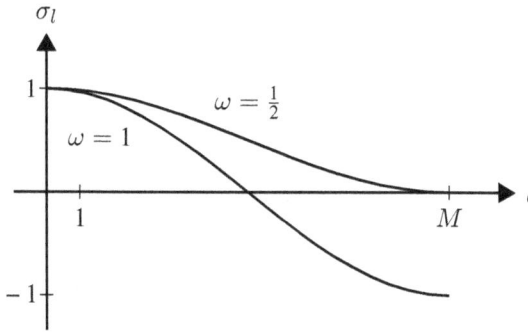

Figure 5.7. Damping factor $\sigma_l(\omega)$ due to (5.40) in dependence upon the frequency $l\pi$, where $l = 1, \ldots, M - 1, Mh = 1$.

case $\omega = 1/2$. As can be seen from the figure, for $\omega = 1$ the low and high frequencies remain roughly preserved, while the medium frequencies are damped. For $\omega = 1/2$, however, the high frequencies are damped, whereas the low frequencies are preserved; this mathematical analysis also explains Figure 5.5, left and right, down to fine detail.

Unfortunately, a comparable spectral analysis for a study of the smoothing properties of the Gauss–Seidel or symmetric Gauss–Seidel iterations is much more complicated, since in this case the eigenfunctions mix. Often such an analysis is done dropping the boundary conditions, which means for the discretized Laplace operator only. For the CG-iteration the same holds. However, a different version of analysis exists, which we will elaborate in the following Section 5.4.3; in this analysis the spectral decomposition only plays an implicit role.

5.4.3 Smoothing Theorems

The convergence results of Section 5.2 are dominated by the largest eigenvalue of the iteration matrix G and therefore describe the convergence rate for $k \to \infty$. This rate depends on the mesh size h and collapses for $h \to 0$. The smoothing of the error, however, is a transient phenomenon, which only appears for small k and is independent of h. In the present section we want to develop a theory that exactly describes this transient starting phase independent of the mesh size. For this advantage we are willing to pay the price that this theory does not supply any deeper insight for $k \to \infty$.

The subsequent theory exploits the fact that the (discrete) energy norm $\| \cdot \|_A$ and the Euclidean norm $\| \cdot \|_2$ reflect the spectrum $\sigma(A)$ of eigenvalues of the matrix A quite differently. For the starting error we know that

$$\|u - u_0\|_A \leq \sqrt{\lambda_{\max}}\|u - u_0\|_2, \tag{5.41}$$

where $\lambda_{\max}$ denotes the largest eigenvalue present in the starting error. From the Poincaré inequality (A.25) $\lambda_{\max}$ is bounded independent of h. For the mathematical modelling of the error smoothing in the course of the iteration one may therefore compare the error norm $\|u - u_k\|_A$ with the initial error norm $\|u - u_0\|_2$.

Before going into details, let us first present an auxiliary result, which is a slight modification of a result in [114] and will be helpful in the sequel.

Lemma 5.14. *Let C be a symmetrizable, positive semidefinite matrix with spectral norm* $\|C\|_2 \leq 4/3$. *Then, for* $k \geq 1$:

$$\|C(I - C)^k\|_2 \leq \frac{1}{k+1}. \tag{5.42}$$

Proof. For the spectrum of C we have by assumption $\sigma(C) \subset \,]0, 4/3]$. From this we may derive an estimate as follows

$$\|C(I - C)^k\|_2 \leq \sup_{\lambda \in \,]0, 4/3]} \varphi(\lambda),$$

where $\varphi(\lambda) = \lambda |1 - \lambda|^k$. First, we check for interior extrema. Taking derivatives yields

$$\varphi'(\lambda) = (1 - \lambda)^{k-1} (1 - (k+1)\lambda) = 0,$$

from which we obtain $\lambda_1 = \frac{1}{k+1}$ and $\lambda_2 = 1$. Insertion into φ supplies

$$\varphi(\lambda_1) = \frac{1}{k+1} \left(\frac{k}{k+1} \right)^k \leq \frac{1}{k+1} \quad \text{and} \quad \varphi(\lambda_2) = 0.$$

Next, we check for a possible violation of (5.42) by boundary extrema. We have $\varphi(0) = 0$ and because of the stronger monotony of $1/3^{k+1}$ compared with $1/(k+1)$ for $k \geq 1$ also

$$\varphi \left(\frac{4}{3} \right) = \frac{4}{3^{k+1}} \leq \frac{1}{k+1}. \qquad \square$$

Symmetrizable Iterative Methods.   To this class belong the (in general damped) Jacobi method and the symmetric Gauss–Seidel method (SGS). By means of Lemma 5.14 we are led to the following results.

Theorem 5.15. *Let the matrix A be spd and satisfy the weak diagonal dominance* (5.8). *Then, for the damped Jacobi method with* $1/2 \leq \omega \leq 2/3$ *and for the symmetric Gauss–Seidel method, the following common estimate holds:*

$$\|u - u_k\|_A \leq \left(\frac{2}{2k+1} \right)^{1/2} \sqrt{\lambda_{\max}(A)} \, \|u - u_0\|_2. \tag{5.43}$$

Proof. We start from a symmetrizable iterative method with

$$G = I - C, \quad C = B^{-1}A \quad \Leftrightarrow \quad A = BC$$

and assume that B, C are spd (w.r.t. some scalar product).

(a) We first have to show for the two methods in question that $\|C\|_2 \le 4/3$. For the damped Jacobi method we have already shown in (5.13) – under the assumption of weak diagonal dominance (5.8) – that

$$\|C\|_2 \le 2\omega$$

which is why the assumption $\omega \le 2/3$ is sufficient. We thus do not cover the classical Jacobi method by this proof. For the SGS-method we use (5.15) and obtain

$$\|C\|_2 = \sup_{v \ne 0} \frac{\langle v, Av \rangle}{\langle v, Bv \rangle} \le 1.$$

(b) With $\|C\|_2 \le 4/3$ assured we first rewrite

$$\|u - u_k\|_A^2 = \langle G^k(u - u_0), AG^k(u - u_0) \rangle = \langle u - u_0, BC(I - C)^{2k}(u - u_0) \rangle.$$

Use of the Cauchy–Schwarz inequality then yields

$$\|u - u_k\|_A^2 \le \|B\|_2 \|C(I - C)^{2k}\|_2 \|u - u_0\|_2^2.$$

By application of Lemma 5.14 (note $2k$ instead of k) we obtain

$$\|u - u_k\|_A^2 \le \left(\frac{\|B\|_2}{2k + 1} \right) \|u - u_0\|_2^2,$$

and finally

$$\|u - u_k\|_A \le \left(\frac{\|B\|_2}{2k + 1} \right)^{1/2} \|u - u_0\|_2.$$

(c) It remains to prove $\|B\|_2$ for the two iterative methods. For the damped Jacobi method we get the estimate

$$\|B\|_2 = \frac{1}{\omega} \|D\|_2 \le \frac{1}{\omega} \|A\|_2,$$

where the last inequality is generated by application of the Rayleigh quotient

$$\|D\|_2 = \max_i a_{ii} = \max_i \frac{\langle e_i, Ae_i \rangle}{\langle e_i, e_i \rangle} \le \sup_{v \ne 0} \frac{\langle v, Av \rangle}{\langle v, v \rangle} = \|A\|_2. \qquad (5.44)$$

The maximum value is obtained for the choice $\omega = 1/2$, whereby we get

$$\|B\|_2 \le 2\|A\|_2 = 2\lambda_{\max}(A). \qquad (5.45)$$

Insertion of this result confirms (5.43) for the damped Jacobi method.

For the symmetric Gauss–Seidel method we have

$$\|B\|_2 = \sup_{v \ne 0} \left[\frac{\langle v, Av \rangle}{\langle v, v \rangle} + \frac{\langle Lv, D^{-1}Lv \rangle}{\langle v, v \rangle} \right] \le \|A\|_2 + \|L^T D^{-1}L\|_2.$$

The second term may be written with the definition $\overline{L} = D^{-1}L$ in the form

$$\|L^T D^{-1} L\|_2 = \sup_{v \neq 0} \frac{\langle \overline{L}^T v, D\overline{L}^T v \rangle}{\langle v, v \rangle} \leq \|D\|_2 \|\overline{L}^T\|_2^2 \leq \|A\|_2,$$

where we have used (5.44) and (5.8). So we finally also obtain (5.45) and thus confirm (5.43) also for this method. $\square$

Gauss–Seidel Method. For a proof of the smoothing property of the GS-method it would be very nice to proceed as in the proof of convergence in Section 5.2.2 and take the detour via the SGS-method. A detailed analysis, however, shows that this is not possible (see Exercise 5.4). If, however, the matrix has *property* $\mathcal{A}$ (see Definition 3.1), then a smoothing theorem can be proved, which we have saved for Exercise 5.5. The reason for this is that the property essentially only holds for matrices from difference methods over uniform meshes including our model problem 5.13. For finite element methods over triangulations, this is only true in very special cases. (To see this, just try to numerate the three nodes of a triangle as in Section 3.1.1 according to the *red/black* scheme.) That is why today in FE-methods, whenever the smoothing property is of particular interest, the SGS-method or the damped Jacobi method, respectively, are preferred among the matrix decomposition methods.

Method of Conjugate Gradients. The error smoothing by the Jacobi or the Gauss–Seidel method originates from their local structure, since only high frequencies are locally well visible. The fact that the nonlocal CG-method exhibits a similar smoothing property, is, at first glance, surprising – also in the direct comparison with the Gauss–Seidel method in Figure 5.6.

The CG-method, however, minimizes the error in the energy norm, wherein high frequencies dominate compared with the L^2-norm and are therefore reduced more significantly. The associated smoothing result dates back to V. P. Il'yin [128].

Theorem 5.16. *In the notation of this section we get*

$$\|u - u_k\|_A \leq \frac{\sqrt{\lambda_{\max}(A)}}{2k + 1} \|u - u_0\|_2. \tag{5.46}$$

Proof. The proof starts as the proof of Theorem 5.8 (see Volume 1, Section 8.3, Theorem 8.17). As u_k is the solution of the minimization problem (5.16), we have for all $w \in U_k$

$$\|u - u_k\|_A \leq \|u - w\|_A.$$

All elements w of the Krylov space U_k possess a representation of the form

$$w = u_0 + P_{k-1}(A)r_0 = u_0 + AP_{k-1}(A)(u - u_0)$$

in terms of a polynomial $P_{k-1} \in \boldsymbol{P}_{k-1}$, such that

$$u - w = u - u_0 - AP_{k-1}(A)(u - u_0) = Q_k(A)(u - u_0),$$

where $Q_k(A) = I - AP_{k-1}(A) \in \boldsymbol{P}_k$ with $Q_k(0) = 1$. Therefore the minimization over w can be replaced by a minimization over all feasible polynomials $Q_k \in \boldsymbol{P}_k$, i.e.,

$$\|u - u_k\|_A \leq \min_{w \in U_k} \|u - w\|_A = \min_{Q_k(A) \in \boldsymbol{P}_k, Q_k(0)=1} \|Q_k(A)(u - u_0)\|_A.$$

If we proceed with the usual estimate

$$\|Q_k(A)(u - u_0)\|_A \leq \|Q_k(A)\|_A \|u - u_0\|_A,$$

we arrive at Theorem 5.8. Instead we follow [128] and continue by

$$\|Q_k(A)(u - u_0)\|_A = \|A^{1/2}Q_k(A)(u - u_0)\|_2 \leq \|A^{1/2}Q_k(A)\|_2 \|u - u_0\|_2.$$

From S. K. Godunov and G. P. Prokopov [103] we take

$$\|A^{1/2}Q_k(A)\|_2 = \max_{\lambda \in \sigma(A)} |\sqrt{\lambda} Q_k(\lambda)| \leq \frac{\sqrt{\lambda_{\max}(A)}}{2k + 1}.$$

Upon collecting all partial results, the statement of the theorem is finally supplied. □

In this context consider also Figure 5.8 w.r.t. the behavior of the CG-method, there without preconditioning.

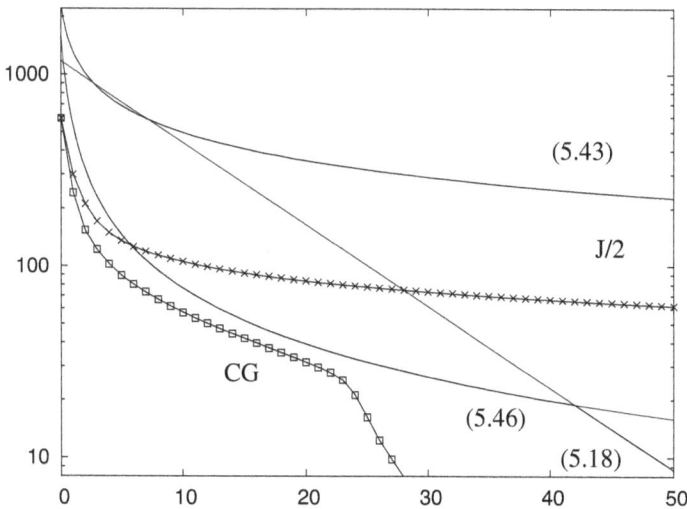

Figure 5.8. Poisson model problem 5.13: Comparison of computed energy errors $\|u - u_k\|_A$ with convergence estimates (5.43) for the damped Jacobi method (J/2) as well as (5.46) and (5.18) for the CG-method (CG).

Interpretation.   We now want to understand why the estimates (5.43) or (5.46) describe a smoothing property of the method. In fact, a meaningful statement of the form

$$\|u - u_k\|_A \approx \left(\frac{1}{k}\right)^{1/2} \|u - u_0\|_A \quad \text{or} \quad \|u - u_k\|_A \approx \frac{1}{2k+1}\|u - u_0\|_A$$

is only obtained, if

$$\|u - u_0\|_A \approx \sqrt{\lambda_{\max}(A)}\,\|u - u_0\|_2,$$

which means that the initial error $u - u_0$ contains a significant portion of eigenvectors corresponding to the largest eigenvalues. These are exactly the "high frequency" eigenfunctions. Smooth initial errors, however, are characterized by the relation

$$\|u - u_0\|_A \approx \sqrt{\lambda_{\min}(A)}\,\|u - u_0\|_2.$$

In this case equation (5.43) supplies, due to $\lambda_{\min} \ll \lambda_{\max}$, for small k nearly no error reduction.

   With this insight we turn to Figure 5.8. The uppermost line showing the upper bound (5.43) for comparison describes the transient behavior of the damped Jacobi method correctly, but exhibits a jump at the beginning: this is just the quotient

$$\frac{\|u - u_0\|_A}{\sqrt{\lambda_{\max}(A)}\,\|u - u_0\|_2} \ll 1.$$

In fact, the example has been constructed in such a way that as many frequencies as possible would enter uniformly into the initial error. For this reason the estimate is qualitatively correct, but quantitatively not sharp. For the CG-method the above estimate (5.46) shows a qualitatively good agreement with the initial error behavior, whereas the upper bound (5.18) only holds roughly asymptotically.

Summary.   In order to assess the smoothing property of iterative methods, one may compare the results (5.43) and (5.46) with (5.41). The comparison suggests that the CG-method smoothes faster than the damped Jacobi method or the symmetric Gauss–Seidel method; this is agreement with Figure 5.5 and 5.6. In addition, the result that the classical Jacobi method does *not* smooth is clearly reflected in Figure 5.5.

5.5   Iterative Hierarchical Solvers

In the preceding Section 5.4 we found out that by repeated application of some of the classical iterative solvers "high" spatial frequencies in the iterative error are damped faster than "lower" ones. From this property we will now draw further advantage. For this purpose, we exploit the fact that the linear equation systems considered here originated from a discretization process. Frequently, a hierarchical system of meshes or nested FE-spaces

$$S_0 \subset S_1 \subset \cdots \subset S_j \subset H_0^1(\Omega)$$

arises. This corresponds to a hierarchical system of linear equations

$$A_k u_k = f_k, \quad k = 0, 1, \ldots, j$$

with symmetric positive definite matrices A_k of dimension N_k. The idea of hierarchical solvers now is to numerically solve not only the original system for $k = j$, i.e., on the finest mesh, but simultaneously all systems for $k = 0, 1, \ldots, j$.

Usually, the system $A_0 u_0 = f_0$ is solved by direct elimination methods. If N_0 is still large, which may well happen in practically relevant problems, one will employ one of the direct sparse solvers that we discussed in Section 5.1 above. For the systems $k > 0$ iterative solvers will be used which are known to damp "high frequency" error components, as shown in the preceding Section 5.4. On a hierarchical mesh, the "height" of a frequency must be seen in relation to the mesh size:

- "high frequency" error components are only "visible" on finer meshes, but invisible on coarser meshes;

- "Low frequency" error components on finer meshes are "high frequency" error components on coarser meshes.

The situation is illustrated in Figure 5.9 for two neighboring grids in one space dimension.

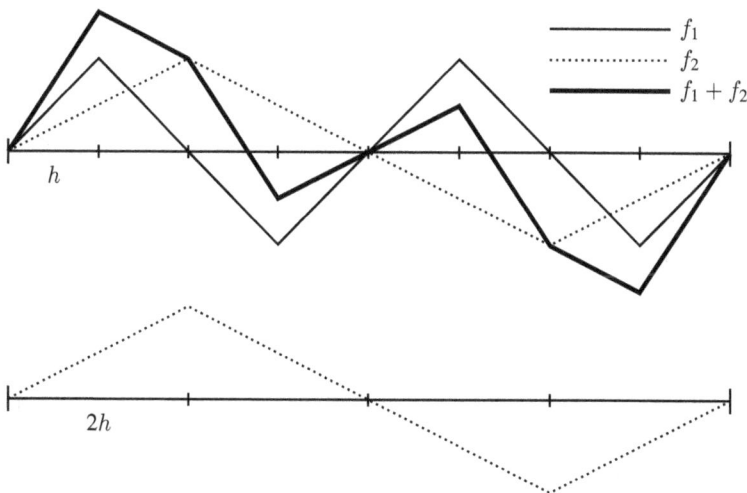

Figure 5.9. Error components on two hierarchical grids ($d = 1$): *Top:* visible component $f_1 + f_2$ on grid $\mathcal{T}_h$. *Bottom:* visible component of $f_1 + f_2$ on grid $\mathcal{T}_{2h}$: obviously, the component f_1 is invisible.

In the following Sections 5.5.1 and 5.5.2 we will present two fundamental ideas on how this insight can be exploited for an algorithmic speed-up.

5.5.1 Classical Multigrid Methods

In this type of method the classical iterative solvers (to iteration matrix G) are colloquially just called "smoothers", since they actually "smooth away" the "high" frequencies in the iterative error. Classical multigrid methods are constructed according to the following basic pattern:

- application of the smoother on a fine grid system (with iteration matrix G_h) damps the high frequency error components there, but for the time being leaves the low frequency components unaltered;

- coarse grid correction: in order to reduce the smooth (low frequency) error components, they are represented on a coarser grid, where they, in turn, contain "high" frequencies that can be reduced by application of a smoother (this time with iteration matrix G_H);

- the coarse grid corrections are recursively performed on successively coarser grids.

One step of a multigrid method on grid level k consists, in general, of three substeps: *presmoothing, coarse grid correction,* and *postsmoothing.* The presmoothing substep realizes an approximate solution $\bar{u}_k \in S_k$ by ν_1-times application of the smoother. The optimal correction on the next coarser grid $\mathcal{T}_{k-1}$ would be given by the best approximation v_{k-1} of $u_k - \bar{u}_k$ in S_{k-1} and satisfies the equation system

$$\langle Av_{k-1}, \varphi \rangle = \langle A(u_k - \bar{u}_k), \varphi \rangle = \langle f - A\bar{u}_k, \varphi \rangle \quad \text{for all } \varphi \in S_{k-1}. \tag{5.47}$$

The trick of *multi*grid methods consists of solving this system only approximately, and this recursively again by some multigrid method. This supplies the coarse grid correction $\hat{v}_{k-1} \in S_{k-1}$. The numerical solution of (5.47) requires establishing the algebraic linear equation $A_{k-1}v_{k-1} = r_{k-1}$ with residual $r_{k-1} = f - A\bar{u}_k \in S_{k-1}^*$. In the postsmoothing substep the coarse grid corrected approximation $\bar{u}_k + \hat{v}_{k-1}$ is subjected ν_2-times to the smoother, which finally produces the result $\hat{u}_k$.

Because of the different bases of the ansatz spaces S_k and S_{k-1} the computation of the *coefficient vectors* requires a change of basis, which is done via a matrix representation of the otherwise trivial embeddings $S_{k-1} \hookrightarrow S_k$ and $S_k^* \hookrightarrow S_{k-1}^*$. For the transitions between neighboring levels two types of matrices are necessary:

- *prolongations* I_{k-1}^k transform the coefficient vectors of *values* $u_h \in S_{k-1}$ on a grid $\mathcal{T}_{k-1}$ to the coefficient vectors associated with S_k on the next finer grid $\mathcal{T}_k$. In the conformal FE-methods (here the only ones considered) the prolongation is realized by *interpolation* of the coarser values on the finer grid;

- *restrictions* I_k^{k-1} transform coefficient vectors of *residuals* $r \in S_k^*$ on some grid $\mathcal{T}_k$ to their coefficient vectors associated with S_{k-1}^* on the next coarser grid $\mathcal{T}_{k-1}$; for conformal FE-methods the relation $I_k^{k-1} = (I_{k-1}^k)^T$ holds.

In contrast to FE-methods, finite difference methods require careful consideration concerning the definition of prolongation and restriction, i.e., of a consistent operator pair (I_k^{k-1}, I_{k-1}^k) (see, e.g, [116]). In both cases, I_{k-1}^k is an (N_k, N_{k-1})-matrix, correspondingly I_k^{k-1} an (N_{k-1}, N_k)-matrix (see also Exercise 5.7 with regard to the transformation of the stiffness matrices A_k and A_{k-1}). Due to the local support of the FE-basis functions, both matrices are sparse. For special FE-bases, their application can be realized by an efficient algorithm on a suitably chosen data structure (cf. Section 7.3.1).

Multigrid Iteration. Having completed these preparations, we are now ready for the recursive formulation of one step MGM_k of a multigrid method. For the description of the smoother we use the representation (5.4) of the iterative error, which includes the unknown solution (e.g., $u_k \in S_k$); strictly speaking, this only covers the linear iterative methods (like the damped Jacobi and symmetric Gauss–Seidel methods); we want, however, to also subsume any PCG-method under this notation.

Algorithm 5.17. *Classical multigrid algorithm.*
$\hat{u}_k := \mathrm{MGM}_k(\nu_1, \nu_2, \mu, r_k)$

  — *Presmoothing* by ν_1-times application of the smoother
  $\bar{u}_k := 0$
  $u_k - \bar{u}_k := G^{\nu_1}(u_k - \bar{u}_k)$
  — *Coarse grid correction* …
  $r_{k-1} := I_k^{k-1}(r_k - A_k \bar{u}_k)$
  if $k = 1$ **then**
      — …by direct solution
      $\hat{v}_0 = A_0^{-1} r_0$
  else
      — …by μ-times recursive application of the multigrid method
      $\hat{v}_{k-1} := 0$
      for $i = 1$ **to** μ **do**
          $\delta v := \mathrm{MGM}_{k-1}(\nu_1, \nu_2, \mu, r_{k-1} - A_{k-1}\hat{v}_{k-1})$
          $\hat{v}_{k-1} := \hat{v}_{k-1} + \delta v$
      end for
  end if
      — *Postsmoothing* by ν_2-times application of the smoother
  $\hat{u}_k := \bar{u}_k + I_{k-1}^k \hat{v}_{k-1}$
  $u_k - \hat{u}_k := G^{\nu_2}(u_k - \hat{u}_k)$

Any multigrid method is therefore characterized by the choice of the smoother (G) as well as by the three parameters (ν_1, ν_2, μ). From the early multigrid developments on, certain intuitive representations of the Algorithm 5.17 have evolved in the form of diagrams, where the initial grid, i.e., the finest grid ($k = j$), is on the top and the coarsest grid ($k = 0$) at the bottom. In Figure 5.10 we show such diagrams for three or four grids, respectively: because of their special form in this representation, one also speaks of "V-cycles" for $\mu = 1$ and of "W-cycles" for $\mu = 2$.

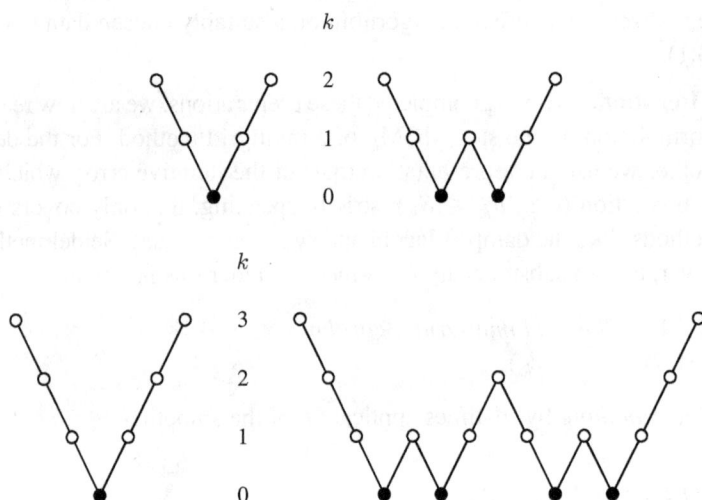

Figure 5.10. Variants of the classical multigrid method 5.17. Descending lines represent restrictions, ascending propagations (i.e., for FE: interpolations). Direct solutions on the coarsest grid are labeled by •. *Top:* three hierarchical grids. *Bottom:* four hierarchical grids. *Left:* V-cycles ($\mu = 1$). *Right:* W-cycles ($\mu = 2$).

Convergence.   One of the fascinating things about multigrid methods is the fact that they permit a proof of *mesh independent linear convergence* with spectral radius

$$\rho(\nu_1, \nu_2, \mu) < 1 \ .$$

In order to demonstrate this, we introduce here the essential core of the classical convergence theory due to W. Hackbusch [113] from 1979. It is based on two pillars.

One of the pillars uses the transfer of the smoothing theorems for iterative solvers that we had presented in Section 5.4.3 for the discrete case; see, e.g., (5.43) for the damped Jacobi method and the symmetric Gauss–Seidel method as well as (5.46) for the CG-method. Upon inserting the nodal basis for finite element norms into the discrete norms applied there, one arrives, after some short intermediate calculation, at an additional scaling factor h and consequently at the *smoothing property* in function space in the form

$$\|G^m(u_h - v_h)\|_A \leq \frac{c}{hm^\gamma}\|u_h - v_h\|_{L^2(\Omega)}. \tag{5.48}$$

By comparison with (5.46) and (5.43) we get $\gamma = 1$ for the CG-method and $\gamma = 1/2$ for the other iterative methods. Furthermore one uses the reduction of the energy norm in the course of smoothing, i.e.,

$$\|Gu_h\|_A \le \|u_h\|_A. \tag{5.49}$$

The second pillar is the *approximation property*[2]

$$\|u - u_h\|_{L^2(\Omega)} \le ch\|u\|_A , \tag{5.50}$$

which holds for the best approximations $u_h \in V_h$ of u in the energy norm.

Theorem 5.18. *Assume that (5.50) and (5.48) hold. Then the multigrid method 5.17 satisfies the contraction property*

$$\|u_k - \hat{u}_k\|_A \le \rho_k \|u_k\|_A \tag{5.51}$$

with the recursively defined contraction factor

$$\rho_k = \frac{c}{v_2^\gamma} + \rho_{k-1}^\mu, \quad \rho_0 = 0. \tag{5.52}$$

For the W-cycle ($\mu = 2$) and under the condition $v_2^\gamma \ge 4c$, the sequence of the ρ_k is bounded independently of k according to

$$\rho_k \le \rho_* = \frac{1}{2} - \sqrt{\frac{1}{4} - \frac{c}{v_2^\gamma}} \le \frac{1}{2}.$$

Proof. The induction starts with $\rho_0 = 0$ due to the direct solution on the coarse grid $\mathcal{T}_0$. Let now $k \ge 1$. With the notation introduced in Algorithm 5.17 we have

$$
\begin{aligned}
\|u_k - \hat{u}_k\|_A &= \|G^{v_2}(u_k - \bar{u}_k - \hat{v}_{k-1})\|_A \\
&\le \|G^{v_2}(u_k - \bar{u}_k - v_{k-1})\|_A + \|G^{v_2}(v_{k-1} - \hat{v}_{k-1})\|_A \\
&\le \frac{c}{hv_2^\gamma}\|(u_k - \bar{u}_k) - v_{k-1}\|_{L^2(\Omega)} + \|v_{k-1} - \hat{v}_{k-1}\|_A.
\end{aligned}
$$

From the second to the third line we arrive by means of (5.49). As v_{k-1} is the exact solution of (5.47), the inequality (5.50) can be applied to the first term. For the second term we know that the multigrid method on level $k - 1$ converges with a contraction factor $\rho_{k-1} < 1$. In summary, we thus obtain

$$\|u_k - \hat{u}_k\|_A \le \frac{c}{v_2^\gamma}\|u_k - \bar{u}_k\|_A + \rho_{k-1}^\mu\|v_{k-1}\|_A \le \left(\frac{c}{v_2^\gamma} + \rho_{k-1}^\mu\right)\|u_k - \bar{u}_k\|_A,$$

---

[2] often also called *Jackson inequality*, in particular in mathematical approximation theory.

where $\|v_{k-1}\|_A \leq \|u_k - \bar{u}_k\|_A$ follows from the Galerkin orthogonality (4.14). Since $u_k - \bar{u}_k = G^{\nu_1} u_k$ from the presmoothing and (5.49), we finally obtain the contraction factor (5.52).

For the verification of the convergence of the W-cycle we merely observe that the sequence $\{\rho_k\}$ is increasing monotonely towards a fixed point ρ_*, which is obtained as a solution of a quadratic equation.   □

Note that presmoothing steps do not show up in the contraction factor (5.52) and therefore are not necessary for the convergence of the W-cycle. They do, however, contribute to the efficiency, as can be seen in Table 5.1.

For the V-cycle ($\mu = 1$), however, we do not get an h-independent convergence result from (5.52) – this case requires some more complex proof technique, which in 1983 has been worked out by D. Braess and W. Hackbusch [40] employing subtle techniques from approximation theory (see also the book [39] of Braess). Here, however, we postpone the proof for the V-cycle to Section 7, where we will offer a more general and at the same time more simple theoretical framework.

Computational Amount.   We start from a mesh independent linear convergence. For simplicity, we assume quasi-uniform triangulations, i.e.,

$$c_0 2^{dk} \leq N_k \leq c_1 2^{dk}, \quad k = 0, \ldots, j.$$

As a measure for the computational amount we take the number of floating point operations (flops); we only want to derive a rough classification and, in particular, the growth behavior in terms of N_j. A meticulous counting of flops is not only cumbersome, but also not very informative, since in present computer architectures the number and distribution of memory accesses is more important for the speed of a program than the number of genuine operations. This is why in special implementations the data structures and the access pattern are of enormous importance. The growth order in terms of N_j is a reasonable quantity for the principal assessment of the efficiency of algorithms, in particular, with growing problem size, since a significant growth cannot be compensated by any efficient data structures. Nevertheless we want to point out that given the problem size the multiplicative constants (assumed to be generic in our proofs) may be more important than the order.

The numerical solution on the coarsest grid is not of importance for an asymptotic consideration. Due to the sparsity of the A_k we count $\mathcal{O}(N_k)$ for a smoothing step, just as for the change of grid levels, so that on each grid level a computational amount $\mathcal{O}(\nu N_k)$ with $\nu = \nu_1 + \nu_2 + 2$ arises. For the V-cycle, we thus obtain per multigrid iteration

$$W^{\mathrm{MG}} = \mathcal{O}\left(\sum_{k=1}^{j} \nu N_k\right) = \mathcal{O}\left(\nu \sum_{k=1}^{j} 2^{dk}\right) = \mathcal{O}\left(\nu 2^{d(j+1)}\right) = \mathcal{O}(\nu N_j) \qquad (5.53)$$

operations. Obviously, the order of this amount is optimal, since even for the evaluation of the solution each of the N_j coefficients must be visited.

The numbers ν_1 and ν_2 of smoothing steps are determined by the minimization of
the *effective computational amount*

$$W_{\text{eff}}^{\text{MG}} = -\frac{W^{\text{MG}}}{\log \rho(\nu_1, \nu_2, \mu)} = \mathcal{O}(N_j), \tag{5.54}$$

which describes the amount per accuracy gain. We will not go into theoretical depth
here, but will only consider an example.

Example 5.19. We return to the Example 5.13 for our Poisson model problem. In
Figure 5.11 we show the behavior of a V-cycle using the damped Jacobi method as
smoother. It is clearly visible that all frequencies are damped to a similar extent, in particular that significantly more high frequency error components are observable would
be expected after a corresponding number of iterations of a classical method (see Figures 5.5 and 5.6). Vice versa, the low frequency error components are significantly
reduced so that, in total, a convergence rate $\rho \approx 0.1$ is achieved. In this way the effective amount is even slightly smaller than that of the CG-method in its asymptotic
convergence phase (cf. 5.4) – with finer grids, the advantage of the multigrid method
becomes more apparent. In Table 5.1 the effective amount for various values of ν_1
and ν_2 is shown. The optimum is attained at $\nu_1 = \nu_2 = 2$, but the efficiency does not
depend too strongly on this choice unless too little smoothing is performed.

Table 5.1. Poisson model problem from Example 5.13. Effective amount of the multigrid
V-cycle of Figure 5.11 for various numbers of pre- and postsmoothing steps.

$\nu_1 \backslash \nu_2$	0	1	2	3	4
0		4.57	3.81	3.77	3.91
1	4.54	3.67	3.54	3.60	3.74
2	3.69	**3.49**	**3.49**	3.58	3.71
3	3.57	3.50	3.53	3.62	3.74
4	3.65	3.59	3.62	3.70	3.82

Remark 5.20. The above shown computational complexity $\mathcal{O}(N_j)$ is based on reach-
ing a *fixed* error reduction. One should, however, also take into account that finer
meshes give rise to smaller discretization errors, in terms of the energy error an order
of magnitude $\|u - u_h\|_a = \mathcal{O}(h_j) = \mathcal{O}(2^{-j})$ for H^2-regular problems. Therefore,
on finer meshes one will have to perform a larger number of multigrid iterations in

k

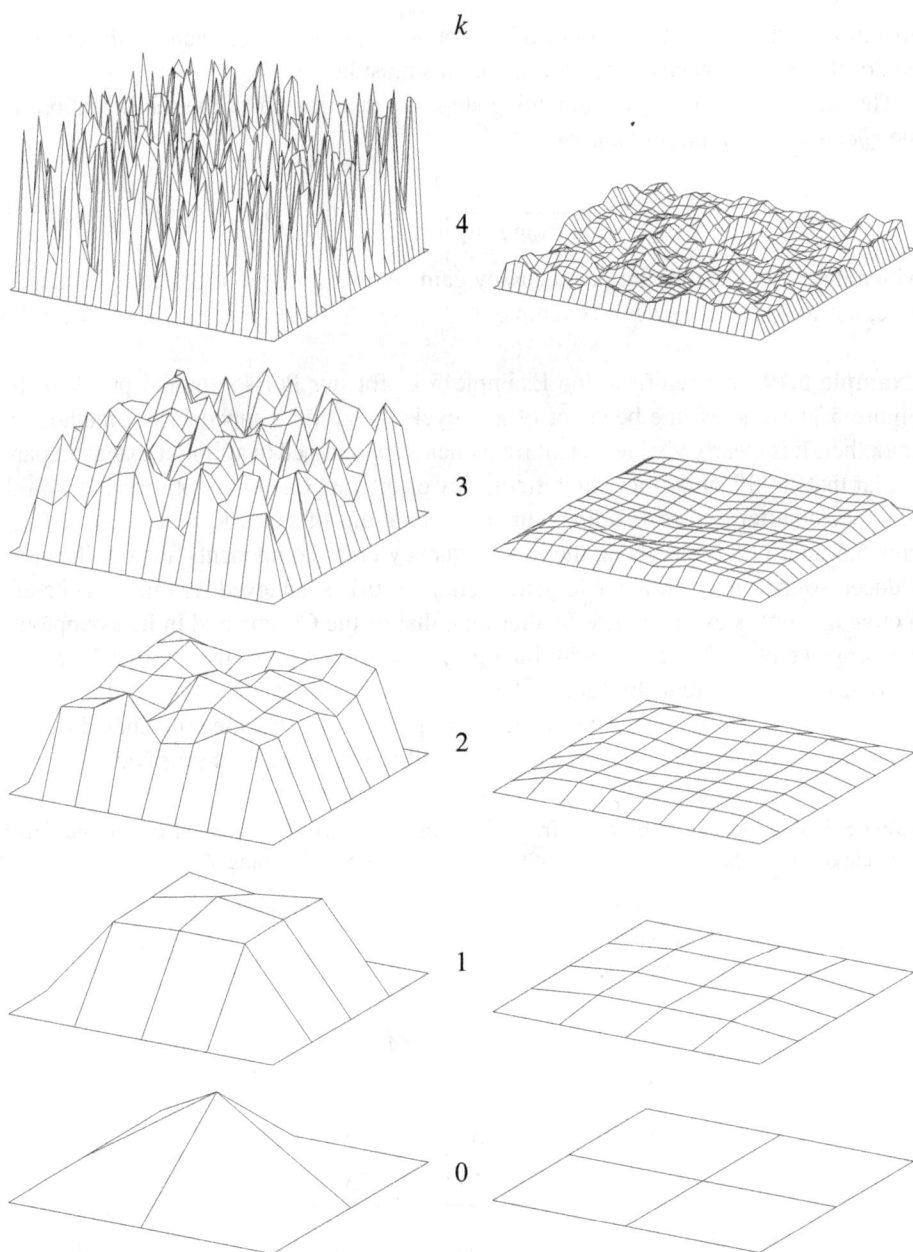

Figure 5.11. Poisson model problem from Example 5.13. Behavior of one V-cycle with smoothing step numbers $v_1 = v_2 = 2$ over five hierarchical grids, i.e., with $j = 4$. *Left:* initial error $u_k - 0$. *Right:* final error $u_k - \hat{u}_k$. For comparison with Figure 5.5 and 5.6: the computational amount corresponds to eight smoothing steps on the finest grid.

order to reach the discretization accuracy. Assuming a contraction factor $\rho_* < 1$ the number k of necessary multigrid iterations is determined by

$$\rho_*^k = \mathcal{O}(2^{-j}) \quad \Rightarrow \quad k = \mathcal{O}(j),$$

which implies a total computational amount of

$$\overline{W}^{MG} = \mathcal{O}(jN_j). \tag{5.55}$$

An algorithmic improvement is to perform more multigrid cycles with less computational amount on coarser mesh levels, such that on the finer levels only the error reduction corresponding to the discretization error improvement from one level to the next finer level has to be achieved. By the usual geometric series argument, the total amount is then seen to be bounded by $\mathcal{O}(N_j)$, which will be proved in Exercise 5.8. This approach is known as *nested iteration* or *F-cycle* and arises most naturally as *cascadic principle* in the context of adaptive mesh refinement (see Chapters 6 and 7 below).

Remark 5.21. From a historical perspective, multigrid methods date back to the Russian mathematician R. P. Fedorenko[3]. Rather early, in connection with a simple weather model in which discretized Poisson problems over an equidistant $(64, 64)$-mesh (i.e., with a $(4096, 4096)$-matrix) had to be repeatedly solved, he had observed and analyzed the different smoothing of error components on different meshes. It seems that his idea was first seen as only having military importance and was not published until 1964 in Russian [86]. An English translation [87] existed in the West after 1973. On this basis the Israeli mathematician A. Brandt introduced "adaptive multilevel methods" in a first publication [43] and has further developed them since then in a series of papers. Independent of this line, "multigrid methods" were detected in 1976 by W. Hackbusch [111], who then extended them in a fast sequence of papers beyond Poisson problems and recessed their theoretical justification. His frequently cited books (see [114, 115]) founded a worldwide "multigrid school". The first comprehensive introduction to multigrid methods, predominantly for finite difference discretizations, appeared in the collection [116] by W. Hackbusch and U. Trottenberg in 1982. First convergence proofs for multigrid methods by W. Hackbusch [113] were theoretically confined to uniform meshes and W-cycles (see, e.g., Theorem 5.18). The breakthrough to a rather general and at the same time transparent convergence theory also for V-cycles is due to the Chinese-American mathematician J. Xu [223] in 1992 (see also the generalization in the beautifully clear survey by H. Yserentant [228]). It is this more recent line of proofs that we will follow in Chapter 7 below.

---

[3] For integral equations the idea dates back to an even earlier paper by H. Brakhage [41] from 1960.

5.5.2 Hierarchical-basis Method

In 1986, H. Yserentant [226] suggested an alternative idea how to exploit the grid hierarchy. In this approach the usual nodal basis $\Phi_h^1 = (\varphi_i)_{i=1,\dots,N}$ from (4.53) is replaced by the so-called *hierarchical* basis (HB)

$$\Phi_h^{HB} = (\varphi_i^{HB})_{i=1,\dots,N} = \bigoplus_{k=0}^{j} \Phi_k^{HB}$$

with

$$\Phi_k^{HB} = \{\varphi_{k,\xi} \in \Phi_k^1 : \xi \in \mathcal{N}_k \setminus \mathcal{N}_{k-1}\}.$$

In Figure 5.12 the two basis systems are represented in one space dimension.

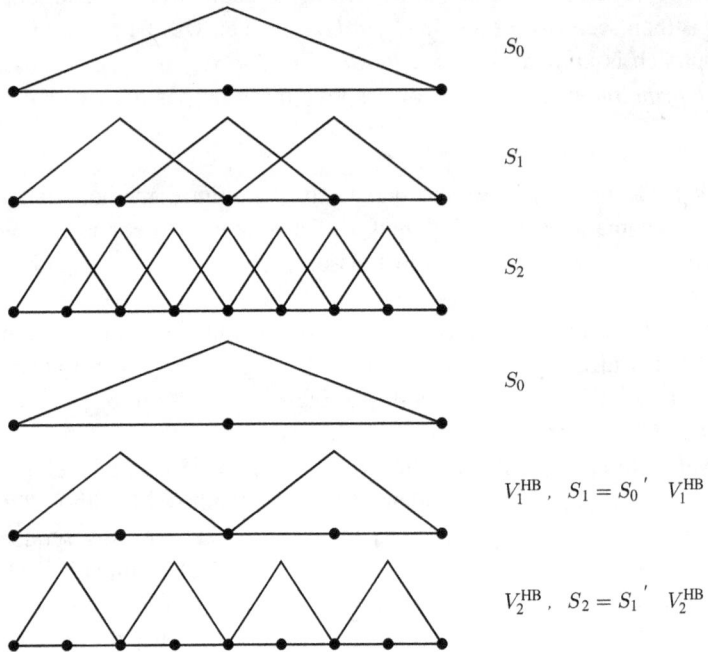

Figure 5.12. FE-basis systems on hierarchical grids ($d = 1$). *Top:* nodal basis. *Bottom:* hierarchical basis (HB).

The knack of this basis change consists in the combination of two properties that are important for the solution of the equation system (4.9):

- the basis change $a_h^{HB} = C a_h$ between the representations

$$u = \sum_{i=1}^{N} a_{h,i} \varphi_i = \sum_{i=1}^{N} a_{h,i}^{HB} \varphi_i^{HB}$$

as well as its transpose C^T are easy to realize algorithmically;

- the condition number of the stiffness matrix $A_h^{\mathrm{HB}} = C A_h C^T$ in the hierarchical basis is significantly smaller than the condition of A_h in the nodal basis.

Both aspects together permit the use of $B^{-1} = C C^T$ as effective preconditioner in a PCG-method from Section 5.3.2 (see also Volume 1, Exercise 8.3).

The connection with the smoothing property of classical iteration methods is slightly more hidden here than in the previous section. The hierarchical basis corresponds to a direct decomposition of the ansatz space

$$S_j = S_0 \oplus \bigoplus_{k=1}^{j} V_k^{\mathrm{HB}}$$

into subspaces V_k^{HB} with different spatial frequency and associated mesh size, so that within V_k^{HB} all represented functions are of the same frequency. The coupling between the subspaces is weak, so that essentially a block diagonal matrix A_h^{HB} arises.

Remark 5.22. The application of the preconditioner $B^{-1} = C I C^T$ realizes one step of smoothing by a simple Richardson iteration method (cf. Exercise 5.2) – independent of each other on the individual subspaces V_k^{HB}. The smoothing property thus reduces the necessarily high frequency error components very effectively. As B is only applied as preconditioner, we may dispense of the choice of a mesh size for the Richardson method (again, see Exercise 5.2).

Convergence. For space dimension $d = 2$ a significant reduction of the condition number is achieved. In the nodal basis we have

$$\mathrm{cond}(A_h) \leq \mathrm{const} / h_j^2 \sim 4^j,$$

whereas in the hierarchical basis we get

$$\mathrm{cond}(A_h^{\mathrm{HB}}) \leq \mathrm{const}(j + 1)^2,$$

see (7.55) in Section 7.2.1, where also the associated proof can be found. As a consequence, a significant reduction of PCG-iterations is obtained if the preconditioning is realized via HB: Let m_j denote the number of iterations in the nodal basis and m_j^{HB} the one in HB. With Corollary 5.1 one then obtains

$$m_j \sim \tfrac{1}{2}\sqrt{\mathrm{cond}_2(A_h)} \sim 2^j, \quad m_j^{\mathrm{HB}} \sim \tfrac{1}{2}\sqrt{\mathrm{cond}_2(A_h^{\mathrm{HB}})} \sim (j + 1)$$

for an error reduction by a fixed factor. For $d = 3$ the condition number reduction is far less marked, the reason for which we will see in Section 7.2.1 below.

Computational Amount. The evaluation of the inner products as well as the matrix-vector multiplications in the PCG-algorithm 5.9 causes an amount of $\mathcal{O}(N_j)$. The evaluation of the precondition is also extremely simple, as we will show briefly.

Due to the Lagrange property $\varphi_i(\xi_l) = \delta_{il}$ for $\xi_l \in \mathcal{N}$ we have

$$a_{h,l} = u(\xi_l) = \sum_{i=1}^{N} a_{h,i}^{\mathrm{HB}} \varphi_i^{\mathrm{HB}}(\xi_l)$$

and thus $C_{il} = \varphi_i^{\mathrm{HB}}(\xi_l)$. We may utilize this simple structure of the hierarchical meshes to realize – without any memory requirement – the application of C and C^T recursively at an amount of $\mathcal{O}(N_j)$.

Example 5.23. For illustration we here give two examples (for space dimensions $d = 1$ and $d = 2$) in Figure 5.13. With the chosen order of nodes we may read off on top ($d = 1$):

$$a_{h,l} = a_{h,i}^{\mathrm{HB}}, \quad l = 1, 2, 3, 4,$$

$$a_{h,5} = a_{h,5}^{\mathrm{HB}} + \frac{1}{2} a_{h,2}^{\mathrm{HB}},$$

$$a_{h,7} = a_{h,7}^{\mathrm{HB}} + \frac{1}{2} a_{h,3}^{\mathrm{HB}},$$

$$a_{h,6} = a_{h,6}^{\mathrm{HB}} + \frac{1}{2} (a_{h,2}^{\mathrm{HB}} + a_{h,3}^{\mathrm{HB}}).$$

At the bottom ($d = 2$) we obtain with $I_0 = (1, 2, 3, 4)$, $\delta I_1 = I_1 - I_0 = (5, 6, 7)$ the explicit evaluation of $a_h^{\mathrm{HB}} = C^T a_h$ in the form

$$a_{h,l}^{\mathrm{HB}} = a_{h,l}, \quad l \in I_0,$$
$$a_{h,l}^{\mathrm{HB}} = a_{h,l} - \tfrac{1}{2}(a_{h,l^-} + a_{h,l^+}), \quad l \in \delta I_1, \text{ with neighboring nodes } l^-, l^+ \in I_0.$$

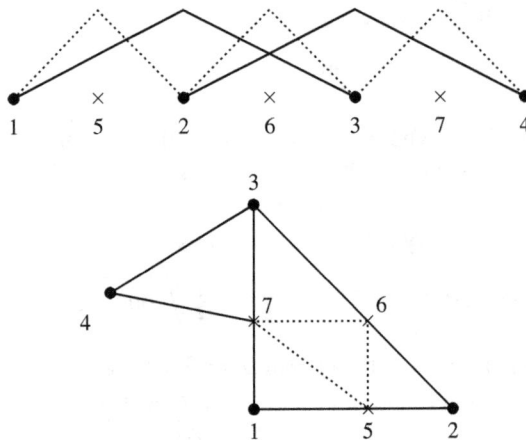

Figure 5.13. Examples illustrating the basis transformation from nodal basis to hierarchical basis (HB). *Top:* $d = 1$. *Bottom:* $d = 2$.

In the *general* case the following recursive evaluation is obtained, where the index sets $I_0, \ldots, I_j$ are defined analogously:

$$a_{h,l}^{\text{HB}} = a_{h,l}, \qquad\qquad\qquad l \in I_0,$$

$$k = 1, \ldots, j: \quad a_{h,l}^{\text{HB}} = a_{h,l} - \tfrac{1}{2}(a_{h,l-} + a_{h,l+}), \quad l \in \delta I_k, \; l^-, l^+ \in I_{k-1}.$$

Obviously, it requires an amount of $\mathcal{O}(N_j)$ in particularly simple form, which can be conveniently realized on a suitable data structure. This leads to an amount of $\mathcal{O}(N_j)$ per step and with Theorem 5.8 an effective amount of

$$W_{\text{eff}}^{\text{HB}} = \mathcal{O}(jN_j) = \mathcal{O}(N_j \log N_j). \tag{5.56}$$

If the discretization error accuracy is taken into account, then one ends up with a total computational amount of

$$\overline{W}^{\text{HB}} = \mathcal{O}(j^2 N_j). \tag{5.57}$$

Comparing (5.56) with (5.53) or (5.57) with (5.55), respectively, only a moderate factor of j is added; this is compensated for by the fact that the HB-method is of utmost simplicity and the multiplicative constant in the amount is rather small.

5.5.3 Comparison with Direct Hierarchical Solvers

For the multigrid solvers presented above we assumed a hierarchy of meshes with corresponding matrices $A_0, \ldots, A_j$ of dimensions $N_0, \ldots, N_j$, where direct methods are only applied to the coarse grid matrix A_0. In principle, the direct solvers presented in Section 5.1 may be applied equally well to each single mesh out of a mesh hierarchy, which means they work on the matrices $A = A_j$ separately.

For full matrices A, i.e., those with $\mathcal{O}(N^2)$ nonvanishing elements, direct elimination methods are known to require a computational amount of $\mathcal{O}(N^3)$ (see, e.g., Volume 1, Section 1). Here, however, we have a sparse matrix A_j with only $\mathcal{O}(N_j)$ nonvanishing elements. In this case the computational amount for direct methods is in general only $\mathcal{O}(N_j^\alpha)$ operations for some $\alpha > 1$, which strongly depends on the space dimension (see the subsequent example in Section 5.6). For classical multigrid methods only $\mathcal{O}(jN_j)$ operations are required if one terminates the multigrid iteration at the level of discretization errors (see (5.55) above). *Asymptotically*, multigrid methods will therefore be quicker. The question is when asymptotics sets in. There will be a separation line at "not too large" problems, which will first of all depend on the space dimension d as well as next on (a) the available computers (and thus shifted with every new computer generation) and (b) the progress of the corresponding algorithms (and thus shifted with every new implementation).

Computational Amount. For space dimension $d = 2$ the present state is that good direct solvers are generally faster than iterative multigrid methods up to a few millions of degrees of freedom, as to be illustrated in the following Section 5.6. For $d = 3$ the mesh nodes are denser connected so that direct methods will produce a larger fill-in. Multigrid methods, however, do not suffer from the higher space dimension, so that they will be faster already above some ten or hundred thousand degrees of freedom.

Array Storage.   On present computers, accessible work storage restricts problem sizes more than computing times with respect to the competitiveness of direct methods. Examples for such restrictions will be given next in Section 5.6 and further below in Section~6.4. That is why for direct methods "problem adapted" meshes are particularly crucial, supplying a prescribed accuracy with as few degrees of freedom as possible. The construction of such *adaptive* meshes will be worked out in detail in the subsequent Chapter 6.

Hybrid Methods.   Of course, all kinds of mixed forms between direct methods and iterative solvers are realized where direct solvers are applied, not only on the coarsest mesh, but also up to a refinement level $k_{dir} > 1$. In Section 6.4 we present such a hybrid method using the example of a "plasmon-polariton waveguide", where direct solvers are used at *every* submesh, i.e., with $k_{dir} = j$.

5.6   Power Optimization of a Darrieus Wind Generator

Wind power plays an important role in any concept of regenerative energy. In order to generate enough energy at justifiable cost, the optimization of wind generators is necessary, which includes in particular the simulation of the air flow around individual turbines and through entire wind parks. In principle, the turbulent air flow around the rotor blade of a wind generator is described by the compressible Navier–Stokes equations (2.19). In this book, however, we have deliberately excluded numerical methods for the computation of turbulent flow (see Section 2.2.2). Instead we restrict ourselves in the following to some simpler model case.

The French engineer Georges Darrieus (1888–1979) suggested a variant of wind generators with *vertical* rotation axis (see Figure 5.14). Its wings parallel to the rotation axis move transversely to the wind thus exploiting the dynamic lift for energy production. The vertical translation invariance permits the application of simpler numerical methods. With this approach we can already study the important question of how the power supplied by the wind generator depends on the rotation velocity.

Mathematical Modelling.   As the simplest model of air flow around the wings we select the *incompressible potential equation* from Section 2.2.1. As mentioned in the beginning, in this way we ignore nearly all practically relevant dynamical effects lsuch as turbulence, stall, and mutual influence of the wings – nevertheless the results give some first insight into the behavior of the wind generator.

In Figure 5.15 the geometric situation is represented schematically. Due to the translation invariance in vertical direction we may restrict ourselves to the two-dimensional horizontal cross section. Depending on the position and the velocity of the wing on its

Figure 5.14. Darrieus rotor for the power supply of the antarctic Neumayer station 1991–2008. Photo: Hannes Grobe/AWI.

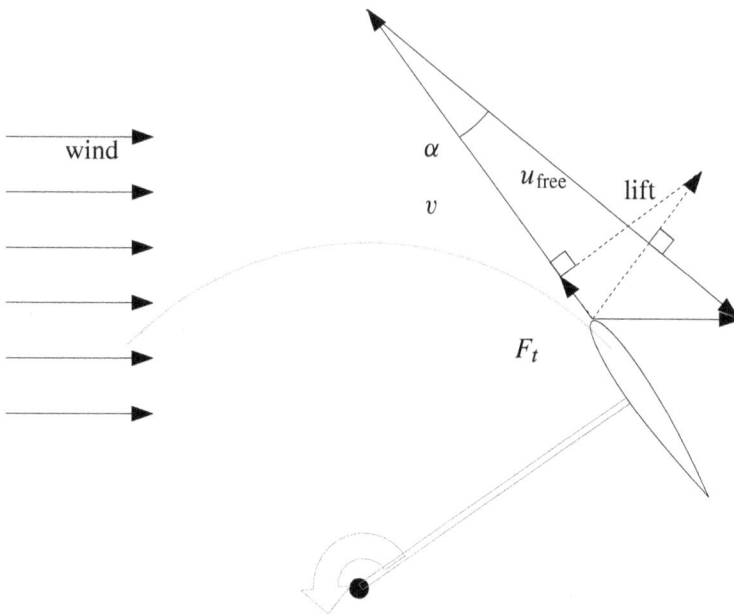

Figure 5.15. Horizontal cross section through a Darrieus rotor.

circuit, the lift angle and the relative flow velocity change and thus also the direction and size of the lift. From the Theorem of Kutta–Joukowski (see, e.g., the classical textbook [50] by A. J. Chorin and J. E. Marsden) the size of the lift, which eventually gives rise to the angular momentum, corresponds to the product of the flow velocity and the so-called *circulation*

$$\Gamma(u) = \int_{\partial \Omega_W} n^T \begin{bmatrix} 0 & 1 \\ -1 & 0 \end{bmatrix} u \, ds$$

of the velocity field u around the *wing*. In not simply connected domains Ω as given here the circulation need not vanish even for curl-free fields, but will do so for gradient fields $u = \nabla \phi$. With the ansatz

$$u = \nabla \phi + \gamma w, \quad w(x, y) = \frac{1}{x^2 + y^2} \begin{bmatrix} -y \\ x \end{bmatrix}, \quad \gamma \in \mathbb{R}$$

we obtain a generalized curl-free potential flow with nonvanishing circulation. The incompressibility then again leads to

$$\operatorname{div} \nabla \phi = -\gamma \operatorname{div} w, \tag{5.58}$$

the *slip condition* (2.16) is given by

$$n^T \nabla \phi = -\gamma n^T w \quad \text{on } \partial \Omega_W.$$

At the artificially introduced *exterior* boundary $\partial \Omega_{\text{ext}}$ we require the same mass flow across the boundary as for the unperturbed constant flow field u_{free} of the wind, i.e.,

$$n^T \nabla \phi = n^T (u_{\text{free}} - \gamma w) \quad \text{on } \partial \Omega_{\text{ext}}. \tag{5.59}$$

Thus ϕ is uniquely determined up to a constant by an inhomogeneous Neumann problem. The necessary compatibility conditions (1.3) follow from Theorem A.3 of Gauss applied to $u_{\text{free}} - \gamma w$. The value of γ is determined by the so-called Kutta condition: The viscosity of real fluids leads to the fact that the fluid flows uniformly at the back edges of the wings and does not flow around the edge (see also [99]). In the situation of Figure 5.15 this means that $e_2^T u(\xi) = 0$ or

$$e_2^T \nabla \phi + \gamma e_2^T w = 0. \tag{5.60}$$

Both the direction and the size of the flow velocity u_{free} only enter into the data of the boundary condition (5.59) (and thus only into the right-hand side of the equation system (5.61) below). Along the circuit of the wing the direction and the size of u_{free} change continuously. For the rotation angle $\beta \in [0, 2\pi]$ and the circuit velocity v we have, at absolute wind with normalized velocity 1 (so that v then is the speed ratio)

$$u_{\text{free}} = \begin{bmatrix} v + \cos\beta \\ -\sin\beta \end{bmatrix}$$

and therefore $|u_{\text{free}}| = \sqrt{1 + 2v\cos\beta + v^2}$ and $\tan\alpha = -\sin\beta/(v + \cos\beta)$. For larger flow angle α the real viscosity of the fluid leads to stall. The arising turbulences reduce the lift significantly and increase the drag. That is why Darrieus rotors run with circuit velocities $v > 1$.

In order to be able to integrate the lift and thus also the angular momentum along the circuit, the circulation component γ_α must be calculated for various right-hand sides f_α according to the flow angle $\alpha \in [\alpha_{\min}, \alpha_{\max}]$ versus the tangential direction. Due to the linearity of (5.61) the flow velocity may be assumed to be 1. The lift acts perpendicular to u_{free}, which implies for the tangential force

$$F_t \sim |u_{\text{free}}|^2 \gamma_\alpha \sin\alpha = |u_{\text{free}}| \gamma_\alpha u_{\text{free},y} \,.$$

In a realistic model the laminar flow of the boundary layer induced by the viscosity must be taken into account. It is proportional to the flow velocity $|u_{\text{free}}|$, so that we end up with

$$F_t \sim |u_{\text{free}}|(\gamma_\alpha u_{\text{free},y} - \eta).$$

Summarizing, the power output of the rotor is proportional to

$$P(v) \sim v \int_{\beta=0}^{2\pi} -|u_{\text{free}}|(\gamma_\alpha \sin\beta - \eta)\, d\beta.$$

Discrete Formulation. In the absence of viscosity, the solutions of the pure Euler equations do not exhibit any Prandtl's boundary layer (see Section 2.2.3), which is why we may dispense with the construction of boundary fitted meshes as in Figure 2.4. Nevertheless the mesh around the wing must be fine enough (see Figure 5.16, left). The discretization of the equations (5.58)–(5.60) for the flow angle α by finite elements leads to a 2×2-block system of the form

$$\begin{bmatrix} A & b \\ c^T & d \end{bmatrix} \begin{bmatrix} \phi_\alpha \\ \gamma_\alpha \end{bmatrix} = \begin{bmatrix} f_\alpha \\ 0 \end{bmatrix}, \tag{5.61}$$

where A is the symmetric positive semidefinite discretization of the Neumann problem and the variable flow angle α only enters into the right-hand side f_α. For reasons of numerical stability it is recommended to assure uniqueness of the discrete solution, e.g., by imposing the condition $\phi(\xi) = 0$ at a selected boundary point ξ.

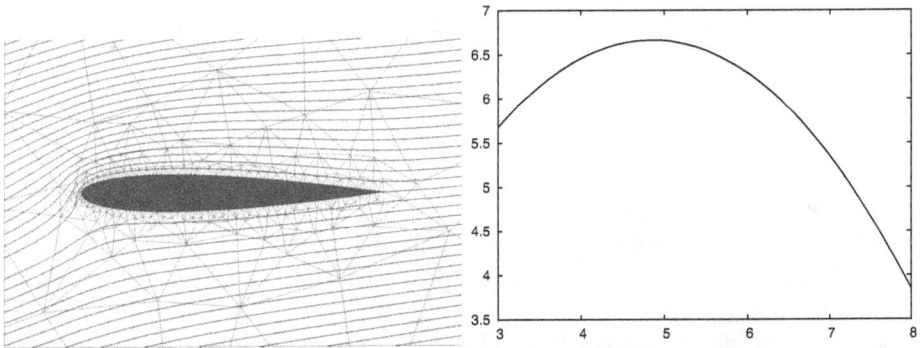

Figure 5.16. *Left:* zoom of the computational domain around the wing profile NACA0012 with coarse mesh and stream lines for a flow angle α of $12°$. *Right:* power output of the Darrieus rotor versus the speed ratio v.

The actual solution of the system may be done directly by block elimination on A, whereby

$$
\begin{bmatrix} A & b \\ 0 & d - c^T A^{-1} b \end{bmatrix} \begin{bmatrix} \phi_\alpha \\ \gamma_\alpha \end{bmatrix} = \begin{bmatrix} f_\alpha \\ -c^T A^{-1} f_\alpha \end{bmatrix}
$$

is obtained. Instead of computing the solution of the Neumann problem $A^{-1} f_\alpha$ for different α each, one will solve the adjoint equation

$$
A^T \lambda = c \tag{5.62}
$$

independent of α. Thus one obtains for γ_α the equation

$$
(d - \lambda^T b)\gamma_\alpha = -\lambda^T f_\alpha
$$

and has to solve only one Neumann problem. However, the potential ϕ_α and thus the flow field is not obtained in this way, but this is not needed for the computation of $P(v)$.

Adjoint equations may reasonably be applied whenever the solution of the equation system requires only the evaluation of a linear functional. In Section 9.2.3 this structure will come up again in the context of error estimators.

Numerical Results. The computation of the flow field u and the power output $P(v)$ in Figure 5.16, right, has been performed on an unstructured triangular mesh by means of the FE package Kaskade 7 by M. Weiser et al. [110] (see Figure 5.16, left). The basic mesh with 451 nodes was generated by the mesh generator Triangle [188]. The nodes are denser close to the wing, so that its shape is well approximated. In order to improve the accuracy of the solution, the mesh has been uniformly refined several times. Thus one obtains the typical shape of the power output curve with a maximum in the range $v \approx 5$, where the exact value of the maximum depends upon the value of the flow resistance η.

Computing Times of Direct versus Iterative Methods.   In Figure 5.17 we show the efficiencies of the different solvers discussed in the present section. All iterative solvers have been terminated at an algebraic tolerance of the order of the discretization error. Obviously, the simple Jacobi method is totally insufficient even for smaller problem sizes. The CG-method, too, is only competitive for relatively few degrees of freedom. For this reason we added a PCG with incomplete Cholesky (IC) preconditioning (see Section 5.3.2), a method rather popular in industry. Here we have used the TAUCS library.

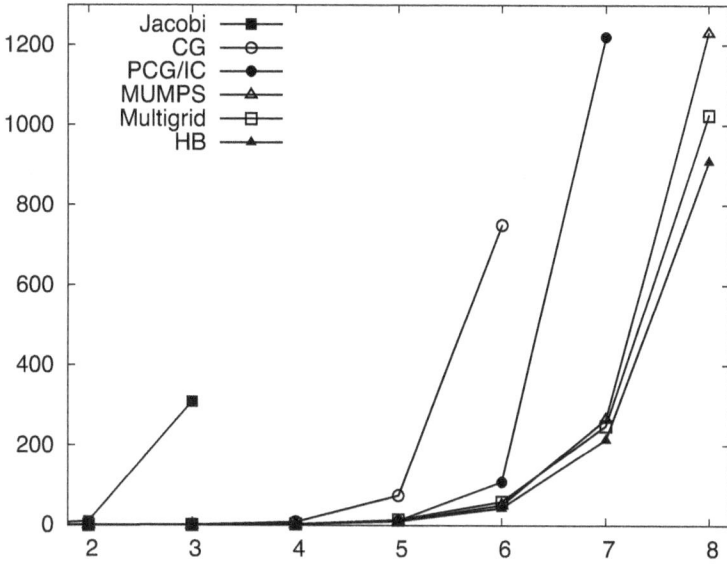

Figure 5.17. Computing times of different methods for the solution of the algebraic adjoint equation (5.62) versus the mesh refinement level $j \sim \log N_j$.

The clear winner among the iterative methods is the classical multigrid method, here realized as a V-cycle with three Jacobi pre- and postsmoothing steps. The direct sparse solver MUMPS is nearly as fast as the multigrid method. Therein the superlinear growth of fill-in elements induces the expected increase of computing time at finer meshes. In total, we measure that direct solution on meshes on different refinement levels exhibit an amount of $\mathcal{O}(N_j^\alpha)$ with $\alpha \approx 1.15$. This effect would be stronger in 3D examples, which would give rise to a higher value of α.

Array Storage.   In Table 5.2 the actual memory required in the course of the program execution is listed for the different methods. As expected, the simple iterative methods require the least storage. In the multigrid methods, however, the prolongation matrices I_k^{k+1} and the projected stiffness matrices $A_k = I_{k+1}^k A_{k+1} I_k^{k+1}$ show up significantly over all refinement levels (geometric series!) by a constant factor of $\mathcal{O}(N_j)$.

Table 5.2. Storage requirement of direct versus iterative solvers when applied to system (5.62). Given is the size before and during the solution process.

refinement	3	4	5	6	7	8
nodes	24 104	95 056	377 504	1 504 576	6 007 424	$24 \cdot 10^6$
CG	33MB	100 MB	425 MB	1.6 GB	6.5 GB	27 GB
PCG/IC	48MB	170 MB	630 MB	2.2 GB	8.9 GB	41 GB
MUMPS	50 MB	205 MB	775 MB	3.4 GB	15 GB	66 GB
multigrid	51 MB	180 MB	700 MB	2.7 GB	10 GB	43 GB

In contrast, the storage requirement of the direct method increases superlinearly. In 3D, due to the denser network of nodes this behavior would show up even stronger.

5.7   Exercises

Exercise 5.1. The bandwidth b_A of a symmetric matrix A is defined as $b_A = \max\{|i - j| : A_{ij} \neq 0\}$. Prove that on a cartesian grid of $n \times n$ nodes in $\mathbb{R}^2$ the stiffness matrix has a bandwidth of at least $n + 1$ for any numbering of the nodes, and $n^2 + n + 1$ for a $n \times n \times n$ grid in $\mathbb{R}^3$.

Exercise 5.2. The simplest linear iterative method is the *Richardson iteration*

$$u_{k+1} - u = u_k - u - \omega(Au_k - b)$$

with $B = \omega^{-1}I$. Show convergence and smoothing property for positive definite A and sufficiently small ω. Determine the optimal choice of the parameter ω, if the extreme eigenvalues $\lambda_{\max}(A), \lambda_{\min}(A)$ are known.

Exercise 5.3. Show that the estimate (5.42) for $k \geq 2$ can be sharpened by a factor of 2. Compute numerically for the Poisson model problem and various matrix decomposition methods sharp upper bounds for the iterative error $\|u - u_k\|_A / \|u - u_k\|_2$ and compare them with the analytic estimates.
   Hint: Generate the iteration matrices G and compute $\|(G^T)^k A G^k\|_2$.

Exercise 5.4. Consider the Gauss–Seidel method with iteration matrix G. Let the matrix A be symmetric positive definite and satisfy the weak diagonal dominance (5.8).

1. Show that under the assumption

$$[G^*, G]_- = G^*G - GG^* = 0 \tag{5.63}$$

the following estimate holds:

$$\|u - u_k\|_A \leq \left(\frac{2}{k+1}\right)^{1/2} \sqrt{\lambda_{\max}(A)}\|u - u_0\|_2.$$

2. Show that the assumption (5.63) cannot be satisfied.

Exercise 5.5. *Smoothing theorem for the Gauss–Seidel method.* Let the matrix A be spd and satisfy the weak diagonal dominance (5.8) as well as the *"property $\mathcal{A}$"*, i.e., there exists a permutation matrix P such that

$$PAP^T = \begin{bmatrix} D_r & L^T \\ L & D_b \end{bmatrix} = \begin{bmatrix} D_r & 0 \\ 0 & D_b \end{bmatrix} \begin{bmatrix} I & \tilde{L}^T \\ \bar{L} & I \end{bmatrix}$$

with positive diagonal matrices D_r, D_b, a rectangular coupling matrix L and the notation $\bar{L} = D_b^{-1}L, \tilde{L}^T = D_r^{-1}L^T$. With $G = G_{GS}, G^*$ from Section 5.2.2 a smoothing theorem will be proved in the following steps:

1. first, show that

$$\|u - u_k\|_A^2 \leq \|AG^{*k}G^k\|_2 \|u - u_0\|_2^2;$$

2. verify the intermediate results

$$G = \begin{bmatrix} 0 & -\tilde{L}^T \\ 0 & \bar{L}\tilde{L}^T \end{bmatrix}, \quad G^k = \begin{bmatrix} 0 & -\tilde{L}^T(\bar{L}\tilde{L}^T)^{k-1} \\ 0 & (\bar{L}\tilde{L}^T)^k \end{bmatrix}$$

as well as

$$G^* = \begin{bmatrix} \tilde{L}^T\bar{L} & 0 \\ -\bar{L} & 0 \end{bmatrix}, \quad G^{*k} = \begin{bmatrix} (\tilde{L}^T\bar{L})^k & 0 \\ -\bar{L}(\tilde{L}^T\bar{L})^{k-1} & 0 \end{bmatrix}$$

and finally

$$AG^{*k}G^k = D \begin{bmatrix} 0 & \tilde{L}^T(\bar{L}\tilde{L}^T)^{2k-1}(-I + \bar{L}\tilde{L}^T) \\ 0 & 0 \end{bmatrix};$$

3. prove the estimate

$$\|u - u_k\|_A \leq \left(\frac{1}{2k}\right)^{1/2} \sqrt{\lambda_{\max}(A)}\|u - u_0\|_2. \tag{5.64}$$

Hint: Introduce the matrices $\overline{C} = D_b^{1/2}\widetilde{L}\widetilde{L}^T D_b^{-1/2}$ as well as $C = I - \overline{C}$ and apply Lemma 5.14.

Exercise 5.6. Transfer Theorem 5.16 on the smoothing property of the CG-method to the PCG-method.

Exercise 5.7. Let $V_H \subset V_h \subset V$ denote two nested FE-spaces with the bases (φ_i^H) and (φ_i^h) and let $A : V \to V^*$ be an elliptic operator. Further let $I_H^h : V_H \to V_h$ denote the matrix representation of the embedding and A_h, A_H the matrix representations of the operator A. Show that

$$A_H = (I_H^h)^T A_h I_H^h.$$

Exercise 5.8. *Nested iteration.* Prove that with nested iteration one can achieve optimal complexity of $\mathcal{O}(N_j)$ instead of (5.55).

Exercise 5.9. Show that the hierarchical basis for $d = 1$ is orthogonal w.r.t. the energy product induced by the Laplace operator and that therefore the PCG-method with hierarchical basis preconditioning converges in *one* iteration.

Exercise 5.10. Instead of the conjugate gradient method for the minimization of the Rayleigh quotient (Section 5.3.4) consider a simple gradient method (which originally had been suggested by other authors). Thus, instead of (5.36), the corrections $p_{k+1} = -g(u_k)$ are used. Develop details of such an algorithm. Show that here, too, a monotone sequence of eigenvalue approximations is obtained. Compare the convergence of this algorithm with that of the variant described in Section 5.3.4.

Chapter 6

Construction of Adaptive Hierarchical Meshes

In the preceding chapters we discussed discretizations on uniform or at least shape regular meshes. In many practically important application problems, however, strongly localized phenomena arise, which can be treated more efficiently on problem adapted hierarchical meshes. Such meshes can be constructed by means of a posteriori estimators of the approximation errors in finite elements.

In many cases, even more important is the determination of the distribution of the actual local discretization errors to assure a sufficient accuracy of the numerical solution.[1] This gives a posteriori error estimators a crucial role in the termination criteria for mesh refinement processes.

In Sections 6.1 and 6.2 we work out the prevalent error estimators in a unified theoretical frame in some detail. On this basis, a subtle strategy of equilibration of discretization errors (Section 6.2.1) supplies local mesh markings, which, by heuristic refinement strategies, are translated into hierarchical meshes (see Section 6.2.2). The thus constructed adaptive meshes can then be combined both with (direct or iterative) solvers of linear algebra and with multigrid methods. For a model adaptive mesh refinement we give an elementary convergence proof (Section 6.3.1). In the subsequent Section 6.3.2 we illustrate how corner singularities are quasi-transformed away by adaptive hierarchical meshes. In the final Section 6.4 we document at a (quadratic) eigenvalue problem for a plasmon-polariton waveguide how far the presented concepts carry over into rough practice.

6.1 A Posteriori Error Estimators

An important algorithmic piece of *adaptive* PDE solvers are error estimators, by means of which *problem adapted hierarchical meshes* can be constructed – a topic that we will treat in the subsequent Section 6.2. In Section 4.4 we presented an approximation theory that described the global discretization error (in the energy norm $\| \cdot \|_a$ or the L^2-Norm $\| \cdot \|_{L^2}$) only on *uniform* or *quasi-uniform* meshes, respectively, via the order w.r.t. a characterizing mesh size h. For realistic problems from engineering and biosciences, however, *problem adapted*, which usually means *nonuniform* meshes are

---

[1] In 1991, the swimming fundament of the oil platform *Sleipner A* leaked and sank. The material broke at a strongly strained point with large mechanical stress, which had not been correctly computed by an FE simulation on a locally too coarse mesh. In the absence of an error estimator the wrong solution had been accepted and the design error had not been corrected. The total damage caused was about 700 million US Dollars [129].

of crucial importance. In this case, a characterization by a single global mesh size h is no longer reasonable; therefore, as in Section 4.3.1, we mean by h some local mesh size, but use an index h in $u_h \in S_h$ as a characterization of the discrete solution.

The error estimators treated in the following are valid for general elliptic boundary value problems. For ease of presentation, however, we pick the example of the Poisson equation with homogeneous Robin boundary conditions

$$-\operatorname{div}(\sigma \nabla u) = f, \quad n^T \sigma \nabla u + \alpha u = 0,$$

in weak formulation written as

$$a(u, v) = \langle f, v \rangle, \quad v \in H^1(\Omega), \tag{6.1}$$

and in discrete weak formulation as

$$a(u_h, v_h) = \langle f, v_h \rangle, \quad v_h \in S_h \subset H^1(\Omega). \tag{6.2}$$

Global Error Estimators.   In actual computation we will only have access to an approximation $\tilde{u}_h$ in lieu of u_h. Consequently, we will have to distinguish the following errors:

- *algebraic error* $\delta_h = \|u_h - \tilde{u}_h\|_a$;
- *discretization error* $\epsilon_h = \|e_h\|_a$ with $e_h = u - u_h$;
- *approximation error* $\tilde{\epsilon}_h = \|\tilde{e}_h\|_a$ with $\tilde{e}_h = u - \tilde{u}_h$.

A usual accuracy check for the numerical solution of the Poisson problem will have the form

$$\tilde{\epsilon}_h \le \text{TOL} \tag{6.3}$$

for given error tolerance TOL. For this special purpose it is sufficient, to compute $\tilde{\epsilon}_h$ approximately, i.e., to *estimate* it approximately. From Theorem 4.3 above we have $a(u_h - \tilde{u}_h, e_h) = 0$, from which we directly obtain

$$\tilde{\epsilon}_h^2 = \epsilon_h^2 + \delta_h^2. \tag{6.4}$$

An efficient possibility to iteratively estimate the *algebraic* error δ_h within a PCG-method was already given in Section 5.3.3. Therefore, we restrict our subsequent attention to an estimation of the discretization error ϵ_h.

In a first seminal paper from 1972 I. Babuška and A. K. Aziz [9] introduced a what is now a generally accepted classification of estimators $[\epsilon_h]$ of discretization errors ϵ_h:

Definition 6.1. An *error estimator* $[\epsilon_h]$ for ϵ_h is called

- *reliable*, if, with a constant $\kappa_1 \ge 1$, one has

$$\epsilon_h \le \kappa_1 [\epsilon_h]. \tag{6.5}$$

If κ_1 is known, then the nonevaluable test (6.3) (with $\delta_h = 0$) can be substituted by the *evaluable check*

$$\kappa_1 [\epsilon_h] \leq \text{TOL},\tag{6.6}$$

from which then, with (6.5), the check (6.3) follows directly;

- *efficient*, if, with some $\kappa_2 \geq 1$, one additionally has

$$[\epsilon_h] \leq \kappa_2 \epsilon_h,$$

which, together with (6.5), can then be combined to

$$\frac{1}{\kappa_1} \epsilon_h \leq [\epsilon_h] \leq \kappa_2 \epsilon_h.\tag{6.7}$$

The product $\kappa_1 \kappa_2 \geq 1$ is also called *efficiency span*, since its deviation from the value 1 is a simple measure of the efficiency of an error estimator;

- *asymptotically exact*, if one additionally has

$$\frac{1}{\kappa_1} \epsilon_h \leq [\epsilon_h] \leq \kappa_2 \epsilon_h \quad \text{with } \kappa_{1,2} \to 1 \text{ for } h \to 0.$$

Localization of Global Error Estimators.   Apart from the obvious purpose of determining the approximation quality of solutions u_h, error estimators can also be used for the construction of problem adapted meshes. This is possible in elliptic problems, as their Green's function (A.13) is strongly localized, i.e., local perturbations of the problem essentially remain local. Vice versa, on the same basis, global errors can be composed from local components, which we will elaborate in the following.

In finite element methods over a triangulation $\mathcal{T}$ there is a natural way to split the global error into its local components via the individual elements $T \in \mathcal{T}$ according to

$$\epsilon_h^2 = \sum_{T \in \mathcal{T}} \epsilon_h(T)^2 = \sum_{T \in \mathcal{T}} a(u - u_h, u - u_h)|_T.\tag{6.8}$$

The exact computation of the local error components $\epsilon_h(T)$ would, similar to the global case, require an amount comparable to the one for the solution of the whole problem. For this reason, one tries to construct local error estimators that are not only "efficient, reliable, and asymptotically exact", but also "cheap": Starting from a given approximation $\tilde{u}_h$ on a given finite element mesh $\mathcal{T}$ one acquires local *a posteriori* error estimates, i.e., error estimates that are locally evaluated *after* the computation of an approximation – in contrast to *a priori* error estimates that are available *before* the computation of an approximation, such as the results in Section 4.4.1.

An alternative option for a localization $\epsilon_h(T)$ to elements $T \in \mathcal{T}$ is via a localization on edges $E \in \mathcal{E}$ (or faces $F \in \mathcal{F}$, respectively) in the form of

$$\epsilon_h^2 = \sum_{E \in \mathcal{E}} \epsilon_h(E)^2$$

is equally useful. However, an interpretation of the kind $\epsilon_h(T)^2 = a(e_h, e_h)|_T$ is then no longer feasible.

Definition 6.2. Local error estimators $[\epsilon_h(T)]$ associated with the local errors $\epsilon_h(T)$ are called *error indicators*, if for two h-independent constants c_1, c_2 the relation

$$\frac{1}{c_1} \epsilon_h(T) \leq [\epsilon_h(T)] \leq c_2 \, \epsilon_h(T), \quad T \in \mathcal{T}.$$

holds. Obviously, this is the localization of the concept of *efficiency* for global error estimators (see (6.7)).

Error Representation.   On some occasions it is useful not only to have an estimate of the *norm* ϵ_h of the error, but also an explicit approximation $[e_h]$ of the discretization error e_h itself. The improved approximation $u_h + e_h$ can, in turn, be used to construct a goal oriented error estimator (see the subsequent Section 6.1.5).

Basic Formula.   In order to estimate the discretization $\epsilon_h = \|e_h\|_a$ we resort to the representation via the dual norm. From the Cauchy–Schwarz inequality

$$a(e_h, v) \leq \|e_h\|_a \|v\|_a \quad \text{for all } v \in H^1(\Omega)$$

we immediately derive the relation

$$\epsilon_h = \sup_{v \in H^1(\Omega), v \neq 0} \frac{a(e_h, v)}{\|v\|_a}. \tag{6.9}$$

The supremum is naturally attained at $v = e_h$, i.e., the exact determination of the supremum would again require the same amount of computation as the exact solution of the original problem (6.1). By some suitable choice of v, however, evaluable bounds of the error can be constructed. Note that the above error representation is localized for now.

6.1.1   Residual Based Error Estimator

This rather popular type of error estimator tries to gain a *global upper bound* for all v in (6.9). The derivation of this bound is a direct consequence of Corollary 4.1 above. There we had introduced the flow condition (4.13); it states that the normal components of the flow for the exact solution u are continuous at the internal faces (or edges) and thus their jumps $[\![n^T \sigma \nabla u]\!]_\Gamma$ vanish. As illustrated in Figure 4.10, this does not hold for the discrete solution u_h. Therefore the jumps of the discrete normal flows will constitute an essential part of the error.

Localization.   This is achieved in two steps: as u_h is differentiable sufficiently often only on isolated elements of the triangulation, we first apply Green's Theorem (A.7) to individual elements $T \in \mathcal{T}$ and sum up as follows:

$$a(e_h, v) = \sum_{T \in \mathcal{T}} a_T(e_h, v) \tag{6.10}$$

$$= \sum_{T \in \mathcal{T}} \left[-\int_T (\operatorname{div}(\sigma \nabla u_h) + f) v \, dx + \int_{\partial T} v \, n^T \sigma \nabla e_h \, ds \right] + \int_{\partial \Omega} \alpha v e_h \, ds.$$

After that we sort the integrals w.r.t. ∂T to obtain a sum of integrals over faces $F \in \mathcal{F}$ (or edges $E \in \mathcal{E}$). Let T_1, T_2 denote two arbitrarily chosen neighboring elements with common face $F \in \mathcal{F}$. Then we arrive at local terms of the form

$$-\int_F v n^T (\sigma \nabla e_h|_{T_1} - \sigma \nabla e_h|_{T_2}) \, ds = \int_F v [\![n^T \sigma \nabla u_h]\!]_F \, ds$$

for internal faces. For boundary faces $F \subset \partial \Omega$ we obtain integrals of the form

$$\int_F v (n^T \sigma \nabla u_h + \alpha u_h) \, ds,$$

so that we may, in order to unify the notation, define "boundary jumps" according to

$$[\![n^T \sigma \nabla u_h]\!]_F = n^T \sigma \nabla u_h + \alpha u_h.$$

Summarizing, we thus obtain the localized error representation

$$a(e_h, v) = \sum_{T \in \mathcal{T}} \int_T (\operatorname{div}(\sigma \nabla u_h) + f) v \, dx + \sum_{F \in \mathcal{F}} \int_F v \, [\![n^T \sigma \nabla u_h]\!]_F \, ds.$$

Applying the Cauchy–Schwarz inequality to all local integrals yields the estimate

$$a(e_h, v) \leq \sum_{T \in \mathcal{T}} \| \operatorname{div}(\sigma \nabla u_h) + f \|_{L^2(T)} \| v \|_{L^2(T)}$$

$$+ \sum_{F \in \mathcal{F}} \| [\![n^T \sigma \nabla u_h]\!]_F \|_{L^2(F)} \| v \|_{L^2(F)}.$$

Due to (4.14) we have $a(e_h, v) = a(e_h, v - v_h)$ for all $v_h \in S_h$, so that the term $\| v \|$ above can be replaced by $\| v - v_h \|$. For the purpose of theory we now choose a v_h such that the contributions $\| v - v_h \|_{L^2(T)}$ and $\| v - v_h \|_{L^2(F)}$ are as small as possible. In fact, for each $v \in H^1(\Omega)$ there exists an (even computable) quasi-interpolant $v_h \in S_h$ with

$$\sum_{T \in \mathcal{T}} h_T^{-2} \| v - v_h \|_{L^2(T)}^2 \leq c^2 |v|_{H^1(\Omega)}^2 \quad \text{and} \quad \sum_{F \in \mathcal{F}} h_F^{-1} \| v - v_h \|_{L^2(F)}^2 \leq c^2 |v|_{H^1(\Omega)}^2, \tag{6.11}$$

where h_T and h_F are again the diameters of T and F and c a constant depending only on the shape regularity of the triangulation (see, e.g., [186]). In view of this we get

$$a(e_h, v) = a(e_h, v - v_h)$$

$$\leq \sum_{T \in \mathcal{T}} h_T \| \operatorname{div}(\sigma \nabla u_h) + f \|_{L^2(T)} \, h_T^{-1} \| v - v_h \|_{L^2(T)}$$

$$+ \sum_{F \in \mathcal{F}} h_F^{1/2} \| [\![n^T \sigma \nabla u_h]\!]_F \|_{L^2(F)} \, h_F^{-1/2} \| v - v_h \|_{L^2(F)}.$$

If we define local error estimators over T, F as

$$\eta_T := h_T \| \operatorname{div}(\sigma \nabla u_h) + f \|_{L^2(T)}, \quad \eta_F := h_F^{1/2} \| [\![n^T \sigma \nabla u_h]\!]_F \|_{L^2(F)},$$

then a second application of the Cauchy–Schwarz inequality, this time in finite dimensions, eventually leads to

$$a(e_h, v) \le \Big(\sum_{T \in \mathcal{T}} \eta_T^2 \Big)^{\frac{1}{2}} \Big(\sum_{T \in \mathcal{T}} h_T^{-2} \| v - v_h \|_{L^2(T)}^2 \Big)^{\frac{1}{2}}$$

$$+ \Big(\sum_{F \in \mathcal{F}} \eta_F^2 \Big)^{\frac{1}{2}} \Big(\sum_{F \in \mathcal{F}} h_F^{-1} \| v - v_h \|_{L^2(E)}^2 \Big)^{\frac{1}{2}}$$

$$\le 2c |v|_{H^1(\Omega)} \Big(\sum_{T \in \mathcal{T}} \eta_T^2 + \sum_{F \in \mathcal{F}} \eta_F^2 \Big)^{\frac{1}{2}}.$$

Due to the positivity of A there exists an $\alpha > 0$ such that $|v|_{H^1(\Omega)} \le \alpha \|v\|_a$. In the above inequality we thus replace $|v|_{H^1(\Omega)}$ by $\|v\|_a$ and swallow the constant α in the generic constant c. After this preparation, dividing by $\|v\|_a$ we can return to the basic formula (6.9) and eventually arrive at the classical error estimator due to I. Babuška and A. Miller [11]:

BM Error Estimator

$$[\epsilon_h] = 2c \Big(\sum_{T \in \mathcal{T}} \eta_T^2 + \sum_{F \in \mathcal{F}} \eta_F^2 \Big)^{\frac{1}{2}} \ge \epsilon_h. \tag{6.12}$$

By derivation, this is a global upper bound of the actual error. In order to algorithmically realize an elementwise localization of this error estimator in the sense of (6.8), the localized jump contributions at the faces $F \in \mathcal{F}$ must be redistributed to the elements $T \in \mathcal{T}$. One therefore defines

$$[\epsilon_h(T)]^2 = 4c^2 \alpha^2 \Big(\eta_T^2 + \sum_{F \in \mathcal{F}, F \subset \partial T} \alpha_{T,F} \, \eta_F^2 \Big), \tag{6.13}$$

where $\alpha_{T_1, F} + \alpha_{T_2, F} = 1$ for two elements T_1 and T_2 adjacent to some F. With the argument that the two adjacent elements must have the same share of the jump on F, mostly $\alpha_{T,F} = 1/2$ is set (see [39]). This argument is, at best, valid for uniform meshes, continuous diffusion coefficients and homogeneous approximations over the whole triangulation $\mathcal{T}$ (cf. the voluminous analysis by M. Ainsworth and J. T. Oden [3], which, however, offers only 1D-arguments for a locally variable choice of these coefficients).

The BM error estimator (6.12) is – in the presence of knowledge about the generic constant c – "reliable" in the sense of (6.5) with $\kappa_1 = 1$. However, the constant c is usually rather inaccessible, so that the check (6.6) performed with the above error estimator is indeed *not reliable* (see [19]). Moreover, the formal reliability of the error estimator (6.12) is based on a sufficiently accurate computation of the terms η_T and η_F. While η_F contains only well-known quantities of the discretization, the computation of η_T requires an integration of f, which is usually realized via numerical quadrature (cf. Volume 1, Section 9). It can be shown, however (see R. Verfürth [209]), that for some class of problems the terms η_T can be dropped. For a characterization of this class we require the following definition, which dates back to W. Dörfler [79].

Definition 6.3 (Oscillation). Let ω_ξ denote some star-shaped domain around a node $\xi \in \mathcal{N}$ and

$$\bar{f}_\xi = \frac{1}{|\omega_\xi|} \int_{\omega_\xi} f \, dx$$

the corresponding average value of f. Then the *oscillation* of f is given by

$$\operatorname{osc}(f; \mathcal{T})^2 = \sum_{\xi \in \mathcal{N}} h_\xi^2 \sum_{T \in \mathcal{T} : \xi \in T} \|f - \bar{f}_\xi\|_{L^2(T)}^2,$$

where h_ξ again denotes the diameter of ω_ξ.

Theorem 6.4. *There exists a constant c independent of the local mesh size h such that for solutions u_h of (6.2) the following result holds:*

$$\sum_{T \in \mathcal{T}} \eta_T^2 \leq c \left(\sum_{F \in \mathcal{F}} \eta_F^2 + \operatorname{osc} \left(\operatorname{div}(\sigma \nabla u_h) + f ; \mathcal{T} \right)^2 \right). \tag{6.14}$$

Proof. The piecewise residual is given by $r = \operatorname{div}(\sigma \nabla u_h) + f \in L^2(\Omega)$ in the interior of each element $T \in \mathcal{T}$. Let $\xi \in \mathcal{N}$ with associated nodal function $\phi_\xi \in S_h$. We define $\mathcal{F}_\xi = \{F \in \mathcal{F} : \xi \in F\}$. Because of (6.2) we have, after an elementary application of the Gaussian theorem,

$$\int_\Omega f \phi_\xi \, dx = \int_\Omega \nabla u_h^T \sigma \nabla \phi_\xi \, dx$$

$$= \sum_{F \in \mathcal{F}_\xi} \int_F \phi_\xi [\![n^T \sigma \nabla u_h]\!]_F \, ds - \int_\Omega \phi_\xi \operatorname{div}(\sigma \nabla u_h) \, dx$$

and, by subsequent application of the Cauchy–Schwarz inequality in finite dimensions, eventually

$$\int_\Omega r \phi_\xi \, dx = \sum_{F \in \mathcal{F}_\xi} \int_F \phi_\xi [\![n^T \sigma \nabla u_h]\!]_F \, ds \leq \sum_{F \in \mathcal{F}_\xi} \|\phi_\xi\|_{L^2(F)} \|[\![n^T \sigma \nabla u_h]\!]_F\|_{L^2(F)}$$

$$\leq \left(\sum_{F \in \mathcal{F}_\xi} h_F^{d-2} \right)^{1/2} \left(\sum_{F \in \mathcal{F}_\xi} h_F \|[\![n^T \sigma \nabla u_h]\!]_F\|_{L^2(F)}^2 \right)^{1/2}$$

$$\leq c h_\xi^{d/2-1} \left(\sum_{F \in \mathcal{F}_\xi} \eta_F^2 \right)^{1/2}. \tag{6.15}$$

By the averaging

$$\bar{r}_\xi = \frac{1}{|\omega_\xi|} \int_{\omega_\xi} r \, dx \leq \frac{\|1\|_{L^2(\omega_\xi)}}{|\omega_\xi|} \|r\|_{L^2(\omega_\xi)} = \frac{\|r\|_{L^2(\omega_\xi)}}{|\omega_\xi|^{1/2}} \tag{6.16}$$

we have (this part of the proof has been sourced out to Exercise 6.1c)

$$\int_{\omega_\xi} 2r(1-\phi_\xi)\bar{r}_\xi \, dx \leq \frac{2\|1-\phi_\xi\|_{L^2(\omega_\xi)}}{|\omega_\xi|^{1/2}} \|r\|_{L^2(\omega_\xi)} \, |\omega_\xi|^{1/2} \bar{r}_\xi \leq \|r\|_{L^2(\omega_\xi)}^2$$

and thus

$$\|\bar{r}_\xi\|_{L^2(\omega_\xi)}^2 \leq \int_{\omega_\xi} (r^2 + 2r(\phi_\xi-1)\bar{r}_\xi + \bar{r}_\xi^2) \, dx = \int_{\omega_\xi} (2r\phi_\xi\bar{r}_\xi + (r-\bar{r}_\xi)^2) \, dx.$$

Finally, we get, due to the L^2-orthogonality of the averaging

$$\|r\|_{L^2(\omega_\xi)}^2 = \|r - \bar{r}_\xi\|_{L^2(\omega_\xi)}^2 + \|\bar{r}_\xi\|_{L^2(\omega_\xi)}^2 \leq 2\|r - \bar{r}_\xi\|_{L^2(\omega_\xi)}^2 + 2\int_{\omega_\xi} r\phi_\xi\bar{r}_\xi \, dx.$$

From (6.16) we obtain

$$\bar{r}_\xi \leq c h_\xi^{-d/2} \|r\|_{L^2(\omega_\xi)},$$

which with (6.15) gives rise to the quadratic inequality

$$\|r\|_{L^2(\omega_\xi)}^2 \leq \beta \|r\|_{L^2(\omega_\xi)} + \gamma,$$

$$\beta = c h_\xi^{-1} \Big(\sum_{F \in \mathscr{F}_\xi} \eta_F^2 \Big)^{1/2}, \quad \gamma = 2\|r - \bar{r}_\xi\|_{L^2(\omega_\xi)}^2.$$

Its roots are given by

$$r_\pm = \tfrac{1}{2}\Big(\beta \pm \sqrt{\beta^2 + 4\gamma} \Big)$$

so that we have for the solutions

$$4\|r\|_{L^2(\omega_\xi)}^2 \leq \big(\beta + \sqrt{\beta^2 + 4\gamma} \big)^2 = \beta^2 + 2\beta\sqrt{\beta^2 + 4\gamma} + \beta^2 + 4\gamma$$

$$\leq 2\beta^2 + 2\beta^2\Big(1 + \frac{2\gamma}{\beta^2} \Big) + 4\gamma = 4(\beta^2 + \gamma)$$

$$\leq c\Big(h_\xi^{-2} \sum_{F \in \mathscr{F}_\xi} \eta_F^2 + \|r - \bar{r}_\xi\|_{L^2(\omega_\xi)}^2 \Big).$$

Multiplication by h_ξ^2 and summation over $\xi \in \mathcal{N}$ then yields (6.14).   □

Simplified BM Error Estimator.   If the oscillation of the right-hand side f can be neglected, then, due to (6.14), the volume contribution in (6.12) can be replaced by the jump contribution. One thus obtains, again by neglecting the generic constants,

$$[\epsilon_h] = \Big(\sum_{F \in \mathscr{F}} \eta_F^2 \Big)^{\frac{1}{2}}. \tag{6.17}$$

This error estimator is indeed reliable in the sense of (6.5). (The associated proof has been postponed until Exercise 7.2.)

Remark 6.5. Strictly speaking, the a priori restriction of the oscillation in Theorem 6.4 is not sufficient as a theoretical basis for the *practical* reliability of the error estimator (6.17): This theorem is based on the fact that u_h satisfies the weak formulation (6.2), which includes the presumption that the integrals $\langle f, v_h \rangle$ have been computed to sufficient accuracy by numerical quadrature. Even for functions f with small oscillation, this can be guaranteed only under additional smoothness assumptions.

Remark 6.6. In [11], I. Babuška and A. Miller presented their theoretical analysis not, as presented here, via the quasi-interpolant (6.11), but by a decomposition into local Dirichlet problems. A modification of such a decomposition will come up again in the immediately following Section 6.1.4, there, however, as an algorithmically realized option.

6.1.2   Triangle Oriented Error Estimators

The unsatisfactorily rough estimate (6.12) originates from the attempt to find a direct upper bound for all v in (6.9) by means of *a priori*-estimates. This insight motivates the alternative, to construct *a posteriori* a special $[e_h] \approx e_h$. For this purpose, the ansatz space S_h must be suitably expanded to a space S_h^+, which in the context of finite elements can be conveniently localized.

Localization.   Following a suggestion of R. Bank and A. Weiser [19] from 1985, we return to the decomposition (6.10), but exploit the information in some way different from the one in the previous Section 6.1.1: in this approach, we may immediately read that the localized error $e_T = e_h|_T$ of the inhomogeneous Neumann problem

$$a_T(e_T, v) = b_T(v; n^T \sigma \nabla e_h) \quad \text{for all } v \in H^1(T) \tag{6.18}$$

with right-hand side

$$b_T(v; g) = \int_T (\operatorname{div}(\sigma \nabla u_h) + f) v \, dx - \int_{\partial T} vg \, ds$$

is all that is needed. This includes that, as wanted, $\epsilon_T^2 = a_T(e_T, e_T)$. For the purpose of error estimation again an approximate computation of e_T is sufficient. Due to the small size of the domain T it can be elementarily realized by some *locally expanded* ansatz space $S_h^+(T) = S_h(T) \oplus S_h^{\oplus}(T) \subset H^1(\Omega)|_T$. Typical examples under application of a polynomial ansatz space $S_h = S_h^p$ are (a) a uniform refinement $S_h^+ = S_{h/2}^p$ or (b) an increase of the polynomial order $S_h^+ = S_h^{p+1}$ (see Figure 6.1 for $p = 1$).

The Neumann problems over the individual triangles $T \in \mathcal{T}$ are, however, unique only up to an additive constant (see Theorem 1.3), and must, in addition, meet a compatibility condition similar to (1.3). Therefore we focus our attention to ansatz spaces

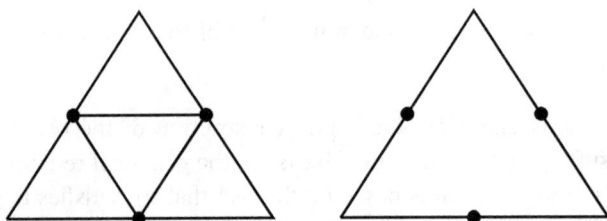

Figure 6.1. Choice of expansion $S_h^{\oplus}(T)$ of the ansatz space for linear finite elements in 2D. *Left:* piecewise linear expansion. *Right:* quadratic expansion.

$S_h^{\oplus}(T)$ that exclude constant functions; thus the bilinear form $a(u,v)$ is elliptic on $S_h^{\oplus}(T)$ (independent of the presence of a Helmholtz term).

Let $e_T^{\oplus} \in S_h^{\oplus}(T)$ be defined as solution of

$$a_T(e_T^{\oplus}, v) = b_T(v; n^T \sigma \nabla e_h) \quad \text{for all } v \in S_h^{\oplus}(T). \tag{6.19}$$

Because of the Galerkin orthogonality (4.14) we immediately obtain the *lower* bound

$$a(e_T^{\oplus}, e_T^{\oplus}) \leq \epsilon_T^2.$$

A direct application of (6.19) to an error estimate fails, since, due to (4.13), the continuous normal flow $n^T g = n^T \sigma \nabla u$ as well as the noncontinuous normal flow $n^T \sigma \nabla e_h = n^T(g - \sigma \nabla u_h)$ are unknown, which requires the construction of an approximation g_h. Just as in the localization of residual based error estimators (see the choice $\alpha_{T,F} = 1/2$ in (6.13)), the simplest, though arbitrary choice of g_h is the *averaging*

$$g_h|_E(x) = \frac{1}{2} \lim_{\epsilon \to 0} \left(\sigma(x + \epsilon n) u_h(x + \epsilon n) + \sigma(x - \epsilon n) u_h(x - \epsilon n) \right). \tag{6.20}$$

This leads to

$$n^T(g_h - \sigma \nabla u_h) = \frac{1}{2} [\![n^T \sigma \nabla u_h]\!].$$

BW Error Estimator.   The local error approximation $[e_T]$ is then defined by

$$a_T([e_T], v) = b_T\left(v; \frac{1}{2} [\![n^T \sigma \nabla u_h]\!]\right) \quad \text{for all } v \in S_h^{\oplus}(T),$$

with the associated local error estimator

$$[\epsilon_T] = \| [e_T] \|_a. \tag{6.21}$$

For linear finite elements this requires a local extension $S_h^{\oplus}(T)$ by quadratic shape functions apart from the evaluation of the jumps along edges and the strong residual $\text{div}(\sigma \nabla u_h) + f$ as well as the solution of local equation systems of dimension $\frac{1}{2} d(d+1)$.

Error Representation.   By composition of the single partial solutions $[e_T]$ one obtains a global approximation $[e_h] \approx e_h$ of the error. One has to observe, however, that in general $[e_h]$ exhibits jumps beyond edges and faces and therefore cannot be contained in $H^1(\Omega)$. A refinement of the solution by the ansatz $u_h + [e_h] \approx u$ would lead to nonconformal solutions.

Reliability.   The restriction to error approximations in a finite dimensional subspace contains the risk to overlook part of the error: It is easy to construct examples, for which $b(v; \frac{1}{2}[n^T \sigma \nabla u_h]) = 0$ holds for all $v \in S_h^{\oplus}(T)$, even though b itself does not vanish. For this reason, the verification of reliability makes the following saturation assumption necessary.

Definition 6.7. *Saturation assumption.*   Let $\hat{S}_h^+ = \Pi_{T \in \mathcal{T}} S_h^+(T)$ and $S_h^+ = \hat{S}_h^+ \cap H^1(\Omega)$ denote an extended conformal ansatz space with corresponding Galerkin solution $u_h^+ \in S_h^+$ and error ϵ_h^+. Then there exists a $\beta < 1$ such that

$$\epsilon_h^+ \leq \beta \epsilon_h. \tag{6.22}$$

It should be explicitly mentioned that this saturation assumption is, in fact, a central *assumption*, since for each given space extension arbitrarily many right-hand sides f may be constructed such that $u_h^+ = u_h$ holds (see Exercise 7.2). Nevertheless it is useful in practice, since for each *fixed* and reasonably smooth right-hand side f it is satisfied on sufficiently fine meshes. In particular, the extension of linear to quadratic elements on simplicial triangulations satisfies the saturation assumption, if the oscillation of f is small, as shown by the following theorem of W. Dörfler and R. H. Nocchetto [80], which we present here without proof.

Theorem 6.8. *Let $S_h = S_h^1$ and $S_h^+ = S_h^2 \subset H_0^1(\Omega)$. Then there exists a constant $\mu < 1$, which only depends on the shape regularity of the triangulation $\mathcal{T}$, so that a small oscillation*

$$\mathrm{osc}(f; \mathcal{T}) \leq \mu \epsilon_h$$

implies the saturation assumption (6.22) *with $\beta = \sqrt{1 - \mu^2}$.*

After employing this assumption, we are now ready to verify that $e_h^{\oplus} = u_h^+ - u_h \in S_h^+$ supplies a reliable error approximation $\epsilon_h^{\oplus} = \|e_h^{\oplus}\|_a$.

Theorem 6.9. *Under the saturation assumption* (6.22) *the following result holds:*

$$\epsilon_h^{\oplus} \leq \epsilon_h \leq \frac{\epsilon_h^{\oplus}}{\sqrt{1 - \beta^2}}. \tag{6.23}$$

Proof. Based on the orthogonality of the error e_h^+ in the ansatz space S_h^+ we have

$$\|e_h\|_a^2 = \|e_h^+\|_a^2 + \|e_h^{\oplus}\|_a^2 \geq \|e_h^{\oplus}\|_a^2,$$

which confirms the lower bound in (6.23). For a verification of the upper bound we
need the saturation assumption:

$$\|e_h\|_a^2 = \|e_h^+\|_a^2 + \|e_h^\oplus\|_a^2 \leq \beta^2 \|e_h\|_a^2 + \|e_h^\oplus\|_a^2.$$

By repositioning of terms we immediately get

$$(1 - \beta^2)\|e_h\|_a^2 \leq \|e_h^\oplus\|_a^2,$$

which is the missing inequality.   □

Now, by comparison with $\epsilon_h^\oplus$, the reliability of the estimator (6.21) can be shown.

Theorem 6.10. *Let, in addition to the assumption* (6.22), *some* $\gamma < 1$ *exist such that
the strengthened Cauchy–Schwarz inequality*

$$a_T(v_h, v_h^\oplus) \leq \gamma \|v_h\|_{a_T} \|v_h^\oplus\|_{a_T}$$

is satisfied for all $T \in \mathcal{T}$ *and* $v_h \in S_h(T)$, $v_h^\oplus \in S_h^\oplus(T)$. *Then*

$$(1 - \gamma^2)\sqrt{1 - \beta^2}\,\epsilon_h \leq [\epsilon_h]. \tag{6.24}$$

Proof. The proof is done in two steps via auxiliary quantities $\hat{e}_T \in S_h^+(T)$ defined by

$$a_T(\hat{e}_T, v) = b_T(I_h^\oplus v) \quad \text{for all } v \in S_h^+(T), \quad \int_T \hat{e}_T \, dT = 0,$$

where we dropped the flow approximation as parameters in b_T. Here $I_h^\oplus : S_h^+(T) \to
S_h^\oplus(T)$ is the projector uniquely defined by the decomposition $S_h^+(T) = S_h(T) \oplus
S_h^\oplus(T)$; in addition, we define $I_h = I - I_h^\oplus$. We first have

$$a_T([e_T], v) = b_T(v) = b_T(I_h^\oplus v) = a_T(\hat{e}_T, v) \quad \text{for all } v \in S_h^\oplus(T).$$

Due to $I_h^\oplus \hat{e}_T \in S_h^\oplus$ we now obtain

$$\|[e_T]\|_{a_T} \|I_h^\oplus \hat{e}_T\|_{a_T} \geq a_T([e_T], I_h^\oplus \hat{e}_T) = a_T(\hat{e}_T, I_h^\oplus \hat{e}_T).$$

We have

$$a_T(\hat{e}_T, I_h \hat{e}_T) = b_T(I_h^\oplus I_h \hat{e}_T) = 0 \tag{6.25}$$

and with the strengthened Cauchy–Schwarz inequality

$$\begin{aligned}
\|[e_T]\|_{a_T} \|I_h^\oplus \hat{e}_T\|_{a_T} &\geq a_T(\hat{e}_T, I_h \hat{e}_T) + a_T(\hat{e}_T, I_h^\oplus \hat{e}_T) \\
&= \|I_h \hat{e}_T\|_{a_T}^2 + 2a_T(I_h \hat{e}_T, I_h^\oplus \hat{e}_T) + \|I_h^\oplus \hat{e}_T\|_{a_T}^2 \\
&\geq \|I_h \hat{e}_T\|_{a_T}^2 - 2\gamma \|I_h \hat{e}_T\|_{a_T} \|I_h^\oplus \hat{e}_T\|_{a_T} + \|I_h^\oplus \hat{e}_T\|_{a_T}^2 \\
&= (\|I_h \hat{e}_T\|_{a_T} - \gamma \|I_h^\oplus \hat{e}_T\|_{a_T})^2 + (1 - \gamma^2)\|I_h^\oplus \hat{e}_T\|_{a_T}^2 \\
&\geq (1 - \gamma^2)\|I_h^\oplus \hat{e}_T\|_{a_T}^2.
\end{aligned}$$

Division by $\|I_h^\oplus \hat{e}_T\|_{a_T}$ leads to the estimate $(1 - \gamma^2)\|I_h^\oplus \hat{e}_T\|_{a_T} \leq \|[e_T]\|_{a_T}$. From (6.25) we may conclude that

$$\|\hat{e}_T\|_{a_T}^2 = a_T(\hat{e}_T, I_h^\oplus \hat{e}_T) \leq \|\hat{e}_T\|_{a_T}\|I_h^\oplus \hat{e}_T\|_{a_T},$$

so that in total we get

$$(1 - \gamma^2)\|\hat{e}_T\|_{a_T} \leq \|[e_T]\|_{a_T} . \tag{6.26}$$

In the second step we turn to global error estimation. Simple summation over the elements $T \in \mathcal{T}$ leads to the global error approximation $\hat{e}_h \in \hat{S}_h^+$ with $\hat{e}_h|_T = \hat{e}_T$ and

$$a(\hat{e}_h, v) = b(I_h^\oplus v) \quad \text{for all } v \in \hat{S}_h^+ \supset S_h^+.$$

Note that the summation of b_T leads directly to the localization (6.10), which is why $a(e_h, v) = b(v)$ holds for all $v \in H^1(\Omega)$, and, with the Galerkin orthogonality $a(e_h^+, v) = 0$ for all $v \in S_h^+$, also

$$a(e_h^\oplus, v) = a(e_h - e_h^+, v) = b(v) \quad \text{for all } v \in S_h^+.$$

Because of $I_h^\oplus e_h^\oplus \in S_h^+$ we thus obtain

$$\|\hat{e}_h\|_a \|e_h^\oplus\|_a \geq a(\hat{e}_h, e_h^\oplus) = b(I_h^\oplus e_h^\oplus) = a(e_h^\oplus, I_h^\oplus e_h^\oplus)$$
$$= a(e_h^\oplus, e_h^\oplus - I_h e_h^\oplus) = a(e_h^\oplus, e_h^\oplus) = \|e_h^\oplus\|_a^2$$

and, after division by $\|e_h^\oplus\|_a$ with Theorem 6.9

$$\|\hat{e}_h\|_a \geq \|e_h^\oplus\|_a \geq \sqrt{1 - \beta^2}\, \epsilon_h. \tag{6.27}$$

The combination of (6.26) and (6.27) then leads to the assertion (6.24). $\qquad\square$

Remark 6.11. The *triangle oriented* error estimator due to R. Bank and A. Weiser [19] from 1985 was the first error estimator in which a localization of the quadratic ansatz space was realized. For a long time it remained the basis of adaptivity in the finite element code PLTMG [16], but has been replaced in newer versions by the gradient recovery treated in the next Section 6.1.3. The auxiliary quantity $\hat{e}_h$ used in the proof of Theorem 6.10 was also considered in the paper [19] as a possible error estimator. It gives rise to a better constant κ_1 in the estimate for the reliability, but has shown to be less efficient in practical tests.

6.1.3   Gradient Recovery

We return to the averaging (6.20) and try to replace this rather simple approximation of the discontinuous discrete flows $g_h = \sigma \nabla u_h \approx \sigma \nabla u = g$ by an improved version called gradient recovery. For that purpose, we project, as shown in Figure 6.2, the gradient $g_h \in S_{h,\text{grad}} = \{\nabla v : v \in S_h\}$ by some suitable projector Q onto an

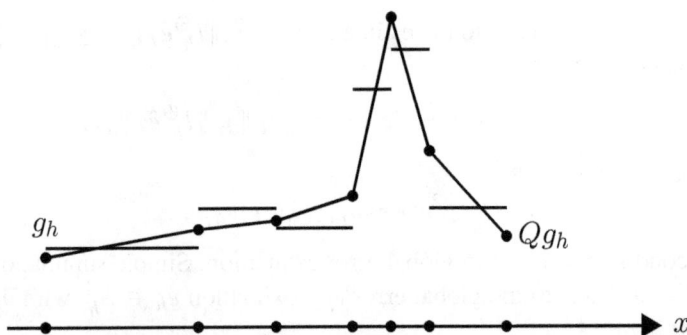

Figure 6.2. Gradient recovery for linear finite elements in $\mathbb{R}^1$ via L^2-projection.

ansatz space $\bar{S}_h$ of *continuous* functions. For the time being, we focus our attention on continuous diffusion coefficients, where both the exact gradients ∇u and the flows g are continuous. So there is no conceptual difference between a projection of the flows and one of the gradients.

We consider the homogeneous Neumann problem with

$$\|e_h\|_a = \|\sigma^{1/2}(\nabla u - \nabla u_h)\|_{L^2(\Omega)}$$

and $\alpha = 0$. Then we replace ∇u by

$$Q\nabla u_h = \arg\min_{q_h \in \bar{S}_h} \|\sigma^{1/2}(\nabla u_h - q_h)\|_{L^2(\Omega)},$$

so that we have for all $q_h \in \bar{S}_h$:

$$\|\sigma^{1/2}(\nabla u_h - Q\nabla u_h)\|_{L^2(\Omega)} \leq \|\sigma^{1/2}(\nabla u_h - q_h)\|_{L^2(\Omega)}$$
$$\leq \|\sigma^{1/2}(\nabla u - \nabla u_h)\|_{L^2(\Omega)} + \|\sigma^{1/2}(\nabla u - q_h)\|_{L^2(\Omega)}$$
$$= \epsilon_h + \|\sigma^{1/2}(\nabla u - q_h)\|_{L^2(\Omega)}. \tag{6.28}$$

First, we neglect the second term $\|\sigma^{1/2}(\nabla u - q_h)\|_{L^2(\Omega)}$ in the sum and obtain

$$[\epsilon_h] = \|\sigma^{1/2}(\nabla u_h - Q\nabla u_h)\|_{L^2(\Omega)}. \tag{6.29}$$

Localization.   From (6.29) we may recognize a retrospective localization immediately. We obtain the error indicators

$$[\epsilon_h(T)] = \|\sigma^{1/2}(\nabla u_h - Q\nabla u_h)\|_{L^2(T)}.$$

Reliability.   The reliability of the error estimator (6.29) is a direct consequence of Theorem 6.4 and of the following theorem.

Theorem 6.12. *There exists a constant c independent of h and f such that for (6.29) the result*

$$\sum_{F \in \mathcal{F}} \eta_F^2 \leq c[\epsilon_h]^2 \tag{6.30}$$

holds.

Proof. For some $\xi \in \mathcal{N}$ we consider the quotient spaces

$$S_\xi = (S_{h,\text{grad}}(\omega_\xi) + \bar{S}_h(\omega_\xi))/\bar{S}_h(\omega_\xi)$$

and on them the seminorm

$$|\nabla u_h|_\xi^2 = \sum_{F \in \mathcal{F}, F \subset \omega_\xi} h_F \|[\![n^T \nabla u_h]\!]_F\|_{L^2(F)}^2$$

as well as the norm

$$\|\nabla u_h\|_\xi = \min_{q_h \in \bar{S}_h} \|\sigma^{1/2}(\nabla u_h - q_h)\|_{L^2(\omega_\xi)}.$$

Because of the bounded dimension of S_ξ there exists a constant c depending on the shape regularity of the elements $T \subset \omega_\xi$, but not on h, with $|\nabla u_h|_\xi \leq c\|\nabla u_h\|_\xi$. Summation over all $\xi \in \mathcal{N}$ leads, due to the bounded overlap of Ω with ω_ξ, to the inequality (6.30).   □

As in the simplified BM error estimator (6.17) the reliability here is only given under the a priori assumption of a small oscillation of f and for exact computation of the scalar products.

Efficiency.   If the neglected term $\|\sigma^{1/2}(\nabla u - q_h)\|_{L^2(\Omega)}$ from (6.28) can be suitably bounded, then the error estimator (6.29) is a lower bound of the error and thus efficient. For this we need a variant of the saturation assumption (6.22).

Theorem 6.13. *Suppose there exists an h-independent constant $\hat{\beta}$ such that*

$$\min_{q_h \in \bar{S}_h} \|\sigma^{1/2}(\nabla u - q_h)\|_{L^2(\Omega)} \leq \hat{\beta}\epsilon_h \tag{6.31}$$

holds. Then $[\epsilon_h] \leq (1 + \hat{\beta})\epsilon_h$.

Proof. Let $w_h = \arg\min_{q_h \in \bar{S}_h} \|\sigma^{1/2}(\nabla u - q_h)\|_{L^2(\Omega)} \in \bar{S}_h$. Then $[\epsilon_h] \leq \epsilon_h + \|\sigma^{1/2}(\nabla u - w_h)\|_{L^2(\Omega)} \leq \epsilon_h + \hat{\beta}\epsilon_h$ holds due to (6.28) and the saturation assumption.   □

For piecewise polynomial finite elements $u_h \in S_h^p$ the gradient ∇u_h lies in

$$S_{h,\text{grad}}^p = \{v \in L^2(\Omega)^d : v|_T \in \mathbb{P}_{p-1} \text{ for all } T \in \mathcal{T}\}.$$

In order to assure the saturation assumption asymptotically for sufficiently smooth u, i.e., for $h \to 0$ with $\kappa_2 \to 1$ in Definition 6.1, the order for $\bar{S}_h = (S_h^p)^d$ is increased – for linear elements this order increase is anyway necessary to achieve continuity.

Error Representation.   As in the residual based error estimator (6.13), we cannot directly deduce an error approximation $[e_h]$ from (6.29). But we have the defect equation

$$- \operatorname{div}(\sigma \nabla e_h) = - \operatorname{div}(\sigma \nabla u - \sigma \nabla u_h) \approx - \operatorname{div}(\sigma(Q \nabla u_h - \nabla u_h)),$$

whose localization to individual elements T similar to (6.18) leads to

$$a_T(e_T, v) = \int_T \left(\operatorname{div}(\sigma \nabla u_h) - \operatorname{div}(\sigma Q \nabla u_h) \right) v \, dx - \int_{\partial T} v n^T \sigma(Q \nabla u_h - \nabla u_h) \, ds. \tag{6.32}$$

The replacement of $f = - \operatorname{div}(\sigma \nabla u)$ by $- \operatorname{div}(\sigma Q \nabla u_h)$ ensures that the compatibility condition (1.3) is satisfied by the flow reconstruction $\sigma Q \nabla u_h$, and thus saves us the arbitrariness of the averaging (6.20). By virtue of the solution of local Neumann problems in the extended local ansatz space $S_h^{p+1}(T)$ one then obtains – up to constants – an approximate error representation $[e_h]$. Alternatively, a hierarchical ansatz space may be applied for the solution of (6.32) just as in the BW error estimator.

Robin Boundary Conditions.   For $\alpha > 0$ an additional term enters into the energy norm

$$\|e_h\|_a^2 = \|\sigma^{1/2} \nabla e_h\|_{L^2(\Omega)}^2 + \|\alpha^{1/2} e_h\|_{L^2(\partial \Omega)}^2 ,$$

which is not covered by the error estimator (6.29). Nevertheless $[\epsilon_h]$ remains formally efficient, as the boundary term is dominated by the domain term.

Theorem 6.14. *There exists a constant* $c > 0$ *with*

$$c \epsilon_h \leq \|\sigma^{1/2} \nabla e_h\|_{L^2(\Omega)} \leq \epsilon_h.$$

Proof. We define the average

$$\bar{e}_h = \frac{1}{|\Omega|} \int_\Omega e_h \, dx$$

and regard it as a constant function in S_h. Then, due to the Galerkin orthogonality (4.14) and the continuity A.10 of the bilinear form a, the following result holds:

$$\epsilon_h^2 \leq \|e_h\|_a^2 + \|\bar{e}_h\|_a^2 = \|e_h - \bar{e}_h\|_a^2 \leq C_a \|e_h - \bar{e}_h\|_{H^1(\Omega)}^2.$$

Form the Poincaré inequality (A.25) we then get

$$\epsilon_h \leq C_P \sqrt{C_a} |e_h|_{H^1} \leq C_P \frac{\sqrt{C_a}}{\inf \sqrt{\sigma}} \|\sigma^{1/2} \nabla e_h\|_{L^2(\Omega)}. \qquad \qquad \square$$

Note, however, that even for asymptotically exact gradient recovery, the error estimator will, in general, not be asymptotically exact, due to the neglection of the boundary terms.

Discontinuous Coefficients. In the case of discontinuous diffusion coefficients σ the *tangential component* of the exact flow – just as the normal component of the gradient – is also discontinuous. The gradient "recovery" by projection to the continuous functions would necessarily lead to a deviation $\nabla u_h - Q\nabla u_h$ of the order of the coefficient jump $[\![\sigma]\!]$ and, correspondingly, to an unnecessarily strong mesh refinement along the discontinuity faces. In these cases a piecewise projection with discontinuities of the recovered gradient $Q\nabla u_h$ is possible. For domains with many interior faces, at which diffusion coefficients are discontinuous, such a concept is rather unhandy, let alone that the recovery effect should, in principle, be achieved by the projection to a continuous space.

ZZ Error Estimator. Due to the h-independent bounded condition number of the mass matrix the computation of the L^2-projection $Q\nabla u_h$ is easily possible by the CG-method, which, however, requires a global computational amount. One of the simplest local recovery formulas is also the oldest one: an averaging of piecewise constant discrete gradients, which dates back to O. C. Zienkiewicz and J. Z. Zhu [231, 232]:

$$(Q\nabla u_h)(\xi) = \frac{1}{|\omega_\xi|} \sum_{T \in \mathcal{T}: \xi \in T} |T|\, \nabla u_h|_T(\xi) \quad \text{for all } \xi \in \mathcal{N}.$$

The asymptotic exactness is based on special *superconvergence results on structured meshes*, which can be extended to unstructured meshes by means of certain smoothing operations (see R. E. Bank and J. Xu [20, 21]). In the survey article [49] C. Carstensen showed that equivalent error estimators can be constructed by means of different but nevertheless simple local averaging operators.

6.1.4   Hierarchical Error Estimators

A rather straightforward way of error estimation does not originate from the localization (6.10), but from the global defect equation

$$a(e_h, v) = a(u - u_h, v) = \langle f, v \rangle - a(u_h, v) \quad \text{for all } v \in H^1(\Omega), \tag{6.33}$$

from which an error approximation $[e_h] \approx u - u_h$ is constructed. This approach dates back to P. Deuflhard, P. Leinen, and H. Yserentant [73] from 1989, who suggested an *edge oriented* error estimator.

Embedded Error Estimator. As before in (6.19), the defect equation (6.33) is solved approximately in some (here) globally extended ansatz space $S_h^+ \supset S_h$. We thus define the error approximation $[e_h] = u_h^+ - u_h \in S_h^+$ as the solution of

$$a([e_h], v) = a(u_h^+ - u_h, v) = \langle f, v \rangle - a(u_h, v) \quad \text{for all } v \in S_h^+ \tag{6.34}$$

and, correspondingly, the embedded error estimator

$$[\epsilon_h] = \|[e_h]\|_a. \tag{6.35}$$

Upon application of the piecewise polynomial ansatz space $S_h = S_h^p$ we are essentially led to the extensions $S_h^+ = S_{h/2}^p$ by uniform refinement and $S_h^+ = S_h^{p+1}$ by order increase (cf. Figure 6.1 for the special case $p = 1$).

Localization and Error Representation.   With the computation of $[e_h] \in H^1(\Omega)$ as Galerkin solution of (6.34) a consistent approximation of the error is readily available. The localization to individual elements $T \in \mathcal{T}$ is obvious:

$$[\epsilon_h(T)] = a_T([e_h], [e_h]).$$

At the same time, with $[e_h]$ a more accurate Galerkin solution $u_h^+ = u_h + [e_h] \in S_h^+$ is available, which might mean that an estimation of the error $\epsilon_h^+ = \|u - u_h^+\|_a$ would be more interesting – which, however, is unavailable. We thus find ourselves in the same kind of dilemma as in the adaptive timestep control for ordinary differential equations (cf. Volume 2, Section 5).

Reliability and Efficiency.   In Theorem 6.9 we already showed that the value $[\epsilon_h]$ from (6.35) supplies an efficient error estimator under the saturation assumption $\beta < 1$; as for the reliability, we again refer to Theorem 6.8. The comparison of the result (6.23) with Definition 6.1 directly leads to the constants $\kappa_1 = 1/\sqrt{1 - \beta^2}$ and $\kappa_2 = 1$.

For asymptotic exactness we require additionally that $\beta \to 0$ for $h \to 0$, which – given sufficient regularity of the solution u – is guaranteed by the higher order extension $S_h^+ = S_h^{p+1}$.

DLY Error Estimator.   We are now ready to introduce the edge oriented error estimator due to [73] for linear finite elements in some detail. For realistic applications, the complete solution of (6.34) is still too costly (cf. Exercise 7.3). We therefore replace the Galerkin solution of (6.34) by the solution of a 'sparsed' equation system: We restrict ourselves to the complement $S_h^\oplus$ and merely use the diagonal of the stiffness matrix so that only local $(1, 1)$-"systems" are to be solved. Let $S_h^\oplus$ be spanned by the basis $\Psi^\oplus$; then we obtain the approximate error representation

$$[e_h] := \sum_{\psi \in \Psi^\oplus} \frac{\langle f - Au_h, \psi \rangle}{\|\psi\|_A^2} \psi \in S_h^\oplus \subset H^1(\Omega)$$

and the associated error estimator

$$[\epsilon_h]^2 := \sum_{\psi \in \Psi^\oplus} \frac{\langle f - Au_h, \psi \rangle^2}{\|\psi\|_A^2}. \tag{6.36}$$

In view of an efficient computation, the basis functions $\psi \in \Psi^\oplus$ should again have a support over few elements $T \in \mathcal{T}$ only. The simplest choice is an extension of linear elements by quadratic "bubbles" on the edges (see Figure 6.3). The localization necessary for an adaptive mesh refinement can then be directly identified from (6.36).

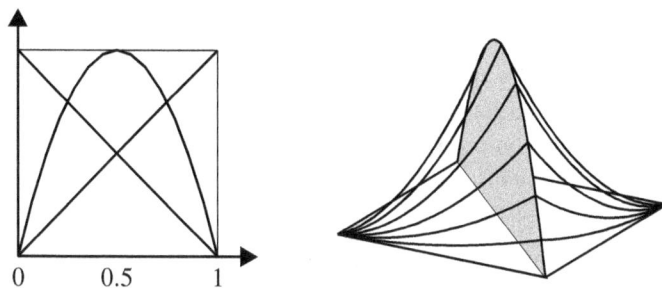

Figure 6.3. Quadratic "bubble" in edge oriented error estimator. *Left:* shape functions on edge (cf. Figure 4.8). *Right:* ansatz function in $\mathbb{R}^2$.

However, in contrast to (6.19), we now get an error estimator localized to individual edges $E \in \mathcal{E}$:

$$[\epsilon_h(E)] = \langle r, \psi_E \rangle / \|\psi_E\|_A. \qquad (6.37)$$

This edge-oriented hierarchical error estimator constitutes the conceptual fundament of the adaptive finite element codes KASKADE (see the software page at the end of the book).

Reliability and Efficiency.   By solving the reduced system (6.34) only approximately, we have not forfeited efficiency, but the efficiency span gets larger and, at the same time, we also lose asymptotic exactness, as shown by the following theorem.

Theorem 6.15. *Under the saturation assumption* (6.22) *the error estimator* (6.36) *is efficient according to* (6.7), *i.e., there exist h-independent constants* $K_1, K_2 < \infty$ *such that*

$$\sqrt{\frac{1 - \beta^2}{K_1}} \, \epsilon_h \leq [\epsilon_h] \leq \sqrt{K_2} \, \epsilon_h.$$

The proof is postponed until Section 7.1.5, where we can work it out much more simply in the context of subspace correction methods.

Connection with Residual-based Error Estimator.   For linear finite elements and piecewise constant data σ and f an interesting connection between the BM error estimator (6.12) and the hierarchical error estimator (7.28) has been found in [37]. For this type of problems the error estimator

$$\sum_{T \in \mathcal{T}} \eta_T^2 + \sum_{F \in \mathcal{F}} \eta_F^2$$

is equivalent to the hierarchical error estimator

$$\sum_{\psi \in \Psi_{TF}^\oplus} \frac{\langle f - Au_h, \psi \rangle^2}{\|\psi\|_A^2},$$

if one extends the ansatz space S_h appropriately:

- for $d = 2$ by locally quadratic "bubbles" on the midpoints of the edges (as in DLY) and by additional cubic "bubbles" in the barycenter of the triangles, whereby the saturation property is satisfied, i.e., the error estimator is efficient; in actual computation this space extension beyond DLY has shown not to pay off;

- for $d = 3$ beyond DLY locally cubic "bubbles" in the barycenters of the triangular faces and quartic "bubbles" in the barycenters of the tetrahedra, whereby, however, the saturation assumption is still not satisfied.

6.1.5 Goal-oriented Error Estimation

Up to now, we have tacitly defined the error ϵ_h as the energy norm of the deviation e_h, which, due to the best approximation property of the Galerkin solution u_h in elliptic PDEs, seemed to be self-evident. In practical applications, however, this is often not the quantity of interest. In structural mechanics, e.g., one might be interested in the *maximal* occurring stresses and strains (see Section 2.3) to be able to assess the risk of a crack or a rupture in material, whereas integral norms conceal the occurrence of stress cusps. In fluid dynamics simulations one is interested in the lift and the drag of wings, whereas an accurate computation of the airflow might be less important. Against this background, R. Rannacher et al. [15, 25] suggested estimating an error, not, as usual, w.r.t. an arbitrary norm, but w.r.t. a *quantity of interest*; this directs us to the concept of *goal-oriented* error estimation.

Let J denote such a quantity of interest. In the simplest case, J is a *linear functional* of the solution u so that we may write the deviation as

$$\epsilon_h = J(u) - J(u_h) = J(u - u_h) = \langle j, e_h \rangle$$

with a unique j due to the Theorem of Fischer–Riesz. The thus defined selectivity may be interpreted geometrically: error components orthogonal to j do not enter into the error estimation and thus need not be resolved by a finer and therefore more expensive discretization. With respect to the quantity of interest, a significantly smaller effort will be sufficient to achieve a prescribed accuracy; of course, the error in other directions may then well be larger.

In case an explicit error approximation $[e_h]$ is available, such as from (7.28), the goal-oriented error can be directly estimated by $[\epsilon_h] = \langle j, [e_h] \rangle$. From this, the localization $[\epsilon_h](T) = |\langle j, [e_h]|_T \rangle|$ can be immediately found. As already mentioned in the introductory section, this form of localization is reasonable only for elliptic problems with spatially smooth j, where, due to the strongly localized Green's function, the error distribution can be simultaneously used for mesh refinement. In other problems, however, e.g., when a discontinuous j occurs or when an explicit error approximation is missing, the goal-oriented error estimator can be more properly expressed by the residual $r_h = f - Au_h$:

$$\epsilon_h = \langle j, e_h \rangle = \langle j, A^{-1} r_h \rangle = \langle A^{-*} j, r_h \rangle$$

In view of this relation we define the *weight function* $z \in H^1(\Omega)$ as the solution of the *dual* problem and thus obtain for the error

$$A^* z = j, \quad \epsilon_h = \langle z, r_h \rangle.$$

Because of this equation the whole approach is called the *dual weighted residual (DWR) method*. Normally z itself needs to be computed again. The elliptic model problems considered in this section are self-adjoint, i.e., we have $A^* = A$, which simplifies the computation of z. Due to the Galerkin orthogonality (4.14), however, an approximation $z_h \in S_h$ would just lead to the useless error estimate $[\epsilon_h] = 0$, so that useful approximations can only be determined from an extended FE space. In fact, approximations $z_h \in S_h^+$ can be obtained by the very methods treated above, such as hierarchical error estimators as described above (see Section 6.1.4), or gradient recovery (see Section 6.1.3).

Remark 6.16. In the course of time, goal-oriented error estimators have been suggested for a series of problem classes, among them for parabolic equations (see Section 9.2.3), for fluid dynamics [23] or for more general optimization problems [24, 212]. Comprehensive surveys are [15, 25].

6.2   Adaptive Mesh Refinement

In this section we present the actual construction of adaptive meshes, i.e., meshes adapted to the problem to be solved. We restrict our attention to the special case of hierarchical simplicial meshes. Starting from a coarse mesh with a moderate number of nodes, a sequence of nested, successively finer meshes is constructed. For that purpose we need two algorithmic modules:

- *marking strategies* on the basis of *local a posteriori error estimators* (see Section 6.1), by means of which we select edges, triangles, or tetrahedra for refinement; as an efficient example we present the method of local error extrapolation in Section 6.2.1;

- *refinement strategies*, i.e., a system of rules, by which we determine how meshes with marked elements should be conformally refined, without generating too obtuse interior angles (cf. Section 4.4.3); in Section 6.2.2 we introduce a selection of efficient techniques for *simplicial elements*, i.e., for triangles ($d = 2$) or tetrahedra ($d = 3$).

The special role of simplicial elements comes from two properties:

- simplices in higher dimensions can be built from those of lower dimensions, e.g., tetrahedra from four triangles; this property is an advantage in the discretization of surfaces of three-dimensional bodies;

- simplices are comparably easy to subdivide into smaller simplices of the same dimension, e.g., triangles into four smaller similar triangles (i.e., with the same interior angles); this property is important for the construction of hierarchical meshes; in Section 6.2.2 we will give suitable illustrative examples.

6.2.1 Equilibration of Local Discretization Errors

Before we turn to methods of mesh refinement, we want to investigate theoretically what an adaptive mesh should look like. The basis for this is a generalization of the error estimates from Section 4.4.1.

Adaptive Mesh Model.   Assertions on nonuniform meshes are hard to make because of their necessarily discrete structure. That is why we turn to a simple model and consider continuous *mesh size functions* $h : \Omega \to \mathbb{R}$. For a number N of mesh nodes we obtain

$$N \approx c \int_{\Omega} h^{-d} \, dx \quad \text{with} \quad c > 0.$$

This continuous model includes a *pointwise* variant of the interpolation error estimate for linear finite elements from Lemma 4.18:

$$|\nabla(u - I_h u)| \le ch|u''| \quad \text{almost everywhere in } \Omega .$$

Integration and application of the Poincaré inequality supplies the global error estimate

$$\|u - I_h u\|^2_{H^1(\Omega)} \le c^2 \int_{\Omega} h^2 |u''|^2 \, dx.$$

Let us now fix the number of nodes and search for a mesh size function h that minimizes this error bound. For this purpose we need to solve the constrained optimization problem

$$\min_{h} \int_{\Omega} h^2 |u''|^2 \, dx \quad \text{subject to} \quad c \int_{\Omega} h^{-d} \, dx = N.$$

According to Appendix A.6 we couple the constraints by some positive Lagrange multiplier $\lambda \in \mathbb{R}$ to the objective functional and thus obtain the saddle point problem

$$\max_{\lambda} \min_{h} \left[\int_{\Omega} h^2 |u''| \, dx - \lambda \left(c \int_{\Omega} h^{-d} \, dx - N \right) \right].$$

For fixed λ differentiation of the integrand supplies the pointwise optimality condition

$$2h|u''|^2 - c\lambda d \, h^{-(d+1)} = 0.$$

The core of this relation is the proportionality

$$h^{d+2} = s^2 |u''|^{-2} \quad \text{with} \quad s^2 = c\lambda d/2 \in \mathbb{R}. \tag{6.38}$$

After this preparation we again turn to discrete meshes and replace the continuous mesh size function h by a piecewise constant function. Accordingly, we replace (6.38) by the approximate local relation

$$h^{d+2} \approx s^2 |u''|^{-2}.$$

For the *local error contributions* on an element $T \in \mathcal{T}$ we then get

$$\|u - I_h u\|_{H^1(T)}^2 \le c \int_T h_T^2 |u''|^2 \, dx = ch_T^{-d} \int_T h_T^{d+2} |u''|^2 \, dx \approx cs^2, \tag{6.39}$$

where we have used $h_T^{-d} \int_T dx = c$ with the usual generic constant c.

Conclusion.   *In an optimal adaptive mesh all local error contributions have (roughly) the same size.*

This means that any strategy for adaptive mesh refinement should aim at an *equilibration of the local discretization errors*: Starting from some given mesh, exactly those elements or edges should be marked whose contribution to the error estimate exceeds some *threshold value*. Various marking strategies differ only in the choice of such a threshold.

In the absence of a uniform mesh size an error estimate as in Lemma 4.18 is inappropriate for adaptive meshes. However, if we choose the number of nodes as a measure, then, on the basis of the above relation (6.39), we obtain the following generalization of Lemma 4.18.

Theorem 6.17. *Let $\Omega \subset \mathbb{R}^d$ be a polytope, $u \in H^2(\Omega)$ and $\mathcal{T}$ a mesh over Ω, which satisfies the equilibration condition*

$$c^{-1}s \le \|u - I_h u\|_{H^1(T)} \le cs \quad \text{for all } T \in \mathcal{T} \tag{6.40}$$

for some $s > 0$ and a constant c depending only on the interior angles of the elements. Then the following error estimate holds:

$$\|u - I_h u\|_{H^1(\Omega)} \le c N^{-1/d} |u|_{H^2(\Omega)}. \tag{6.41}$$

Proof. First we define a function z, piecewise constant on $\mathcal{T}$, by

$$z^2|_T = |T|^{-1} \|u\|_{H^2(T)}^2 \quad \text{where} \quad |T| = ch_T^d.$$

Then Lemma 4.18, which we here apply locally on each element $T \in \mathcal{T}$, yields

$$s \le c\|u - I_h u\|_{H^1(T)} \le ch_T \|u\|_{H^2(T)} \le ch_T^{1+d/2} z.$$

From this, we obtain for the number $M = |\mathcal{T}|$ of elements

$$M = \sum_{T \in \mathcal{T}} 1 = c \sum_{T \in \mathcal{T}} \int_T h_T^{-d} \, dx = c \int_\Omega h^{-d} \, dx \leq c \int_\Omega \left(\frac{s}{z} \right)^{-\frac{2d}{2+d}} dx$$

and thus

$$s \leq c \left(M^{-1} \int_\Omega z^{\frac{2d}{2+d}} \, dx \right)^{\frac{2+d}{2d}}.$$

Because of the equilibration condition (6.40) the total error can be estimated as

$$\|u - I_h u\|_{H^1(\Omega)}^2 = s^2 M \leq c M^{1 - \frac{2+d}{d}} \left(\int_\Omega z^{\frac{2d}{2+d}} \, dx \right)^{\frac{2+d}{d}} \leq c M^{-2/d} \|z\|_{L^q(\Omega)}^2$$

where $q = 2d/(2+d) < 2$. With $M \geq cN$ and $\|z\|_{L^q(\Omega)} \leq c\|z\|_{L^2(\Omega)} = |u|_{H^2(\Omega)}$ we finally arrive at the assertion (6.41).   □

Attentive readers may have observed that the estimate

$$\|z\|_{L^q(\Omega)} \leq c\|z\|_{L^2(\Omega)}$$

at the end of the above proof is rather generous. In fact, the result (6.41) can be shown for less regular functions $u \in W^{2,q}(\Omega)$, so that domains with reentrant corners are also covered (see our corresponding Example 6.23 in the subsequent Section 6.3.2). However, we do not want to go deeper into this theory, but instead refer to [162].

Based on the Galerkin optimality (4.15), Theorem 6.17 immediately supplies the corresponding generalization of Theorem 4.19.

Corollary 6.1. *Under the assumptions of Theorem* 6.17 *the energy error of an FE solution u_h of*

$$a(u_h, v_h) = \langle f, v_h \rangle, \quad v_h \in S_h$$

satisfies the error estimate

$$\|u - u_h\|_a \leq c N^{-1/d} |u|_{H^2(\Omega)} = c N^{-1/d} \|f\|_{L^2(\Omega)}. \tag{6.42}$$

Local Error Extrapolation.   In what follows we introduce an iterative algorithmic strategy for realizing the desired equilibration approximately; this was suggested in 1978 by I. Babuška and W. C. Rheinboldt [13] for elliptic problems. In Volume 1, Section 9.7, we have already presented this strategy using the simple example of numerical quadrature.

As a basis we require one of the local error estimators worked out in Section 6.1 above. For ease of presentation, we first treat finite elements in 1D and later consider the difference in higher space dimensions for different estimators $[\epsilon(T)]$, defined on triangles T, and $[\epsilon(E)]$, defined on edges. We want to construct a hierarchical refinement strategy aiming at an equilibration of the discretization error over the elements

$T \in \mathcal{T}$ of a given mesh. Let $T_0 = (t_l, t_m, t_r)$ denote a selected element with left and right boundary nodes t_l, t_r and a midpoint node t_m. By subdivision of this element subelements T_l and T_r are generated, where

$$T_l := \left(t_l, \frac{t_l + t_m}{2}, t_m \right) \quad \text{and} \quad T_r := \left(t_m, \frac{t_r + t_m}{2}, t_r \right).$$

Upon refining twice, we obtain a *binary tree*, depicted in Figure 6.4.

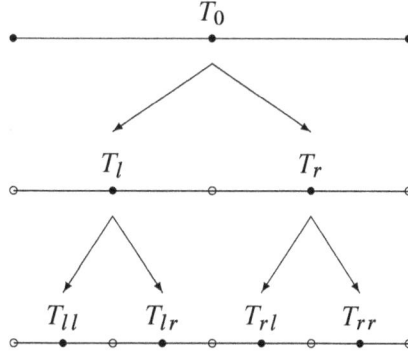

Figure 6.4. Double refinement of edge $T_0 = (t_l, t_m, t_r)$.

The figure selects three levels of a hierarchical refinement, which we will denote by $\mathcal{T}^-, \mathcal{T}, \mathcal{T}^+$ so that

$$T_0 \in \mathcal{T}^-, \quad \{T_l, T_r\} \in \mathcal{T}, \quad \{T_{ll}, T_{lr}, T_{rl}, T_{rr}\} \in \mathcal{T}^+.$$

With this notation, we now turn to the local error estimation. In Section 4.4.1 the error behavior of linear finite elements over a triangulation with mesh size h was given: Theorem 4.19 supplied $\mathcal{O}(h)$ for the error in the energy norm, Theorem 4.21 $\mathcal{O}(h^2)$ in the L^2-norm; in both theorems, however, the usually unrealistic assumption $u \in H^2(\Omega)$ was set, which would require convex domains Ω or those with smooth boundaries. For general domains, however, the local error behavior is hard to predict: in the interior of a domain one will expect $\mathcal{O}(h)$ for the local energy error, close to reentrant corners $\mathcal{O}(h^{1/\alpha - \epsilon})$ with a factor determined by the interior angle $\alpha \pi > \pi$, and finally $\mathcal{O}(h^\beta)$ for *rough* right-hand sides f with an a priori unknown exponent β. In this rather unclear situation I. Babuška and W. C. Rheinboldt [13] suggested making the ansatz

$$\epsilon(T) \doteq C h^\gamma, \quad T \in \mathcal{T} \tag{6.43}$$

for the discretization error, where h denotes the local mesh size, γ a local order and C a local problem dependent constant. The two unknowns C, γ can be determined from the numerical results of $\mathcal{T}^-$ and $\mathcal{T}$. Similar to (6.43) one then obtains for the edge level $\mathcal{T}^-$

$$\epsilon(T^-) \doteq C(2h)^\gamma, \quad T^- \in \mathcal{T}^-$$

as well as for level $\mathcal{T}^+$

$$\epsilon(T^+) \doteq C(h/2)^\gamma, \quad T^+ \in \mathcal{T}^+.$$

Without being forced to estimate the errors on the finest level $\mathcal{T}^+$ explicitly, these three relations permit, by *local extrapolation*, to gain the look-ahead estimate

$$[\epsilon(T^+)] \doteq \frac{[\epsilon(T)]^2}{[\epsilon(T^-)]}.$$

This gives us an idea in advance what effect a refinement of the elements in $\mathcal{T}$ would have. As the threshold value, above which we refine an element, we take the maximum local error that we would get by *global uniform refinement*, i.e., by refinement of *all* elements. This leads to the definition

$$\kappa^+ := \max_{T^+ \in \mathcal{T}^+} [\epsilon(T^+)]. \tag{6.44}$$

The situation is illustrated in Figure 6.5, which shows that the errors towards the

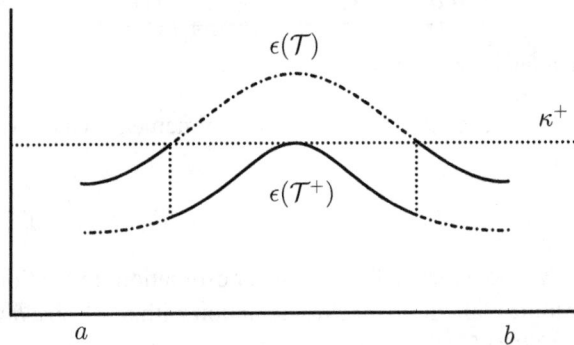

Figure 6.5. Iterative step towards discretization error equilibration: error distribution $\{\epsilon(\mathcal{T})\}$, extrapolated distribution $\{\epsilon(\mathcal{T}^+)\}$ for (not realized) uniform refinement, expected distribution (fat line) for realized nonuniform refinement.

right and left boundary already lie below the level that would be expected by further refinement; therefore we do not need to refine there. Only in the center part might a refinement pay off. So we arrive at the following *refinement rule*: Refine only such elements $T \in \mathcal{T}$, for which

$$[\epsilon(T)] \geq \kappa^+, \quad T \in \mathcal{T}.$$

Application of this rule supplies an expected error distribution, depicted in Figure 6.5 by thick lines. Compared with the original distribution, we recognize that this strategy has brought us one step closer to the desired equilibration of the local errors – assuming that the actual local errors arising in the next refinement step roughly behave as expected. Upon repetition of the procedure in subsequent refinement steps the distribution will get successively "smaller", i.e., we will iteratively approach the equilibration up to a width of 2^γ (see the analysis in Volume 1, Section 9.7.2).

The result of local error extrapolation is a *marking* of elements which, in view of the nested structure of hierarchical meshes, directly carries over to higher space dimensions. For $d > 1$, however, the *edges* of a triangulation may be not completely nested, since by subdivision of elements new edges may be generated not contained in the edge set on coarser levels (cf. the subsequent section). When using the edge-oriented DLY error estimator (6.36) for local error extrapolation, the question of how to deal with such edges arises. The simplest way of dealing with parentless edges has turned out to be to skip them in the maximum search (6.44), which will merely lead to a possibly slightly lower threshold value κ^+.

6.2.2   Refinement Strategies

In this section we start, as an example, from a *triangulation with marked edges*. The aim now is to generate a refined mesh which is both *conformal* and *shape regular*. In two or three space dimensions this task is all but trivial, since the local subdivision of an element will, in general, lead to a nonconformal mesh. To construct a conformal mesh, the neighboring elements have to be suitably subdivided such that this process does not spread over the whole mesh. At the same time, one must take care that the approximation properties of the FE ansatz spaces on the meshes do not deteriorate, i.e., that the maximal interior angles remain bounded (cf. (4.65) in Section 4.4.3). In some cases, a constraint on the minimal interior angles is of interest (see Section 7.1.5). In view of an extension of multigrid methods from Section 5.5 to adaptive meshes one is also interested in the construction of nested mesh hierarchies.

Definition 6.18. *Mesh hierarchy.* A family $(\mathcal{T}_k)_{k=0,\dots,j}$ of meshes over a common domain $\Omega \subset \mathbb{R}^d$ is called *mesh hierarchy* of depth j, if for each $T_k \in \mathcal{T}_k$ there exists a set $\hat{T} \subset \mathcal{T}_{k+1}$ with $T_k = \bigcup_{T \in \hat{T}} T$. The *depth* of an element $T \in \bigcup_{k=0,\dots,j} \mathcal{T}_k$ is given by $\min\{k \in \{0,\dots,j\} : T \in \mathcal{T}_k\}$; the depth for faces, edges, or nodes is defined in a similar way.

An interesting property of mesh hierarchies is the embedding of the nodal set: $\mathcal{N}_k \subset \mathcal{N}_{k+1}$ (see Exercise 7.7). In view of the objective criteria conformity, shape regularity, and mesh hierarchy essentially two strategies have evolved, the bisection and so-called red-green refinement.

Bisection

The simplest method, already worked out in 1984 by M.-C. Rivara [175], is the bisection of marked edges coupled with a decomposition into two neighboring simplices (see Figure 6.6).

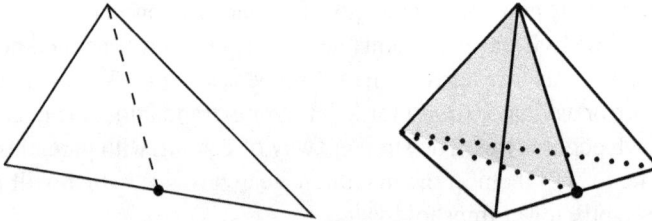

Figure 6.6. Bisection of triangles (*left*) and tetrahedra (*right*).

In order to maintain shape regularity, the longest edge is usually selected for bisection. The refinement is continued as long as hanging nodes still exist (see Figure 6.7).

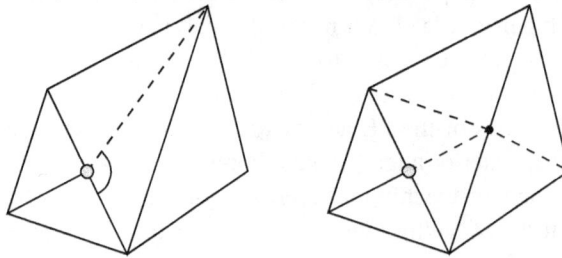

Figure 6.7. Conformal mesh refinement by continued bisection in 2D.

In general, the refinement process terminates before the mesh has been uniformly refined (cf. Exercise 6.2): If a hanging node is situated on a "longest" edge, then further bisection does not generate another new node. Otherwise a new hanging node is generated on a longer edge. Hence, the continuation of the refinement can only affect elements with longer edges.

Red-green refinement

A more tricky method dates back to R. E. Bank et al. [18] from 1983, who had introduced some "red-green" refinement strategy in his adaptive FE package PLTMG (in 2D).[2]

---

[2] R. E. Bank started, however, from the algorithmic basis of triangle oriented error estimators in 2D.

Red Refinement.   In this kind of refinement (named as such by R. E. Bank) the elements are subdivided into 2^d smaller simplices, where exactly the midpoints of all edges are introduced as new nodes. In 2D, this yields a unique decomposition of a triangle into four geometrically similar triangles; as the interior angles do not change, the shape regularity is preserved (see Figure 6.8, left).

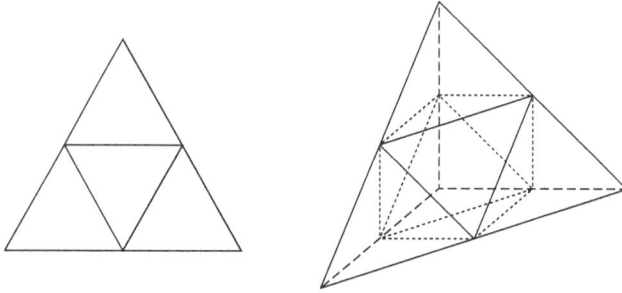

Figure 6.8. Red refinement of triangles (*left*) and tetrahedra (*right*).

In 3D, things are more complex geometrically: here one first gets a unique decomposition into four tetrahedra at the vertices of the original tetrahedron as well as an octahedron in the center (see Figure 6.8, right). By selecting one of the diagonals as the common new edge, the octahedron may be decomposed into four further tetrahedra, which, however, are no longer similar to the original tetrahedron, so that the shape regularity will now depend on the refinement depth. In fact, the selection of the diagonal to be subdivided must be done with great care in order to be able to prove that shape regularity is preserved. As a rule, the shortest diagonal is selected.

In order to preserve conformity of the triangulation, elements neighboring already subdivided elements must also be subdivided. In such a strategy, however, the red refinement would inevitably spread over the whole mesh and end up with a global uniform refinement. That is why a remedy has been thought of early: Red refinements are performed locally only, if (in 2D) *at least two* edges of a triangle are marked. By this rule, the refinement process will spread, but will remain sufficiently local.

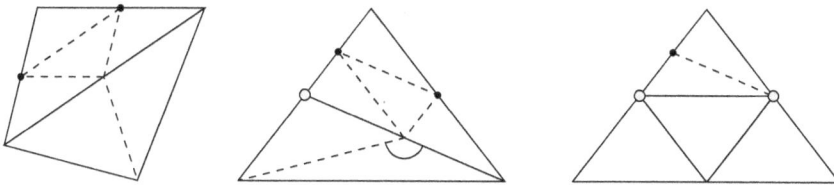

Figure 6.9. Red-green refinement strategy in 2D. *Left:* combination of a red and a green refinement at a marked edge with only one hanging node. *Center:* increase of the interior angle at continued green refinement. *Right:* shape regularity by removal of green edges before refinement.

Green Refinement.   We now assume (for illustration again in 2D) that in a local triangle only one edge has been marked for refinement. In this case we introduce a new so-called *green* edge (again named by R. Bank) and subdivide the triangle into two triangles (see Figure 6.9, left). In this way, the refinement is not continued any further, which is why the green edges are called *green completion*. The interior angles of triangles from green refinements may be both larger and smaller than the ones of the original triangle. Hence, by continuation of green refinements, the shape regularity might suffer (see Figure 6.9, center). In order to assure shape regularity nevertheless, all green completions are removed from the mesh before the next refinement, possibly adding green edges once red refinement has been performed (see Figure 6.9, right). Along this line, shape regularity can be preserved.

In 3D, the process runs in a similar way, but requires three different types of green completions (see [36]), depending on whether one, two, or three edges are marked (see Figure 6.10). With four marked edges red refinement can again take place.

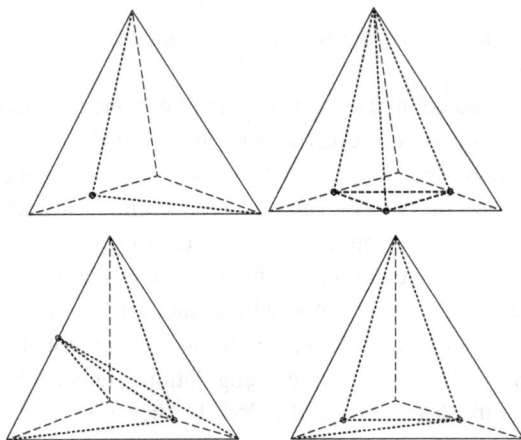

Figure 6.10. Four different types of green completions with tetrahedra [36]. *Top left:* one edge marked. *Top right:* three edges of a face marked. *Bottom:* two edges marked. In all other cases red refinement is performed.

Remark 6.19. The preservation of shape regularity by red refinement of simplices of arbitrary space dimension in finitely many similarity classes was already proven in 1942 by H. Freudenthal [92], for the adaptive mesh refinement in 3D, however, rediscovered only 50 years later [27, 102]. The stability of selecting the shortest diagonal has been shown in [136, 230].

Remark 6.20. If in the course of the refinement of meshes simultaneously both the mesh size h and the local order p of the finite elements are adapted, one speaks of hp-refinement. We have skipped this efficient method, since throughout most of this beek we treat elements of lower order. We refer interested readers to the book of C. Schwab [183].

6.2.3 Choice of Solvers on Adaptive Hierarchical Meshes

Up to this point we have elaborated how to construct adaptive hierarchical meshes. Let us now briefly outline the kind of solvers that can be combined with this kind of meshes. As in Section 5.5 above, we build upon a mesh hierarchy and nested FE spaces:

$$S_0 \subset S_1 \subset \cdots \subset S_j \subset H_0^1(\Omega).$$

For boundary value problems this corresponds to a hierarchical system of linear equations

$$A_k u_k = f_k, \quad k = 0, 1, \ldots, j \qquad (6.45)$$

with symmetric positive definite matrices A_k of dimension N_k. In contrast to the situation in Section 5.5, where we started from a fixed fine mesh $k = j$, for *adaptive* hierarchical meshes the order from coarse mesh $k = 0$ to fine mesh $k = j$ is now compulsory. In particular, the linear equation systems (6.45) have to be solved on *each* level k in turn, in order to enable the adaptive construction of the mesh hierarchy.

This successive solution can be performed in two principal ways:

1. by methods from numerical linear algebra, such as the *direct elimination methods* presented in Section 5.1, or iterative eigenvalue solvers (see, e.g., Volume 1, Section 8.5), which we here will only demonstrate in a nontrivial example (see the subsequent Section 6.4);

2. by adaptive multigrid methods, which we will discuss in the subsequent Chapter 7; there it is sufficient to reduce the algebraic error $\|u_j - \hat{u}_j\|$ down to the size of the discretization error $\|u - u_j\|$ – an obviously reasonable complexity reduction. Note that a nested iteration approach for multigrid as discussed in Exercise 5.8 comes in quite natural in the context of adaptive mesh refinement. As usual, we suppose that at least the system $A_0 u_0 = f_0$ can be solved by direct elimination methods. Of course, all kinds of intermediate forms are realized where direct solvers are used up to refinement level $0 < k_{\max} < j$.

As a general rule, direct elimination methods, due to their superlinear growth of the computational amount are the methods of choice for "smaller" problems, whereas iterative methods with linearly growing amount are inevitable for "large" problems (see Section 5.5.3 above). As already mentioned there, adaptive meshes are particularly crucial for direct methods. This is illustrated in the technological example given in the subsequent Section 6.4.

6.3 Convergence on Adaptive Meshes

Up to now, we have tacitly assumed that the meshes refined by means of localized error estimators actually generate a sequence $\{\mathcal{T}_k\}$ of meshes with corresponding FE solutions $u_k \in S_k = S_h^1(\mathcal{T}_k)$ that essentially realize the optimal convergence order

$$\|u - u_k\|_A \leq c N_k^{-1/d} \|f\|_{L^2(\Omega)}$$

from Corollary 6.1. Despite convincing results, this had not been theoretically confirmed for a long time – not even the pure convergence $u_k \to u$. After 1996 W. Dörfler [79] managed to prove convergence of adaptive solvers by introducing his concept of *oscillation* (see Definition 6.3). The *optimal* convergence rate was shown in 2004 by P. Binev, W. Dahmen, and R. De Vore [29], although by application of not very practicable coarsening steps. In 2007 R. Stevenson [193] was able to prove convergence for the typical algorithms in common use.

In the following Section 6.3.1 we give such a convergence proof. In the subsequent Section 6.3.2 we illustrate the computational situation by an elementary 2D example with reentrant corners.

6.3.1   A Convergence Proof

In order to work out the main aspects of a convergence proof for adaptive methods, we first define a model algorithm for hierarchical refinement.

Adaptive Model Algorithm.   We focus our attention to an idealized situation with linear finite elements S_h^1 on a triangular mesh ($d = 2$). For the marking of edges we use a hierarchical error estimator. As extension $S_h^\oplus$ instead of quadratic "bubbles" as in the DLY estimator (see Figure 6.1, right), we here use piecewise linear "bubbles" (see Figure 6.1, left). As refinement rule we combine red refinement with hanging nodes: Whenever a hanging node $\xi \in \mathcal{N}$ is midpoint of an edge (ξ_l, ξ_r) that on one side has an unrefined, on the other side a refined triangle, then we use interpolation by linear finite elements and require that $u_h(\xi) = (u_h(\xi_l) + u_h(\xi_r))/2$. Ee allow at most one hanging node per edge. The thus generated meshes are – up to the missing green completions – identical with red-green refined meshes.

By this model of an adaptive algorithm, we ensure, on the one hand, that the generated sequence $\{S_k\}$ of ansatz spaces is nested, i.e., that $S_k \subset S_{k+1}$ holds, and, on the other hand, that the ansatz functions applied for error estimation are contained in the ansatz space of the next level of mesh refinement. In order to construct a convergent method in this frame, it seems reasonable to impose a refinement of those triangles with the largest error indicator. However, not too many triangles should be refined, in particular those with especially small error indicator, since then the fast growing number N_k of nodes would not coincide with an associated error reduction and thus the convergence rate would be suboptimal only. We therefore choose as limiting case in each refinement step an edge $E \in \mathcal{E}$ with maximum error indicator, i.e., where $[\epsilon_h(E)] \geq [\epsilon_h(e)]$ holds for all $e \in \mathcal{E}$, and mark exactly the triangles incident with E for refinement. Hence, due to the locality of the refinement, the number of additional nodes is bounded by some constant m.

For this idealized case we are now able to show convergence of the FE solution $u_k \in S_k$. For the proof, we employ the *saturation* instead of the *oscillation* assumption, both of which were shown to be equivalent by W. Dörfler and R. H. Nocchetto [80] in 2002 (see Theorem 6.8).

Theorem 6.21. *Let the saturation assumption (6.22) be satisfied. Then the FE solution u_j converge to u and there exist constants $c, s > 0$ such that*

$$\|u - u_j\|_A \leq c N_j^{-s} \|u - u_0\|_A, \quad j = j_0, \dots.$$

Proof. We use the notation of Section 6.1.4. From the error localization (6.37) and the Galerkin orthogonality we obtain due to $\psi_E \in S_{k+1}$ the following result for the edge oriented error estimator with linear finite elements ($k = 0, \dots, j$)

$$[\epsilon_k(E)] \|\psi_E\|_A = \langle f - Au_k, \psi_E \rangle = \langle A(u - u_k), \psi_E \rangle = \langle A(u_{k+1} - u_k), \psi_E \rangle$$
$$\leq \|u_{k+1} - u_k\|_A \|\psi_E\|_A,$$

which also holds for the modulus, of course. Thus we can estimate the error reduction. Again due to the Galerkin orthogonality we have

$$\|u - u_k\|_A^2 = \|u - u_{k+1}\|_A^2 + \|u_{k+1} - u_k\|_A^2 \geq \|u - u_{k+1}\|_A^2 + [\epsilon_k(E)]^2,$$

which immediately supplies

$$\epsilon_{k+1}^2 \leq \epsilon_k^2 - [\epsilon_k(E)]^2. \tag{6.46}$$

For the following let $c > 0$ again denote a generic constant. As $[\epsilon_k(E)]$ is maximal, Theorem 6.15 yields

$$[\epsilon_k(E)]^2 \geq \frac{1}{|\mathcal{E}_k|} \sum_{e \in \mathcal{E}_k} [\epsilon_k(e)]^2 \geq c \frac{[\epsilon_k]^2}{N_k} \geq c \frac{\epsilon_k^2}{N_k},$$

where we have bounded the number of edges by the number of nodes. We further get $N_{k+1} \leq N_k + m$, i.e., $N_{k+1} \leq (k+1)m + N_0$, as well as $N_0 \geq 1$, which leads to

$$[\epsilon_k(E)]^2 \geq c \frac{\epsilon_k^2}{k+1}.$$

Upon inserting this into (6.46) we obtain after some brief calculation

$$\epsilon_j^2 \leq \left(1 - \frac{c}{j}\right) \epsilon_j^2 \leq \prod_{k=0}^{j-1} \left(1 - \frac{c}{k+1}\right) \epsilon_0^2 \leq c j^{-c} \epsilon_0^2 \leq c N_j^{-s} \epsilon_0^2$$

and thus convergence as stated above.   □

For the model algorithm studied here, Theorem 6.21 supplies convergence, but the convergence rate remains unknown. With considerably more effort in the proof, one

may also obtain convergence rates that are optimal in the sense of principal approxima-
bility of the solution u (see, e.g., [193]). However, even with optimal convergence rate,
the method described does not achieve optimal computational amount: As in each refinement step, only a bounded number of new degrees of freedom is added, the number
of refinement steps is very large so that the total amount W_j is of the order $\mathcal{O}(N_j^2)$. In
order to keep an optimal amount of the order $\mathcal{O}(N_j)$, in practice sufficiently many triangles are refined, which, however, requires slightly more complex proof techniques.

Remark 6.22. A theoretical concept as described here underlies the adaptive finite
element code ALBERTA (for 2D and 3D), written by ALfred Schmidt and KuniBERT
Siebert[3] and published in [180]; this program uses the residual based error estimator
(see Section 6.1.1), coupled with local error extrapolation due to [13] and *bisection*
for refinement (see Section 6.2.2).

6.3.2 An Example with a Reentrant Corner

We have repeatedly indicated that the asymptotic behavior of the discretization error
with uniform meshes of mesh size h changes at reentrant corners due to the corner
singularities occurring there. For $d = 2$ we characterize the interior angle by $\alpha\pi$ with
$\alpha > 1$. From Remark 4.20 we take that the error behavior for *Dirichlet problems* in
the *energy norm* as

$$\|u - u_h\|_a \leq c h^{\frac{1}{\alpha}-\epsilon}, \quad \epsilon > 0. \tag{6.47}$$

For one-sided *homogeneous Neumann problems* the value α in the above exponent has
to be substituted by 2α, if the problem can be extended symmetrically to a Dirichlet
problem. Analogously, from Remark 4.22 we have for the error behavior in the L^2-
norm

$$\|u - u_h\|_{L^2(\Omega)} \leq c h^{\frac{2}{\alpha}-\epsilon}, \quad \epsilon > 0. \tag{6.48}$$

Here, too, one-sided Neumann problems would require the factor α in the above exponent to be replaced by 2α. For an illustration of adaptive meshes with reentrant corners
we modify an example that originally was originally constructed by R. E. Bank [16].

Example 6.23. *Poisson model problem on slit domain.* Consider the simple Poisson
equation

$$-\Delta u = 1$$

on a circular domain with slit along the positive axis, as depicted in Figure 6.11. On
the left, homogeneous Dirichlet boundary conditions are prescribed on both sides of
the slit. On the right, however, homogeneous Neumann boundary conditions are prescribed at the lower side of the slit (observe the orthogonality of the level lines), while

---

[3] Originally, the program was called ALBERT, which, however, caused some legal objection from economy.

on the upper side homogeneous Dirichlet boundary conditions are prescribed. For both cases, the numerical solution is represented by level lines (see bottom row of Figure 6.11). In the rows above the meshes $\mathcal{T}_j$ generated adaptively by linear finite elements are shown for $j = 0, 5, 11$. The circular boundary is approximated by successively finer polygonal boundaries. For the generation of the adaptive meshes the DLY error estimator from Section 6.1.4, local extrapolation from Section 6.2.1 and red-green refinement from Section 6.2.2 have been selected. The numerical solution was done by one of the multigrid solvers which will be described in Section 7.3, but which does not play a role in the present context.

For this example, approximation theory (see (6.47) and (6.48)) supplies the following error behavior ($\epsilon \approx 0$):

- *Dirichlet* problem: $\alpha = 2$, i.e.,

$$\text{ERRD-EN} := \|u - u_h\|_a^{(D)} \sim \mathcal{O}(h^{\frac{1}{2}}), \quad \text{ERRD-L2} := \|u - u_h\|_{L^2}^{(D)} \sim \mathcal{O}(h);$$

- *Neumann* problem: $2\alpha = 4$, i.e.,

$$\text{ERRN-EN} := \|u - u_h\|_a^{(N)} \sim \mathcal{O}(h^{\frac{1}{4}}), \quad \text{ERRN-L2} := \|u - u_h\|_{L^2}^{(N)} \sim \mathcal{O}(h^{\frac{1}{2}}).$$

The behavior of these four error measures can be nicely observed in double-logarithmic scale in Figure 6.12, top.

In Theorem 6.17 we presented a theory for *adaptive* meshes, the core of which was the characterization of the FE approximation error vs. the number N of unknowns instead of a characterization versus a mesh size h, which really works only in the uniform case. This theory also holds for reentrant corners. To apply it, we consider the above Poisson problem *without* reentrant corners, no matter whether with Dirichlet or with Neumann boundary conditions. For this case, *uniform* meshes as in Section 4.4.1 with $N = ch^d, d = 2$ would have led to an energy error

$$\|u - u_h\|_a \sim \mathcal{O}(h) \sim \mathcal{O}(N^{-\frac{1}{2}})$$

and an L^2-error

$$\|u - u_h\|_{L^2} \sim \mathcal{O}(h^2) \sim \mathcal{O}(N^{-1}).$$

In fact, this same behavior is observed if one draws the *adaptive* meshes vs. $N = N_j$, as in Figure 6.12, bottom. The errors ERRD-EN and ERRN-EN only differ by some constant, which exactly reflects assertion (6.42). We have not given an analogous proof for the L^2-norm; nevertheless, the same behavior can be also seen for the error measures ERRD-L2 and ERRN-L2.

In conclusion, from these figures we obtain:

> *Adaptive meshes "remove" corner singularities by some implicit transformation.*

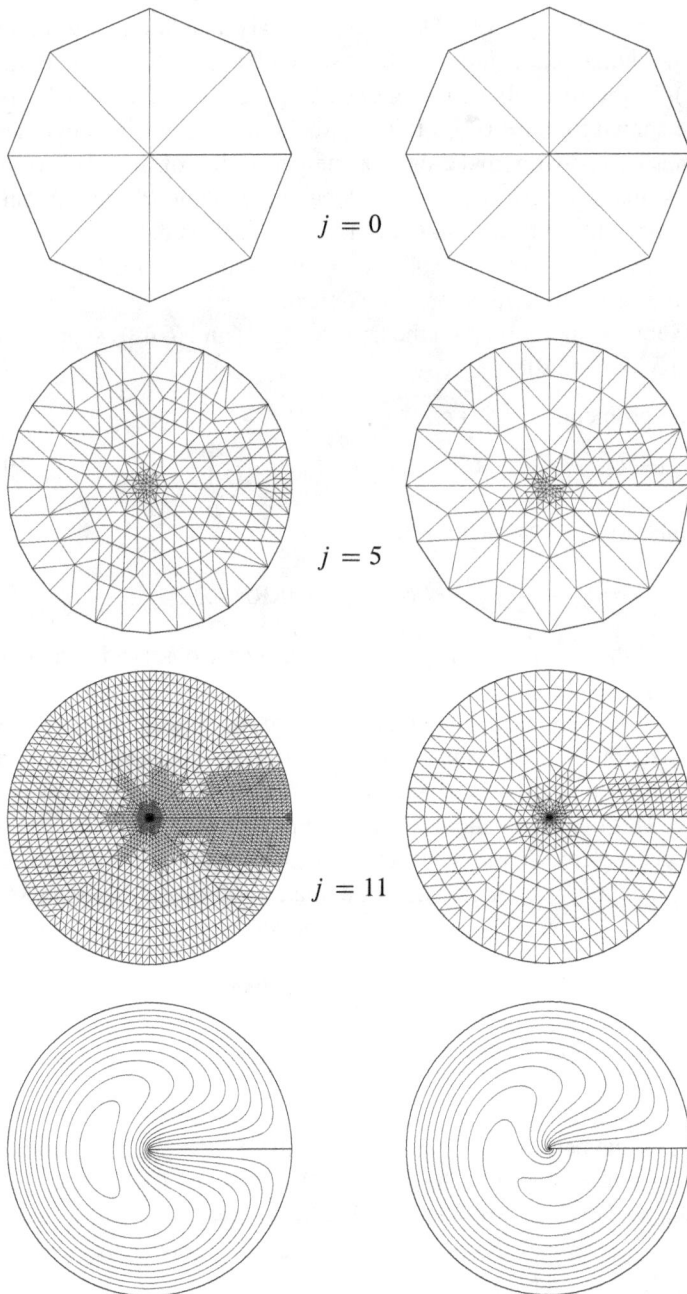

Figure 6.11. Example 6.23: adaptive mesh refinement history. *Left:* Dirichlet boundary conditions; coarse mesh with $N_0 = 10$ nodes, selected refined meshes with $N_5^{(D)} = 253$, $N_{11}^{(D)} = 2152$, level lines for the solution. *Right:* Neumann boundary conditions; same coarse mesh, refined meshes with $N_5^{(N)} = 163$, $N_{11}^{(N)} = 594$, level lines for the solution.

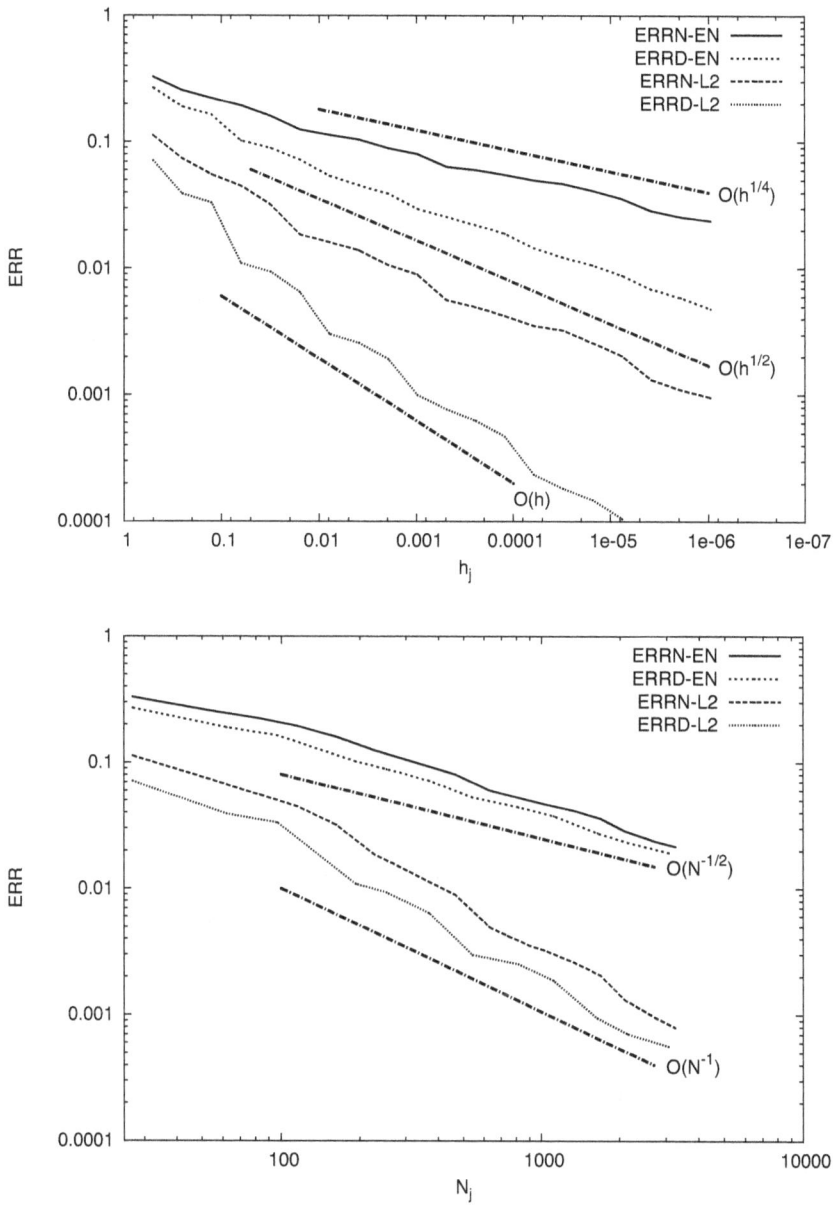

Figure 6.12. Example 6.23: adaptive mesh refinement history. *Top:* error measures vs. smallest mesh size $h = h_j$. *Bottom:* error measures versus number of nodes $N = N_j$. Cf. Theorem 6.17.

6.4 Design of a Plasmon-Polariton Waveguide

This type of semiconductor device has only rather recently become a focus of interest. It serves as biosensor as well as interconnect on integrated semiconductor chips. The nontrivial example presented here due to [46] will give a glimpse into the problem world of nanotechnology. As will turn out, we are led to a quadratic eigenvalue problem.

Figure 6.13 presents a scheme of such a special optical wave guide by showing some 2D cross section (see [26]): on a dielectric material 1 (with relative dielectric constant $\varepsilon_{r,1} = 4.0$) there lies a very thin silver strip with thickness $h = 24.5$ nm and width $b = 1$ μm. A dielectric material 2 (with $\varepsilon_{r,2} = 3.61$) covers the device from above. The geometry is invariant in z-direction, i.e., the device is modelled as infinitely long.

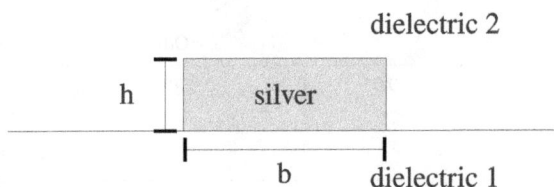

Figure 6.13. Schematic representation of a plasmon-polariton waveguide.

It is of technological interest that light fields spread along the z-direction in this structure but remain localized at the surface of the thin silver strip. As already introduced in Section 2.1.2, such fields are called *waveguide modes*. If light spreads on a metal surface, one speaks of *plasmon-polariton waveguides*.

For our present case we have $\omega = 2\pi/\lambda_0 \, c$ with $\lambda_0 = 633$ nm and the light velocity c. For this choice of ω one has in silver (chemical notation: Ag) the relative dielectric constant $\varepsilon_{r,\text{Ag}} = -19 + 0.53i$. The large real part and nonvanishing imaginary part characterize the damping of the optical wave.

Mathematical Modelling. A hierarchy of optical models was given in Section 2.1.2, among them also a special model for waveguides. In the present case we require a variant of the model introduced there. For this purpose, we recall the *time-harmonic* Maxwell equations. We start from equation (2.5), merely exchanging $(E, \epsilon) \leftrightarrow (H, \mu)$. If we write the differential operator componentwise as vector product of the gradient (cf. Exercise 2.3), then we come up with the following equation for the electric field E:

$$\begin{bmatrix} \partial_x \\ \partial_y \\ \partial_z \end{bmatrix} \times \frac{1}{\mu} \begin{bmatrix} \partial_x \\ \partial_y \\ \partial_z \end{bmatrix} \times E(x,y,z) - \omega^2 \varepsilon(x,y,z) E(x,y,z) = 0.$$

The dielectric constant depends solely on the cross section coordinates, which means that $\varepsilon = \varepsilon(x, y)$. In our specific example we had $\mu_r = 1$, but in the general case the relative magnetic permeability would be $\mu_r = \mu_r(x, y)$.

For the waveguide modes we make an ansatz just as in (2.9), i.e.,

$$E(x, y, z) = \hat{E}(x, y)e^{ik_z z}$$

with a propagation coefficient k_z. As determining equation for the eigenvalue k_z we then get

$$\begin{bmatrix} \partial_x \\ \partial_y \\ ik_z \end{bmatrix} \times \frac{1}{\mu} \begin{bmatrix} \partial_x \\ \partial_y \\ ik_z \end{bmatrix} \times \hat{E}(x, y) - \omega^2 \varepsilon(x, y)\hat{E}(x, y) = 0. \qquad (6.49)$$

Upon expanding the above terms (which we leave for Exercise 6.4) we find that the eigenvalue occurs both linearly and quadratically. Thus we encounter a *quadratic eigenvalue problem*. By some elementary transformation (also left for Exercise 6.4) this problem can be reformulated as a linear eigenvalue problem of double dimension (see, e.g., the survey article by F. Tisseur and K. Meerbergen [199]).

Discrete Formulation. In principle, the eigenvalue problem is stated in the whole $\mathbb{R}^2$. For its numerical discretization the problem must be restricted to some finite domain of interest. The aim is to compute the waveguide mode with the strongest localization around the silver strip, the so-called *fundamental mode*. For this purpose, it is enough to allocate homogeneous Dirichlet boundary conditions on a sufficiently large section around the silver strip such that the "truncation error" can be neglected. For discretization, *edge elements* of second order were applied, which we have skipped in this book (see J. C. Nédélec [158]). Along this line a very large algebraic linear eigenvalue problem with some nonsymmetric matrix arises. For its numerical solution the "inverse shifted Arnoldi method" is applied, a well-proven method of numerical linear algebra (see [143]). During the course of this algorithm sparse linear equations of up to several million unknowns have to be solved, which is done by the program package PARDISO [178].

Numerical Results. The design of a waveguide requires obviously the selection of an eigenvalue via some property of the corresponding eigenfunction: it should be "strongly localized" around the silver strip. In order to approach this goal, first 100 or so discrete eigenvalue problems have been simultaneously approximated on relatively coarse meshes and the localization of their eigenfunctions quantified. From this set, successively fewer eigenvalues have been selected on finer and finer meshes. The eigenvalue finally left over by this iterative procedure is, for technological reasons, required to high accuracy, i.e., it needs to be computed on rather fine meshes. In Figure 6.14 we show isolines of the computed intensity $\|E\|^2$, in logarithmic scale, the representation preferred by chip designers. One recognizes the desired localization

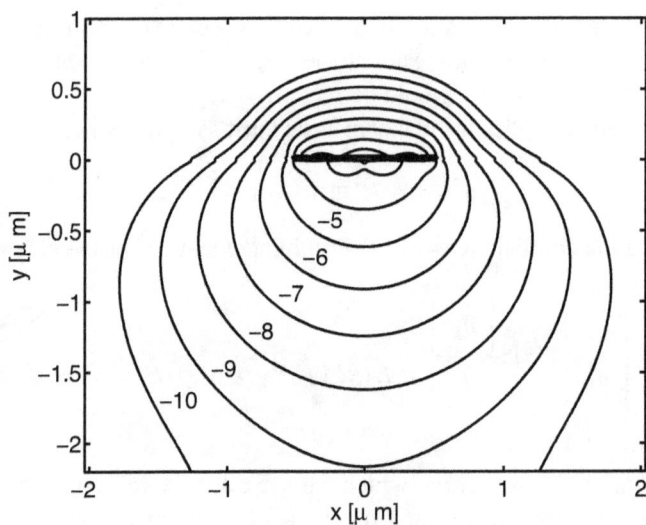

Figure 6.14. Plasmon-polariton waveguide: isolines of the field intensity $\|E\|^2$ (in logarithmic scale).

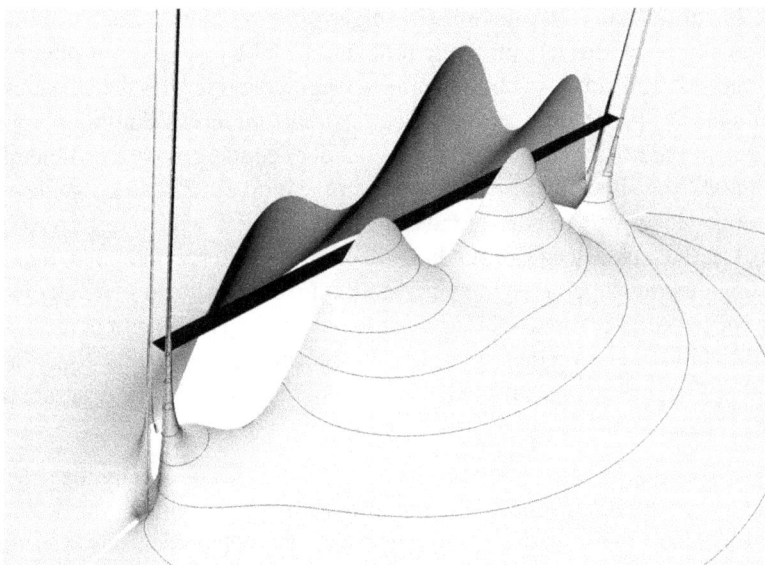

Figure 6.15. Plasmon-polariton waveguide: dominant y-component of the field E. The black line marks the silver strip. The singularities at the ends of the silver strip have been cut off.

around the silver strip. Moreover, the field exhibits an unpleasant number of *singularities* at the ends of the silver strip, which has been made especially visible in Figure 6.15.

Adaptive Meshes.   Because of the singularities involved, an adaptive refinement of the FE mesh is of utmost importance. In [46], the residual based error estimator from Section 6.1.1 has been chosen; even though it is only determined up to some constant, it is very well suited as an error indicator in the presence of singularities.

In Table 6.1 the computed eigenvalues k_z vs. the number N of unknowns are listed. In practical applications, one needs the real part of k_z up to the fourth decimal place after the dot. The imaginary part should be computed up to two significant decimal digits.

Table 6.1. Plasmon-polariton waveguide: eigenvalue approximations k_z vs. adaptive mesh refinement ($k_0 = 2\pi/\lambda_0$).

N	$\Re(k_z)/k_0$	$\Im(k_z)/k_0$
1 830	2.003961e+00	1.031e−03
4 150	2.004239e+00	1.125e−03
8 757	2.003659e+00	1.071e−03
18 955	2.003339e+00	1.034e−03
38 329	2.003155e+00	1.011e−03
74 574	2.003061e+00	9.992e−04
148 872	2.003012e+00	9.929e−04
271 281	2.002986e+00	9.894e−04
528 178	2.002973e+00	9.875e−04
1 010 541	2.002965e+00	9.864e−04
1 903 730	2.002962e+00	9.858e−04
3 518 031	2.002960e+00	9.855e−04

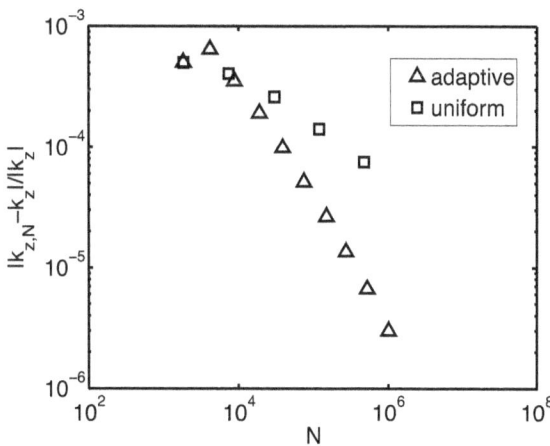

Figure 6.16. Plasmon-polariton waveguide: comparison of convergence of eigenvalue approximations k_z for adaptive versus uniform meshes.

From this, a bound for the relative discretization error of 10^{-6} is obtained. Figure 6.16 demonstrates a comparison of the discretization errors from the generated adaptive FE meshes compared with those from uniform FE meshes. Even with roughly one million unknowns, the uniform mesh refinement does not achieve the desired accuracy. In Table 6.1 the number of unknowns increases to even more than 3.5 million. The memory (32 GB) of the computer used for this part of the simulation did not permit any further uniform refinement beyond 4 million nodes. In view of the comparison of direct vs. iterative solvers discussed in Section 5.5.3 we find that for direct solvers in particular, due to their rapidly increasing array storage, the application of adaptive meshes is reasonable.

6.5 Exercises

Exercise 6.1. Show that for nodal functions φ_ξ over simplicial triangulations the estimate $2\|1 - \varphi_\xi\|_{L^2(\omega_\xi)} \le |\omega_\xi|^{1/2}$ holds.

Exercise 6.2. Construct a nontrivial mesh which by subdivision of one element by means of bisection refines globally in the attempt to restore conformity by continued refinement.

Exercise 6.3. *Local extrapolation.* Let u denote a function on $[0, 1]$ and u^j an associated approximation over the mesh

$$\mathcal{T}^j = \{t_i^j : 0 \le i < n_j\} \quad \text{with} \quad t_i^j :=]x_i^j, x_{i+1}^j[$$

with $0 = x_0^j < x_1^j < \cdots < x_{n_j}^j = 1$. As an initial mesh let $\mathcal{T}^0 = \{]0, 1/2[,]1/2, 1[\}$ be given, and let the mesh $\mathcal{T}^1$ be generated by uniform refinement. Let the other meshes $\mathcal{T}^j$, $j > 1$ be suggested by bisection in combination with a refinement strategy on the basis of local extrapolation (see Section 6.2.1). Suppose η_i^j are the local error indicators, defined by

$$\eta_i^j = \|u - u^j\|_{t_i^j}^2.$$

Two model assumptions will be compared:

1. *Same exponent, different prefactor.* Let the local error on a subinterval $t_h^j :=$ $]z, z + h[$ satisfy the condition

$$\|u - u^j\|_{t_h^j}^2 = \begin{cases} c_1 h^\gamma & \text{for } t_h^j \subset]0, 1[, \\ c_2 h^\gamma & \text{for } t_h^j \subset]1/2, 1[\end{cases}$$

with $0 < c_1 < c_2$. Show that there exists a $j_0 > 0$ at which the error is "nearly uniform" in the following sense: for $j > j_0$ two further refinement steps lead to exactly one refinement in each subinterval.

2. *Same prefactor, different exponent.* Let the local error on a subinterval t_h^j satisfy the condition

$$\|u - u^j\|_{t_h^j}^2 = \begin{cases} ch^{\gamma_1} & \text{for } t_h^j \subset \,]0, 1/2[, \\ ch^{\gamma_2} & \text{for } t_h^j \subset \,]1/2, 1[\end{cases}$$

with $0 < \gamma_1 < \gamma_2$. What changes in this case?

Exercise 6.4. We return to equation (6.49) for the plasmon-polariton waveguide. Expand the differential operators and write the quadratic eigenvalue problem in weak formulation in terms of matrices. Perform a transformation of the kind $v = k_z u$ to obtain a linear eigenvalue problem of double dimension. Two principal reformulations exist. What properties do the arising matrices of double dimension have?

Chapter 7

Adaptive Multigrid Methods for Linear Elliptic Problems

In the preceding chapter we treated the numerical solution of elliptic boundary value problems after they had been discretized, i.e., in the form of elliptic grid equations. In weak formulation, the infinite dimensional space H_0^1 was replaced with some subspace $S_h \subset H_0^1$ with fixed dimension $N = N_h$. As a consequence, the iterative methods presented in Sections 5.2 and 5.3 could only exploit the *algebraic* structure of the arising (N, N)-matrices A.

In contrast, the present chapter will consider elliptic boundary value problems throughout as equations in infinite dimensional space, which means that we see discretization and numerical solution as a unity. The infinite dimensional space is modelled by some *hierarchy* of subspaces $S_k, k = 0, \dots, j$ with increasing dimension $N_0 < N_1 < \dots < N_j$. As a general rule, we assume these subspaces to be nested, i.e., we have $S_0 \subset S_1 \subset \dots \subset S_j \subset H_0^1$. Correspondingly, we here construct *hierarchical solvers*, which will exploit the *hierarchical geometric* structure of the nested FE triangulations $\mathcal{T}_k$. In this way a hierarchy of (N_k, N_k)-matrices A_k arises whose redundance can be turned into a reduction of the required computing times. As a realization of this fundamental concept we discuss here only *adaptive* multigrid methods, i.e., multigrid methods on problem adapted hierarchical meshes. The construction of such meshes has been extensively elaborated in Chapter 6.

In Section 5.5.1 above we introduced first convergence proofs for *nonadaptive* multigrid methods; however, they were only applicable to W-methods on uniform meshes. In Section 5.5.2 we introduced the method of hierarchical bases completely without any convergence proof. In order to fill this gap, we therefore deal with abstract subspace correction methods in Sections 7.1 and 7.2, which will permit transparent convergence proofs in the cases left open. On this basis, we then discuss in Section 7.3 several classes of adaptive multigrid methods for elliptic boundary value problems; we start with *additive* and *multiplicative* ones, two quite closely related variants, as well as a compromise between these two variants, the *cascadic* multigrid methods. Finally, in Section 7.5, we present two multigrid variants for linear elliptic eigenvalue problems.

7.1 Subspace Correction Methods

In Section 5.5.1 we already presented convergence proofs for classical multigrid methods (W-cycle only), although under the assumption of *uniform* meshes. Quite obviously, this assumption does not hold for the adaptive meshes generated in our Example 6.23, as can be seen from Figure 6.11. In Table 7.1, we summarize this information

Table 7.1. Example 6.23: comparison of nodal growth in uniform refinement $N_{j,\text{uniform}}$ and in adaptive refinement for Neumann boundary conditions $N_{j,\text{adaptive}}^{(N)}$ vs. Dirichlet boundary conditions $N_{j,\text{adaptive}}^{(D)}$; cf. also Figure 6.11.

j	$N_{j,\text{uniform}}$	$N_{j,\text{adaptive}}^{(N)}$	$N_{j,\text{adaptive}}^{(D)}$
0	10	10	10
1	27	27	27
2	85	53	62
3	297	83	97
4	1 105	115	193
5	4 257	163	253
6	16 705	228	369
7	66 177	265	542
8	263 225	302	815
9	1 051 137	393	1 115

in a different way: in uniform refinement, the number of nodes (here for $d = 2$) asymptotically behaves like

$$N_{j+1,\text{uniform}} \sim 4\, N_{j,\text{uniform}}$$

independent of the boundary conditions. In adaptive refinement, the difference of the boundary conditions is nicely observable (cf. Figure 6.11, bottom left and right) – associated with the different structure of the solution.

Therefore, in Sections 7.3.1 and 7.3.2, we will present convergence proofs for multigrid methods (including V-cycles) that also hold for *nonuniform* mesh hierarchies from adaptive methods. For this purpose, we need the abstract frame of subspace correction methods, which we will build up in the following.

7.1.1 Basic Principle

We begin with an example from 1870. In that year H. A. Schwarz (1843–1921) invented a constructive existence proof for harmonic functions on composed complex domains that was based on a principle of *domain decomposition* [184]. In Figure 7.1 we show one of the domains considered by Schwarz. Harmonic functions in $\mathbb{C}$ are just solutions of the Laplace equation in $\mathbb{R}^2$, so that we are back in the context of elliptic boundary value problems.

Example 7.1. We consider the Poisson model problem

$$-\Delta u = f \quad \text{in } \Omega,$$
$$u = 0 \quad \text{on } \partial\Omega \tag{7.1}$$

on the domain Ω, shown in Figure 7.1. This domain is the union of a circular domain Ω_1 and a rectangular domain Ω_2. Schwarz started from the fact that the analytical solutions of the Laplace equation on the circular domain as well as on the rectangular domain were given in the form of series representations. He posed the question of how to find an analytical solution on the whole of Ω from the given partial solutions. Schwarz' idea was to replace this problem on Ω by a sequence of alternating problems to be solved on the subdomains Ω_i. Of course, the two partial solutions must eventually agree on the intersection $\Omega_1 \cap \Omega_2$.

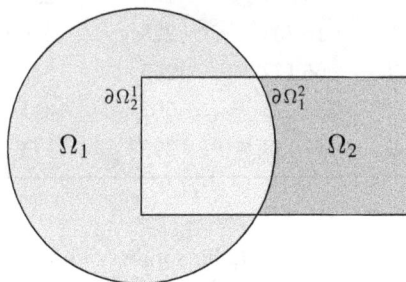

Figure 7.1. Poisson model problem (7.1); domain decomposition due to H. A. Schwarz[1] [184].

This method, now called the *alternating Schwarz method*, iteratively constructs a sequence $\{u_i\}$ of approximate solutions

$$u_i(x) = \begin{cases} u_i^1(x), & x \in \Omega_1 \setminus \Omega_2, \\ u_i^2(x), & x \in \Omega_2 \end{cases}$$

defined by

$$-\Delta u_i^1 = f \quad \text{in } \Omega_1, \qquad\qquad -\Delta u_i^2 = f \quad \text{in } \Omega_2,$$
$$u_i^1 = u_{i-1}^2 \text{ on } \partial\Omega_1^2 = \partial\Omega_1 \cap \Omega_2, \qquad u_i^2 = u_i^1 \text{ on } \partial\Omega_2^1 = \partial\Omega_2 \cap \Omega_1,$$
$$u_i^1 = 0 \quad \text{on } \partial\Omega_1 \setminus \partial\Omega_1^2, \qquad\qquad u_i^2 = 0 \quad \text{on } \partial\Omega_2 \setminus \partial\Omega_2^1.$$

As we will show later, this sequence converges *linearly* towards the solution of (7.1) with a convergence rate depending on the size of the overlap $\Omega_1 \cap \Omega_2$. The limiting cases are clear in advance: (i) for full overlap, i.e., for $\Omega_1 = \Omega_2 = \Omega$, the exact solution is available after one step; (ii) without overlap, i.e., for $\Omega_1 \cap \Omega_2 = \emptyset$, the method will not converge at all, since on the separation line $\partial\Omega_1 \cap \partial\Omega_2$ we always have $u_i = 0$.

---

[1] This classic drawing serves as the logo of the international conference series on domain decomposition.

The alternating Schwarz method can be directly extended to an arbitrary number of subdomains Ω_i. In Section 7.1.4 below we will derive a convergence theory for this general case on the basis of Sections 7.1.2 and 7.1.3. In Section 7.1.5 we will give a further application of this abstract theory to finite element methods of higher order.

Abstract Subspace Correction Methods. A rather flexible framework for a convergence analysis of both domain decomposition methods and multigrid methods has been established by *subspace correction methods* (here abbreviated as: SC). We start from the variational problem for $u \in V$,

$$\langle Au, v \rangle = \langle f, v \rangle \quad \text{for all } v \in V,$$

where A is no longer a symmetric positive definite matrix, but a corresponding operator $A : V \to V^*$, and $f \in V^*$ the associated right-hand side. This formulation is equivalent to the minimization of the variational functional

$$J(u) = \tfrac{1}{2}\langle Au, u \rangle - \langle f, u \rangle, \quad u \in V.$$

Before we turn to specific subspace correction methods in later sections, let us first work out an abstract convergence theory. Hereby we essentially follow the pioneering paper of J. Xu [223] and the beautifully clear survey article of H. Yserentant [228]. The basic ideas underlying all subspace correction methods are:

- *decomposition of ansatz space*

$$V = \sum_{k=0}^{m} V_k, \tag{7.2}$$

  where in general the sum need not be direct;

- *approximate solution* on the subspaces V_k

$$v_k \in V_k : \quad \langle B_k v_k, v \rangle = \langle f - A\tilde{u}, v \rangle \quad \text{for all } v \in V_k; \tag{7.3}$$

  here $B_k : V_k \to V_k^*$ is a symmetric positive definite "preconditioner" and $\tilde{u}$ an approximate solution.

Example 7.2. *Matrix ecomposition methods.* We begin with a space decomposition w.r.t. the *nodal basis* $\{\varphi_k\}$ in an FE discretization; it induces a direct decomposition into one-dimensional subspaces $V_k = \text{span } \varphi_k$, so that the "preconditioner" can even be chosen to be exact:

$$B_k = \langle A\varphi_k, \varphi_k \rangle / \langle \varphi_k, \varphi_k \rangle \in \mathbb{R}.$$

In the matrix notation of Section 5.2, the B_k are just the positive diagonal elements a_{kk}. In this way the *Gauss–Seidel method* from Section 5.2.2 may be interpreted as an SC-method, wherein the scalar equations (7.3) are solved *sequentially*. But also the classical *Jacobi method* from Section 5.2.1 may be regarded as an SC-method, where, however, the subproblems (7.3) are solved independently from each other, i.e., *in parallel*.

The example already shows that we will have to consider (at least) two different variants of subspace correction methods. By generalization of the Gauss–Seidel method we arrive at *sequential subspace correction* methods (in short: SSC). Correspondingly, from the Jacobi method we are led to *parallel subspace correction* methods (in short: PSC).

7.1.2   Sequential Subspace Correction Methods

As a generalization of the Gauss–Seidel method we get

Algorithm 7.3. *Sequential subspace correction method.*
$u^{\nu_{\max}} := \text{SSC}(u^0)$

> **for** $\nu = 0$ **to** $\nu_{\max} - 1$ **do**
> > $w_{-1} := u^\nu$
> > **for** $k = 0$ **to** m **do**
> > > find $v_k \in V_k$ with $\langle B_k v_k, v \rangle = \langle f - A w_{k-1}, v \rangle$ for all $v \in V_k$
> > > $w_k := w_{k-1} + v_k$
> > **end for**
> > $u^{\nu+1} := w_m$
> **end for**

In Figure 7.2 we give a geometric interpretation of this algorithm for the case $m = 1$. For the iterative error we obtain the following simple representation.

Lemma 7.4. *With* $T_k = B_k^{-1} A : V \to V_k$ *the following relation holds for the SSC-Algorithm 7.3:*

$$u - u^{\nu+1} = (I - T_m)(I - T_{m-1}) \cdots (I - T_0)\, (u - u^\nu). \tag{7.4}$$

First of all, let us mention that T_k is well-defined: because of $V_k \subset V$ we have $V^* \subset V_k^*$, with the consequence, however, that functionals distinguishable in V^* might be identical in V_k^*. Thus the restriction $V^* \subset V_k^*$ is not injective. The expression $B_k^{-1} A$ is to be understood in this sense.

Proof. With the above introduced notation we have here

$$\langle B_k v_k, v \rangle = \langle f - A w_{k-1}, v \rangle = \langle A u - A w_{k-1}, v \rangle \quad \text{for all } v \in V_k$$

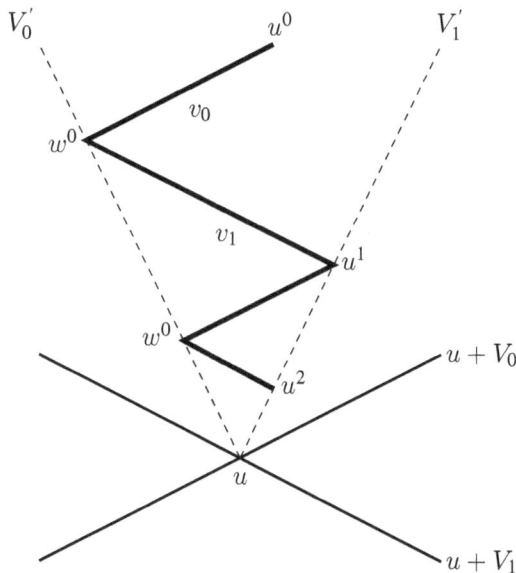

Figure 7.2. Sequential subspace correction method (Algorithm 7.3 SSC). Geometric interpretation for two subspaces V_0, V_1: sequential orthogonal projection to $V_0^\perp$, $V_1^\perp$ (dashed lines). The errors $u - u^\nu$ are, of course, A-orthogonal to the V_k, but depicted, for reasons of geometric representation, as Euclidean orthogonal. Cf. Figure 7.4 for PSC-methods.

and therefore $v_k = B_k^{-1} A(u - w_{k-1}) = T_k(u - w_{k-1})$. By induction, we get

$$u - w_k = u - (w_{k-1} + v_k) = u - w_{k-1} - T_k(u - w_{k-1})$$
$$= (I - T_k)(u - w_{k-1}) = (I - T_k)(I - T_{k-1}) \cdots (I - T_0)(u - w_{-1}). \qquad \square$$

Due to the *product* representation (7.4) sequential subspace correction methods are mostly called *multiplicative* methods.

Theoretical Assumptions. For an abstract convergence theory we need the following three assumptions:

(A0) *local convergence:* there exists some $\lambda < 2$ such that

$$\langle Av_k, v_k \rangle \le \lambda \langle B_k v_k, v_k \rangle \quad \text{for all } v_k \in V_k;$$

(A1) *stability:* there exists a positive constant $K_1 < \infty$ such that, for all $v \in V$, there is a decomposition $v = \sum_{k=0}^{m} v_k$ with $v_k \in V_k$ and

$$\sum_{k=0}^{m} \langle B_k v_k, v_k \rangle \le K_1 \langle Av, v \rangle;$$

(A2) *strengthened Cauchy–Schwarz inequality:* there exists a genuinely upper triangular matrix $\gamma \in \mathbb{R}^{(m+1)\times(m+1)}$ with $K_2 = \|\gamma\| < \infty$ such that

$$\langle Aw_k, v_l \rangle \leq \gamma_{kl} \langle B_k w_k, w_k \rangle^{1/2} \langle B_l v_l, v_l \rangle^{1/2}$$

holds for $k < l$, all $w_k \in V_k$, and $v_l \in V_l$.

(i)　Assumption (A0) means that the preconditioners B_k establish convergent fixed point iterations for the subproblems (7.3), since

$$\lambda_{\max}(B_k^{-1} A) \leq \lambda < 2$$

follows immediately: Due to the positivity of A as well as of B_k we then get $\rho(I - B_k^{-1} A) < 1$ for the spectral radius ρ of the iteration operator. Compare also assumption (5.5) in Theorem 5.1 for symmetrizable iteration methods.

(ii)　Assumptions (A0) and (A2) assure that the B_k are not "too small", so that the subspace corrections $B_k^{-1}(f - Au^\nu)$ do not shoot too far off the target. In contrast, assumption (A1) assures that the B_k are not "too large", so that the subspace corrections perform a significant step towards the solution. Figure 7.3 illustrates that this is directly connected with the "angles" between the subspaces V_k.

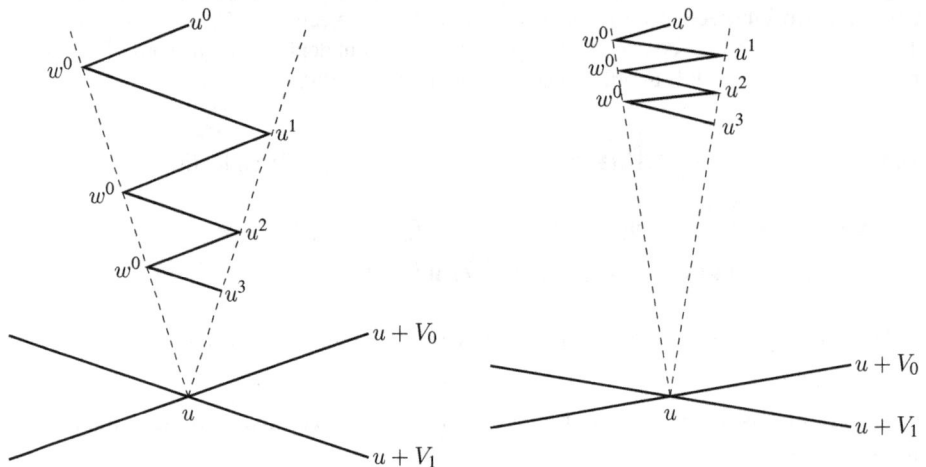

Figure 7.3. Dependence of convergence rate of sequential subspace correction methods on the stability of the space decomposition into the subspaces V_k.

(iii)　Assumption (A2) with $K_2 = \lambda \sqrt{m}$ follows directly from (A0) and the Cauchy–Schwarz inequality. The strengthening here consists in showing that $K_2 \ll \lambda \sqrt{m}$.

(iv)　Note that assumption (A2) depends on the *ordering of the decomposition*, which structurally agrees with sequential subspace correction methods – cf., for example our presentation of the Gauss–Seidel method in Section 5.2.2. Below we will introduce a slightly stronger assumption (A2'), which is independent of the ordering and thus structurally agrees with parallel subspace correction methods. On the other hand, a slightly weaker version of assumption (A2) is sufficient for sequential subspace corrections (see, e.g., [228]), but this will not be required here.

Convergence.   With the assumptions (A0)–(A2) the convergence of the SSC-algorithm 7.3 can now be confirmed.

Lemma 7.5. *Under the assumption* (A1) *the following result holds for* $v = \sum_{k=0}^m v_k$ *with* $v_k \in V_k$ *and for arbitrary* $w_0, \ldots, w_m \in V$:

$$\sum_{k=0}^m \langle v_k, A w_k \rangle \leq \sqrt{K_1 \langle Av, v \rangle} \left(\sum_{k=0}^m \langle A T_k w_k, w_k \rangle \right)^{1/2}.$$

Proof. Upon using the Cauchy–Schwarz inequality for the scalar product induced by B_k^{-1} on V_k^* we get

$$\sum_{k=0}^m \langle v_k, A w_k \rangle = \sum_{k=0}^m \langle B_k^{-1} B_k v_k, A w_k \rangle = \sum_{k=0}^m \langle B_k v_k, A w_k \rangle_{B_k^{-1}}$$

$$\leq \sum_{k=0}^m \langle B_k^{-1} B_k v_k, B_k v_k \rangle^{1/2} \langle B_k^{-1} A w_k, A w_k \rangle^{1/2}.$$

Let $\xi, \eta \in \mathbb{R}^{m+1}$ with $\xi_k = \langle B_k v_k, v_k \rangle^{1/2}$ and $\eta_k = \langle A T_k w_k, w_k \rangle^{1/2}$. Then the above inequality reads

$$\sum_{k=0}^m \langle v_k, A w_k \rangle \leq \xi^T \eta \leq |\xi| \, |\eta|,$$

where we have used the Cauchy–Schwarz inequality for the discrete Euclidean scalar product and the Euclidean vector norm $|\cdot|$. With (A1) the assertion follows.   □

Theorem 7.6. *Let the assumptions* (A0), (A1), *and* (A2) *be satisfied. Then the following result holds for the SSC-Algorithm 7.3:*

$$\|u - u^{\nu+1}\|_A^2 \leq \left(1 - \frac{2 - \lambda}{K_1 (1 + K_2)^2} \right) \|u - u^\nu\|_A^2. \tag{7.5}$$

Proof. We do the proof in two steps.
  (I) Let $v = \sum_{k=0}^m v_k$ denote a decomposition according to (A1) and

$$E_k = (I - T_k) \cdots (I - T_0), \quad k = 0, \ldots, m, \quad E_{-1} = I.$$

Then

$$\langle Av, v \rangle = \sum_{l=0}^m \langle Av, v_l \rangle = \sum_{l=0}^m \langle A E_{l-1} v, v_l \rangle + \sum_{l=1}^m \langle A (I - E_{l-1}) v, v_l \rangle, \tag{7.6}$$

where, due to $E_{-1} = I$, the term for $l = 0$ in the right-hand sum drops out. The first sum can be estimated by means of Lemma 7.5:

$$\sum_{k=0}^{m} \langle A E_{k-1} v, v_k \rangle \leq \sqrt{K_1 \langle Av, v \rangle} \Big(\sum_{k=0}^{m} \langle A T_k E_{k-1} v, E_{k-1} v \rangle \Big)^{1/2}. \qquad (7.7)$$

For the second sum we obtain the recursion

$$E_l = (I - T_l) E_{l-1} = E_{l-1} - T_l E_{l-1} = E_{-1} - \sum_{k=0}^{l} T_k E_{k-1},$$

i.e.,

$$I - E_{l-1} = \sum_{k=0}^{l-1} T_k E_{k-1}.$$

Using $w_k = T_k E_{k-1} v$ in (A2) we get the estimate

$$\sum_{l=1}^{m} \langle A(I - E_{l-1}) v, v_l \rangle = \sum_{l=1}^{m} \sum_{k=0}^{l-1} \langle A w_k, v_l \rangle$$

$$\leq \sum_{l=1}^{m} \sum_{k=0}^{l-1} \gamma_{kl} \langle B_k w_k, w_k \rangle^{1/2} \langle B_l v_l, v_l \rangle^{1/2}$$

$$\leq K_2 \Big(\sum_{k=0}^{m} \langle A T_k E_{k-1} v, E_{k-1} v \rangle \Big)^{1/2} \Big(\sum_{l=0}^{m} \langle B_l v_l, v_l \rangle \Big)^{1/2}.$$

The stability condition (A1) then leads to

$$\sum_{l=1}^{m} \langle A(I - E_{l-1}) v, v_l \rangle \leq \sqrt{K_1 \langle Av, v \rangle} \, K_2 \Big(\sum_{k=0}^{m} \langle A T_k E_{k-1} v, E_{k-1} v \rangle \Big)^{1/2}. \qquad (7.8)$$

Insertion of (7.7) and (7.8) into (7.6), collection of equal terms, taking the square, and division by $\langle Av, v \rangle$ finally yields

$$\langle Av, v \rangle \leq K_1 (1 + K_2)^2 \sum_{k=0}^{m} \langle A T_k E_{k-1} v, E_{k-1} v \rangle. \qquad (7.9)$$

After this preparation we turn to the original estimate (7.5). With the starting point $E_k = (I - T_k) E_{k-1}$ we get, for all $v \in V$, the recursion

$$\langle A E_m v, E_m v \rangle$$

$$= \langle A E_{m-1} v, E_{m-1} v \rangle - 2 \langle A T_m E_{m-1} v, E_{m-1} v \rangle + \langle A T_m E_{m-1} v, T_m E_{m-1} v \rangle$$

$$= \langle Av, v \rangle - \sum_{k=0}^{m} (2 \langle A T_k E_{k-1} v, E_{k-1} v \rangle - \langle A T_k E_{k-1} v, T_k E_{k-1} v \rangle).$$

By insertion of $w = E_{k-1}v \in V_k$ we may apply (A0) to obtain

$$\langle AT_k w, T_k w \rangle \leq \lambda \langle B_k T_k w, T_k w \rangle = \lambda \langle AT_k w, w \rangle.$$

Insertion of this relation supplies

$$\langle A E_m v, E_m v \rangle \leq \langle Av, v \rangle - (2 - \lambda) \sum_{k=0}^{m} \langle AT_k E_{k-1}v, E_{k-1}v \rangle.$$

The right-hand sum can be estimate from below by virtue of (7.9). Hence, due to $2 - \lambda > 0$, we get

$$\|E_m v\|_A^2 \leq \left(1 - \frac{2 - \lambda}{K_1(1 + K_2)^2}\right) \|v\|_A^2.$$

In particular, for $v = u - u^\nu$, we obtain the statement (7.5) of the theorem. □

The contraction rate in (7.5) depends, just as (A2), on the ordering of the subspace corrections. With a slightly stronger assumption, a contraction rate independent of the ordering can be verified (see Exercise 7.9).

7.1.3 Parallel Subspace Correction Methods

As a generalization of the Jacobi method we obtain

Algorithm 7.7. *Parallel subspace correction method.*
$u^{\nu_{\max}} := \mathrm{PSC}(u^0)$

> **for** $\nu = 0$ **to** $\nu_{\max}$ **do**
> > **for** $k = 0$ **to** m **do**
> > > determine $v_k \in V_k$ with $\langle B_k v_k, v \rangle = \langle f - Au^\nu, v \rangle$ for all $v \in V_k$
> > **end for**
> > $u^{\nu+1} := u^\nu + \sum_{k=0}^{m} v_k$
> **end for**

In Figure 7.4 we give a geometric interpretation of this algorithm for $m = 1$. For the iterative error we obtain the following simple representation.

Lemma 7.8. *With $T_k = B_k^{-1}A : V \to V_k$ it holds for the PSC-Algorithm 7.7 that*

$$u - u^{\nu+1} = \left(I - \sum_{k=0}^{m} T_k\right)(u - u^\nu) = (I - B^{-1}A)(u - u^\nu), \tag{7.10}$$

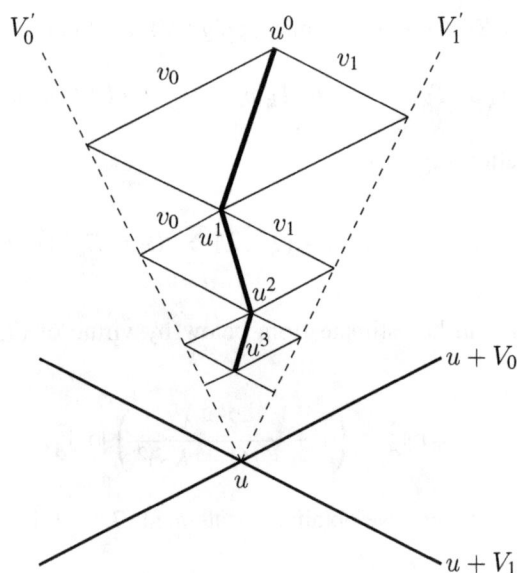

Figure 7.4. Parallel subspace correction methods (Algorithm 7.7 PSC). Geometric interpretation for two subspaces V_0, V_1: parallel orthogonal projection to $V_0^\perp, V_1^\perp$ (dashed lines) with parallelogram addition. Of course, the single components of the errors $u - u^\nu$ are A-orthogonal to the V_k, for reasons of geometric representation, however, they are depicted as Euclidean orthogonal. Cf. Figure 7.2 for SSC-methods.

wherein the inverse

$$B^{-1} = \sum_{k=0}^{m} B_k^{-1} \tag{7.11}$$

of the preconditioner $B : V \to V^*$ is symmetric and positive.

Proof. Upon interpreting B_k^{-1} as in Lemma 7.4 we get for the iterative error

$$\langle B_k v_k, v \rangle = \langle f - A u^\nu, v \rangle \quad \text{for all } v \in V_k.$$

Let therefore $v_k = B_k^{-1} A(u - u^\nu) = T_k(u - u^\nu)$. Then

$$u - u^{\nu+1} = u - u^\nu - \sum_{k=0}^{m} v_k = \left(I - \sum_{k=0}^{m} T_k \right)(u - u^\nu)$$

holds. Because of the symmetry of the B_k^{-1}, their sum B^{-1} is itself symmetric. The ellipticity of the B_k leads to the continuity of B^{-1}. Moreover, due to the positivity of the B_k^{-1} we get that

$$0 = \langle B^{-1} w, w \rangle = \sum_{k=0}^{m} \langle B_k^{-1} w, w \rangle$$

implies $w = 0 \in V_k^*$ for all k, i.e., $\langle w, v_k \rangle = 0$ for all $v_k \in V_k$. Because of $V = \sum_{k=0}^{m} V_k$ then also $\langle w, v \rangle = 0$ for all $v \in V$ and thus $w = 0$, which confirms the positivity of B^{-1}.   □

Due to the *sum* representation (7.10) parallel subspace correction methods are mostly called *additive* methods.

Theoretical Assumptions. For an abstract convergence theory this time just two assumptions will do, where (A1) is already known from sequential subspace correction methods:

(A1) *stability:* there exists a positive constant $K_1 < \infty$, such that for every $v \in V$ a decomposition $v = \sum_{k=0}^{m} v_k$ with $v_k \in V_k$ and

$$\sum_{k=0}^{m} \langle B_k v_k, v_k \rangle \le K_1 \langle Av, v \rangle$$

exists;

(A2′) *strengthened Cauchy–Schwarz inequality (modified):* there exists a symmetric matrix $\gamma \in \mathbb{R}^{(m+1)\times(m+1)}$ such that

$$\langle A w_k, v_l \rangle \le \gamma_{kl} \| w_k \|_{B_k} \| v_l \|_{B_l} \tag{7.12}$$

holds for all $w_k \in V_k$ and $v_l \in V_l$. For a given disjoint index splitting

$$\mathcal{J}_0 \cup \cdots \cup \mathcal{J}_r = \{0, \ldots, m\} \tag{7.13}$$

with $\gamma(\mathcal{J}_l) = (\gamma_{ij})_{i,j \in \mathcal{J}_l} \in \mathbb{R}^{|\mathcal{J}_l| \times |\mathcal{J}_l|}$ let the following bound be satisfied:

$$K_2 = \sum_{l=0}^{r} \| \gamma(\mathcal{J}_l) \| \le \infty. \tag{7.14}$$

Compared with the three assumptions for SSC-methods, we here can dispense with (A0). Moreover, assumption (A2), which depends on the ordering of the decomposition, has been replaced by assumption (A2′), which is ordering independent and structurally agrees with PSC-methods. This again assures that the B_k are "not too small" , whereas (A1) again states the B_k are "not too large".

Convergence. Let us first consider the trivial example $m = 1$ with $V_0 = V_1 = V$. Obviously, in this case the *direction* of the iterative step is perfect, but not its *length* (it is a factor of 2 too large), so that the simple PSC-Algorithm 7.7 would diverge. However, we may hope to produce a convergent method by a subsequent stepsize control. In fact, PSC-methods should most reasonably not be regarded as independent iterative methods, but as a preconditioners B for a gradient or a CG-method. For their convergence, the condition number estimate in the following theorem is sufficient.

Theorem 7.9. *Let the assumptions* (A1) *and* (A2′) *be satisfied. Then the following result holds for the parallel subspace correction method Algorithm 7.7*

$$\lambda_{\min}(B^{-1}A) \geq \frac{1}{K_1}, \quad \lambda_{\max}(B^{-1}A) \leq K_2.$$

Consequently,

$$\kappa(B^{-1}A) \leq K_1 K_2 \tag{7.15}$$

holds. If $K_2 < 2$, *then Algorithm 7.7 converges.*

Proof. The preconditioned operator $T = B^{-1}A : V \to V$ is symmetric w.r.t. the scalar product $(\cdot, \cdot)_A = \langle A\cdot, \cdot \rangle$ induced by A on V. Its condition number is determined by the quotient of the eigenvalues $\lambda_{\max}(T)/\lambda_{\min}(T)$.

(a) We begin with an estimate of $\lambda_{\max}(T)$. With the index splitting underlying (A2′) we have

$$T(\mathcal{J}_j) = \sum_{k \in \mathcal{J}_j} B_k^{-1}A, \quad T = \sum_{j=0}^{r} T(\mathcal{J}_j).$$

Application of the strengthened Cauchy–Schwarz condition (7.12) then yields

$$(T(\mathcal{J}_j)v, T(\mathcal{J}_j)v)_A = \sum_{k,l \in \mathcal{J}_j} \langle AT_k v, T_l v \rangle$$

$$\leq \sum_{k,l \in \mathcal{J}_j} \gamma_{kl} \langle B_k T_k v, T_k v \rangle^{1/2} \langle B_l T_l v, T_l v \rangle^{1/2}$$

$$= \sum_{k,l \in \mathcal{J}_j}^{m} \gamma_{kl} \langle AT_k v, v \rangle^{1/2} \langle AT_l v, v \rangle^{1/2} = \xi^T \gamma(\mathcal{J}_j)\xi$$

with $\xi \in \mathbb{R}^{|\mathcal{J}_j|}$, $\xi_k = (T_k v, v)_A^{1/2}$. Now

$$\xi^T \gamma(\mathcal{J}_j)\xi \leq \|\gamma(\mathcal{J}_j)\| \, |\xi|^2 = \|\gamma(\mathcal{J}_j)\| \sum_{k \in \mathcal{J}_j} (T_k v, v)_A = \|\gamma(\mathcal{J}_j)\| \, (T(\mathcal{J}_j)v, v)_A.$$

Collecting all estimates then leads to

$$(T(\mathcal{J}_j)v, T(\mathcal{J}_j)v)_A \leq \|\gamma(\mathcal{J}_j)\| \, (T(\mathcal{J}_j)v, v)_A,$$

which immediately verifies that $\lambda_{\max}(T(\mathcal{J}_j)) \leq \|\gamma(\mathcal{J}_j)\|$. Summation over all index sets $\mathcal{J}_0, \ldots, \mathcal{J}_r$ in (A2′) with $T = \sum_{j=0}^{r} T(\mathcal{J}_j)$ finally confirms $\lambda_{\max}(T) \leq K_2$.

(b) For the minimal eigenvalue we consider a decomposition of v according to (A1). With $w_k = v$ in Lemma 7.5 we obtain $(v, v)_A \leq K_1(Tv, v)_A$ and thus $\lambda_{\min}(T) \geq K_1^{-1}$. Therefore, the condition is bounded by

$$\kappa(T) = \lambda_{\max}(T)/\lambda_{\min}(T) \leq K_1 K_2,$$

which confirms statement (7.15).

(c) The spectral radius ρ of the iteration operator $I - T$ of Algorithm 7.7 is bounded by $\rho \leq \max(|1 - \lambda_{\max}(T)|, |1 - \lambda_{\min}(T)|)$. With

$$\frac{1}{K_1} \leq \lambda_{\min}(T), \quad \lambda_{\max}(T) \leq K_2, \quad K_1 K_2 \geq 1$$

we then conclude $\rho \leq K_2 - 1$. Hence, for $\rho < 1$ the relation $K_2 < 2$ is sufficient. $\quad \square$

Summarizing, the convergence theory for PSC-methods is significantly simpler than the one for SSC-methods (fewer assumptions, shorter proofs); moreover, they are easier to parallelize.

7.1.4 Overlapping Domain Decomposition Methods

In domain decomposition methods large problems are *spatially* decomposed, i.e., the domain Ω is split into subdomains Ω_i, which can be distributed to different compute nodes, where the computer nodes are only loosely coupled. Then each node must essentially handle only a discretization of the associated subdomain and compute the corresponding solution independently. An efficient decomposition into subdomains that belong to one computer node is an interesting problem of its own, mainly from computer science and graph theory. Communication between the computer nodes is only necessary for the exchange of data in the overlapping domain. The aim is to distribute data as well as speeding up computation, which is why it is advantageous to solve the subproblems *in parallel* and independently from each other. Only by this procedure can massively parallel high performance computers with distributed memory be effectively used to solve PDEs. As a theoretical frame for a convergence theory abstract parallel subspace correction methods (PSC) are well-suited, which have been established in the previous Section 7.1.3.

PSC-methods. We start from some domain Ω, decomposed into $m + 1$ subdomains Ω_k, i.e.,

$$\Omega = \text{int}\left(\bigcup_{k=0}^{m} \overline{\Omega_k} \right),$$

where all subdomains satisfy the interior cone condition A.13. Let the maximum diameter of the subdomains be denoted by

$$H = \max_{k=0,\ldots,m} \max_{x,y \in \Omega_k} \|x - y\|.$$

In this section we treat the case that the subdomains overlap: For the overlap we define subdomains extended by some margin of width ρH,

$$\Omega'_k = \bigcup_{x \in \Omega_k} B(x; \rho H) \cap \Omega,$$

where $B(x; \rho H)$ denotes a ball with center $x \in \Omega_k$ and radius ρH. We focus on problems with *Robin boundary conditions*, so that the linear operator $A : H^1(\Omega) \to H^1(\Omega)^*$ is continuous with continuous inverse. For simplicity, we choose the ansatz space $V = H^1(\Omega)$ here. Over the extended subdomains Ω'_k we define subspaces

$$V_k = \left\{ v \in V : \text{supp}\, v \subset \overline{\Omega'_k} \right\} \subset H^1(\Omega'_k).$$

Thus a spatial decomposition (7.2) is defined. We assume that none of the Ω'_k covers the whole domain Ω, so that all V_k satisfy *homogeneous Dirichlet boundary conditions*, at least on some part of $\partial \Omega'_k$ (cf. our presentation of the historical Example 7.1 by H. A. Schwarz; see also Figure 7.1).

In order to study the PSC-method on this domain decomposition, we have to verify the two assumptions (A1) and (A2$'$) of Theorem 7.9. For simplicity, we start from exact solutions of the subproblems, i.e., we have $B_k = A|_{V_k}$. We begin with the strengthened Cauchy–Schwarz condition (A2$'$). For this, we require the concept of neighborhood of two subdomains. Two domains Ω_k and Ω_l are called *neighbors*, if their extensions Ω'_k and Ω'_l intersect. Correspondingly, we define the *neighborhood sets*

$$\mathcal{N}(k) = \{ l \in 0, \ldots, m : \Omega'_k \cap \Omega'_l \neq \emptyset \}$$

and $N = \max_{k=0,\ldots,m} |\mathcal{N}(k)|$, where $N - 1$ is the maximal number of genuine neighbors of a subdomain. Note that this definition depends on the width ρH of the margin.

Lemma 7.10. *For the overlapping domain decomposition method as just described, the strengthened Cauchy–Schwarz condition (A2$'$) holds with $K_2 = N - 1$.*

Proof. We have $\langle Aw_k, v_l \rangle \leq \gamma_{kl} \langle Aw_k, w_k \rangle^{1/2} \langle Av_l, v_l \rangle^{1/2}$ for all $w_k \in V_k$, $v_l \in V_l$ and $\gamma_{kl} = 1$, if Ω_k and Ω_l are neighbors, otherwise $\gamma_{kl} = 0$. As each subdomain has at most $N - 1$ genuine neighbors, we get $\sum_{l=0, l\neq k}^m |\gamma_{kl}| \leq N - 1$ for all k. From the Theorem of Gerschgorin the spectral radius of γ is bounded by the off-diagonal row sum, here $N - 1$. $\square$

The proof of the stability condition (A1) is harder, in particular for smooth functions. For a better understanding, we go ahead by constructing an example in space dimension $d = 1$.

Example 7.11. Let us consider the Laplace problem on $\Omega = \,]0, 1[$ with Robin boundary conditions

$$\Delta u = 0, \quad -u'(0) + u(0) = 0 = u'(1) + u(1).$$

The corresponding variational functional (4.11) from Theorem 4.2 has the form

$$\langle Au, u \rangle = \int_0^1 u'^2 \, dx + u^2(0) + u^2(1).$$

Let us define the subdomains as uniform subdivision $\Omega_k =]kH, (k+1)H[$ with diameter $H = 1/(m+1)$ and overlap $\rho H < H/2$. As a prototype of a smooth function we choose, for simplicity again, the function $v = 1$. This supplies $\langle Av, v \rangle = 2$ independent of the subdivision. For this function we may define the decomposition $v = \sum_{k=0}^{m} v_k$ according to

$$v_k(x) = \begin{cases} \frac{1}{2} + \frac{x-kH}{2\rho H}, & x \in]kH - \rho H, kH + \rho H[, \\ 1, & x \in]kH + \rho H, (k+1)H - \rho H[, \\ \frac{1}{2} + \frac{(k+1)H - x}{2\rho H}, & x \in](k+1)H - \rho H, (k+1)H + \rho H[\end{cases}$$

for $k = 1, \ldots, m-1$. Because of $\Omega_0 \cap \Omega_m = \emptyset$ for $m > 1$ the functions v_0 and v_m on the boundary intervals Ω_0' and Ω_m' are thus uniquely determined (see Figure 7.5).

for some fixed $\sigma \geq \lambda_{\min}(A)$. By virtue of the Friedrichs inequality (A.26) we may then conclude

$$\sum_{k=0}^{m} \langle B_k v_k, v_k \rangle = \sum_{k=0}^{m} \langle A v_k, v_k \rangle \geq \alpha_A \sum_{k=0}^{m} |v_k|_{H^1(\Omega'_k)}^2$$

$$\geq \frac{\alpha_A}{(1+2\rho)^2 H^2} \sum_{k=0}^{m} \|v_k\|_{L^2(\Omega'_k)}^2 \geq \frac{\alpha_A}{N(1+2\rho)^2 H^2} \|v\|_{L^2(\Omega)}^2$$

$$\geq \frac{\alpha_A}{\sigma N(1+2\rho)^2 H^2} \|v\|_A^2 = \mathcal{O}(m^{2/d}) \|v\|_A^2. \tag{7.17}$$

So we have proved that $K_1 = \mathcal{O}(m^{2/d})$. The m-dependence of the condition number estimate is therefore unavoidable.

Compared with the above one-dimensional example, the case $\rho \to 0$ seems to be harmless. This stems from the fact that for fixed $H > 0$ not every smooth function achieves the expected effect (see Exercise 7.11). As we have only proved a lower bound for K_1, there could still be functions for which the condition $\rho \geq \rho_0 > 0$ would be strictly necessary. As an example we choose again $v = 1$ on an interior subdomain Ω_i with the decomposition $v = \sum_{k=0}^{m} v_k$. As will be shown in Exercise 7.15, there exists a constant $c > 0$ such that

$$|v_i|_{H^1(\Omega'_i)} \geq \frac{c}{\rho}.$$

On the basis of this example we thus have $K_1 \to \infty$ for $\rho \to 0$ in general. Hence, for *robust* domain decomposition methods an overlap will be necessary.

Incorporation of a Coarse Grid Space.   As elaborated above, the unsatisfactory dependence of the condition number estimate on m stems from the representation of smooth functions as a sum of local functions, which inevitably exhibit large gradients at the subdomain boundaries. In order to also be able to represent spatially low frequency functions in a stable way, one extends the space V by some *coarse grid space* W. As an example of such an extension we want to introduce an additive *agglomeration method* [131] here.

Motivated by the one-dimensional example, we construct a coarse grid space W from a basis that represents a positive *partition of unity* with the following properties:

(a) supp $\phi_k \subset \overline{\Omega'_k}$,

(b) $0 \leq \phi_k \leq 1$,

(c) $\sum_{k=0}^{m} \phi_k = 1$.

This can be achieved, e.g., by the following construction

$$\hat{\phi}_k(x) = \max\left(0, 1 - \frac{\text{dist}(x, \Omega_k)}{\rho H}\right), \quad \phi_k = \frac{\hat{\phi}_k}{\sum_{k=0}^{m} \hat{\phi}_k}. \tag{7.18}$$

For $d = 1$ and uniform subdivision this is, without fixing the scaling, exactly Figure 7.5. As each point $\xi \in \Omega$ lies in the completion of at least one subdomain Ω_k, the denominator is always strictly positive, i.e.,

$$s(\xi) = \sum_{k=0}^{m} \hat{\phi}_k(\xi) > 0. \tag{7.19}$$

Hence, the ansatz functions ϕ_k exist and we can define the elements of the coarse grid space by

$$w = \sum_{k=0}^{m} \alpha_k \phi_k \in W. \tag{7.20}$$

With this extension of the space decomposition by the coarse grid space at hand, we again examine the assumptions (A1) and (A2′) in Theorem 7.9. As before, the strengthened Cauchy–Schwarz condition (A2′) is easier to verify.

Lemma 7.12. *For the overlapping domain decomposition with coarse grid extension $V = \sum_{k=0}^{m} V_k + W$ the strengthened Cauchy–Schwarz condition (A2′) holds with the constant $K_2 = N$.*

Proof. For the index splitting we choose $\mathcal{J}_m = \{0, \dots, m\}$ and $\mathcal{J}_W$. Then with $\gamma(\mathcal{J}_W) = 1$ and, due to Lemma 7.10, with $\|\gamma(\mathcal{J}_m)\| \le N - 1$ we directly obtain $K_2 = N$. □

Some more work is needed to show the stability condition (A1).

Lemma 7.13. *Let ϕ_k, $k = 0, \dots, m$ denote the coarse grid ansatz functions. Then*

$$\|\nabla \phi_k\|_{L^\infty(\Omega)} \le \frac{N}{\rho H}. \tag{7.21}$$

Proof. First of all, $\hat{\phi}_k$ is Lipschitz-continuous with Lipschitz constant $L = (\rho H)^{-1}$. Now let $x \in \overline{\Omega}_i$. With (7.19) we get for all $y \in S(x, \rho H) \subset \Omega_i'$

$$
\begin{aligned}
|\phi_j(x) - \phi_j(y)| &= \left| \frac{\hat{\phi}_j(x)}{s(x)} - \frac{\hat{\phi}_j(y)}{s(y)} \right| \\
&\le \frac{|\hat{\phi}_j(x) - \hat{\phi}_j(y)|}{s(x)} + \left| \frac{\hat{\phi}_j(y)}{s(x)} - \frac{\hat{\phi}_j(y)}{s(y)} \right| \\
&\le L\|x - y\| + \hat{\phi}_j(y) \frac{|s(y) - s(x)|}{s(x)s(y)} \\
&\le L\|x - y\| + \sum_{k=0}^{m} |\hat{\phi}_k(y) - \hat{\phi}_k(x)|.
\end{aligned}
$$

Each term in the above sum, for which $\Omega_k \neq \Omega_i$ are neighbors of Ω_i, is positive, all others vanish. Thus at most $N - 1$ terms arise. Hence,

$$|\phi_j(x) - \phi_j(y)| \leq NL\|x - y\|. \tag{7.22}$$

Moreover, the ϕ_j are continuously differentiable apart from a set of measure zero, the subdomain boundaries, and thus weakly differentiable on the whole of Ω. From the Lipschitz bound (7.22) we then can confirm the assertion (7.21).   □

Let us now consider the decomposition

$$v = \sum_{k=0}^{m} v_k + w, \quad v_k \in V_k$$

by an explicit construction of v_k. In view of (7.17) and

$$\|v - w\|_{L^2(\Omega)} \leq \sum_{k=0}^{m} \|v_k\|_{L^2(\Omega)}$$

we will choose the decomposition such that the upper bound is as small as possible. This can be achieved by the approximate L^2-projections (see Exercise 7.16)

$$v_k = (v - \bar{v}_k)\phi_k \quad \text{with} \quad \bar{v}_k = |\Omega'_k|^{-1} \int_{\Omega'_k} v\, dx, \ k = 0, \dots, m. \tag{7.23}$$

With the representation (7.20) and the partition of unity we then have

$$v = \sum_{k=0}^{m} v_k + w = \sum_{k=0}^{m}(v - \bar{v}_k)\phi_k + \sum_{k=0}^{m}\alpha_k(v)\phi_k = v + \sum_{k=0}^{m}(\alpha_k(v) - \bar{v}_k)\phi_k,$$

from which we immediately obtain the identity $\alpha_k(v) = \bar{v}_k$. After these preparations we are now ready to state the following lemma.

Lemma 7.14. *For the coarse grid projection $Q : V \to W$ with*

$$Qv = \sum_{k=0}^{m} \bar{v}_k \phi_k \tag{7.24}$$

there exists a constant c independent of H and N such that

$$|Qv|_{H^1(\Omega)} \leq \frac{cN^3}{\rho}|v|_{H^1(\Omega)}.$$

Proof. For arbitrary constants c_k, the linearity of Q and the result (7.21) lead to

$$|Qv|^2_{H^1(\Omega)} \leq \sum_{k=0}^{m} |Qv|^2_{H^1(\Omega'_k)} = \sum_{k=0}^{m} |Qv - c_k|^2_{H^1(\Omega'_k)} \leq \sum_{k=0}^{m} |Q(v - c_k)|^2_{H^1(\Omega'_k)}$$

$$\leq \sum_{k=0}^{m} \Big\| \sum_{j \in \mathcal{N}_k} \alpha_j (v - c_k) \nabla \phi_j \Big\|^2_{L^2(\Omega'_k)}$$

$$\leq \Big(\frac{N}{\rho H}\Big)^2 \sum_{k=0}^{m} |\mathcal{N}_k| \sum_{j \in \mathcal{N}_k} \|\alpha_j(v - c_k)\|^2_{L^2(\Omega'_k \cap \Omega'_j)}.$$

If we define $\hat{\Omega}_k = \bigcup_{j \in \mathcal{N}_k} \Omega'_j$ and

$$c_k = \frac{1}{|\hat{\Omega}_k|} \int_{\hat{\Omega}_k} v \, dx,$$

then, with the Cauchy–Schwarz inequality and the Poincaré inequality (A.25), we may derive from Theorem A.16

$$\|\alpha_j(v - c_k)\|_{L^2(\Omega'_j)} = |\Omega'_j|^{1/2} |\alpha_j(v - c_k)| \leq |\Omega'_j|^{-1/2} \|v - c_k\|_{L^2(\Omega'_j)} \|1\|_{L^2(\Omega'_j)}$$

$$\leq \|v - c_k\|_{L^2(\hat{\Omega}_k)} \leq C_p H |v - c_k|_{H^1(\hat{\Omega}_k)} = C_p H |v|_{H^1(\hat{\Omega}_k)}.$$

Since $\hat{\Omega}_k$ covers at most $(N+1)^2$ subdomains Ω_j, we obtain

$$|Qv|^2_{H^1(\Omega)} \leq \frac{N^2(N+1)}{(\rho H)^2} \sum_{k=0}^{m} \sum_{j \in \mathcal{N}(k)} C_p^2 H^2 |v|^2_{H^1(\hat{\Omega}_k)}$$

$$\leq \frac{C_p^2 N^2 (N+1)^4}{\rho^2} |v|^2_{H^1(\Omega)}. \qquad \square$$

Remark 7.15. In the case of homogeneous Dirichlet boundary conditions the coarse grid space W will be restricted to those functions that satisfy the boundary conditions. Thus, subdomains incident to the Dirichlet boundary will be excluded from the construction of W. Consequently, the proof of Lemma 7.14 will have to apply the Friedrichs inequality (A.26) instead of the Poincaré inequality (A.25) at the boundary domains.

After these preparations we are now ready to bound the stability constant K_1 independently of m.

Theorem 7.16. *There exists a constant $c < \infty$ independent of m, ρ, and N such that for the decomposition of $v \in V$ according to (7.18), (7.23), and (7.20) the assumption (A1) is satisfied with*

$$K_1 = c N^6 \rho^{-2}.$$

Proof. Due to the ellipticity of A (see A.10) and $B_k = A|_{V_k}$ and the (at least on parts of the boundary valid) Dirichlet boundary conditions of the v_k we have, after application of the Friedrichs inequality (A.26),

$$\sum_{k=0}^{m} \langle B_k v_k, v_k \rangle + \langle B_W w, w \rangle \le C_a C_F \sum_{k=0}^{m} |v_k|^2_{H^1(\Omega)} + C_a \|w\|^2_{H^1(\Omega)}. \quad (7.25)$$

For $k = 0, \ldots, m$ the result (7.23) and Lemma 7.13 imply

$$
\begin{aligned}
|v_k|_{H^1(\Omega)} &= \|\nabla((v - \bar{v}_k)\phi_k)\|_{L^2(\Omega'_k)} \\
&\le \|(v - \bar{v}_k)\nabla\phi_k\|_{L^2(\Omega'_k)} + \|\nabla(v - \bar{v}_k)\phi_k\|_{L^2(\Omega'_k)} \\
&\le \frac{N}{\rho H} \|v - \bar{v}_k\|_{L^2(\Omega'_k)} + \|\nabla v\|_{L^2(\Omega'_k)}. \quad (7.26)
\end{aligned}
$$

The first term in the sum can be estimated by means of the Poincaré inequality (A.25) so that we get

$$|v_k|_{H^1(\Omega)} \le \left(\frac{N C_P H}{\rho H} + 1 \right) |v|_{H^1(\Omega'_k)} \le \frac{cN}{\rho} |v|_{H^1(\Omega'_k)} \quad \text{for } k = 0, \ldots, m.$$

In addition, we have $\|v_k\|_{L^2(\Omega)} \le \|v\|_{L^2(\Omega'_k)}$. The corresponding estimate for w is obtained from Lemma 7.14. Insertion into (7.25) then leads to

$$
\begin{aligned}
\sum_{k=0}^{m} \langle B_k v_k, v_k \rangle + \langle B_W w, w \rangle &\le C_a \left(\frac{c^2 N^6}{\rho^2} |v|^2_{H^1(\Omega)} + \frac{cN^2}{\rho^2} \sum_{k=0}^{m} |v|^2_{H^1(\Omega'_k)} \right) \\
&\le \frac{C_a}{\rho^2} \left(cN^6 + cN^3 \right) |v|^2_{H^1(\Omega)} \le \frac{C_a}{\alpha_a} \frac{cN^6}{\rho^2} \langle Av, v \rangle
\end{aligned}
\quad (7.27)
$$

due to the ellipticity of A. □

Remark 7.17. Apart from the here presented overlapping domain decomposition methods there are also *nonoverlapping* domain decomposition methods, usually called *iterative substructuring methods*; we do not treat them here, since they lead to saddle point problems. Instead we refer interested readers to the comprehensive monographs by A. Toselli and O. B. Widlund [201] or A. Quarteroni and A. Valli [173].

7.1.5 Higher-order Finite Elements

In this section we give two related examples of subspace correction methods, where the decomposition may be regarded on the one hand as a spatial domain decomposition and on the other hand as a decomposition of the *function space* with respect to frequencies. For this purpose, we apply the abstract convergence theory (from Section 7.1.3) for PSC-methods to finite element methods with polynomial elements of higher order (cf. Section 4.3.3).

Hierarchical Error Estimator. In Section 6.1.4 we had not given any proof of the reliability of the DLY error estimator (6.36), for which we want to make up here. We may regard the error approximation

$$[e_h] = B^{-1}(f - Au_h) = \sum_{\psi \in \Psi^\oplus} \frac{\langle f - Au_h, \psi \rangle}{\|\psi\|_A^2} \psi \in S_h^+ \subset H^1(\Omega) \qquad (7.28)$$

as *one* step of the PSC-Algorithm 7.7 over the direct space decomposition

$$S_h^+ = S_h \oplus \bigoplus_{\psi \in \Psi^\oplus} \text{span } \psi,$$

where $\Psi^\oplus$ denotes the set of quadratic "bubbles" on the edges, which span the complement $S_h^\oplus$. The preconditioner B is obtained from the exact solution of the so defined subspaces, i.e., $B_k = A$. In this framework we now want to prove Theorem 6.15, which we repeat for the convenience of the readers.

Theorem 7.18. *Under the saturation assumption* (6.22) *the error estimator* (6.36), *i.e.,*

$$[\epsilon_h]^2 := \|[e_h]\|_B^2 = \sum_{\psi \in \Psi^\oplus} \frac{\langle f - Au_h, \psi \rangle^2}{\|\psi\|_A^2},$$

is efficient: There exist two constants $K_1, K_2 < \infty$ *independent of h such that*

$$\frac{[\epsilon_h]}{\sqrt{K_2}} \le \epsilon_h \le \sqrt{\frac{K_1}{1 - \beta^2}} [\epsilon_h].$$

Proof. In order to be able to apply Theorem 7.9 for parallel subspace correction methods, we have to check for assumptions (A1) and (A2′). Let us begin with the stability assumption (A1), in which the decomposition of $v \in S_h^+$ into

$$v = v_h + v_h^\oplus = v_h + \sum_{\psi \in \Psi^\oplus} v_\psi \psi$$

is unique. Because of $\psi(\xi) = 0$ for all $\psi \in \Psi^\oplus$ and $\xi \in \mathcal{N}$ the relation $v_h = I_h v$ is given by interpolation at the mesh points. Correspondingly, we obtain $v_h^\oplus = (I - I_h)v$ and also, over all elements $T \in \mathcal{T}$, due to Lemma 4.18 and the fixed dimension of the polynomial spaces

$$\|v_h^\oplus\|_{H^1(T)} \le ch_T \|v\|_{H^2(T)} \le c\|v\|_{H^1(T)}. \qquad (7.29)$$

Here the generic constants c do not depend on h_T, but only on the interior angles of T. Summation over $T \in \mathcal{T}$ supplies $\|v_h^\oplus\|_{H^1(\Omega)} \le c\|v\|_{H^1(\Omega)}$. From the triangle inequality and the ellipticity of the operator A we then conclude

$$\|v_h\|_A \le c\|v\|_A. \qquad (7.30)$$

By the equivalence of norms in finite dimensional spaces we obtain from (7.29) also

$$\sum_{\psi \in \Psi^{\oplus} : T \subset \text{supp} \, \psi} v_{\psi}^2 \|\psi\|_{H^1(T)}^2 \leq c \|v_h^{\oplus}\|_{H^1(T)}^2.$$

Again the constant c does not depend on h_T, but only on the interior angles. Summation over all $T \in \mathcal{T}$ leads, with (7.29) and the ellipticity of A, to

$$\sum_{\psi \in \Psi^{\oplus}} \|v_{\psi} \psi\|_A^2 \leq c \|v_h^{\oplus}\|_A^2 \leq c \|v\|_A^2. \tag{7.31}$$

Upon combining (7.30) and (7.31), we get

$$\|v_h\|_A^2 + \sum_{\psi \in \Psi^{\oplus}} \|v_{\psi} \psi\|_A^2 \leq K_1 \|v\|_A^2$$

with an h-independent constant K_1.

Let us now turn to the strengthened Cauchy–Schwarz condition (A2'). Here only those quadratic "bubbles" enter, whose supports overlap:

$$\langle \psi_i, A \psi_j \rangle \leq \gamma_{ij} \|\psi_i\|_A \|\psi_j\|_A \quad \text{with } \gamma_{ij} = \begin{cases} 1, & \exists T \in \mathcal{T} : T \subset \text{supp} \, \psi_i \cap \text{supp} \, \psi_j, \\ 0, & \text{else.} \end{cases}$$

With bounded interior angles of the elements T the number of neighboring ψ is bounded independent of h, so that each row of the matrix γ contains only a bounded number of entries. Then, from the Theorem of Gerschgorin, we may conclude that $\|\gamma\|_2 \leq c$. In the index splitting (7.13) the subspace S_h will be considered separately so that we obtain the h-independent constant $K_2 = c + 1$.

Application of Theorem 7.9 then yields, on the one hand,

$$\|[e_h]\|_B^2 = \langle [e_h], B[e_h] \rangle = \langle A(u_h^+ - u_h), B^{-1} A(u_h^+ - u_h) \rangle \leq K_2 \|u_h^+ - u_h\|_A^2$$

and, on the other hand,

$$\|[e_h]\|_B^2 \geq \frac{1}{K_1} \|u_h^+ - u_h\|_A^2.$$

Finally, by the saturation assumption, the proof of the theorem is completed. □

Remark 7.19. Instead of (6.36) one might be tempted to define an alternative error estimator according to $[\epsilon_h] = \|[e_h]\|_A$, which, however, is not recommended: on the one hand, its evaluation is more expensive, on the other hand, the efficiency span increases. This can be seen by applying Theorem 7.9 again, which yields:

$$\frac{\sqrt{1 - \beta^2}}{K_1} \|e_h\|_A \leq \|[e_h]\|_A = \|B^{-1} A e_h\|_A \leq K_2 \|e_h\|_A.$$

Preconditioner for Higher-order Finite Elements. Here we want to construct a preconditioner which is independent both from the mesh size h and from the ansatz order $p > 1$. We consider (possibly strongly nonuniform) triangular or tetrahedral meshes $\mathcal{T}$ and choose the ansatz space

$$V = S_h^p = \{v \in H^1(\Omega) : v|_T \in \mathbb{P}_p \text{ for all } T \in \mathcal{T}\}.$$

The overlapping local subdomains Ω_k' are formed in an intuitive manner as unions of neighboring elements $T \in \mathcal{T}$. As in the preceding section a coarse grid space is needed to be able to represent smooth functions by a stability constant K_1 independent of the number of subdomains and thus also of the mesh size. With this goal, the choice of the space $W = S_h^1$ of *linear* finite elements seems to be natural.

For the specific choice of subdomains essentially two simple constructions will come into mind (see Figure 7.6): the stars (i) $\Omega_i' = \omega_{\xi_i}$ around nodes $\xi_i \in \mathcal{N}$ or (ii) $\Omega_i' = \omega_{E_i}$ around edges $E_i \in \mathcal{E}$ (cf. Definition 4.12). In both cases it is clear that the number N of neighboring domains is bounded by the *minimal interior angle*. Hence, similar to Lemma 7.12, we here also obtain the constant $K_2 = N$ from assumption (A2′).

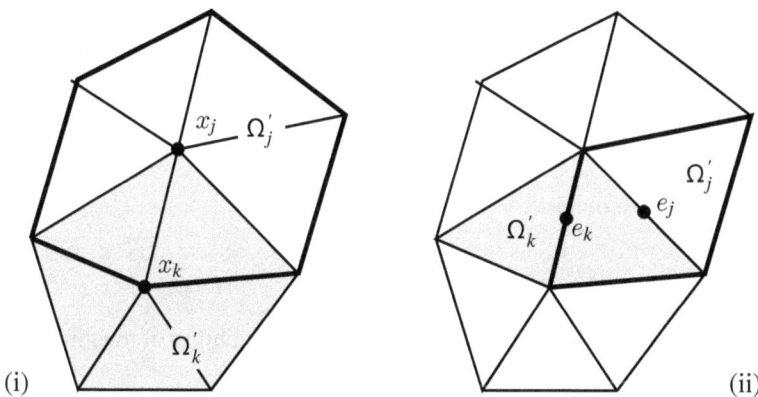

Figure 7.6. Possible constructions of overlapping subdomains for $d = 2$.

But the stability of the decomposition $v = \sum_{k=0}^m v_k + w$ depends on the construction of the subdomains. For choice (ii) we necessarily have $w(\xi_i) = v(\xi_i)$ for all nodes $\xi_i \in \mathcal{T}$, and thus $w \in S_h^1$ is uniquely determined by interpolation. However, since $H^1(\Omega) \not\subset C(\Omega)$ (cf. Theorem A.14), the stability constant K_1 will then depend on the order p (see Exercise 7.12). Hence, we here choose

$$\Omega_k' := \omega_{\xi_k} = \bigcup_{T \in \mathcal{T} : \xi_k \in T} T \tag{7.32}$$

and define, as before, the decomposition $v = \sum_{k=0}^m v_k + w$ by (7.23) and (7.20). The required partition of unity is ensured by the nodal basis ϕ_k of S_h^1, as we had worked out in Section 4.3.2.

Let H_k denote the length of the longest edge in Ω'_k. For some ρ with $0 < \rho \le 1$, which due to the minimal interior angle is bounded away from 0, we construct a ball $B(x_k; \rho H_k) \subset \Omega'_k$. Then

$$\|\nabla \phi_k\|_{L^\infty(\Omega)} \le (\rho H_k)^{-1}. \tag{7.33}$$

As an equivalent to Lemma 7.14 we thus obtain the following lemma.

Lemma 7.20. *Let again $Q : V \to W$ denote the coarse grid projection (7.24). Then there exists a constant c independent of p, ρ, and H_k such that*

$$|Qv|_{H^1(\Omega)} \le \frac{c}{\rho^2} |v|_{H^1(\Omega)}.$$

Proof. The proof follows the lines of the proof of Lemma 7.14, so that we use the same notation but restrict our attention to the differences. From (7.33) we get

$$|Qv|^2_{H^1(\Omega)} \le \sum_{k=0}^m \rho^{-2} H_k^{-2} \sum_{k=0}^m |\mathcal{N}_k| \sum_{j \in \mathcal{N}_k} \|\alpha_j (v - c_k)\|^2_{L^2(\Omega'_k \cap \Omega'_j)}.$$

The diameter of $\hat{\Omega}_k$ is bounded by $2 H_k (1 + 1/\rho)$, which is why application of the Poincaré inequality (A.25) yields

$$\|\alpha_j (v - c_k)\|_{L^2(\Omega'_j)} \le 2 C_p H_k \frac{\rho + 1}{\rho} |v|_{H^1(D_k)}.$$

Correspondingly, we obtain

$$|Qv|^2_{H^1(\Omega)} \le 4 C_p^2 N^3 \frac{(1 + \rho)^2}{\rho^4} |v|^2_{H^1(\Omega)}. \qquad \square$$

After this preparation we are now ready to prove the stability of the thus constructed decomposition.

Theorem 7.21. *There exists a constant $c < \infty$ independent of ρ, p, and H_k such that, using the nodal basis ϕ_k for the decomposition of $v \in V$ according to (7.23) and (7.20), the assumption (A1) holds with*

$$K_1 = \frac{c}{\rho^4}.$$

Proof. Here we have due to (7.33), similar to (7.26),

$$|v_k|_{H^1(\Omega)} \le \frac{1}{\rho H_k} \|v - \bar{v}_k\|_{L^2(\Omega'_k)} + \|\nabla v\|_{L^2(\Omega'_k)}.$$

With the Poincaré inequality (A.25) we eventually obtain

$$|v_k|_{H^1(\Omega)} \le \frac{c}{\rho} |v|_{H^1(\Omega'_k)} \quad \text{for } k = 0, \ldots, m.$$

Together with Lemma 7.20 this leads, in analogy to (7.25) and (7.27), to the result

$$\sum_{k=0}^{m} \langle B_k v_k, v_k \rangle + \langle B_W w, w \rangle \leq \frac{C_a}{\alpha_a} \frac{c}{\rho^4} \langle Av, v \rangle. \qquad \square$$

We have tacitly assumed that, as in the preceding section, the solution of the sub-problems is exact. On the local subdomains, this is rather reasonable due to the bounded problem size. For finer meshes, however, the exact computation of the coarse grid component is rather costly, which is why here approximate solutions, i.e., preconditioners $B_W \neq A$, come into play. For this purpose, *additive* multigrid methods (see Section 7.3.1), are particularly well-suited.

Remark 7.22. The kind of preconditioning for finite elements of higher order presented here was suggested by L. Pavarino for spectral methods on Cartesian meshes [167] and only recently transferred to unstructured triangular or tetrahedral meshes by J. Schöberl et al. [181]. As elaborated above, the generous geometric overlap (7.32) with the nodes of the triangulation in the interior of the subdomains Ω'_k is necessary in order to avoid a definition of the coarse grid part by interpolation (Sobolev Lemma). As a consequence, however, a large computational and storage amount will come in, since many medium size and mostly dense local stiffness matrices need to be stored and factorized (for $d = 2$ a size of $3p(p-1) + 1$ each, for $d = 3$ even larger). However, by some finer space decomposition [181], one may significantly reduce the *algebraic* overlap, i.e., the dimension of the ansatz spaces on the subdomains and thus the size of the local stiffness matrices.

7.2 Hierarchical Space Decompositions

Here again we build upon the theoretical fundament of abstract subspace correction methods prepared in Section 7.1.2 and 7.1.3. In contrast to the preceding sections we decompose the ansatz space hierarchically according to

$$S_j = \sum_{k=0}^{j} V_k \quad \text{and} \quad v = \sum_{k=0}^{j} v_k, \quad v_k \in V_k \qquad (7.34)$$

into subspaces V_k with different *spatial frequencies*, which have been constructed by adaptive refinement using error estimators. As subspace solvers B_k we resort to *smoothing iterative methods* from Section 5.4. Again we focus on simplicial mesh hierarchies $(\mathcal{T}_k)_{k=0,\dots,j}$, which we assume to be generated by red-green refinement[2] and to be *shape regular*. That is why the refinement process for the generation of these meshes will not play a role in this section – the mesh hierarchy can be uniquely fixed by

---

[2] The convergence results derived in the following hold equally well for bisection, but some of the estimates are a bit more technical and include different constants in the proofs.

the coarse initial mesh $\mathcal{T}_0$ and the fine final mesh $\mathcal{T}_j$. In contrast to Section 5.5, where we had assumed uniform meshes throughout, we here assume adaptive, nonuniform meshes.

As has been elaborated in the preceding Sections 7.1.2 and 7.1.3, proofs by virtue of subspace correction methods first of all are based on a clever decomposition of subspaces. In Sections 7.2.1 and 7.2.2 we therefore want to examine subtle space decompositions in which the total computational complexity caused by the smoothers B_k on all subspaces V_k remains $\mathcal{O}(N_j)$. Based only on this, we will examine two types of multigrid methods in more detail, i.e., the additive ones (Section 7.3.1) and the multiplicative ones (Section 7.3.2).

7.2.1   Decomposition into Hierarchical Bases

Particularly simple is the disjoint decomposition of the nodal set $\mathcal{N}_j$ presented in Section 5.5.2 and shown in Figure 7.7 according to

$$\mathcal{N}_j = \bigcup_{k=0}^{j} \mathcal{N}_k^{\mathrm{HB}}, \quad \mathcal{N}_k^{\mathrm{HB}} = \mathcal{N}_k \setminus \mathcal{N}_{k-1}, \quad \mathcal{N}_{-1} = \emptyset.$$

It corresponds to the direct space decomposition

$$S_j = \bigoplus_{k=0}^{j} V_k^{\mathrm{HB}}, \quad V_k^{\mathrm{HB}} = \mathrm{span}\{\varphi_{k,\xi} : \xi \in \mathcal{N}_k^{\mathrm{HB}}\},$$

where the $\varphi_{k,\xi}$ again are the nodal basis functions of S_k.

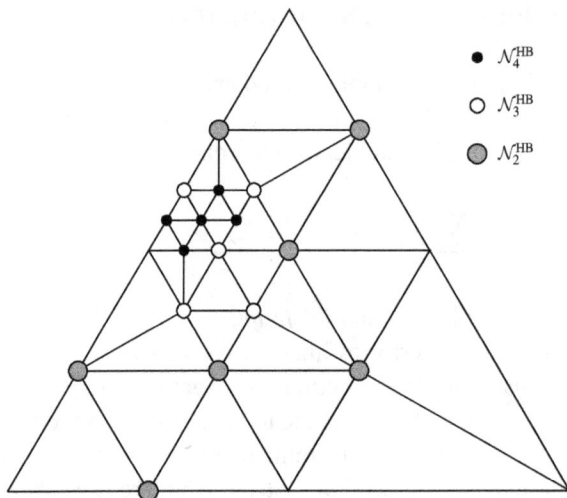

Figure 7.7. Disjoint decomposition of the nodal set $\mathcal{N}_4$ into the nodal sets $\mathcal{N}_k^{\mathrm{HB}}$ of the hierarchical basis.

With the help of interpolation operators $I_k : C(\overline{\Omega}) \rightarrow S_k$, defined by the property $(I_k u)(\xi) = u(\xi)$ for all $\xi \in \mathcal{N}_k$, this decomposition can be represented using the notation (7.34) in the form

$$v = \sum_{k=0}^{j} v_k = I_0 v + \sum_{k=1}^{j} (I_k - I_{k-1}) v \in S_j. \tag{7.35}$$

Let us now apply the apparatus of the abstract subspace correction methods to this decomposition. If not stated otherwise, we will restrict our attention to linear finite elements.

Stability Assumption (A1). First we want to check assumption (A1), which we need both for sequential (SSC) and for parallel (PSC) subspace correction methods. By means of a result by H. Yserentant [226] we will be able to find a result that in two space dimensions is not optimal, but suboptimal.

Lemma 7.23. *With the notations as introduced here, there exists a generic constant $C < \infty$ such that, in space dimension $d = 2$,*

$$|I_k v|^2_{H^1(\Omega)} \le C \, (j - k + 1) \, |v|^2_{H^1(\Omega)} \tag{7.36}$$

$$\|I_k v\|^2_{L^2(\Omega)} \le C \, (j - k + 1) \, \|v\|^2_{H^1(\Omega)} \tag{7.37}$$

for all $v \in S_j$.

Proof. Both assertions are based on an estimate valid for $d = 2$ that we just quote here without proof from [226]:

$$\frac{\|u\|_{L^1(B(0;r))}}{\pi r^2} \le \sqrt{\frac{\log \frac{R}{r} + \frac{1}{4}}{2\pi}} \, |u|_{H^1(B(0;R))} \quad \text{for all } u \in H^1_0(B(0; R)). \tag{7.38}$$

The logarithmic growth factor in terms of the quotient R/r corresponds to the Green's function for the Poisson equation for $d = 2$, which exactly induces the H^1-seminorm (cf. (A.13) in Appendix A.4).

Let us first pick certain triangles $T \in \mathcal{T}_k$. Let $B(\bar{x}; h_T/2)$ denote the circumcircle of T and $u \in H^1(T)$, then u may be continued to an extended function on the double circumcircle $B(\bar{x}; h_T)$, where $|u|_{H^1(B(\bar{x};h_T))} \le c_1 \|u\|_{H^1(T)}$ with a constant c_1 depending only on the minimal interior angle of T (see Exercise 7.13). Let $t \in \mathcal{T}_j$, $t \subset T$ be a triangle with diameter h_t. Then, following (7.38), we get

$$\frac{4\|u\|_{L^1(t)}}{\pi h_t^2} \le c_1 \sqrt{\frac{\log \frac{2h_T}{h_t} + \frac{1}{4}}{2\pi}} \, \|u\|_{H^1(T)}.$$

Now let v be linear on t. Then, due to the equivalence of norms in finite dimensional spaces, there exists a constant c_2 such that

$$\|v\|_{L^\infty(t)} \le c_2 h_t^{-2} \|v\|_{L^1(t)} \le c_1 c_2 \frac{\pi}{4} \sqrt{\frac{\log \frac{2h_T}{h_t} + \frac{1}{4}}{2\pi}} \, \|v\|_{H^1(T)} \, .$$

Again c_2 only depends on the minimal interior angle of t. If $v \in S_j$, then v is linear on all triangles $t \in \mathcal{T}_j$. Due to red-green refinement we have $h_t \geq 2^{k-j} h_T$, which immediately implies

$$\log \frac{2h_T}{h_t} + \frac{1}{4} \leq j - k + 1 + \frac{1}{4} \leq \frac{5}{4}(j - k + 1) \, .$$

If we take the maximum over all $t \in \mathcal{T}_j$, $t \subset T$, then we obtain (without estimating too sharply)

$$\|v\|_{L^\infty(T)} \leq c_3 \sqrt{j - k + 1} \, \|v\|_{H^1(T)}.$$

We are now ready to turn to the interpolation. For $v \in S_j$ we have

$$|I_k v|_{H^1(T)} \leq 2 \min_{z \in \mathbb{R}} \|I_k v - z\|_{L^\infty(T)} \leq 2 \min_{z \in \mathbb{R}} \|v - z\|_{L^\infty(T)} \tag{7.39}$$

$$\leq c_3 \sqrt{j - k + 1} \, \min_{z \in \mathbb{R}} \|v - z\|_{H^1(T)}.$$

With the Poincaré inequality (A.25) we conclude that

$$\min_{z \in \mathbb{R}} \|v - z\|_{L^2(T)} \leq C_P h_T |v|_{H^1(T)}$$

and therefore

$$|I_k v|^2_{H^1(T)} \leq c_3^2(j - k + 1)(1 + C_P^2 h_T^2)|v|^2_{H^1(T)}.$$

Summation over all triangles $T \in \mathcal{T}_k$ leads to (7.36). In order to confirm (7.37), we proceed similarly, using, however, instead of (7.39) the inequality

$$\|I_k v\|_{L^2(\Omega)} \leq h_T \|I_k v\|_{L^\infty(T)},$$

thus dispensing of the Poincaré inequality. $\square$

Lemma 7.24. *For $d = 2$ there exists a j-independent constant $C < \infty$ such that for all $v \in S_j$*

$$\|v_0\|^2_{H^1(\Omega)} + \sum_{k=1}^{j} 4^k \|v_k\|^2_{L^2(\Omega)} \leq C(j + 1)^2 \|v\|^2_{H^1(\Omega)}. \tag{7.40}$$

Proof. Let us first, for $T \in \mathcal{T}_{k-1}$, consider the finite dimensional space $V_T^k = \{v_k|_T : v_k \in V^k\}$. In particular, the vertices of T are contained in $\mathcal{N}_{k-1}$ such that $v_k(\xi) = 0$ for all $\xi \in \mathcal{N}_{k-1}$ and $v_k \in V_T^k$. From the general Poincaré inequality (A.27) we now get $4^k \|v_k\|^2_{L^2(T)} \leq c |v_k|^2_{H^1(T)}$ for all $v_k \in V_T^k$. As we assumed shape regularity, the constant depends only on the size of the triangle T, but not on its interior angles. Summation over all triangles in $\mathcal{T}_{k-1}$ immediately leads to

$$4^k \|v_k\|^2_{L^2(\Omega)} \leq c |v_k|^2_{H^1(\Omega)} \quad \text{for all } v_k \in V^k,$$

where, for simplicity, we may assume $c \geq 1$. Upon application of (7.36) we obtain

$$|v_0|^2_{H^1(\Omega)} + \sum_{k=1}^{j} 4^k \|v_k\|^2_{L^2(\Omega)} \leq |I_0 v|^2_{H^1(\Omega)} + c \sum_{k=1}^{j} |I_k v - I_{k-1} v|^2_{H^1(\Omega)}$$

$$\leq |I_0 v|^2_{H^1(\Omega)} + 2c \sum_{k=1}^{j} (|I_k v|^2_{H^1(\Omega)} + |I_{k-1} v|^2_{H^1(\Omega)})$$

$$\leq 4c \sum_{k=0}^{j} |I_k v|^2_{H^1(\Omega)}$$

$$\leq 4cC \sum_{k=0}^{j} (j - k + 1)|v|^2_{H^1(\Omega)}$$

$$\leq 2cC(j+1)^2 |v|^2_{H^1(\Omega)}.$$

For $k = 0$, (7.37) supplies the assertion (7.40). □

After these rather technical preparations we are now ready to confirm the stability assumption (A1) for subspace correction methods in an elementary way.

Theorem 7.25. *Let A denote an $H^1(\Omega)$-elliptic operator on $\Omega \subset \mathbb{R}^2$. For the preconditioner, defined according to $B_0 = A$ and $B_k : V_k^{HB} \to (V_k^{HB})^*$, assume that*

$$c^{-1} 2^k \|v_k\|_{L^2(\Omega)} \leq \|v_k\|_{B_k} \leq c 2^k \|v_k\|_{L^2(\Omega)} \quad \text{for all } v_k \in V_k^{HB}. \tag{7.41}$$

Then there exists a constant $K_1^{HB} = c(j + 1)^2$ such that for the decomposition (7.35) into hierarchical bases the following result holds:

$$\sum_{k=0}^{j} \langle B_k v_k, v_k \rangle \leq K_1^{HB} \langle A v, v \rangle. \tag{7.42}$$

Proof. Assumption (7.41) and Lemma 7.25 immediately supply

$$\sum_{k=0}^{j} \langle B_k v_k, v_k \rangle \leq c \left(\|v_0\|^2_{H^1(\Omega)} + \sum_{k=1}^{j} 4^k \|v_k\|^2_{L^2(\Omega)} \right) \leq cC(j+1)^2 \|v\|^2_{H^1(\Omega)}.$$

The ellipticity of A then yields (7.42). □

In the framework of subspace correction methods we retain the result

$$K_1^{HB} = \mathcal{O}(j^2) \tag{7.43}$$

as essential.

Strengthened Cauchy–Schwarz Inequality (A2'). As the second assumption for subspace correction methods we check (7.12). Let us start from the extended Poisson problem with spatially varying diffusion coefficient $\sigma(x)$.

Theorem 7.26. *In addition to the assumptions of Theorem 7.25 let σ be piecewise constant on the coarse grid $\mathcal{T}_0$. Then there exists some constant depending only on $\mathcal{T}_0$ and the operator A*

$$K_2^{\mathrm{HB}} = \mathcal{O}(1),$$

such that the strengthened Cauchy–Schwarz condition (A2') holds.

Proof. For $v_k \in V_k^{\mathrm{HB}}$ and $w_l \in V_l^{\mathrm{HB}}$, $k \le l \le j$ we first demonstrate the auxiliary result

$$\langle Av_k, w_l \rangle \le \gamma_{kl} |v_k|_{H^1(\Omega)} \|w_l\|_{L^2(\Omega)} \quad \text{with} \quad \gamma_{kl} = c\sqrt{2}^{k-l}2^l \tag{7.44}$$

as well as a j-independent constant c. From (7.14) we recall that

$$K_2 = \sum_{l=0}^{r} \|(\gamma_{kl})\|_2.$$

In order to derive an estimate for K_2, we consider (7.44) on a triangle $T \in \mathcal{T}_k \setminus \mathcal{T}_{k-1}$ and define

$$\chi(x) = \max(0, 1 - 2^l \operatorname{dist}(x, \partial T)).$$

Clearly, the support of χ is a strip of width 2^{-l} along ∂T with volume $|\operatorname{supp} \chi| \le 2^{-l} h_T^{d-1}$.

Because of $1 - \chi = 0$ on ∂T Green's Theorem A.5 supplies

$$\langle Av_k, (1-\chi)w_l \rangle_T = -\int_T (1-\chi)w_l \operatorname{div}(\sigma \nabla v_k)\, dx = 0,$$

since v_k is linear and σ is constant. To the remaining part we apply the Cauchy–Schwarz inequality

$$\langle Av_k, w_l \rangle_T = \langle Av_k, \chi w_l \rangle_T = \int_{\operatorname{supp}\chi} \nabla v_k^T \sigma \nabla \chi w_l\, dx \le c|v_k|_{H^1(\operatorname{supp}\chi)}|\chi w_l|_{H^1(T)}.$$

As ∇v_k is constant on T, we have

$$|v_k|_{H^1(\operatorname{supp}\chi)} \le \sqrt{\frac{|\operatorname{supp}\chi|}{|T|}}|v_k|_{H^1(T)} \le c\sqrt{2}^{-k-l}|v_k|_{H^1(T)},$$

where the constant c depends on the minimal interior angle of T. The other factor may be estimated by the inverse inequality (7.79) as

$$|\chi w_l|_{H^1(T)} \le \|\nabla\chi w_l\|_{L^2(T)} + \|\chi\nabla w_l\|_{L^2(T)}$$
$$\le 2^l \|w_l\|_{L^2(T)} + \|\nabla w_l\|_{L^2(T)} \le c2^l \|w_l\|_{L^2(T)}.$$

Then

$$\langle Av_k, w_l \rangle_T \le c\sqrt{2}^{k-l} |v_k|_{H^1(T)} 2^l \|w_l\|_{L^2(T)} \tag{7.45}$$

holds. For $T \in \mathcal{T}_k \cap \mathcal{T}_{k-1}$ we trivially also get (7.45) due to $v_k|_T = 0$.

The Cauchy–Schwarz inequality in finite dimension then yields

$$\langle Av_k, w_l \rangle = \sum_{T \in \mathcal{T}_k} \langle Av_k, w_l \rangle_T = c\sqrt{2}^{k-l} 2^l \sum_{T \in \mathcal{T}_k} |v_k|_{H^1(T)} \|w_l\|_{L^2(T)}$$

$$\le c\sqrt{2}^{k-l} 2^l \left(\sum_{T \in \mathcal{T}_k} |v_k|^2_{H^1(T)} \right)^{1/2} \left(\sum_{T \in \mathcal{T}_k} \|w_l\|^2_{L^2(T)} \right)^{1/2}$$

$$= c\sqrt{2}^{k-l} 2^l |v_k|_{H^1(\Omega)} \|w_l\|_{L^2(\Omega)}.$$

Upon applying the inverse inequality (7.79) again, this time for v_k, we are led to

$$\langle Av_k, w_l \rangle \le c\sqrt{2}^{k-l} 2^k \|v_k\|_{L^2(\Omega)} 2^l \|w_l\|_{L^2(\Omega)}.$$

Together with (7.41) and (7.44) we get

$$\langle Av_k, w_l \rangle \le c\sqrt{2}^{k-l} 2^k (4^{-k}\langle B_k v_k, v_k \rangle)^{1/2} 2^l (4^{-l}\langle B_l w_l, w_l \rangle)^{1/2}$$

$$= c\sqrt{2}^{k-l} \|v_k\|_{B_k} \|w_l\|_{B_l}$$

with a constant c independent of k and l. So we can satisfy the strengthened Cauchy–Schwarz inequality (7.12) with $\gamma_{kl} = c\sqrt{2}^{-|k-l|}$.

Finally, the theorem of Gerschgorin yields a j-independent bound for the spectral radius of γ, i.e., an upper bound for K_2. $\qquad\square$

Choice of Suitable Smoothers.   Up to now we have not specified the choice of the preconditioner B_k, for which we want to make up now.

Lemma 7.27. *The Jacobi smoother B_k^J with*

$$\langle B_k^J \varphi_{k,l}, \varphi_{k,m} \rangle = \delta_{lm} \langle A\varphi_{k,l}, \varphi_{k,m} \rangle \tag{7.46}$$

for all basis functions $\varphi_{k,l}, \varphi_{k,m}$ of V_k^{HB} satisfies (7.41).

Proof. For $v_k \in V_k^{HB}$ we get

$$\langle B_k^J v_k, v_k \rangle = \sum_{\xi, \nu \in \mathcal{N}_k^{HB}} v_k(\xi) v_k(\nu) \langle B_k^J \varphi_{k,\xi}, \varphi_{k,\nu} \rangle = \sum_{\xi \in \mathcal{N}_k^{HB}} v_k(\xi)^2 \langle A\varphi_{k,\xi}, \varphi_{k,\xi} \rangle. \tag{7.47}$$

Beyond that we have

$$c^{-1} 2^{(d-2)k} \le \langle A\varphi_{k,\xi}, \varphi_{k,\xi} \rangle \le c 2^{(d-2)k}, \tag{7.48}$$

where the generic constant c only depends on the interior angles of the triangles in the support of $\varphi_{k,\xi}$ and on the constants C_a, α_a of the operator A (see Exercise 4.14).

For nodal basis functions $\varphi_{k,\xi}$ of V_k^{HB} associated with nodes $\xi \in \mathcal{N}_k^{\mathrm{HB}}$ the support $\operatorname{supp}\varphi_{k,\xi} = \omega_{k,\xi}$ has a diameter $h_{k,\xi}$ with

$$c^{-1}2^{-k} \leq h_{k,\xi} \leq c2^{-k}.$$

Hence,

$$c^{-1}2^{-dk} \leq \|\varphi_{k,\xi}\|_{L^2(\Omega)}^2 \leq c2^{-dk},$$

and with (7.47) and (7.48) we conclude that

$$c^{-1}2^{2k} \sum_{\xi \in \mathcal{N}_k^{\mathrm{HB}}} v_k(\xi)^2 \|\varphi_{k,\xi}\|_{L^2(\Omega)}^2 \leq \|v_k\|_{B_k^j}^2 \leq c2^{2k} \sum_{\xi \in \mathcal{N}_k^{\mathrm{HB}}} v_k(\xi)^2 \|\varphi_{k,\xi}\|_{L^2(\Omega)}^2.$$

Because of the bounded condition number of the mass matrix (see Exercise 4.14) we then obtain

$$c^{-1}2^k \|v_k\|_{L^2(\Omega)} \leq \|v_k\|_{B_k^j} \leq c2^k \|v_k\|_{L^2(\Omega)}. \qquad \square$$

For $d = 3$, however, the strengthened Cauchy–Schwarz inequality (A2') also holds, but not the stability result (A1). This is due to an underlying deeper structure: the HB-decomposition (7.35) builds upon the *interpolation operator*, which requires $u \in C(\Omega)$, whereas we can only assume $u \in H^1(\Omega)$. By Definition A.12 and the Sobolev embedding Theorem A.14 we recognize that this only holds in general, if $d < 2$, so that the HB-decomposition is only stable in $d = 1$ independently of j – for this case, it even diagonalizes the Laplace operator. The limiting case $d = 2$ can be captured by means of Lemma 7.23 by the logarithmic singularity of the Green's function (of the Poisson equation), whereas for $d = 3$ (and higher) the Green's function exhibits too strong a singularity – see, e.g., (A.13). As a consequence, the stability constant K_1^{HB} for $d = 3$ no longer grows logarithmically, but linearly with the mesh size. In order to cure this deficiency, we present an alternative in the next section, which, however, requires some larger computational amount.

7.2.2 L^2-orthogonal Decomposition: BPX

In the preceding Section 7.2.1 we worked out why the HB-decomposition implies an algorithmic restriction to the case $d = 2$. In this we present an extension of the space decomposition suggested in 1989 by J. Xu in his dissertation [222] and published a year later by J. H. Bramble, J. E. Pasciak, and J. Xu in [42] – hence the now commonly used abbreviation BPX. For uniform refined meshes they suggested the extension $V_k = S_k$ and defined the decomposition by

$$v = \sum_{k=0}^{j} v_k = Q_0 v + \sum_{k=1}^{j}(Q_k - Q_{k-1})v \qquad (7.49)$$

with L^2-orthogonal projectors $Q_k : H^1 \to S_k$ instead of the interpolation operators I_k in (7.35). For *adaptive meshes*, however, this maximal extension of $\mathcal{N}_k^{\text{HB}}$ no
longer exhibits optimal complexity. Therefore we will follow the lines of a notice-
able improvement of the algorithm (in the adaptive case) by H. Yserentant [227], also
from 1990.[3]

In fact, the decomposition (7.49) allows to bound the stability constant K_1^{BPX} also
for $d > 2$. In the first papers [42, 222] a result of the kind

$$K_1^{\text{BPX}} = \mathcal{O}(j), \quad \text{for } d \geq 2, \tag{7.50}$$

has been shown, already an improvement beyond K_1^{HB} from (7.43), which additionally
is only valid for $d = 2$. In 1991, P. Oswald [166] managed to improve the above
result to

$$K_1^{\text{BPX}} = \mathcal{O}(1) \quad \text{for } d \geq 2 \tag{7.51}$$

by employing Besov spaces, a generalization of Sobolev spaces. Shortly after, W. Dahmen and A. Kunoth came up with a simpler proof of this result [60]. Unfortunately, all
of these proofs require theoretical tools far beyond the scope of the present volume,
which is why we will be content here to prove the slightly weaker result (7.50).

Subspace Extension. As already mentioned, the papers [42, 222] extended the subspaces V_k to the full S_k. Here, however, we want to follow the more subtle suggestion
in [227]: starting from the decomposition into hierarchical bases, we perform a "cautious" extension, as illustrated in Figure 7.8, by those nodal basis functions that are

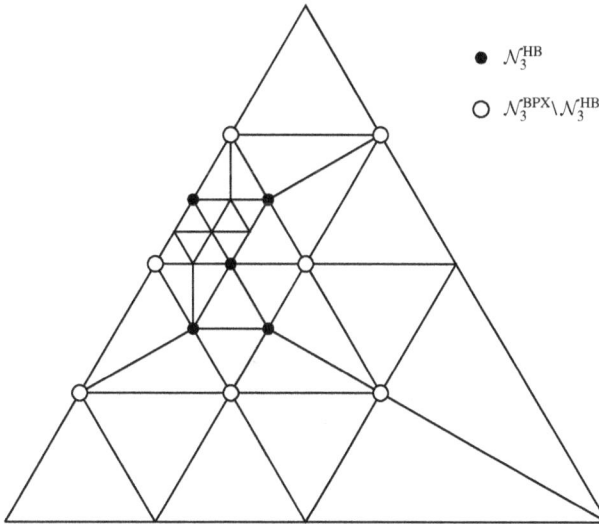

Figure 7.8. Extension of the nodal set $\mathcal{N}_3^{\text{HB}}$ to $\mathcal{N}_3^{\text{BPX}}$.

---

[3] This algorithm might therefore be rightly named BPXY-algorithm.

associated with neighboring nodes of $\mathcal{N}_k^{\mathrm{HB}}$ in $\mathcal{T}_k$:

$$\mathcal{N}_k^{\mathrm{BPX}} = \mathcal{N}_k \cap \bigcup_{T \in \mathcal{T}_k : T \cap \mathcal{N}_k^{\mathrm{HB}} \neq \emptyset} T, \quad V_k^{\mathrm{BPX}} = \mathrm{span}\{\varphi_{k,\xi} | \xi \in \mathcal{N}_k^{\mathrm{BPX}}\}, \quad k < j$$

$$\mathcal{N}_j^{\mathrm{BPX}} = \mathcal{N}_j, \qquad\qquad\qquad V_k^{\mathrm{BPX}} = S_j.$$

Hence, the relation $\mathcal{N}_k^{\mathrm{HB}} \subset \mathcal{N}_k^{\mathrm{BPX}}$ is clear. As the number of elements incident with a node is bounded by the minimal interior angle, there exists a constant c for $k < j$ only dependent on the coarse grid with $|\mathcal{N}_k^{\mathrm{BPX}}| \leq c|\mathcal{N}_k^{\mathrm{HB}}|$. Thus we have

$$\sum_{k=0}^{j} |\mathcal{N}_k^{\mathrm{BPX}}| \leq c N_j,$$

and one step of the subspace correction method has the optimal computational amount $\mathcal{O}(N_j)$.

Strengthened Cauchy–Schwarz Inequality (A2'). Theorem 7.26 and Lemma 7.27 hold for the BPX-decomposition in the same way as for the HB-decomposition. The proofs run totally analogously, except that the supports of the nodal basis functions of V_k^{BPX} are slightly larger – however, due to the cautious extension, only by a fixed factor, so that all estimates hold, though with slightly different constants.

Stability Assumption (A1). For the construction of a stable decomposition we define the averaging operators $M_k : S_j \to V_k^{\mathrm{BPX}}$ by

$$(M_k v)(\xi) = \frac{\langle v, \varphi_{k,\xi} \rangle}{\langle 1, \varphi_{k,\xi} \rangle}, \quad \xi \in \mathcal{N}_k,$$

and decompose $v \in S_j$ via

$$v = M_0 v + \sum_{k=1}^{j-1} (M_k - M_{k-1}) v + (I_j - M_{j-1}) v. \tag{7.52}$$

For $\xi \in \mathcal{N}_k \setminus \mathcal{N}_k^{\mathrm{BPX}}$ we get, because of the identical stars around ξ in the meshes $\mathcal{T}_k$ and $\mathcal{T}_{k-1}$, just $\varphi_{k,\xi} = \varphi_{k-1,\xi}$ and thus $((M_k - M_{k-1})v)(\xi) = 0$. As a consequence, we have $(M_k - M_{k-1})v \in V_k^{\mathrm{BPX}}$, and the decomposition (7.52) is useful for the stability assumption (A1).

Lemma 7.28. *There exists a constant c depending only on the coarse grid $\mathcal{T}_0$ such that*

$$\|M_0 v\|_{L^2(\Omega)} \leq c\|v\|_{L^2(\Omega)}, \tag{7.53}$$

$$|(M_k - M_{k-1})v|_{H^1(\Omega)} \leq c|v|_{H^1(\Omega)},$$

$$|(I_j - M_j)v|_{H^1(\Omega)} \leq c|v|_{H^1(\Omega)}$$

for all $v \in S_j$.

Proof. Due to the bounded condition number of the mass matrix (see Exercise 4.14) and the Cauchy–Schwarz inequality we have

$$\|M_0 v\|^2_{L^2(\Omega)} \le c \sum_{\xi \in \mathcal{N}_0} \frac{\langle v, \varphi_{0,\xi} \rangle^2}{\langle 1, \varphi_{0,\xi} \rangle^2} \|\varphi_{0,\xi}\|^2_{L^2(\Omega)} \le c \sum_{\xi \in \mathcal{N}_0} \|v\|^2_{L^2(\omega_\xi)} \frac{\|\varphi_{0,\xi}\|^4_{L^2(\Omega)}}{\langle 1, \varphi_{0,\xi} \rangle^2}.$$

In ω_ξ there are only finitely many elements $T \in \mathcal{T}_0$ and occur multiply in the sum. This is why we get with $\|\varphi_{0,\xi}\|_{L^2(\Omega)} \le \langle 1, \varphi_{0,\xi} \rangle$ the result (7.53).

Let now $T \in \mathcal{T}_k$. Since M_k reproduces constants exactly, we obtain for all $\bar{v} \in \mathbb{R}$

$$|M_k v|_{H^1(T)} = |M_k(v - \bar{v})|_{H^1(T)} \le \frac{c}{h_T} \|M_k(v - \bar{v})\|_{L^2(\Omega)},$$

where in the second step of the calculation we used the fact that in finite dimensional spaces all norms are equivalent. The arising constant c depends only on the interior angles of T, but not on the diameter h_T. With $U(T) = \bigcup_{\xi \in \mathcal{N}_k \cap T} \omega_\xi$ and the Poincaré inequality (A.25) we then get

$$|M_k v|_{H^1(T)} \le \frac{c}{h_T} \|v - \bar{v}\|_{L^2(U(T))} \le c\, C_P |v|_{H^1(U(T))}.$$

Taking squares and summing over $T \subset \mathcal{T}_k$ then supplies a constant c with

$$|M_k v|^2_{H^1(\Omega)} \le c |v|^2_{H^1(\Omega)}, \tag{7.54}$$

since again each T is covered only be finitely many $U(T)$. The remaining two statements of the lemma follow from (7.54) by the triangle inequality.   □

With Lemma 7.28 we now have the basis to prove the stability condition (A1).

Theorem 7.29. *Let A be an $H^1(\Omega)$-elliptic operator. Let $B_0 = A$ and $B_k : V_k^{\mathrm{BPX}} \to (V_k^{\mathrm{BPX}})^*$ satisfy*

$$c^{-1} 2^k \|v_k\|_{L^2(\Omega)} \le \|v_k\|_{B_k} \le c 2^k \|v_k\|_{L^2(\Omega)} \quad \text{for all } v_k \in V_k^{\mathrm{HB}}.$$

Then there exists a constant K_1^{BPX} depending only on the coarse grid and the operator A such that for all $v = \sum_{k=0}^{j} v_k \in S_j$ the following result holds:

$$\sum_{k=0}^{j} \langle B_k v_k, v_k \rangle \le K_1^{\mathrm{BPX}} \langle Av, v \rangle \quad \text{with} \quad K_1^{\mathrm{BPX}} = c(j + 1).$$

Proof. The proof is in full analogy to those of Lemma 7.24 and Theorem 7.25, with the only difference that now Lemma 7.28 replaces Lemma 7.23.   □

Obviously, the theorem confirms the result (7.50) already mentioned above.

7.3   Multigrid Methods Revisited

In Section 5.5 we have already introduced *nonadaptive* multigrid methods on fixed given hierarchical meshes as well as the related method of hierarchical bases. The proofs worked out there only covered W-methods on uniform meshes and required H^2-regular problems. In this section we want to fill this gap by discussing the two classes of additive and multiplicative multigrid methods. Moreover, in the subsequent Section 7.4, we present a compromise between these two variants, the cascadic multigrid methods.

7.3.1   Additive Multigrid Methods

Additive multigrid methods are multigrid realizations of the *parallel* subspace correction Algorithm 7.7 based on a hierarchical space decomposition. We will restrict our attention to the two most important ones, the HB- and the BPX-decomposition, which we have studied in the preceding sections. As the approximate subspace solver B_k we choose the simplest variant, the Jacobi preconditioner (i.e., the diagonal of the spd-matrix) (see Section 5.2.1). As outer iteration adaptive PCG-methods might be used (see Sections 5.3.2 and 5.3.3).

HB-preconditioning

This special additive multigrid method realizes a preconditioner B^{HB} such that, for $d = 2$, the condition number is bounded according to

$$\kappa((B^{\mathrm{HB}})^{-1}A) \le c(j+1)^2. \tag{7.55}$$

As a consequence, an HB-preconditioned PCG-method will require $m_j^{\mathrm{HB}} \sim \mathcal{O}(j)$ iterations only to converge. This property and the extremely simple computation of the preconditioner are the reason why for $d = 2$ and at a realistic mesh depth the HB-preconditioner will be the method of choice.

Summation Formula.   In what follows we will develop an explicit representation of the application of the HB-preconditioner to a residual $r = f - Au$. First of all, from (7.11) we have

$$(B^{\mathrm{HB}})^{-1}r = \sum_{k=0}^{j} v_k = \sum_{k=0}^{j} (B_k^J)^{-1}r,$$

where $v_k = (B_k^J)^{-1}r \in V_k^{\mathrm{HB}}$ satisfies the Galerkin condition

$$\langle B_k^J v_k, \varphi_{k,\xi} \rangle = \langle r, \varphi_{k,\xi} \rangle \quad \text{for all } \xi \in \mathcal{N}_k^{\mathrm{HB}}.$$

Insertion of the representation into the nodal basis

$$v_k = \sum_{\nu \in \mathcal{N}_k^{\mathrm{HB}}} v_{k,\nu} \varphi_{k,\nu}$$

leads after (7.46) to

$$\langle A\varphi_{k,\xi}, \varphi_{k,\xi} \rangle v_{k,\xi} = \langle r, \varphi_{k,\xi} \rangle.$$

Thus we obtain as an explicit representation

$$(B^{\mathrm{HB}})^{-1} r = B_0^{-1} r + \sum_{k=1}^{j} \sum_{\xi \in \mathcal{N}_k^{\mathrm{HB}}} \frac{\langle r, \varphi_{k,\xi} \rangle}{\langle A\varphi_{k,\xi}, \varphi_{k,\xi} \rangle} \varphi_{k,\xi}. \qquad (7.56)$$

Restriction. In general, the values $\langle r, \varphi_{k,\xi} \rangle$ and $\langle A\varphi_{k,\xi}, \varphi_{k,\xi} \rangle$ required above are not directly available, since the assembling should be performed with local ansatz functions. Instead one has only the values $\langle r, \varphi_{j,\xi} \rangle$ and $\langle A\varphi_{k,\nu}, \varphi_{k,\xi} \rangle$ for all $\xi, \nu \in \mathcal{N}_j$, so that the evaluation of (7.56) (and below of (7.57)) still requires a basis change – a restriction well-known from Section 5.5.1. The foundation for an efficient computation is the recursive representation of ansatz functions $\varphi_{k,\xi}$ on the next finer mesh $\mathcal{T}_{k+1}$ (see Figure 7.9):

$$\varphi_{k,\xi} = \varphi_{k+1,\xi} + \tfrac{1}{2} \sum_{\nu \in \mathcal{N}_{k,\xi}} \varphi_{k+1,\nu} \quad \text{with} \quad \mathcal{N}_{k,\xi} = \mathrm{int}\, \omega_{k,\xi} \cap \mathcal{N}_{k+1} \backslash \{\xi\}.$$

Then we immediately get

$$r_{k,\xi} := \langle r, \varphi_{k,\xi} \rangle = \langle r, \varphi_{k+1,\xi} \rangle + \tfrac{1}{2} \sum_{\nu \in \mathcal{N}_{k,\xi}} \langle r, \varphi_{k+1,\nu} \rangle$$

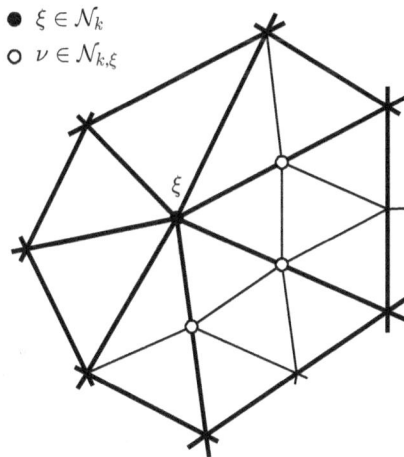

Figure 7.9. Neighborhood set $\mathcal{N}_{k,\xi}$.

and

$$a_{k,\xi} := \langle A\varphi_{k,\xi}, \varphi_{k,\xi} \rangle = \langle A\varphi_{k+1,\xi}, \varphi_{k+1,\xi} \rangle + \sum_{\nu \in \mathcal{N}_{k,\xi}} \langle A\varphi_{k+1,\xi}, \varphi_{k+1,\nu} \rangle$$

$$+ \frac{1}{4} \sum_{\nu,\mu \in \mathcal{N}_{k,\xi}} \langle A\varphi_{k+1,\mu}, \varphi_{k+1,\nu} \rangle.$$

The conversion of the residuals is possible in-place for a vector $(r_\xi)_{\xi \in \mathcal{N}_j}$ indexed with the nodes $\xi \in \mathcal{N}_j$, whereas the computation of the energy products of the hierarchical basis functions is somewhat more laborious. Fortunately they do not depend on the residuals, which change during the CG-iteration, so that they can be computed off-line in advance. As an especially simple alternative, which for smooth diffusion coefficients is even equivalent, the following scaled form seems to be appropriate:

$$a_{k,\xi} = 2^{(d-2)(k-j)} \langle A\varphi_{j,\xi}, \varphi_{j,\xi} \rangle .$$

Prolongation. Eventually, the solution represented in terms of the hierarchical basis must be backtransformed into the nodal basis of S_j. This is the prolongation we became acquainted with in Section 5.5.1. As for the residuals, this can be performed by in-place overwriting the coefficients – this time in the reverse direction by the mesh hierarchy. With these preparations we are now ready to write down an informal HB-algorithm.

Algorithm 7.30. HB-preconditioner
$v := \mathrm{HB}(r)$

> **for** $k = j$ **to** 1 **step** -1 **do**
> — smoother on level k
> **for** $\xi \in \mathcal{N}_k^{\mathrm{HB}}$ **do**
> $v_\xi := r_\xi / a_{k,\xi}$
> **end for**
> — restriction of residual
> **for** $\xi \in \{\eta \in \mathcal{N}_{k-1} : \mathcal{N}_{k-1,\eta} \neq \emptyset\}$ **do**
> $r_\xi := r_\xi + \frac{1}{2} \sum_{\nu \in \mathcal{N}_{k,\xi}} r_\nu$
> **end for**
> **end for**
>
> — direct coarse grid correction on $\mathcal{T}_0$
> $(v_\xi)_{\xi \in \mathcal{N}_0} = B_0^{-1}(r_\xi)_{\xi \in \mathcal{N}_0}$
>
> — prolongation of the correction
> **for** $k = 1$ **to** j **do**

for $\xi \in \{\eta \in \mathcal{N}_{k-1} : \mathcal{N}_{k-1,\eta} \neq \emptyset\}$ **do**
 for $\nu \in \mathcal{N}_{k-1,\xi}$ **do**
  $v_\nu := v_\nu + \frac{1}{2} v_\xi$
 end for
end for
end for

The only nontrivial part of this algorithm is the construction of an efficient data structure for the neighborhood sets $\mathcal{N}_{k,\xi}$ and $\{\xi \in \mathcal{N}_k : \mathcal{N}_{k,\xi} \neq \emptyset\}$, which requires some programming care.

BPX-preconditioning

For $d = 3$ the BPX-preconditioner is more efficient than the HB-preconditioner, since the stability constant K_1^{BPX} remains bounded (see (7.51)). In principle, this holds for $d = 2$ as well. But the computational amount, compared with HB, is much larger, so that this advantage only comes into play for rather fine meshes.

Summation Formula. The extension of the subspaces of V_k^{HB} in the direction of V_k^{BPX} leads to an extension of the HB-summation formula (7.56) according to

$$(B^{\text{BPX}})^{-1} r = B_0^{-1} r + \sum_{k=1}^{j} \sum_{\xi \in \mathcal{N}_k^{\text{BPX}}} \frac{\langle r, \varphi_{k,\xi} \rangle}{\langle A\varphi_{k,\xi}, \varphi_{k,\xi} \rangle} \varphi_{k,\xi}. \tag{7.57}$$

In contrast to the HB-preconditioner, the BPX-preconditioner cannot be realized by direct overwriting of coefficients, so that a slightly larger memory is required. The similarity of the summation formulas (7.57) and (7.56) belies the rather different computational complexity when implementing the two formulas.

Algorithm 7.31. BPX-preconditioning.
$v := \text{BPX}(r)$

 for $k = j$ **to** 1 **step** -1 **do**
  — Jacobi smoother
  for $\xi \in \mathcal{N}_k^{\text{BPX}}$ **do**
   $v_{k,\xi} := r_\xi / a_{k,\xi}$
  end for
  — restriction of the residual
  for $\xi \in \{\eta \in \mathcal{N}_{k-1} : \mathcal{N}_{k-1,\eta} \neq \emptyset\}$ **do**
   $r_\xi := r_\xi + \frac{1}{2} \sum_{\nu \in \mathcal{N}_{k,\xi}} r_\nu$
  end for
 end for

— direct solution on $\mathcal{T}_0$

$$(v_\xi)_{\xi \in \mathcal{N}_j} = 0$$
$$(v_\xi)_{\xi \in \mathcal{N}_0} = B_0^{-1}(r_\xi)_{\xi \in \mathcal{N}_0}$$

— prolongation of the corrections
for $k = 1$ **to** j **do**
    for $\xi \in \{\eta \in \mathcal{N}_{k-1} : \mathcal{N}_{k-1,\eta} \neq \emptyset\}$ **do**
        for $\nu \in \mathcal{N}_{k-1,\xi}$ **do**
            $v_\nu := v_\nu + \frac{1}{2}v_\xi$
        end for
    end for
    for $\xi \in \mathcal{N}_k^{\text{BPX}}$ **do**
        $v_\xi := v_\xi + v_{k,\xi}$
    end for
end for

Remark 7.32. In 1989, P. Deuflhard, P. Leinen, and H. Yserentant [73] published the concept of the code KASKADE[4], which appeared to be the first fully adaptive implementation of an additive multigrid method. It realized an adaptive hierarchical mesh generator in 2D based on the DLY error estimator for linear finite elements, the local extrapolation method due to [13], and the red-green mesh refinement due to R. Bank, A. Sherman, and A. Weiser [18]. For the solution of the arising algebraic linear systems a PCG-method with the hierarchical basis preconditioner as applied. In 1993, an extension of the code to 3D was worked out by F. A. Bornemann, B. Erdmann, and R. Kornhuber [36]. It implemented a PCG-method with the above described BPX-preconditioner and a 3D extension of the red-green mesh refinement (see Figure 6.10).

7.3.2   Multiplicative Multigrid Methods

Multiplicative multigrid methods are a multigrid realization of the *sequential* subspace correction Algorithm 7.3 based on the hierarchical space decomposition introduced in Section 7.2. The HB-decomposition from Section 7.2.1 gives rise to the hierarchical basis multigrid method due to R. E. Bank, T. Dupont, and H. Yserentant [17]. Even though this algorithm, just like its additive counterpart, exhibits only nearly optimal convergence rate, it is, for $d = 2$ and realistic refinement levels j, the most efficient among the multiplicative multigrid algorithms. Since 1988 it has been realized in the program package PLTMG. For details we refer to the original paper [17]. In the following we will instead elaborate the BPX-decomposition with $V_k = S_k$ from

---

[4] This name was suggested by W. Mackens.

Section 7.2.2, which covers the classical multigrid algorithm including the adaptive variant considered here.

Convergence. In Section 5.5.1 we worked out a convergence proof for the classical multigrid method that, however, is sufficient only for W-cycles and uniform meshes. The in practice more often used V-cycles were not covered. This gap will now be filled by applying the sequential subspace correction methods from Section 7.1.2. In comparison with additive multigrid methods we now need to verify the three assumptions (A0), (A1), and (A2′), instead of just two. The proof of the assumptions (A1) and (A2′) in the BPX-decomposition was already done and discussed in Section 7.2.2. The only thing missing is the confirmation of assumption (A0), which we repeat here for the convenience of the reader: *there exists a $\lambda < 2$, such that*

$$\langle Av_k, v_k \rangle \leq \lambda \langle B_k v_k, v_k \rangle \quad \text{for all } v_k \in V_k.$$

This assumption is equivalent to the assumption (5.5) from Theorem 5.1, to be proved for smoothers in finite dimensional subspaces V_k. For the *damped* Jacobi method this is shown by Theorem 5.3, if we identify the above λ with 2ω so that $\lambda < 2$ is identical with $\omega < 1$. For the *symmetric* Gauss–Seidel method (SGS) we employ the result (5.15) from the proof of Theorem 5.5. A corresponding proof for the (unsymmetric) Gauss–Seidel method (GS) is not assured under the general assumption of spd-matrices, which is why SGS has more and more prevailed over GS as a smoother within classical multigrid methods.[5] For the CG-method the assumption is not as easily shown, since this method is nonlinear. However, in view of Theorem 5.16 we may – purely theoretically for the purpose of convergence analysis – formally define an spd-matrix B_k on each level a posteriori, the inverse of which reproduces the effect of the CG-method and which satisfies the conditions (A0)–(A2′). For this hypothetical linear method a convergence rate can be shown. By construction, the multigrid method with CG-smoother supplies the same theoretical contraction rate.

Nested Multigrid Methods. The realization of adaptive multiplicative multigrid methods is slightly more complicated than the one of additive methods. As can be seen from Algorithm 7.3, the residual $r_k = f - Aw_k$ on mesh level k must be computed after the addition of the correction $v_k = B_k^{-1} r_{k-1}$. This requires the availability of the whole matrix A_k instead of just the diagonal entries as in the additive methods. The algorithmic structure corresponds largely to the one of the classical multigrid method 5.17. For a really efficient implementation, however, quite complex data structures need to be realized. In general, the V-cycle we met in Section 5.5.1 is replicated, this time starting from the coarse mesh. Thus, in combination with adaptive mesh refinement, a *nested iteration* is obtained (see Figure 7.10).

---

[5] Nevertheless, multigrid convergence can also be shown for unsymmetric smoothers (see N. Neuss [159]).

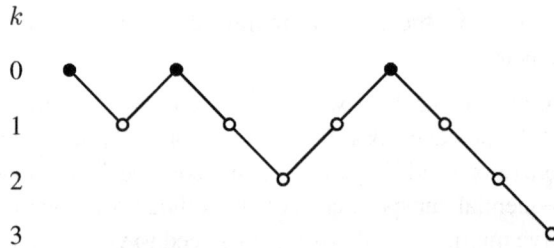

Figure 7.10. Nested V-cycle: coupling of multigrid method with adaptive mesh refinement. This time the coarse mesh $k = 0$ as initial mesh lies on top, while the finest mesh lies at the bottom – the V "stands upside down". In contrast to Figure 5.10, the descending lines now represent prolongations (i.e., for FE: interpolations), ascending ones restrictions. Direct solutions on the initial mesh are again labeled by •, and ○ denotes the fixed prescribed numbers (v_1, v_2) of pre- or postsmoothing steps.

Computational Complexity of Nested Multigrid Method.   In order to reach the discretization error of $\epsilon_k \approx cN_k^{-1/d}$ in step k, according to Corollary 6.1 an error reduction of

$$\frac{\epsilon_k}{\epsilon_{k-1}} \approx \left(\frac{N_{k-1}}{N_k}\right)^{1/d}$$

is required. Let us assume that the mesh refinement yields a geometric progression of nodes with $1 < q \leq N_k/N_{k-1} \leq 2^d$. Then we have $\epsilon_k/\epsilon_{k-1} \geq 2^{-d}$. From a mesh-independent contraction rate $\rho_* < 1$ of a multigrid method in refinement step k we may infer that

$$\rho^{m_k} \leq 2^{-d} \quad \Rightarrow \quad m_k \leq cd.$$

Thus a constant number m_k of iterations is sufficient. Hence, the computational amount in step k is then $\mathcal{O}(N_k)$. Summing up concering all refinement steps up to j yields an optimal total computational amount of $\mathcal{O}(N_j)$ due to the boundedness of the geometric series (cf. also Exercise 5.8).

Remark 7.33. The question of whether multiplicative or additive multigrid algorithms are faster cannot be easily answered. On the one hand, it was proven in [22] for simple 1D model problems that the contraction rates in the multiplicative case are smaller, i.e., that these methods require *fewer iterations*; this analysis is illustrated by numerical tests at simple 2D problems, both on uniform and on adaptive meshes. On the other hand, however, additive multigrid algorithms require *less computing time per iteration*, so that the total computing times are often shorter (see, e.g., [67, Table 1.1]). Thus both types of multigrid methods are roughly equivalent with respect to computing times. Moreover, for sufficiently complex application problems and problem adapted meshes the bulk of the work of numerical computation is spent in the assembling of the matrices. The decision on which variant to choose should therefore be made on the grounds of other criteria (see, e.g., Section 7.5.2, where a multiplicative method assures the monotonicity of eigenvalue approximations, whereas this could not have been achieved by an additive method).

Remark 7.34. Multiplicative multigrid methods are well-suited for numerous further problem classes far beyond the boundary value problems discussed here. Interested readers are referred to [204] by U. Trottenberg, C. Oosterlee, and A. Schüller. Efficient multigrid methods for the Navier-Stokes equations have been suggested by G. Wittum [218, 219]. Multigrid methods also play an increasing role in the field of nondifferentiable optimization; for this topic see the survey article by C. Gräser and R. Kornhuber [106].

7.4 Cascadic Multigrid Methods

In 1993, P. Deuflhard tried to compare the actual computing times of adaptive additive multigrid methods with different preconditioners at various test problems, in particular with HB versus BPX, both for $d = 2$ and for $d = 3$, to relate the theoretical complexity results with practical experience. In order to scale the comparison, he used the simple Jacobi preconditioning (i.e., only diagonal elements of A_h). To the surprise of nearly all the experts, the simple variant yielded the by far shortest computing times [66]. The core idea of this method, however, was an adaptation of the number ν_k of smoothing steps on coarser levels k such that errors to show up later on finer levels would already be eliminated on the coarser levels. The author first used the corresponding PCG iteration as smoother within the adaptive multigrid methods and derived this variant as a doubly-nested Galerkin method. For this reason, he called the method *cascadic conjugate gradient method*, or CCG-method. Upon replacing the PCG-iteration by some more general smoother, one is led to "cascadic" multigrid methods, in short CMG-methods.

A first convergence proof for the CCG-method was given by the Russian mathematician V. Shaidurov [187] in 1994. He oriented the scheme of his proof along the lines for classical multigrid methods on uniform meshes worked out here in Section 5.5.1. His central theoretical tool was the smoothing Theorem 5.16 by V. P. Il'yin [128]. On this basis, F. A. Bornemann found a generalization of the proof for general smoothers, still under the assumption of (quasi-)uniform meshes. The extension of this proof to adaptive meshes as well as the derivation of a ν_k-control was published by F. A. Bornemann and P. Deuflhard [35] in 1996.

7.4.1 Theoretical Derivation

Cascadic multigrid methods may be interpreted from two dual points of view:

- as *additive* multigrid methods with "economized" preconditioner:
  BPX → HB → Jacobi;

k

0 •

1 ◦

2 ◦

3 ◦

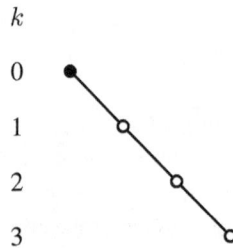

Figure 7.11. Cascadic multigrid method: as in Figure 7.10 the coarsest mesh $k = 0$ as initial mesh lies on top, the finest mesh at the bottom. In contrast to Figure 7.10 only descending lines occur (in FE: interpolations), coarse grid corrections are totally missing. Direct solutions on the initial mesh are again labeled by • and ◦ denotes a level dependent number ν_k of smoothing steps.

• as *multiplicative* multigrid methods of V-cycle type, but *without* any coarse grid correction (see the scheme in Figure 7.11), to be compared with the adaptive V-cycle in Figure 7.10.

For the analysis which follows we introduce the usual notation: we again assume that adaptive mesh refinement in j refinement steps generates a mesh hierarchy $\{\mathcal{T}_k\}$, $k = 0, \ldots, j$ with FE spaces $V_k = S_h^1(\mathcal{T}_k)$. Let $u_k \in V_k$ denote the (exact) FE solutions defined by

$$a(u_k, v_k) = \langle f, v_k \rangle \quad \text{for all } v_k \in V_k.$$

The cascadic multigrid method produces approximate solutions $\hat{u}_k \in V_k$ by ν_k-times application of the smoother G:

$$u_k - \hat{u}_k = G^{\nu_k}(u_{k-1} - \hat{u}_{k-1}).$$

This corresponds exactly to the graphical representation in Figure 7.11, for which reason CMG is also known as "backslash algorithm".

Convergence

In contrast to additive or multiplicative multigrid methods a classical convergence result is not possible here. Instead, our aim is to reduce the total error, i.e., discretization error and algebraic error, to a prescribed tolerance TOL. Obviously, this is a reasonable option for any adaptive algorithm. In what now follows we will take a perspective slightly different from the original one given in [35] both in terms of the theoretical derivation and the algorithmic realization.

In the computation of $\hat{u}_j$ by some smoother, starting from $\hat{u}_{j-1}$, the accuracy of $\hat{u}_{j-1}$ clearly plays a big role for the number of required iterations and thus for the computational amount. We are therefore recursively interested in the question of how accurate the $\hat{u}_k$ should be computed and how much total work is involved.

Basic Assumptions.   To begin with, we collect all theoretical assumptions that we will use in the subsequent analysis.

- *Monotonicity of the smoother*:

$$\|G(u_k - v)\|_A \leq \|u_k - v\|_A \quad \text{for all } v \in V_k. \tag{7.58}$$

  This property is equally satisfied by the Jacobi, the symmetric Gauss–Seidel, and the CG-methods.

- *Smoothing property*:

$$\|G^{\nu_k}(u_k - u_{k-1})\|_A \leq \frac{c_G}{\nu_k^{\gamma}}\|u_k - u_{k-1}\|_A \quad \text{for } k = 1, \ldots, j \tag{7.59}$$

  with $\gamma = 1/2$ (damped Jacobi and SGS-method) or $\gamma = 1$ (CG-method). For *uniform* mesh hierarchies this property can be immediately derived from the smoothing property (5.48) and the contraction property (5.51). For *adaptive* nonuniform mesh hierarchies it can be justified, at least for the CG-method, under rather natural assumptions on the mesh refinement process (see [35]).

- *Approximation property*:

$$0 < \underline{c_f} \leq \frac{\|u - u_k\|_A N_k^{1/d}}{\|f\|_{L^2(\Omega)}} \leq \overline{c_f}. \tag{7.60}$$

  The upper bound follows for H^2-regular problems from Theorem 6.17. The lower bound describes the generic case – in particular, the trivial case $u \in V_k$ is thus excluded.

- *Sufficiently fast mesh refinement*:

$$1 < \underline{q} \leq \frac{N_k}{N_{k-1}} \leq \overline{q} \leq 2^d \quad \text{for } k = 1, \ldots, j. \tag{7.61}$$

Basic Idea.   The maximal refinement level j is not known at the beginning, but only implicitly determined by the constraint

$$\|u - \hat{u}_j\|_A \leq \text{TOL}$$

and by the minimization of the computational amount. As in (6.4) we split the total error into the algebraic and the discretization error according to

$$\|u - \hat{u}_j\|_A^2 = \|u - u_j\|_A^2 + \|u_j - \hat{u}_j\|_A^2$$

and demand that, for some factor $0 < \beta < 1$ to be determined:

$$\|u_j - \hat{u}_j\|_A \leq \beta \, \text{TOL}, \quad \|u - u_j\|_A \leq \sqrt{1 - \beta^2} \, \text{TOL} < \|u - u_{j-1}\|_A. \tag{7.62}$$

The basic idea is now to set the tolerances for the algebraic error $\delta_k := \|u_k - \hat{u}_k\|_A$ on refinement level $k < j$ such that the total amount for achieving the error level TOL is minimized. For this purpose, we need a work model, which we will elaborate and minimize in the following.

Work Minimization

As in all multigrid methods the algorithmic work is characterized by

$$W_j = \sum_{k=1}^{j} \nu_k N_k.$$

The dimension sequence $\{N_k\}$ is obtained via the adaptive mesh construction, where above we assumed sufficient growth. The required number $\{\nu_k\}$ of smoothing steps may be estimated by the following lemma (ignoring the integer nature of the ν_k).

Lemma 7.35. *Under the assumption* $\delta_k > \delta_{k-1}$ *and with the notation introduced above, the following result holds:*

$$\nu_k^\gamma \leq \frac{c_G \overline{c_f}\, \overline{q}^{1/d} \|f\|_{L^2(\Omega)}}{(\delta_k - \delta_{k-1}) N_k^{1/d}}. \tag{7.63}$$

Proof. Step by step, we use the triangle inequality, the smoothing property (7.59), and the monotonicity property (7.58), and thus arrive at the estimate

$$\begin{aligned}
\delta_k = \|u_k - \hat{u}_k\|_A &= \|G^{\nu_k}(u_k - \hat{u}_{k-1})\|_A \\
&\leq \|G^{\nu_k}(u_k - u_{k-1})\|_A + \|G^{\nu_k}(u_{k-1} - \hat{u}_{k-1})\|_A \\
&\leq \frac{c_G}{\nu_k^\gamma}\|u_k - u_{k-1}\|_A + \|u_{k-1} - \hat{u}_{k-1}\|_A \\
&\leq \frac{c_G}{\nu_k^\gamma}\|u - u_{k-1}\|_A + \delta_{k-1}.
\end{aligned}$$

Application of the approximation property (7.60) in the adaptive case and subsequent introduction of the refinement factors (7.61) then supplies

$$0 < \delta_k - \delta_{k-1} \leq \frac{c_G}{\nu_k^\gamma}\overline{c_f}\|f\|_{L^2(\Omega)} N_{k-1}^{-1/d} \leq \frac{c_G \overline{c_f}\, \overline{q}^{1/d}}{\nu_k^\gamma}\|f\|_{L^2(\Omega)} N_k^{-1/d}$$

and thus assertion (7.63).   □

The total work for the computation of $\hat{u}_j$ may thus be estimated by

$$W_j \leq \overline{W}_j := c \sum_{k=0}^{j} (\delta_k - \delta_{k-1})^{-1/\gamma} N_k^{\frac{d\gamma-1}{d\gamma}}.$$

We now want to minimize the upper bound $\overline{W}_j$ of the total amount W_j by an optimal sequence $\{\nu_k\}$ of the number of smoothing steps. By means of the above lemma this can be implicitly achieved by an optimal choice $\{\bar{\delta}_k\}$ for the algebraic errors $\{\delta_k\}$. For the subsequent consideration we replace the δ_k by their differences $s_k = \delta_k - \delta_{k-1}$ (assumed to be positive). With $\delta_0 = \bar{\delta}_0 = 0$ (direct solution on the initial mesh) we immediately get

$$\delta_k = \sum_{i=1}^{k} s_i, \quad \delta_j \leq \bar{\delta}_j = \beta \, \text{TOL}.$$

Note that the last refinement level j is only implicitly defined here. This leads us to the constrained minimization problem

$$\min_{s_k} \sum_{k=0}^{j} s_k^{-1/\gamma} N_k^{\frac{d\gamma-1}{d\gamma}} \quad \text{subject to} \quad \sum_{k=1}^{j} s_k = \beta \, \text{TOL},$$

whose optimal solution $\{\bar{s}_k\}$, $\{\bar{\delta}_k\}$ is summarized in the following theorem.

Theorem 7.36. *With the optimal choice of $\bar{\delta}_k$ for the algebraic error δ_k the total amount of work for the cascadic multigrid method is constrained by*

$$W_j \leq c \begin{cases} N_j, & d\gamma > 1, \\ j^{\frac{1+\gamma}{\gamma}} N_j, & d\gamma = 1, \end{cases}$$

where c is a j-independent generic constant.

Proof. Due to Appendix A.6 we couple the constraints by a (positive) Lagrange multiplier λ to the objective and thus obtain the equations

$$-\frac{1}{\gamma} \bar{s}_k^{-\frac{1+\gamma}{\gamma}} N_k^{\frac{d\gamma-1}{d\gamma}} + \lambda = 0.$$

From this we derive the representation

$$\bar{s}_k = (N_k^{\frac{d\gamma-1}{d\gamma}} (\lambda\gamma)^{-1})^{\frac{\gamma}{1+\gamma}}.$$

Upon taking the quotients $\bar{s}_k/\bar{s}_j$, the Lagrange multipliers cancel out, and we obtain

$$\bar{s}_k = \bar{s}_j \left(\frac{N_k}{N_j}\right)^{\frac{d\gamma-1}{d(1+\gamma)}}. \tag{7.64}$$

After insertion into the formula for the work a short intermediate calculation yields

$$\overline{W}_j = c \, \bar{s}_j^{-1/\gamma} N_j^{\frac{d\gamma-1}{d\gamma}} \sum_{k=1}^{j} \left(\frac{N_k}{N_j}\right)^{\frac{d\gamma-1}{d(1+\gamma)}}$$

as well as the side condition

$$\beta \, \text{TOL} = \bar{s}_j \sum_{k=1}^{j} \left(\frac{N_k}{N_j}\right)^{\frac{d\gamma-1}{d(1+\gamma)}}.$$

At this stage we use the growth of the number of nodes as required in (7.61). So we obtain, for the time being,

$$\beta \, \text{TOL} \leq \bar{s}_j \sum_{k=1}^{j} q^{\frac{(k-j)(d\gamma-1)}{d(1+\gamma)}} .$$

In order to simplify the calculation, we introduce the quantity $b = q^{-\frac{d\gamma-1}{d(1+\gamma)}}$. We obviously have

$$d\gamma = 1 \quad \Rightarrow \quad b = 1, \; d\gamma > 1 \quad \Rightarrow \quad b < 1,$$

while the case $d\gamma < 1$ does not occur for neighboring meshes. Thus we obtain

$$\beta \, \text{TOL} \leq \begin{cases} \bar{s}_j/(1-b), & d\gamma > 1, \\ j\bar{s}_j, & d\gamma = 1, \end{cases}$$

which immediately implies

$$\bar{s}_j \geq \begin{cases} \beta \, \text{TOL}(1-b), & d\gamma > 1, \\ \beta \, \text{TOL} /j, & d\gamma = 1. \end{cases}$$

In passing we note that in both cases $\bar{s}_j > 0$ holds, from which with (7.64) directly $\bar{s}_k > 0$, $k = 0, \ldots, j-1$ follows; thus the assumption $\delta_k > \delta_{k-1}$ in Lemma 7.35 is assured at least for the optimal choice $\{\delta_k\}$. For $d\gamma > 1, b < 1$ the amount can be estimated via the geometric series as

$$\overline{W}_j \leq c(\beta \, \text{TOL}(1-b))^{-1/\gamma} N_j^{\frac{d\gamma-1}{d\gamma}} (1-b)^{-1}.$$

From the approximations assumption (7.60) and the accuracy requirement (7.62) now follows

$$\sqrt{1-\beta^2} \, \text{TOL} \geq \frac{c_f \|f\|_{L^2(\Omega)}}{N_j^{1/d}} = c N_j^{-1/d}$$

and thus finally (with generic constants independent of β and j)

$$\overline{W}_j \leq c \left(\frac{\sqrt{1-\beta^2}}{\beta(1-b)} \right)^{1/\gamma} \frac{N_j}{1-b}, \tag{7.65}$$

i.e., the first assertion of the theorem. For $d\gamma = 1, b = 1$ we obtain similarly

$$\overline{W}_j \leq c \left(\frac{j\sqrt{1-\beta^2}}{\beta} \right)^{1/\gamma} jN_j .$$

Collecting the exponents of j then supplies the second assertion of the theorem.   □

For $d\gamma > 1$, i.e., for the CG-method in $d = 2, 3$ as well as the Jacobi and the Gauss–Seidel methods in $d = 3$, the above theorem supplies *optimal complexity* $\mathcal{O}(N_j)$. For the limiting case $d\gamma = 1$ only suboptimal complexity $\mathcal{O}(j^{(1+\gamma)/\gamma} N_j)$ is obtained. Therefore we will restrict the following study to $d\gamma > 1$. For this case we may exploit Theorem 7.36 even further to determine the hitherto unknown parameter β by minimization of the amount of work.

Theorem 7.37. *For $d\gamma > 1$ the following result holds*

$$W_j \leq c\, \beta^{-1/\gamma}(1 - \beta^2)^{\frac{1-d\gamma}{2\gamma}}$$

with a constant c independent of β and j. The minimum of the upper bound is achieved for the value

$$\beta_{\min} = \frac{1}{\sqrt{d\gamma}}. \qquad (7.66)$$

Proof. In order to get the pure dependence of the work on the parameter β, we merely replace the factor N_j in (7.65) by virtue of (7.61), (7.60), and (7.62) by

$$N_j \leq \bar{q} N_{j-1} \leq \bar{q} \left(\frac{\overline{c_f}\|f\|_{L^2(\Omega)}}{\|u - u_{j-1}\|_A} \right)^d \leq \bar{q} \left(\frac{\overline{c_f}\|f\|_{L^2(\Omega)}}{\sqrt{1 - \beta^2}\, \text{TOL}} \right)^d$$

and obtain

$$W_j \leq c \left(\frac{\sqrt{1 - \beta^2}}{\beta} \right)^{1/\gamma} \left(\frac{1}{\sqrt{1 - \beta^2}} \right)^d = c\,(1 - \beta^2)^{\frac{1-d\gamma}{2\gamma}} \beta^{-1/\gamma} =: \varphi(\beta).$$

We thus have to minimize the function $\varphi(\beta)$. With $\varphi'(\beta) = 0$ as well as the fact that $\varphi(0)$ and $\varphi(1)$ are positive and unbounded, we may directly derive the relation (7.66). □

The value $\beta_{\min}$ given here has to be inserted into the accuracy requirement (7.62).

7.4.2 Adaptive Realization

The algorithmic realization of the cascadic multigrid method relies on computable estimators $[\delta_k] \leq \delta_k$, like the ones introduced for the adaptive CG-method in Section 5.3.3. In order to algorithmically determine an optimal refinement depth j at which the prescribed accuracy TOL for the total error is achieved, we require for the estimated algebraic error

$$[\delta_k] \leq \bar{\delta}_k = \sum_{i=1}^{k} \bar{s}_i \quad \Rightarrow \quad [\delta_j] \leq \bar{\delta}_j = \beta\, \text{TOL}.$$

Of course, we want to guide the sequence $\{[\delta_k]\}$ as close as possible to the optimal sequence $\{\bar{\delta}_k\}$ to avoid unnecessary smoothing steps, in particular on the finest levels. However, the $\bar{s}_i$ still depend on the N_j (see (7.64)), i.e., they cannot be evaluated explicitly. For the derivation of an upper bound that can be evaluated within the algorithm we want to calculate as much as possible explicitly and return to approximations as late as possible. In a first step, we obtain from the proof of Theorem 7.36

$$\bar{\delta}_k = \bar{\delta}_k \frac{\beta \, \text{TOL}}{\bar{\delta}_j} = \frac{\beta \, \text{TOL} \sum_{i=1}^{k} \bar{s}_i}{\sum_{i=1}^{j} \bar{s}_i} = \frac{\beta \, \text{TOL} \, \bar{s}_j \sum_{i=1}^{k} (\frac{N_i}{N_j})^\alpha}{\bar{s}_j \sum_{i=1}^{j} (\frac{N_i}{N_j})^\alpha} = \beta \, \text{TOL} \frac{\sum_{i=1}^{k} N_i^\alpha}{\sum_{i=1}^{j} N_i^\alpha}$$

with $\alpha = (d\gamma - 1)/(d(1 + \gamma))$. Introducing the notations

$$y_k := \sum_{i=1}^{k} N_i^\alpha, \quad z_k := \sum_{i=k+1}^{j} N_i^\alpha,$$

the above relation can be written in the form

$$\bar{\delta}_k = \beta \, \text{TOL} \frac{y_k}{y_k + z_k}. \tag{7.67}$$

In step k of the algorithm the term y_k can be evaluated, but not the term z_k. Therefore this term must be substituted by some estimate $[z_k]$. For this purpose, we assume, in the absence of more accurate knowledge, that the actual mesh refinement rate is also valid for the next refinement steps. So we obtain from the finite geometric series

$$z_k = N_k^\alpha \sum_{i=1}^{j-k} \left(\frac{N_{k+i}}{N_k} \right)^\alpha \approx N_k^\alpha \sum_{i=1}^{j-k} q_k^{i\alpha} = N_k^\alpha \frac{q_k^{(j-k)\alpha} - 1}{1 - q_k^{-\alpha}} \quad \text{with} \quad q_k := \frac{N_k}{N_{k-1}}.$$

The only unknown in this expression is q_k^{j-k}. This can be estimated by

$$q_k^{j-k} \approx \frac{N_j}{N_k} = q_k^{-1} \frac{N_j}{N_{k-1}} \doteq q_k^{-1} \left(\frac{\|u - u_{k-1}\|_A}{\sqrt{1 - \beta^2} \, \text{TOL}} \right)^d$$

and the discretization error $\|u - u_{k-1}\|_A$ there by the error estimator $[\epsilon_{k-1}]$ already used for the mesh refinement from $\mathcal{T}_{k-1}$ to $\mathcal{T}_k$. In total we obtain

$$z_k :\approx N_k^\alpha \frac{(\frac{[\epsilon_{k-1}]}{\sqrt{1-\beta^2} \, \text{TOL}})^{d\alpha} - q_k^\alpha}{q_k^\alpha - 1} =: [z_k].$$

The application of this estimator in lieu of $\bar{\delta}_k$ in (7.67) leads to the implementable truncation criterion

$$[\delta_k] \le [\bar{\delta}_k] := \frac{\beta \, \text{TOL} \, y_k}{y_k + [z_k]} \quad \Rightarrow \quad [\delta_j] \le [\bar{\delta}_j] = \beta \, \text{TOL}.$$

Since the initial phase of the CG iteration is relevant for the termination here, we use the error estimator from Section 5.4.3, which considers the initial smoothing phase separately.

Improved Error Estimation within the PCG-method. In Section 5.3.3 we derived an error estimation technique for the finite dimensional PCG-method based on Galerkin orthogonality. However, if one tries to extract information about the asymptotic convergence rate Θ in (5.23) from current iterates, the error may be significantly underestimated, with the result that the iteration is terminated too early. As can be seen from Figure 5.8, a biphasic error behavior is rather typical, of which the asymptotic linear convergence is only one part. That is why we want to work out some heuristics here based on the estimates (5.46) and (5.18). On this background, we are led to the ansatz

$$\epsilon_k \approx \min\left(\frac{a}{2k+1}, \epsilon_0 \Theta^k\right).$$

In practical examples the actual error behavior appeared to be better modelled by replacing the first term in the form

$$\epsilon_k \approx \min\left(\frac{a}{\sqrt{k}+1}, \epsilon_0 \Theta^k\right). \tag{7.68}$$

With Galerkin orthogonality we are therefore led to

$$[\epsilon_k]_n^2 = \epsilon_k^2 - \epsilon_n^2 \approx \frac{a^2}{(\sqrt{k}+1)^2} - b^2. \tag{7.69}$$

The model parameters a^2 and b^2 may be determined by solving the linear least squares problem

$$\min_{a^2, b^2} \sum_{k=0}^{n-1} \frac{1}{(k+1)^2}\left(\frac{a^2}{(\sqrt{k}+1)^2} - b^2 - [\epsilon_k]_n^2\right)^2,$$

in which the chosen prefactor $(k+1)^{-2}$ gives a larger weight to earlier iterative steps. For the asymptotic convergence phase we may then set

$$\epsilon_k = \epsilon_0 \Theta^k \tag{7.70}$$

and thus obtain

$$\Theta^{2k} \geq \frac{\epsilon_k^2}{\epsilon_0^2} = \frac{[\epsilon_k]_n^2 + \epsilon_n^2}{[\epsilon_0]_n^2 + \epsilon_n^2}. \tag{7.71}$$

With $\epsilon_n^2 = \Theta^{2n}\epsilon_0^2 = \Theta^{2n}([\epsilon_0]_n^2 + \epsilon_n^2)$ we arrive at

$$\epsilon_n^2 = \Theta^{2n}\frac{[\epsilon_0]_n^2}{1 - \Theta^{2n}}.$$

Insertion into (7.71) eventually leads to

$$\Theta^{2k} = \frac{[\epsilon_k]_n^2 + \Theta^{2n}([\epsilon_0]_n^2 - [\epsilon_k]_n^2)}{[\epsilon_0]_n^2},$$

where, for each $k = 1, \ldots, n-1$, some value Θ_k may be computed fast and conveniently via some fixed point iteration. As a cautious option we decided to take $\Theta = \max_{k=1,\ldots,n-1} \Theta_k$.

Thus the only thing left to do is to find out when the asymptotic linear convergence phase actually sets in. A necessary condition for this is that the error (7.70) lies significantly below the error (7.69). The crossover point between the two terms may be defined by some index k_c according to

$$\Theta^{2k_c} \frac{[\epsilon_0]_n^2}{1 - \Theta^{2n}} = \frac{a^2}{(\sqrt{k_c} + 1)^2} + \frac{[\epsilon_0]_n^2}{1 - \Theta^{2n}} - a^2.$$

Herewe have shifted the curve (7.69) such that for $k = 0$ it is consistent with (7.70). In a semilogarithmic diagram, the mapping (7.69) is convex, whereas (7.70) is linear, so that exactly one nontrivial cross point k_c occurs. For sufficiently large n, say $\rho n \geq k_c$ with a safety factor $\rho < 1$, we may assume to have reached the asymptotic convergence phase. As previously,in Section 5.3.3, we run the scheme over a number of indices to obtain some robustness of the heuristic estimates.

Algorithm.   After these preparations we have all necessary pieces of an adaptive cascadic multigrid method available.

Algorithm 7.38. *Cascadic conjugate gradient method (CCG).*
$\hat{u}_j := \text{CCG}(f, \text{TOL})$

> $\hat{u}_0 := A_0^{-1} f_0 = u_0$
> **for** $k = 1, \ldots$ **do**
>> compute error estimator $[\epsilon_{k-1}]$
>> **if** $[\epsilon_{k-1}] \leq \sqrt{1 - \beta^2}\, \text{TOL}$ **then**
>>> $\hat{u}_j := \hat{u}_{k-1}$
>>> **exit**
>> **end if**
>> refine mesh according to $[\epsilon_{k-1}]$
>> $\hat{u}_k := \text{CG}(A_k, f_k, I_{k-1}^k \hat{u}_{k-1}; \beta\, \text{TOL}\, y_k / (y_k + [z_k]))$
> **end for**

The adaptive CCG-algorithm worked out above slightly differs in some details and in the underlying theoretical concept from the one published in [35].

Numerical Experiments.   In order to illustrate the CCG method, we perform numerical experiments for a Poisson problem on a slit domain similar to our previous Example 6.23. The results are depicted in Figure 7.12. Note that the CCG version presented here slightly differs from the one published in [35].

In Figure 7.12, left, the typical behavior of the iterative number of smoothing steps is shown: following the strategy worked out above, an intermediate bump occurs that asymptotically merges into one iteration; note that we have added six CG iterations per step used to estimate the CG error according to (7.68). In Figure 7.12, right, the

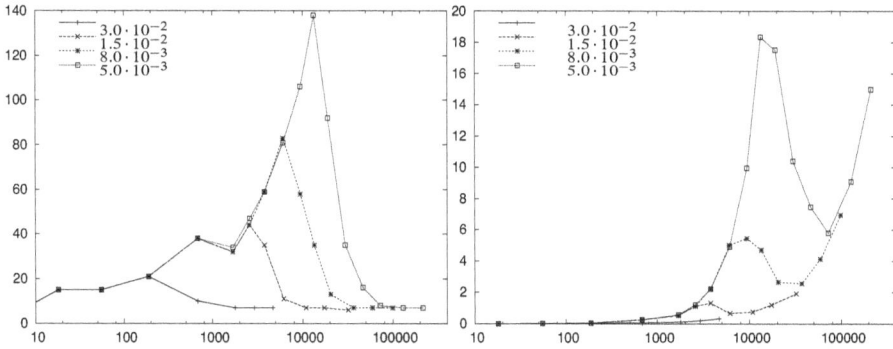

Figure 7.12. CCG method for prescribed error tolerances TOL $= \{3.0 \cdot 10^{-2}, \dots, 5.0 \cdot 10^{-3}\}$. Horizontal axis: numbers N_k of nodes (logarithmic scale) at iterative refinement level k. *Left:* iterative number ν_k of smoothing steps. In contrast to the original paper [35], six iterations used for the CG error estimation as a standard option are included, which explains the asymptotic constant. *Right:* iterative computational work $10^{-5} \nu_k N_k$.

Table 7.2. CCG method: dependence of number N_k of nodes on prescribed error tolerance TOL, as observed from Figure 7.12.

TOL	$3.0 \cdot 10^{-2}$	$1.5 \cdot 10^{-2}$	$8.0 \cdot 10^{-3}$	$5.0 \cdot 10^{-3}$
N_k	5 208	20 384	77 129	279 910
TOL $N_k^{1/2}$	2.165	2.142	2.221	2.645

iterative work at refinement level k is given; due to the semilogarithmic scale the asymptotic linear behavior shows up as exponential behavior.

On both sides of Figure 7.12, the dependence on the prescribed error tolerance TOL is clearly visible: a slight decrease of TOL induces a drastic increase in the necessary number of nodes and thus of computational work. This effect can be directly read from Table 7.2. In the last row the necessary number of degrees of freedom is seen to nicely follow the convergence law for adaptive meshes derived in Section 6.3.1 (cf. Figure 6.12, which roughly states TOL $\sim N_k^{-1/2}$ for the energy norm). This effect would be even more marked in space dimension $d = 3$, where TOL $\sim N_k^{-1/3}$ would be obtained.

Conclusion. Compared to additive or multiplicative multigrid methods, cascadic multigrid methods stand out by their *extreme simplicity*, which can be seen both from the above algorithm and also in Figure 7.10. That is why they are favored as subroutines in particularly complex software systems (e.g., in climate modelling [30]). However, the advantages of a simpler implementation and a smaller computational amount are balanced by the fact that the accuracy TOL to be achieved must be chosen *in advance*; a mere continuation of the computation, should the result not meet the expectations of the user, is not possible. Moreover, the theory of cascadic multigrid methods only assures a reduction of the *energy error*. An extension of the theory to a reduction of the L^2-error, which exists in the multiplicative or the BPX-preconditioned additive multigrid methods, is impossible (see F. A. Bornemann and R. Krause [38]).

7.5   Eigenvalue Problem Solvers

In Section 4.1.3 we discussed weak solutions of linear self-adjoint eigenvalue problems, in Section 4.4.2 we elaborated the corresponding approximation theory for finite elements. Compared to the previous sections, we now deal with a *sequence* of finite dimensional eigenvalue problems of growing dimension, which is why we slightly change the notation in order to reduce the number of indices.

To begin with, we only consider the minimal eigenvalue λ with corresponding eigenfunction $u \in H_0^1(\Omega)$. Let $\lambda_k, u_k \in S_k, k = 0, \dots, j$ denote the exact eigenvalues and eigenfunctions in the FE-spaces $S_k \subset H_0^1$. The associated discrete eigenvalue problems read

$$a(u_k, v_h) = \lambda_k \langle u_k, v_h \rangle \quad \text{for all} \quad v_h \in S_k \in H_0^1(\Omega).$$

From (1.15) we then conclude by means of the Rayleigh quotient

$$\lambda = \min_{v \in H_0^1} R(v) = R(u).$$

Transition to the weak formulation yields

$$\lambda_k = \min_{v_h \in S_k} R(v_h) = R(u_k).$$

As $S_0 \subset S_1 \subset \cdots S_j \subset H_0^1$, the minimum property immediately supplies:

$$\lambda \le \lambda_j \le \lambda_{j-1} \le \cdots \le \lambda_0.$$

By insertion of a basis, we obtain the algebraic eigenvalue problems of dimension $N_k = \dim(S_k)$

$$A_k u_k = \lambda_k M_k u_k, \tag{7.72}$$

where now the u_k are to be interpreted as the coefficient vectors with respect to the FE-basis and M_k as positive definite (N_k, N_k)-mass matrix. This is in agreement with the algebraic representation

$$\lambda_k = \min_{v_h \in S_k} \frac{v_h^T A_k v_h}{v_h^T M_k v_h} = \frac{u_k^T A_k u_k}{u_k^T M_k u_k} = R(u_k)$$

which we will prefer to use in the present subsection, dropping the index h for ease of presentation.

For the numerical solution of linear eigenvalue problems there exist two principal approaches, a linear one and a nonlinear one, both of which we will discuss now.

7.5.1 Linear Multigrid Method

This algorithmic approach was suggested by W. Hackbusch [112] in 1979 and further developed in his monograph [114].

Let us start with explaining the *basic idea*, for the time being without the multigrid context. Linear eigenvalue problems may be regarded as *singular* linear equation systems

$$(A_k - \lambda_k M_k)\, u_k = 0, \quad u_k \neq 0 \tag{7.73}$$

in terms of a parameter λ_k, which, given the eigenfunctions u_k, can be computed from the Rayleigh quotient as

$$\lambda_k = R(u_k). \tag{7.74}$$

In principle, one might construct a fixed point algorithm alternating between (7.73) and (7.74), starting with approximations $\tilde{\lambda}_k, \tilde{u}_k$. In the course of the solution of (7.73) there arises a nonvanishing *residual*

$$r_k = (A_k - \tilde{\lambda}_k M_k)\, \tilde{u}_k.$$

In order to gain corrections from this relation, we define

$$\tilde{u}_k = u_k + \delta u_k, \quad \tilde{\lambda}_k = \lambda_k + \delta \lambda_k.$$

Insertion into the above expression supplies (in linearized approximation)

$$(A_k - \lambda_k M_k)\, \delta u_k = r_k + \delta \lambda_k M_k (u_k + \delta u_k) \doteq r_k + \delta \lambda_k M_k u_k. \tag{7.75}$$

Obviously, this approximate system of equations is only solvable if the right-hand side lies in the image of $A_k - \lambda_k M_k$ and is thus orthogonal to the nullspace $\mathcal{N}_k$. In the simplest case, λ_k is simple and $\mathcal{N}_k = \{u_k\}$. Therefore, in order to generate a solvable system, the right-hand side must be projected into $\mathcal{N}_k^{\perp}$:

$$r_k \rightarrow P_k^{\perp}(r_k + \delta \lambda_k M_k u_k) = P_k^{\perp} r_k = r_k - \frac{u_k^T M_k r_k}{u_k^T M_k u_k}\, u_k.$$

As the exact eigenvector u_k is not known, one will replace it by the available approximation $\tilde{u}_{k-1}$ and solve the *perturbed singular* linear equation system

$$(A_k - \tilde{\lambda}_k M_k)\, v_k = \tilde{P}_k^{\perp} r_k := r_k - \frac{\tilde{u}_k^T M_k r_k}{\tilde{u}_k^T M_k \tilde{u}_k}\, \tilde{u}_k. \tag{7.76}$$

The matrix in this system is nearly singular, i.e., ill-conditioned. Nevertheless the solution of this equation system is well-conditioned, since the right-hand side lies in the perturbed degenerate column space of the matrix. In order to recall this circumstance, we pick Example 2.33 from Volume 1.

Example 7.39. Consider an equation system $Ax = b$ with two different right-hand sides b_1 and b_2

$$A := \begin{bmatrix} 1 & 1 \\ 0 & \varepsilon \end{bmatrix}, \quad b_1 = \begin{pmatrix} 2 \\ \varepsilon \end{pmatrix}, \quad b_2 = \begin{pmatrix} 0 \\ 1 \end{pmatrix},$$

where $0 < \varepsilon \ll 1$ is regarded as input. The condition number of the matrix is

$$\kappa(A) = \|A^{-1}\|_\infty \|A\|_\infty \doteq \frac{2}{\varepsilon} \gg 1.$$

As corresponding solutions we obtain

$$x_1 = A^{-1}b_1 = \begin{pmatrix} 1 \\ 1 \end{pmatrix}, \quad x_2 = A^{-1}b_2 = \begin{pmatrix} 1/\varepsilon \\ 1/\varepsilon \end{pmatrix}.$$

The solution x_1 is independent of ε, i.e., for this right-hand side the equation system is well-conditioned for all ε, even in the strictly singular limit case $\varepsilon \to 0$. The solution x_2, however, does not exist in the limit case (only its direction). This worst case, which, however, does not occur in our context, is described by the condition number of the matrix.

Following [114], the approximate solution of the perturbed system (7.76) is done by some "nested " multigrid method as exemplified in Figure 7.10 for an adaptive V-cycle. On each mesh refinement level the residual must be suitably projected. On the coarsest level ($k = 0$) one of the iterative solvers can be applied that we have presented in Volume 1 for the direct solution of eigenvalue problems. For $k > 0$ one will define the starting value by prolongation from the next coarser grid as

$$\tilde{u}_k = I_{k-1}^k \tilde{u}_{k-1} . \tag{7.77}$$

For the coarse grid correction the conjugate transformation applies, but with the approximate projection $\tilde{P}_{k-1}^\perp$.

Due to the linearization in (7.75) and the difference of the projections $P_k^\perp$ and $\tilde{P}_k^\perp$, $\tilde{P}_{k-1}^\perp$ the convergence of the fixed point iteration is guaranteed only for "sufficiently good" starting values $\tilde{u}_k$. This is made precise in the convergence theory given in [112]. It is based on the classical theory for boundary value problems presented in Section 5.5.1. Just as for boundary value problems, here, too, a restriction to at least W-cycles is required; in addition, an argument is developed for why postsmoothing should be totally avoided. However, whether or not the choice (7.77) produces a "sufficiently good" starting value depends on the fineness of the mesh on level $k - 1$. By this method, S. Sauter and G. Wittum [177] computed the eigenmodes of Lake Constance, starting from an initial grid that resolves the shoreline to rather fine detail.

In summary, this approach requires relatively fine coarse grids and is therefore not sufficiently robust, as we will illustrate below at Example 7.41. A more robust method will be presented in the next subsection.

7.5.2 Rayleigh Quotient Multigrid Method

The second approach is along the line of minimizing the Rayleigh quotient. A first multigrid algorithm on this basis as worked out in 1989 by J. Mandel and S. Mc-Cormick [151] and further pursued in a wider context in [154]. We present here a version which uses the CG-variant of B. Döhler [78] as smoother (see Section 5.3.4). Our presentation will follow along the lines of the dissertation of T. Friese [95].

The *monotony* (5.39) of the RQCG-method from Section 5.3.4 quite obviously depends on the iterative *space extension*, the confirmation of which can be formally written as

$$\lambda_{k+1} = \min_{u \in \tilde{U}_{k+1}} R(u) \leq R(u_k) = \lambda_k \quad \text{with} \quad \tilde{U}_{k+1} := \{u_k, p_{k+1}\}.$$

This principle can be extended in a natural way to nested hierarchical finite element spaces $S_0 \subset S_1 \subset \cdots \subset S_j \subset H_0^1$.

Let the multigrid method be adaptively realized, as depicted schematically in Figure 7.10 for the V-cycle.[6] On the coarse mesh $k = 0$ we solve the eigenvalue problem

$$(A_0 - \lambda_0 M_0)u_0 = 0$$

directly. Let us next consider the refinement level $k > 0$. As *smoother* we naturally use the RQCG-method in $\mathbb{R}^{N_k}$ described above. Strictly speaking, there is still no proof that the RQCG-method actually is a smoother. This is assured only for the asymptotic case of the CG-method. Beyond this, there is also overwhelming experimental evidence in the case of the RQCG. We thus apply Algorithm 5.12 both for ν_1-times presmoothing and for ν_2-times postsmoothing, where we can find suitable starting values from the multigrid context. It remains to be noted that all smoothing steps lead to a monotone sequence of eigenvalue approximations.

The construction of a *coarse grid correction* requires more subtle investigation. For this we define the subspaces

$$\tilde{U}_{k-1} := \text{span}\{\tilde{u}_k\} \oplus S_{k-1} \subset S_k$$

as well as the associated $(N_k, N_{k-1} + 1)$-matrices

$$U_{k-1} := [\tilde{u}_k, I_{k-1}^k],$$

---

[6] The method works equally well on a given hierarchical grid, starting from the finest grid.

defined in terms of suitable bases of S_{k-1}. In analogy to the one-dimensional case we make the ansatz

$$\hat{u}_k(\xi) := \hat{\sigma}_k(\tilde{u}_k + I_k^{k-1}\alpha) = \hat{\sigma}\, U_{k-1}\xi \quad \text{with} \quad \xi^T := (1, \alpha^T), \ \alpha \in \mathbb{R}^{N_{k-1}}.$$

The corresponding projected Rayleigh quotient is

$$R_{k-1}^+(\xi) := \frac{\xi^T A_{k-1}^+ \xi}{\xi^T M_{k-1}^+ \xi}$$

with the projected $(N_{k-1}+1, N_{k-1}+1)$-matrices

$$A_{k-1}^+ = U_{k-1}^T A_k U_{k-1}, \quad M_{k-1}^+ = U_{k-1}^T M_k U_{k-1}.$$

As we had assumed a representation via FE-bases, the matrix A_{k-1}^+ has the partitioned shape

$$A_{k-1}^+ = \begin{bmatrix} \tilde{u}_k^T A_k \tilde{u}_k & (\tilde{u}_k^T A_k \tilde{u}_k)^T \\ \tilde{u}_k^T A_k \tilde{u}_k & (I_{k-1}^k)^T A_k I_{k-1}^k \end{bmatrix} = \begin{bmatrix} \tilde{\lambda}_k & \gamma_{k-1}^T \\ \gamma_{k-1} & A_{k-1} \end{bmatrix},$$

where we have used Exercise 5.7 and $(I_{k-1}^k)^T = I_k^{k-1}$ as follows

$$\tilde{u}_k^T A_k \tilde{u}_k = \tilde{\lambda}_k, \quad I_k^{k-1} A_k I_{k-1}^k = A_{k-1}, \quad \gamma_{k-1} = I_k^{k-1} A_k \tilde{u}_k.$$

For the realization of the restriction of $A_k \tilde{u}_k$ to $\gamma_{k-1} \in \mathbb{R}^{N_{k-1}}$, we refer to Section 5.5.2, where an efficient computation via the mesh data structure is described in detail. For the matrix M_{k-1}^+ a comparable partitioning holds.

These two matrices enter as input into the RQCG-Algorithm 5.12. As starting values for the eigenfunction approximations we take

$$\hat{u}_k^{(0)}(\xi) := \tilde{u}_k, \quad \alpha^{(0)} = 0, \quad \xi^T := (1, 0).$$

Let $\hat{v}_{k-1} = I_k^{k-1}\hat{\alpha}_{k-1} \in S_{k-1}$ denote the coarse grid correction generated after several applications of the RQCG-algorithm. Then, due to (5.39), *monotonicity* arises in the form

$$\hat{\lambda}_k = R(\hat{u}_k) = R(\tilde{u}_k + \hat{v}_{k-1}) < R(\tilde{u}_k) = \tilde{\lambda}_k. \tag{7.78}$$

In this substep of the algorithm we have *reduced* the dimension N_k of the original problem to the dimension $N_{k-1}+1$. Recursive repetition of the just elaborated algorithmic substeps eventually leads to a sequence of eigenvalue problems with successively lower dimension down to the directly solvable eigenvalue problem. With the notation just introduced we finally obtain the following multigrid algorithm for the minimization of the Rayleigh quotient.

Algorithm 7.40. RQMGM-*Algorithm for* FEM.
$(\tilde{\lambda}, \tilde{u}) = \text{RQMGM}_k(A, M, u^{(0)}, \nu_1, \nu_2)$

$(A_0 - \lambda_0 M_0)u_0 = 0 \quad \tilde{u}_0 = u_0.$
for $k = 1$ **to** $k_{\max}$ **do**
$\qquad \tilde{u}_k^{(0)} = I_{k-1}^k \tilde{u}_{k-1}$
$\qquad (\tilde{\lambda}_k, \tilde{u}_k) = \text{RQCG}(A_k, M_k, \tilde{u}_k^{(0)}, \nu_1)$
$\qquad (\hat{\lambda}_k, \hat{u}_k) = \text{RQMGM}_{k-1}(A_{k-1}^+, M_{k-1}^+, \tilde{u}_k, \nu_1, \nu_2)$
$\qquad (\tilde{\lambda}_k, \tilde{u}_k) = \text{RQCG}(A_k, M_k, \hat{u}_k, \nu_2)$
end for
$(\tilde{\lambda}, \tilde{u}) = (\tilde{\lambda}_k, \tilde{u}_k)$

Extension to Eigenvalue Cluster.   If several eigenvalues $\lambda_1, \ldots, \lambda_q$ are sought simultaneously, then the multigrid method presented here can be extended in a way similar to the one presented at the end of Section 5.3.4: an orthogonal basis of the corresponding invariant subspace is computed (see (4.21)) which is constructed by minimization of the trace (4.22) instead of that of a single Rayleigh quotient. An extension to nonselfadjoint eigenvalue problems is oriented on the Schur canonical form (see T. Friese et al. [95, 96]).

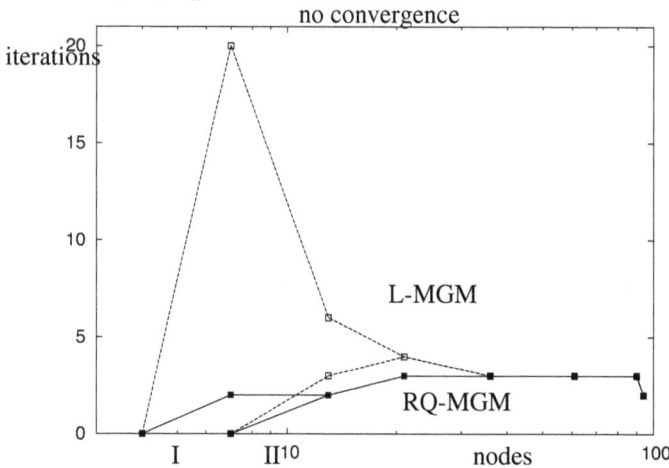

Figure 7.13.   Example 7.41: comparison of linear multigrid method (L-MGM) with the Rayleigh quotient multigrid method (RQ-MGM) for two differently fine initial coarse meshes I and II.

Example 7.41.   In order to compare the Rayleigh quotient multigrid method with the linear multigrid method presented above in Section 7.5.1, we work out an elementary example in one space dimension [70]. To be solved is the eigenvalue problem

$$-u_{xx} - \epsilon u = \lambda u, \quad \text{in } \Omega = \,]-3, 3[, \quad u'(-3) = u'(3) = 0,$$

where the material parameter ϵ exhibits a jump:

$$\epsilon(x) = \begin{cases} (3.2k_0)^2, & x \in]-3, -1[\cup]1, 3[, \\ (3.4k_0)^2, & x \in]-1, 1[. \end{cases}$$

With a wave number $k_0 = 2\pi/1.55$ (in the concrete units of [70]) this corresponds to a wavelength of light of $1.55\,\mu m$ in vacuum. Obviously, the problem is a Helmholtz problem in the critical case (see Section 1.5). Both multigrid methods can handle this type of problem, in principle. As initial coarse mesh the two variants

Coarse mesh I: $\{-3, -1, 1, 3\}$,   Coarse mesh II: $\{-3, -2, -1, 0, 1, 2, 3\}$

have been chosen. In both multigrid methods, the SGS-method has been used for smoothing. As error indicator for the generation of adaptive meshes the relative change of the Rayleigh quotient has been selected which is in any case computed in both methods. In Figure 7.13, the number of multigrid iterations for both coarse meshes is compared. For the coarser mesh I, the linear multigrid method does not converge at the first attempt; it was therefore terminated after 20 multigrid iterations and restarted at this iterate (i.e., by direct solution on more nodes than in coarse mesh I). For the finer coarse mesh II, both multigrid methods converge with a comparable amount of work, since they asymptotically merge (cf. Section 5.3.4). In order to obtain the lowest eigenvalue $\lambda = -188.2913\ldots$ to seven decimal digits, an adaptive mesh with 95 nodes has been constructed.

7.6 Exercises

Exercise 7.1. Given piecewise continuous f or $f \in H^1(\Omega)$. Show that for the oscillation (see Definition 6.3), the following relation holds:

$$\mathrm{osc}(f; \mathcal{T}) = \mathcal{O}(h^s)$$

Give the order s for each of the two cases.

Exercise 7.2. Given an equidistant mesh over $\Omega =]0, 1[$ with meshsize h. Construct right-hand sides $f_h \neq 0$ such that for exact error estimators (6.17), (6.36), and (6.29) the equation $[\epsilon_h] = 0$ holds.

Exercise 7.3. Consider a uniform discretization of a Poisson problem for $d = 1, 2, 3$ with linear and quadratic finite elements, respectively. Estimate the quotient N_Q/N_L of the numbers of degrees of freedom and the quotient $\mathrm{nnz}_Q/\mathrm{nnz}_L$ of nonzeros in the stiffness matrices. Quantify the memory and computational savings achieved by the DLY error estimator compared to a full hierarchic error estimator.

Exercise 7.4. *Inverse inequality.* Show the existence of a constant $c < \infty$ depending only on the minimal interior angle such that

$$\|u_h\|_{H^1(\Omega)} \le ch^{-1}\|u_h\|_{L^2(\Omega)} \quad \text{for all } u_h \in S_h^1 \tag{7.79}$$

holds. Here h is the minimal diameter of the elements in $\mathcal{T}$.

Exercise 7.5. Why can't the L^2-error of an FE-solution on an adaptive mesh due to Theorem 6.17 be estimated in analogy to Theorem 4.21?

Exercise 7.6. Calculate the weight function z of the goal oriented error estimation for the quantity of interest $J(u) = u(\xi)$, $\xi \in \Omega$, given the Poisson problem $-\Delta u = f$ with Dirichlet boundary conditions on $\Omega =]0, 1[$. With this result show the interpolation property of linear finite elements on the grid $0 = x_0 < \cdots < x_n = 1$: $u_h(x_i) = u(x_i)$.

Exercise 7.7. Show the nesting of the nodal sets $\mathcal{N}_k \subset \mathcal{N}_{k+1}$ in hierarchical meshes.

Exercise 7.8. Let the Green's function $u(s) = -\log|x|$ of the Poisson problem in $\mathbb{R}^2$ on the unit ball $B(x; 1)$ be interpolated with linear finite elements by *adaptive* refinement from a coarse mesh. Let j denote the refinement depth and N_j the number of nodes of the resulting mesh. Determine the order of growth of

$$W(j) = N_j^{-1} \sum_{k=0}^{j} N_k$$

with respect to j. Compare the result with the one for *uniform* refinement.

Exercise 7.9. Show that under the assumption (A2′) instead of (A2) the bound (7.5) for the contraction rate of the multiplicative subspace correction method is independent of the ordering of the subspaces.

Exercise 7.10. Prove the convergence of the Jacobi and Gauss–Seidel methods by means of the techniques for subspace correction methods.

Exercise 7.11. For fixed subdomain diameter $H > 0$ construct a smooth function $v = \sum_{k=0}^{m} v_k$ such that the stability constant

$$K(v) = \frac{\|v\|_A^2}{\sum_{k=0}^{m} \|v_k\|_A^2}$$

is independent of the relative width ρ of the overlap.
   Hint: The constant σ that defines the smoothness $\|v\|_A \le \sigma\|v\|_{L^2(\Omega)}$ may depend on H.

Exercise 7.12. Show that the domain decomposition preconditioners for higher order finite elements can only be independent of the order p, if the mesh nodes x_k lie in the interior of the subdomains Ω'_k.

Hint: Use the approximation of discontinuous functions $u \in H^1(\Omega)$ with small energy by finite element functions $u_h \in S_h^p$ and their interpolation $u_h^1 \in S_h^1$ for the construction of examples that do not permit a p-independent stable decomposition (see Figure 7.14).

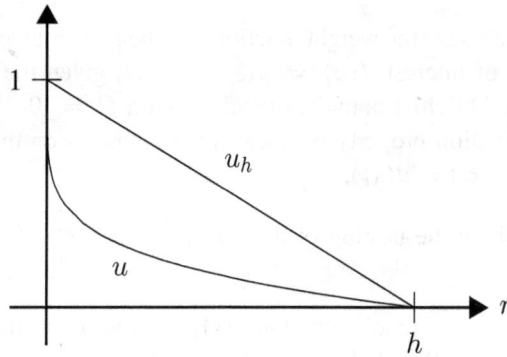

Figure 7.14. Exercise 7.12: cross section of linear interpolation of the radially symmetric function $u(r) = 1 - (r/h)^t$ with $0 < t \le 1$. For $d = 2$ one has $|u|_{H^1}^2 = \pi t$ and $|u_h|_{H^1}^2 = \pi$.

Exercise 7.13. Let $T \subset \mathbb{R}^2$ denote a triangle with circumcircle $B(x; h_T/2)$ and $u \in H^1(T)$. Show that u can be continued to $u \in H_0^1(B(x; h_T))$ such that

$$|u|_{H^1(B(x;h_T))} \le c \|u\|_{H^1(T)},$$

where the constant c only depends on the shape regularity, i.e., on the minimal interior angle of T.

Hint: Embed T in a regular mesh of congruent triangles and continue u by reflection at the edges.

Exercise 7.14. For uniformly refined mesh hierarchies as well as space decomposition $V_k = S_k$ and $v \in H^2(\Omega)$ show that the stability constant K_1 in assumption (A1) does not depend on the refinement depth j.

Hint: Use an H^1-orthogonal decomposition of v_h.

Exercise 7.15. Let $\Omega \subset \mathbb{R}^d$ satisfy the interior cone condition A.13 as well as a corresponding exterior cone condition and let

$$\Omega_s = \bigcup_{x \in \partial\Omega} B(x; \rho).$$

Show the existence of a ρ-independent constant $c > 0$ such that for all $v \in H^1$ with $v = 0$ on $\partial\Omega_s \cap (\mathbb{R}^d \backslash \Omega)$ and $v = 1$ on $\partial\Omega_s \cap \Omega$ the relation

$$|v|_{H^1} \geq \frac{c}{\rho} \qquad\qquad (7.80)$$

holds. *Hint:* In a first step, show the relation (7.80) for polygonal or polyhedral domains Ω.

Exercise 7.16. Derive the relation (7.23) from the minimization of $\|v_k\|_{L^2(\Omega)}$.

Chapter 8

Adaptive Solution of Nonlinear Elliptic Problems

Nonlinear elliptic differential equations in science and engineering often arise in the context of some *minimization problem*

$$f(u) = \min, \quad f : D \subset X \to \mathbb{R},$$

where we assume f to be a *strictly convex* functional over some Banach space X. This problem is equivalent to the nonlinear elliptic PDE

$$F(u) := f'(u) = 0, \quad F : D \to X^*. \tag{8.1}$$

The corresponding Fréchet derivative $F'(u) = f''(u) : X \to X^*$ is, due to the strict convexity of f, a *symmetric positive* operator.

To solve equation (8.1) we have two principal possibilities: *discrete* or *continuous* Newton methods, the latter often also called *function space oriented* Newton methods; they are shown in a common diagram in Figure 8.1. In the case of *fixed* discretization in space the diagram commutes. For adaptive discretizations, however, the order of discretization and linearization does play an important role.

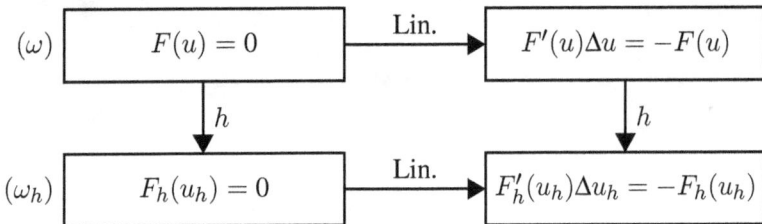

Figure 8.1. *Discrete* Newton methods: first discretization (h), then linearization. Top left → bottom left → bottom right. *Continuous* or also *function space oriented* Newton methods: first linearization, then discretization (h). Top left → top right → bottom right.

If we discretize the problem first in space, we obtain a finite dimensional nonlinear equation system $F_h(u_h) = 0$. Then in each Newton iteration a linear equation system with a symmetric positive definite matrix has to be solved. Adaptivity nevertheless turns up in the form of a stepsize damping strategy and a termination criterion for an inner PCG iteration. This variant will be presented in Section 8.1. If, however, we first linearize the operator equation, then we obtain a continuous Newton iteration. There in each step a linear elliptic boundary value problem has to be solved, which has to be discretized. As additional adaptive algorithmic pieces, one has the adaptive mesh generation and, on this basis, adaptive multigrid methods. This variant will be worked out below in Section 8.2. In the final Section 8.3 we give the nontrivial problem of operation planning in cranio-maxillo-facial surgery; as will be seen, we will have to leave the simple context of convex optimization there.

Remark 8.1. For experts we want to set out the theoretical frame somewhat more precisely: first, we assume that the Banach space is *reflexive*, i.e., $X^{**} = X$, which is the basic assumption for the *existence* of a minimum of f. Moreover, the Newton iterates usually do not exhaust the full Banach space X, but remain in some "smoother" subspace, which will be denoted by $\underline{X} \subset X$. In this subspace, $F'(u)$ is a symmetric strictly positive operator. Moreover, the (subsequently) constructed local energy norms are then equivalent to the corresponding locally weighted H^1-norms. For simplicity we will not distinguish here between $\underline{X}$ and X. However, the structure will show up in the context of the numerical examples, in which we will illustrate the adaptive concepts in Section 8.2.

8.1 Discrete Newton Methods for Nonlinear Grid Equations

In this section we presuppose that the nonlinear elliptic PDE has already been discretized on a fixed mesh with dimension N, i.e., we deal with nonlinear grid equations. For their numerical solution we treat *adaptive Newton methods*, which in each step require the solution of systems of linear grid equations. Where these linear systems are solved by direct elimination methods, we speak of exact Newton methods (see Section 8.1.1); if they are attacked by iterative methods from linear algebra, then so-called inexact Newton methods arise (see Section 8.1.2). To simplify the notation, we drop the suffix h, i.e., F_h, u_h is written as F, u. We start from a discrete minimization problem

$$f(u) = \min, \quad f : D \subset \mathbb{R}^N \to \mathbb{R}, \tag{8.2}$$

where f is *convex*. This problem is equivalent to the nonlinear equation system

$$F(u) := f'(u)^T = 0, \quad F : D \to \mathbb{R}^N.$$

The corresponding Jacobian matrix $F'(u) = f''(u)$ is, due to the convexity of f, certainly *symmetric positive semidefinite*. If we further assume that the functional f is *strictly* convex, then $F'(u)$ is symmetric positive definite.

Affine Invariance. In Volume 1, Section 4.2, we required invariance of the general nonlinear systems $F(u) = 0$ under the transformation $F \to G = AF$ for arbitrary matrix A. Here, however, we have a more special case, as the equation system originates from a convex minimization problem. Suppose we transform the domain space of f according to

$$g(\bar{u}) = f(C\bar{u}) = \min, \quad u = C\bar{u},$$

where C denotes an arbitrary nonsingular matrix, then we are led to the transformed nonlinear equation system

$$G(\bar{u}) = C^T F(C\bar{u}) = 0$$

and the transformed associated Jacobian matrix

$$G'(\bar{u}) = C^T F'(u)C.$$

Obviously, the affine transformation of the Jacobian matrix is *conjugate*, which has inspired the naming *affine conjugate*. With $F'(u)$ simultaneously all G' are symmetric positive definite.

Not only for reasons of mathematical aesthetics, but also in the adaptive algorithms we only want to use terms that are invariant under affine conjugation, i.e., terms that are independent of the choice of the matrix C (see [68]). Such terms are, e.g., the values of the functional f itself, but also so-called *local energy products*

$$\langle w, z \rangle_{F'(u)} = w^T F'(u)z, \quad u, w, z \in D.$$

Their invariance under the choice of C can be seen as follows: If we transform

$$u, w, z \rightarrow u = C\bar{u}, \ w = C\bar{w}, \ z = C\bar{z},$$

then

$$\langle \bar{w}, \bar{z} \rangle_{G'(\bar{u})} = \bar{w}^T C^T F'(u)C\bar{z} = \langle w, z \rangle_{F'(u)}$$

follows. The energy products induce *local energy norms*, which we may write as

$$\|w\|_{F'(u)} = (\langle w, w \rangle_{F'(u)})^{1/2} = (w^T F'(u)w)^{1/2}.$$

Thus, purely on the basis of invariance considerations, we have obtained exactly those scalar products which we had used anyway in the numerical treatment of discretized linear elliptic PDEs. In this framework the *affine conjugate Lipschitz condition* fits nicely, which we will introduce belo in Sections 8.1.1 and 8.1.2; it is of central importance for the construction of *adaptive* Newton methods. A more detailed representation of the affine invariant theory and the associated adaptive Newton algorithms can be found in the monograph [68].

8.1.1   Exact Newton Methods

For the solution of finite-dimensional nonlinear systems, we already introduced affine invariant Newton methods with adaptive damping strategy in Volume 1, Section 4.2, in our context written in the form ($k = 0, 1, \ldots$)

$$F'(u^k)\Delta u^k = -F(u^k), \quad u^{k+1} = u^k + \lambda_k \Delta u^k, \quad \lambda_k \in \ ]0, 1]. \tag{8.3}$$

Of course, here Δ does not denote the Laplace operator, but instead represents the Newton correction, defined as the solution of a linear equation system. If the above equation system is solved by direct elimination methods (see Section 5.1), then (8.3) may be called an *exact* Newton method. Due to the underlying minimization problem (8.2) we will end up with a damping strategy different from the one in Volume 1, Section 4.2.

We begin with a shortened version of Theorem 3.23 from [68].

Theorem 8.2. *Let $f : D \subset \mathbb{R}^N \to \mathbb{R}$ be some strictly convex C^2-functional. Let the nonlinear grid equation system to be solved, $F(u) = f'(u)^T = 0$, have a symmetric positive definite Jacobian matrix $F'(u) = f''(u)$. For arguments $u, v \in D$ let the affine conjugate Lipschitz condition*

$$\| F'(u)^{-1}(F'(v) - F'(u))(v - u) \|_{F'(u)} \le \omega \| v - u \|^2_{F'(u)} \tag{8.4}$$

hold with some Lipschitz constant $0 \le \omega < \infty$. For an iterate $u^k \in D$ let the following quantities be defined:

$$\epsilon_k = \| \Delta u^k \|^2_{F'(u^k)}, \quad h_k = \omega \| \Delta u^k \|_{F'(u^k)}. \tag{8.5}$$

Let $u^k + \lambda \Delta u^k \in D$ for $\lambda \in [0, \lambda^k_{\max}]$ with

$$\lambda^k_{\max} = \frac{4}{1 + \sqrt{1 + 8h_k/3}} \le 2.$$

Then

$$f(u^k + \lambda \Delta u^k) \le f(u^k) - t_k(\lambda)\epsilon_k, \tag{8.6}$$

where

$$t_k(\lambda) = \lambda - \tfrac{1}{2}\lambda^2 - \tfrac{1}{6}\lambda^3 \, h_k. \tag{8.7}$$

As optimal damping factor in the sense of the estimate one obtains

$$\lambda^k_{\text{opt}} = \frac{2}{1 + \sqrt{1 + 2h_k}} \le 1. \tag{8.8}$$

Proof. We drop the iteration index k. Application of the mean value theorem (in its integral form) supplies after some short calculation

$$f(u + \lambda \Delta u) - f(u)$$

$$= -\lambda\epsilon + \tfrac{1}{2}\lambda^2\epsilon + \lambda^2 \int_{s=0}^{1} \int_{t=0}^{1} s \langle \Delta u, (F'(u + st\lambda\Delta u) - F'(u))\Delta u \rangle \, dt \, ds.$$

Upon applying the Lipschitz condition (8.4) and the Cauchy–Schwarz inequality we get

$$f(u + \lambda \Delta u) - f(u) + \left(\lambda - \tfrac{1}{2}\lambda^2\right)\epsilon \leq \lambda^3 \int_{s=0}^{1}\int_{t=0}^{1} \omega s^2 t \|\Delta u\|_{F'(u)}^3 \, dt \, ds = \tfrac{1}{6}\lambda^3 h\epsilon.$$

This confirms (8.6) with (8.7). Minimization of t_k by virtue of $t_k' = 0$ and solution of the arising quadratic equation then yields (8.8). Finally, the cubic equation

$$t_k = \lambda\left(1 - \frac{1}{2}\lambda - \frac{1}{6}\lambda^2 h_k\right) = 0$$

has only on *positive* root, $\lambda_{\max}^k$.   □

Thus we have found a *theoretically optimal* damping strategy, which, however, contains the computationally inaccessible quantities h_k, i.e., it cannot be directly realized. In passing we mention the special case of the *ordinary* Newton method ($\lambda_k = 1$), for which, under suitable additional assumptions, quadratic convergence in the form

$$\|u^{k+1} - u^k\|_{F'(u^k)} \leq \frac{1}{2}\omega\|u^k - u^{k-1}\|_{F'(u^{k-1})}^2 \tag{8.9}$$

can be proved (again, see [68]).

Adaptive Damping Strategy.   We can nevertheless develop a damping strategy on the basis of Theorem 8.2. Its underlying idea is conceivably simple. We replace the inaccessible affine conjugate quantity $h_k = \omega\epsilon_k^{1/2}$ with some computationally accessible, also affine conjugate estimate $[h_k] = [\omega]\epsilon_k^{1/2}$, which we insert into (8.8) to obtain an estimated optimal damping factor:

$$[\lambda_{\text{opt}}^k] = \frac{2}{1 + \sqrt{1 + 2[h_k]}} \leq 1. \tag{8.10}$$

For an estimate $[\omega]$ of the Lipschitz constant ω we can only use local information, which induces that $[\omega] \leq \omega$, and, consequently,

$$[h_k] \leq h_k \quad \Rightarrow \quad [\lambda_{\text{opt}}^k] \geq \lambda_{\text{opt}}^k.$$

A thus determined damping factor might well be "too large", which implies that we have to develop both a *prediction strategy* and a *correction strategy*. For $\lambda = [\lambda_{\text{opt}}^k]$ the following reduction of the functional is guaranteed:

$$f(u^k + \lambda \Delta u^k) \leq f(u^k) - \frac{1}{6}\lambda(\lambda + 2)\epsilon_k.$$

It is exactly this "monotonicity test" that we will use to decide whether a suggested damping factor λ is accepted or not.

Computational estimation of Lipschitz constant. For the definition of $[h_k] = [\omega]\epsilon_k^{1/2}$ we start from the inequality

$$E(\lambda) = f(u^k + \lambda\Delta u^k) - f(x^k) + \lambda\left(1 - \frac{1}{2}\lambda\right)\epsilon_k \leq \frac{1}{6}\lambda^3 h_k\epsilon_k. \qquad (8.11)$$

In the evaluation of E some care must be taken to avoid extinction of leading digits. Whenever $E(\lambda) \leq 0$, the Newton method behaves locally better or at least as good as predicted by the quadratic model for the purely linear case $h_k = 0$; in this case we leave the old value $[\lambda_{\text{opt}}^k]$ unchanged. Let therefore $E(\lambda) > 0$. For a given value of λ we define

$$[h_k] = \frac{6\,E(\lambda)}{\lambda^3\epsilon_k} \leq h_k.$$

Insertion of this value into (8.10) for $[\lambda_{\text{opt}}^k]$ requires at least one test value for λ or for $f(u^k + \lambda\Delta u^k)$, respectively. Thus this estimate delivers a *correction strategy* for $i \geq 0$:

$$\lambda_k^{i+1} = \frac{2}{1 + \sqrt{1 + 2[h_k(\lambda_k^i)]}}. \qquad (8.12)$$

In order to find a theoretically backed starting value λ_k^0, we may use the relation

$$h_{k+1} = \left(\frac{\epsilon_{k+1}}{\epsilon_k}\right)^{1/2} h_k,$$

which directly emerges from Definition (8.5). Correspondingly, we may define an estimate

$$[h_{k+1}^0] = \left(\frac{\epsilon_{k+1}}{\epsilon_k}\right)^{1/2} [h_k^*],$$

where $[h_k^*]$ denotes the latest computed estimate (8.12) from step k. This leads us to the *prediction strategy* for $k \geq 0$:

$$\lambda_{k+1}^0 = \frac{2}{1 + \sqrt{1 + 2[h_{k+1}^0]}} \leq 1. \qquad (8.13)$$

We still have to set the value λ_0^0, for which we can get no information along the sketched line. As standard option, $\lambda_0^0 = 1$ for "weakly nonlinear" problems is chosen, while for "highly nonlinear" problems the value $\lambda_0^0 = \lambda_{\min} \ll 1$ is set. The thus defined damping strategy requires, as a rule, only one or two tries per iteration, even in difficult application problems; the theoretical basis shines through by the efficiency of the method.

Termination of the Newton Iteration.   If the problem does not have a solution or the starting value u^0 is "too bad", then the above damping strategy will lead to suggestions $\lambda_k < \lambda_{\min}$ and thus to a truncation of the iteration without a solution.

In the convergent case the iteration will merge into the ordinary Newton method with $\lambda_k = 1$. In this case, the iteration will be terminated as soon as the "energy error" is "sufficiently small", i.e., whenever one of the two criteria

$$\epsilon_k^{1/2} \leq \text{ETOL} \quad \text{or} \quad f(u^k) - f(u^{k+1}) \leq \frac{1}{2} \text{ETOL}^2 \tag{8.14}$$

is met for a user prescribed absolute accuracy parameter ETOL.

Software.   The adaptive Newton method described here has been implemented in the program NLEQ-OPT (see the software list at the end of the book).

8.1.2   Inexact Newton-PCG Methods

The linear equation systems to be solved in each iterative step of the Newton method are frequently so large that it is advisable to solve them iteratively themselves; this leads to the so-called *inexact* Newton methods. In this setting the Newton iteration is realized as an *outer iteration* ($k = 0, 1, \ldots$) according to

$$F'(u^k)\delta u^k = -F(u^k) + r^k, \quad u^{k+1} = u^k + \lambda_k \delta u^k, \quad \lambda_k \in \,]0, 1], \tag{8.15}$$

or, written in equivalent form,

$$F'(u^k)(\delta u^k - \Delta u^k) = r^k, \quad u^{k+1} = u^k + \lambda_k \delta u^k, \quad \lambda_k \in \,]0, 1]. \tag{8.16}$$

Here Δu^k denotes the exact (not computed) Newton correction and δu^k the inexact (actually computed) Newton correction. The outer iteration is coupled with an *inner iteration* ($i = 0, 1, \ldots, i_k$), in which the residuals $r^k = r_i^k|_{i=i_k}$ are computed by means of a PCG-iteration (see Section 5.3.3). Then, from Theorem 4.3, a *Galerkin condition* holds, which in our context reads:

$$\langle \delta u_i^k, \delta u_i^k - \Delta u^k \rangle_{F'(u^k)} = \langle \delta u_i^k, r_i^k \rangle = 0. \tag{8.17}$$

Recalling Section 5.3.3 we define the relative energy error

$$\delta_i^k = \frac{\|\Delta u^k - \delta u_i^k\|_{F'(u^k)}}{\|\delta u_i^k\|_{F'(u^k)}}.$$

If we start the inner iteration with $\delta u_0^k = 0$, the quantity δ_i^k decreases monotonically with increasing index i (cf. (5.24)); in the generic case we thus get

$$\delta_{i+1}^k < \delta_i^k,$$

which means that in the course of the PCG-iteration the correction norms will eventually fall below a prescribed upper bound $\bar{\delta}_k < 1$, i.e.,

$$\delta_i^k|_{i=i_k} \leq \bar{\delta}_k \quad \text{or in short} \quad \delta^k \leq \bar{\delta}_k .$$

This defines the residual $r^k = r_i^k|_{i=i_k}$ in the outer Newton iteration (8.15) or (8.16).

After these preparations we are ready to examine the convergence of the inexact Newton iteration for some damping parameter $\lambda \in]0, 2]$ in analogy to the exact Newton iteration, where we follow [68, 76] in our presentation here.

Theorem 8.3. *The assertions of Theorem* 8.2 *hold for the inexact Newton PCG-method as well if only the exact Newton corrections* Δu^k *are replaced by the inexact Newton corrections* δu^k *and the quantities* ϵ_k, h_k *by*

$$\begin{aligned}
\epsilon_k^\delta &:= \|\delta u^k\|_{F'(u^k)}^2 = \epsilon_k/(1 + \delta_k^2), \\
h_k^\delta &:= \omega\|\delta u^k\|_{F'(u^k)} = h_k/\sqrt{1 + \delta_k^2}.
\end{aligned} \tag{8.18}$$

Proof. Let us drop the iteration index k. Then the first line of the proof of Theorem 8.2 is modified as follows:

$$f(u + \lambda\delta u) - f(u)$$

$$= -\lambda\epsilon^\delta + \frac{1}{2}\lambda^2\epsilon^\delta + \lambda^2\int_{s=0}^1 s\int_{t=0}^1 \langle\delta x, (F'(u + st\lambda\delta u) - F'(u))\delta u\rangle\mathrm{d}t\,\mathrm{d}s + \langle\delta u, r\rangle.$$

The fourth term on the right-hand side vanishes due to the Galerkin condition (8.17), so that indeed only Δu must be replaced by δu to be able to perform the proof for the inexact case.   □

Damping Strategy for the Outer Newton Iteration.   The adaptive damping strategy worked out in Section 8.1.1 for the exact Newton method can be directly carried over replacing the terms Δu^k by δu^k, ϵ_k by ϵ_k^δ, and h_k by h_k^δ. In actual computation the orthogonality condition (8.17) can be spoiled by rounding errors arising from the evaluation of the scalar products within the PCG-method. Therefore an expression for $E(\lambda)$ slightly different from (8.11) should be evaluated in the form

$$E(\lambda) = f(u^k + \lambda\delta u^k) - f(u^k) - \lambda\langle F(u^k), \delta u^k\rangle - \frac{1}{2}\lambda^2\epsilon_k^\delta \tag{8.19}$$

with $\epsilon_k^\delta = \langle\delta u^k, F'(u^k)\delta u^k\rangle$. We have analogously

$$E(\lambda) \leq \frac{1}{6}\lambda^3 h_k^\delta \epsilon_k^\delta,$$

from which, for $E(\lambda) \geq 0$, the estimate

$$[h_k^\delta] = \frac{6\,E(\lambda)}{\lambda^3\epsilon_k^\delta} \leq h_k^\delta \tag{8.20}$$

can be gained. On the basis of this estimate both the prediction strategy (8.13) and the correction strategy (8.12) can be appropriately modified.

Accuracy Matching Strategy for Inner PCG-iteration. In order to implement the above damping strategy, we still have to discuss the adaptive choice of $\bar{\delta}_k$. From [68] we take for the damping phase (reasons given there)

$$\bar{\delta}_k = 1/4, \quad \text{if} \quad \lambda_k < 1.$$

After merging into the phase of the ordinary Newton method ($\lambda_k = 1$) several possibilities are open, which will only be sketched here (a more elaborate presentation can be found in [68, 75]). For exact ordinary Newton iteration ($\lambda_k = 1$) we have quadratic convergence, which can be most simply read from the inequality

$$h_{k+1} \le \frac{1}{2} h_k^2 \quad \text{with} \quad h_k = \omega \| \Delta u^k \|_{F'(u^k)}.$$

An analogous derivation, which we here quote without proof from [75], leads to the estimate

$$h_{k+1}^{\delta} \le \frac{1}{2} h_k^{\delta} \left(h_k^{\delta} + \bar{\delta}_k \left(h_k^{\delta} + \sqrt{4 + (h_k^{\delta})^2} \right) \right).$$

Depending on the choice of $\bar{\delta}_k$ we can asymptotically (i.e., for $h_k^{\delta} \to 0$) achieve either *linear* or *quadratic* convergence. In [68, 75] a simple complexity model was used to show that for finite-dimensional nonlinear grid equation systems asymptotically the *quadratic* convergence mode is more efficient. With the special choice

$$\bar{\delta}_k = \rho \frac{h_k^{\delta}}{h_k^{\delta} + \sqrt{4 + (h_k^{\delta})^2}} \tag{8.21}$$

we indeed obtain

$$h_{k+1}^{\delta} \le \frac{1+\rho}{2} (h_k^{\delta})^2,$$

which means quadratic convergence; the factor $\rho > 0$ is still open to be chosen, with $\rho < 1$ as a rule. In order to obtain a computationally accessible quantity $[\bar{\delta}_k]$ from (8.21), we substitute the quantity h_k^{δ} by its numerical estimate $[h_k^{\delta}]$. With $\delta^k \le [\bar{\delta}_k]$ we finally obtain a common adaptive strategy for damping as well as for accuracy matching.

Termination of the outer Newton-iteration.   Instead of the criterion (8.14) we terminate the inexact Newton PCG-iteration, subject to (8.18), once the following relation holds:

$$\left((1 + (\delta^k)^2)\epsilon_k^{\delta}\right)^{1/2} \leq \text{ETOL} \quad \text{or} \quad f(u^k) - f(u^{k+1}) \leq \frac{1}{2}\text{ETOL}^2. \qquad (8.22)$$

Software.   The adaptive Newton method described here is implemented in the program GIANT-PCG (see the software list at the end of the book).

8.2   Inexact Newton-Multigrid Methods

In this section we again treat the numerical solution of nonlinear elliptic PDEs. In contrast to the preceding Section 8.1 we want to directly take on elliptic problems as operator equations in function space, i.e., *before discretization*, where we will draw upon Section 8.3 of the monograph [68]. Just as at the beginning of Chapter 7 for the linear special case, we will be guided by the idea that a function space is characterized by the asymptotic behavior in the process of successive exhaustion by finite dimensional subspaces, e.g., in the limiting process $h \to 0$. Following this idea it will not be sufficient to replace a function space with a single subspace of finite, even though of high dimension. Instead we want to represent it by a *sequence of finite dimensional subspaces of growing dimension*. In this view, adaptive multigrid methods come into play not only for efficiency reasons, but also as means to analyze the problem to be solved.

8.2.1   Hierarchical Grid Equations

Let us return to the convex minimization problem

$$f(u) = \min,$$

where $f : D \subset X \to \mathbb{R}$ is a *strictly convex* functional defined over a *convex* subset D of a reflexive Banach space X. We have to decide about X. While linear elliptic PDEs live in the function space $H^1(\Omega)$, we need $X = W^{1,p}(\Omega)$ with $1 < p < \infty$ for nonlinear elliptic PDEs; this more general choice is occasionally necessary for the existence of a solution of the nonlinear minimization problem. Moreover we assume that a solution in D exists; otherwise the Newton method would not converge. Because of the strict convexity the solution will then be unique. As in the finite dimensional case, which we had treated in Section 8.1, this minimization problem is equivalent to the nonlinear operator equation

$$F(u) = f'(u) = 0, \quad u \in D \subset W^{1,p}(\Omega). \qquad (8.23)$$

This is a nonlinear elliptic PDE problem. For simplicity, we again assume that this PDE is *strictly* elliptic, so that the symmetric Fréchet derivative $F'(u) = f''(u)$ is *strictly positive.* Then, for not too pathological arguments $u \in \underline{X} \subset X$ (cf. Remark 8.1), there also exist local energy products $\langle \cdot, F'(u) \cdot \rangle$ in function space. They again induce *local energy norms*

$$\| \cdot \|_{F'(u)} = \langle \cdot, F'(u) \cdot \rangle^{1/2},$$

which are equivalent to locally weighted H^1-norms.

Asymptotic Mesh Independence.   The Newton iteration in $W^{1,p}$ associated with (8.23) reads (without damping)

$$F'(u^k) \Delta u^k = -F(u^k), \quad u^{k+1} = u^k + \Delta u^k, \quad k = 0, 1, \dots. \tag{8.24}$$

In each Newton step a linearization of the nonlinear operator equation must be solved, i.e., a linear elliptic PDE problem. Now let ω denote the corresponding affine conjugate Lipschitz constant, defined in analogy to (8.4) in Theorem 8.2. Similar as in (8.9), we obtain local quadratic convergence in the form

$$\| u^{k+1} - u^k \|_{F'(u^k)} \le \frac{\omega}{2} \| u^k - u^{k-1} \|^2_{F'(u^{k-1})}.$$

Note that different local norms arise on the two sides of the above estimate, which means that a convergence concept based on this estimate will differ from a convergence in $W^{1,p}(\Omega)$. Instead we will focus algorithmically on an iterative reduction of the functional f.

In actual computation we will have to solve discretized nonlinear PDEs of finite dimension, i.e., in the case considered here, a sequence of nonlinear grid equation systems

$$F_j(u_j) = 0, \quad j = 0, 1, \dots,$$

where F_j denotes the finite dimensional nonlinear mapping on level j of a hierarchical mesh. The associated ordinary Newton method reads

$$F'_j(u_j^k) \Delta u_j^k = -F_j(u_j^k), \quad u_j^{k+1} = u_j^k + \Delta u_j^k, \quad k = 0, 1, \dots.$$

In each Newton step a linear grid equation system has to be solved. Now let ω_j be the affine conjugate Lipschitz constant corresponding to the mapping F_j, as defined in (8.4). Then we obtain, as in (8.9), local quadratic convergence in the form

$$\| u_j^{k+1} - u_j^k \|_{F'_j(u_j^k)} \le \frac{\omega_j}{2} \| u_j^k - u_j^{k-1} \|^2_{F'_j(u_j^{k-1})}.$$

Without proof we quote from [68, Lemma 8.4] or [215], respectively, the asymptotic result

$$\omega_j \le \omega + \sigma_j, \quad \lim_{j \to \infty} \sigma_j = 0. \tag{8.25}$$

For increasingly fine meshes the grid equation systems become more and more similar, which is reflected in the (also observable) fact that the convergence behavior of the associated Newton methods is asymptotically mesh independent.

Nonlinear Multigrid Methods. There are two principal variants of multigrid methods for nonlinear problems which exploit the above characterized redundancy of asymptotic mesh independence:

- *Multigrid-Newton methods.* In this variant the multigrid methods are the outer loop, and finite dimensional Newton methods the inner loop. The Newton methods operate simultaneously on all equation systems $F_j(u_j) = 0$ for $j = 0, \ldots, l$, i.e., on fixed given hierarchical triangulations $\mathcal{T}_0, \ldots, \mathcal{T}_l$ with fixed index l. This variant does not permit an adaptive construction of meshes. It has been extensively elaborated in the book [115] of Hackbusch or in the more recent survey article [134] by Kornhuber.

- *Newton-multigrid methods.* In this variant the Newton iteration is the outer loop, the multigrid method the inner loop. The Newton iteration is directly realized on the operator equation, which means that the arising linear operator equations, i.e., linear elliptic PDE problems, are solved by adaptive multigrid methods for linear problems. This variant permits to construct hierarchical meshes *adaptively* on the basis of the convergence theory for $j = 0, 1, \ldots$, i.e., from the coarse mesh to successively finer, in general nonuniform meshes. It has been derived in detail in [68, Section 8.3] (see also [75, 76]).

Because of its possible combination with adaptive mesh refinement we will only work out the Newton-multigrid variant here.

8.2.2 Realization of Adaptive Algorithm

The Newton-multigrid algorithm to be presented now realizes three nested loops:

- *outer loop*: function space inexact Newton method (with adaptive damping strategy);

- *middle loop*: adaptive mesh refinement;

- *inner loop*: PCG-method with multilevel preconditioner and with adaptive truncation criterion (see Section 5.3.3).

For comparison: the inexact Newton-PCG method in Section 8.1.2 realizes only an outer loop (finite dimensional Newton iteration) and an inner loop (PCG-iteration), the meshes remain fixed.

Control of Discretization Error. A numerical solution of the operator equation (8.23) cannot be realized without an *approximation error*. This means that, instead of the exact Newton method (8.24) in Banach space, we must choose an *inexact* Newton method in Banach space as the theoretical frame. On level j of a hierarchical mesh we therefore define

$$F'(u_j^k)\, \delta u_j^k = -F(u_j^k) + r_j^k,$$

or equivalently

$$F'(u_j^k)(\delta u_j^k - \Delta u^k) = r_j^k.$$

The inevitable discretization errors are visible both in the residuals r_j^k and in the discrepancy between the (actually computed) inexact Newton corrections δu_j^k and the (only formally introduced) exact Newton corrections Δu^k. Among the discretization methods we restrict our attention to *Galerkin methods*, which are known to satisfy (cf. Theorem 4.3)

$$\langle \delta u_j^k, \delta u_j^k - \Delta u^k \rangle_{F'(u_j^k)} = \langle \delta u_j^k, r_j^k \rangle = 0. \tag{8.26}$$

In perfect analogy to Section 8.1.2 we define the relative energy error

$$\delta_j^k = \frac{\| \Delta u^k - \delta u_j^k \|_{F'(u_j^k)}}{\| \delta u_j^k \|_{F'(u_j^k)}}.$$

This error depends on the level j of the hierarchical mesh that has been generated in the course of an adaptive multigrid method for the linear elliptic problem. In our present context we choose $j = j_k$ such that

$$\delta^k = \delta_j^k |_{j=j_k} \le 1. \tag{8.27}$$

For a detailed justification of the upper bound 1 we refer again to [68].

Ordinary Newton Method.   First we look at the ordinary Newton method, which converges only locally. In [68, 75] a simplified complexity model was employed to show that for Newton-multigrid methods asymptotically the *linear* convergence mode is efficient. This agrees nicely with our condition (8.27) above. To illustrate this, we first present an example in some detail.

Example 8.4. As a modification of an example given by R. Rannacher in [174] we consider the functional in space dimension $d = 2$

$$f(u) = \int_\Omega \left((1 + |\nabla u|^2)^{p/2} - gu \right) dx, \quad \Omega \subset \mathbb{R}^2, \ u \in W_D^{1,p}(\Omega),$$

which for $p > 1$ is strictly convex. In particular we choose here $g \equiv 0$. The functional f is continuous and coercive in the affine subspace $W_D^{1,p}(\Omega)$ given by Dirichlet conditions on parts of the boundary. For this functional we obtain the following derivative expressions of first and second order (in the sense of dual pairing)

$$\langle F(u), v \rangle = \int_\Omega \left(p(1 + |\nabla u|^2)^{p/2-1} \nabla u^T \nabla v - gv \right) dx,$$

$$\langle w, F'(u)v \rangle = \int_\Omega p(1 + |\nabla u|^2)^{p/2-1} \nabla w^T \left(I + (p-2) \frac{\nabla u \nabla u^T}{1 + |\nabla u|^2} \right) \nabla v \, dx.$$

Note that for $1 < p < 2$ the diffusion tensor is nearly everywhere in Ω positive definite, but only *uniformly* positive definite, if ∇u is bounded. This is obviously a restriction to a "smoother" subspace $\underline{X} \subset X$, as discussed above in Remark 8.1. For convex domains Ω and full smooth Dirichlet boundary conditions this situation actually occurs. Here, however, we choose some slightly more challenging nonconvex domain and with $p = 1.4$ an arbitrary value in the critical range.

The inexact Newton method is implemented in the FE-package KASKADE, in which the arising linear elliptic problems in each Newton step are solved by an adaptive multigrid method. The ordinary Newton method has been controlled in the linear convergence mode.

In Figure 8.2 the starting value u^0 on the coarse mesh $\mathcal{T}_0$ (with $j_0 = 1$) over the given domain Ω is compared with the Newton iterate u^3 on the finer mesh $\mathcal{T}_3$ (with $j_3 = 14$). The finer mesh comprises $n_{14} = 2\,054$ nodes. It is visibly denser at the two critical points on the boundary of the domain Ω, the reentrant corner and the discontinuity point of the boundary conditions; this illustrates the efficiency of the applied local error estimators. For comparison: On level $j = 14$ a uniformly refined mesh would require $\bar{n}_{14} \approx 136\,000$ nodes; as multigrid methods require $\mathcal{O}(n)$ operations, the computational amount would be a factor of $\bar{n}_{14}/n_{14} \approx 65$ higher.

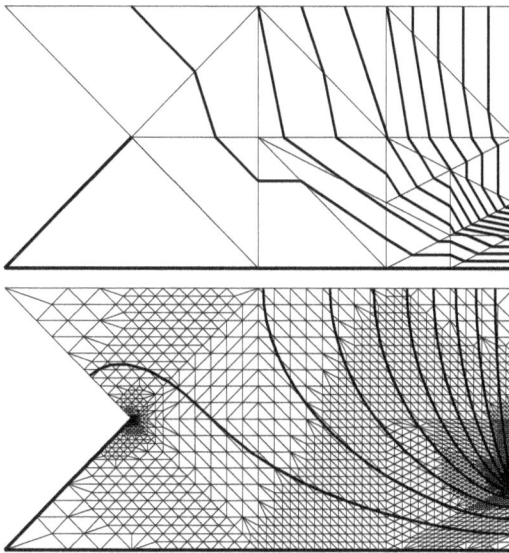

Figure 8.2. Nonlinear elliptic problem 8.4 with $p = 1.4$: Inexact Newton-multigrid method. Inhomogeneous Dirichlet boundary conditions: thick boundary lines, homogeneous Neumann boundary conditions: thin boundary lines. *Top*: level lines of the Newton starting value u^0 on triangulation $\mathcal{T}_0$ at refinement level $j_0 = 1$. *Bottom*: level lines of Newton iterate u^3 on triangulation $\mathcal{T}_3$ at refinement level $j_3 = 14$.

Adaptive Damping Strategy.   In the following we consider the globally convergent inexact Newton-Galerkin method

$$u^{k+1} = u^k + \lambda_k \delta u^k, \quad \lambda_k \in \,]0, 1]$$

with iterates $u^k \in W^{1,p}$, inexact Newton corrections δu^k and damping factors λ_k, whose choice we want to study now. In Section 8.1.2 we had already discussed the finite dimensional version, the global Newton-PCG methods. If in Theorem 8.3 we replace the finite dimensional Galerkin condition (8.17) by its infinite dimensional counterpart (8.26), we directly arrive at the following convergence theorem.

Theorem 8.5. *Notation as just introduced. Let the functional $f : D \to \mathbb{R}$ be strictly convex, to be minimized over an open convex domain $D \subset W^{1,p}$. Let $F'(x) = f''(x)$ exist and be strictly positive in the weak sense. For $u, v \in D$ let the affine conjugate Lipschitz condition*

$$\|F'(u)^{-1}(F'(v) - F'(u))(v - u)\|_{F'(u)} \le \omega \|v - u\|_{F'(u)}^2 \qquad (8.28)$$

hold with $0 \le \omega < \infty$. For $u^k \in D$ let the local energy norms be defined as

$$\epsilon_k = \|\Delta u^k\|_{F'(u^k)}^2, \quad \epsilon_k^\delta = \|\delta u^k\|_{F'(u^k)}^2 = \frac{\epsilon_k}{1 + \delta_k^2},$$

$$h_k = \omega \|\Delta u^k\|_{F'(u^k)}, \quad h_k^\delta = \omega \|\delta u^k\|_{F'(u^k)} = \frac{h_k}{\sqrt{1 + \delta_k^2}}.$$

Furthermore let $u^k + \lambda \delta u^k \in D$ for $0 \le \lambda \le \lambda_{\max}^k$ with

$$\lambda_{\max}^k := \frac{4}{1 + \sqrt{1 + 8h_k^\delta / 3}} \le 2.$$

Then

$$f(u^k + \lambda \Delta u^k) \le f(u^k) - t_k(\lambda)\epsilon_k^\delta \qquad (8.29)$$

with

$$t_k(\lambda) = \lambda - \tfrac{1}{2}\lambda^2 - \tfrac{1}{6}\lambda^3 h_k^\delta.$$

The optimal choice of the damping factor is

$$\overline{\lambda}_k = \frac{2}{1 + \sqrt{1 + 2h_k^\delta}} \le 1.$$

In order to generate a theoretically backed algorithmic strategy from this theorem, we replace, as in the finite dimensional case, h_k^δ and δ_k by computationally conveniently accessible estimates $[h_k^\delta]$ and $[\delta_k]$. For this purpose, we again compute the expressions $E(\lambda)$ from (8.19) and thus obtain from (8.20) the perfectly analogous formula

$$[h_k^\delta] = \frac{6\,E(\lambda)}{\lambda^3\epsilon_k^\delta} \leq h_k^\delta.$$

The difference to (8.20) merely consists in the meaning of the terms. On this basis we realize a correction strategy (8.12), where h_k is replaced by $[h_k^\delta]$, and a prediction strategy (8.13), where 8.3 h_{k+1} is replaced by $[h_{k+1}^\delta]$. In all of these formulas the iterative estimate δ_k of the discretization error according to (8.27) enters, which implies that the damping strategy and the control of the discretization error (i.e., also the iterative control of the refinement level of the triangulations) are coupled.

Termination Criterion. For the termination of the inexact Newton-multigrid iteration we take from (8.22) the criterion

$$((1 + (\delta_j^k)^2)\epsilon_k^\delta)^{1/2} \leq \text{ETOL} \quad \text{or} \quad f(u_j^k) - f(u_j^{k+1}) \leq \tfrac{1}{2}\,\text{ETOL}^2.$$

8.2.3 An Elliptic Problem Without a Solution

For an illustration of the above derived algorithm we next present a critical example in some detail.

Example 8.6. We return to Example 8.4, this time, however, for the critical value $p = 1$. Note that the Banach space $W^{1,1}$ is not reflexive, which is already why the existence of a solution of the minimization problem is not guaranteed; it depends on the domain and the boundary conditions.

In Figure 8.3 we give the specifications for two different cases: in the left case (Example 8.6a) there is a unique solution, which we have computed but do not represent here; the example is only for comparison (see Table 8.1, bottom). Our main interest

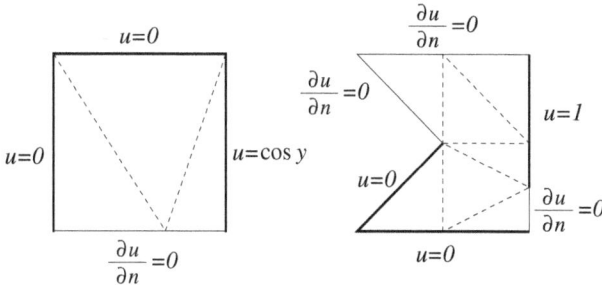

Figure 8.3. Nonlinear elliptic problem 8.4 in the critical case $p = 1$. Inhomogeneous Dirichlet boundary conditions: thick boundary lines; homogeneous Neumann boundary conditions: thin boundary lines. *Left*: problem with unique solution (Example 8.6a). *Right*: problem without a solution (Example 8.6b).

focuses on the right Example 8.6b, where *no (physical) solution exists*. As initial value u^0 for the Newton iteration we choose the prescribed Dirichlet boundary data and zero otherwise.

In Example 8.6b we want to compare the different behavior of two Newton algorithms:

- our *function space based* approach that we worked out here as adaptive inexact Newton-multigrid method;

- the *finite dimensional* approach, as it is typically implemented in multigrid-Newton methods, but with a multigrid method as outer loop.

In the finite dimensional approach the discrete nonlinear finite element problem is successively solved on each refinement level j of a hierarchical mesh; the damping factors will run towards the value 1, once the problem has been solved on that level. In contrast to this, our function space based approach aims at a solution of the operator equation, where information from all refinement levels simultaneously can be exploited. If a unique solution of the operator equation exists, then the damping parameters will run towards the asymptotic value 1 simultaneously with the mesh refinement process; if, however, a unique solution does not exist, then the damping parameters will tend towards the value 0.

Such a behavior is illustrated synoptically in Figure 8.4. The multigrid-Newton method seems to suggest after 60 iterations that the problem is uniquely solvable. In contrast, the Newton-multigrid method terminates after 20 iterations

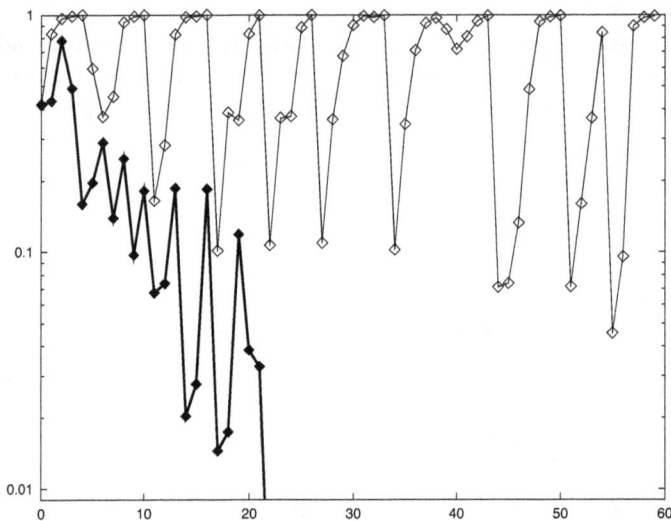

Figure 8.4. Nonlinear elliptic problem 8.6b (without solution): damping strategies for adaptive multigrid-Newton iteration ($\diamond$) compared with adaptive Newton-multigrid method ($\blacklozenge$).

Table 8.1. Estimates $[\omega_j]$ of the affine conjugate Lipschitz constants on refinement level j. Example 8.6a: unique solution exists, asymptotic mesh independence (cf. (8.25)). Example 8.6b: no unique solution exists, increase of Lipschitz constants.

	Example 8.6a		Example 8.6b	
j	# unknowns	$[\omega_j]$	# unknowns	$[\omega_j]$
0	4	1.32	5	7.5
1	7	1.17	10	4.2
2	18	4.55	17	7.3
3	50	6.11	26	9.6
4	123	5.25	51	22.5
5	158	20.19	87	50.3
6	278	19.97	105	1 486.2
7	356	9.69	139	2 715.6
8	487	8.47	196	5 178.6
9	632	11.73	241	6 837.2
10	787	44.21	421	12 040.2
11	981	49.24	523	167 636.0
12	1 239	20.10	635	1 405 910.0

with $\lambda < \lambda_{\min} = 0.01$. For a deeper understanding, we compare in Table 8.1 (from [68, 215]) the numerical estimates of the affine conjugate Lipschitz constants $[\omega_j] \leq \omega_j$ for both example problems. In Example 8.6a (with unique solution) we observe the above elaborated *asymptotic mesh independence*, whereas in Example 8.6b (without solution) the Lipschitz estimates increase with the refinements of the mesh. Note that an increase of the lower bounds $[\omega_j]$ in Table 8.1 at the same time implies an increase of the Lipschitz constants ω_j – for this phenomenon the local estimates just lie on the "correct" side.

Interpretation. The results of the algorithm seem to suggest that in this example the Lipschitz constants behave like

$$\lim_{j \to \infty} \omega_j = \infty.$$

For the convergence radii $\rho_j \sim 2/\omega_j$ from Volume 1, Theorem 4.10, we then have correspondingly

$$\lim_{j \to \infty} \rho_j = 0.$$

In fact, in the course of the process of mesh refinement the convergence radii (estimated) shrink from $\rho_1 \sim 1$ down to $\rho_{22} \sim 10^{-6}$. Hence, even though the problem has a unique finite dimensional solution on each refinement level j, the operator equation nevertheless does not have a solution in $W^{1,1}$.

Remark 8.7. The class of problems in $W^{1,1}$ described here was carefully analyzed in [84, Chapter V]. There a generalized solution concept was suggested, which, however, does not guarantee that the Dirichlet boundary conditions are satisfied. This is only assured under restrictive assumptions on the smoothness of the domain and the boundary conditions. Obviously, these conditions are violated in Example 8.6b.

Software.   The here described adaptive Newton method is implemented in the finite element package KASKADE (see the software list at the end of the book).

8.3   Operation Planning in Facial Surgery

Patients with congenital bone malfunctions in the maxilla domain (see Figure 8.5, left) generally suffer from low social acceptance due to their appearance. Often connected with it are functional limitations such as breathing discomfort, susceptibility to infection, and digestion problems due to the their defective chewing capability. Such malfunctions can be corrected with *cranio-maxillo-facial* surgery: in the course of several hours of the operation, both upper and lower jaw can be sawed and shifted a few centimeters. From a medical point of view, there usually are several therapeutically advisable and surgically performable positionings, which might much differ quite with respect to their postoperative appearance. If a patient is asked *before* the operation, how he or she would like to look *after* the operation, then this leads to a mathematical problem where the efficient numerical solution of PDEs is of crucial importance [77, 216]. The following three-stage paradigm has evolved in virtual medicine (cf., e.g., [67]):

1. a sufficiently accurate geometric 3D-model of the individual patient together with the "material properties" of bone and soft tissue is established in the computer, the so-called *virtual patient*;

2. on this patient specific model simulations are performed, i.e., PDEs are solved numerically and various surgical options are studied;

3. any of the surgical options found in the *virtual OR* can be realized in the actual operation.

Based on this knowledge, supported by visualizations close to reality, the operation can be planned by the surgeon *together* with the patient and decided together.

Mathematical Modelling.   The prediction of the postoperative appearance, i.e., the deformation of the skin from the dislocation of the bone fragments, is a problem arising from the *elastomechanics* of the soft tissue (see Section 2.3). The bone dislocations enter as Dirichlet boundary conditions. Of interest is only the postoperative state, but not any movement during the operation. Therefore we may restrict ourselves to the computation of the *stationary* state. The deformations of the elastic soft tissue Ω are, especially close to the bones, most often so large that *linear* elastomechanics as represented in Section 2.3.2 will not yield a good approximation. Therefore the soft tissue – muscles, skin, fat, and connective tissue – should be modelled as hyperelastic material, usually isotropic in the absence of concrete data about the fiber direction. A comprehensive survey on material models of *biomechanics* can be found in [98]. Generally speaking, biomechanics may be subsumed under *nonlinear* elastomechanics, which was discussed in Section 2.3.1.

The postoperative appearance in cranio-maxillo-facial surgery operations is thus described by a variational problem of the type

$$\min_{u \in H^1(\Omega)} f(u), \quad u|_{\Gamma_D} = u_D \quad \text{with} \quad f(u) = \int_\Omega \Psi(u_x)\,dx, \qquad (8.30)$$

where we have exchanged the notation W for the internal energy from Section 2.3.1 with the notation f for reasons of connectivity with the preceding Section 8.2.

Modification of Adaptive Newton Multigrid Method. At first glance, the problem above (8.30) appears to be just one of the convex minimization problems treated so far in this chapter, so that we could apply the affine conjugate Newton-multigrid method worked out in Section 8.2.2. Unfortunately, we know from Section 2.3.1 that in *nonlinear* elastomechanics f *cannot be convex* (see also Exercise 8.3). From this we immediately conclude that the second derivative $f'' = F'$ is still symmetric, but no longer positive; hence, it does not induce any norm and is not suitable for the adaptive control of the method. The essential consequence of nonconvexity is that the Newton direction, if it exists at all, need not be a direction of descent – the Newton method may converge towards a saddle point or a local maximum of the energy instead of a local minimum. Only close to a solution one may rely on the fact that a damped Newton method reaches its goal.

Therefore we will sketch here an adaptive Newton-multigrid method due to [213] which asymptotically merges into the algorithm from Section 8.2.2 and has proven to work in our application context. As for the second derivative, we use the fact that the operator $F'(0)$ corresponding to *linear* elastomechanics, i.e., the linearization around the initial configuration $u = 0$, is symmetric positive and thus induces an affine conjugate norm $\|\cdot\|_{F'(0)}$. With this background, we modify Definition (8.28) of the affine conjugate Lipschitz constant ω such that

$$\|F'(0)^{-1}(F'(v) - F'(u))(v - u)\|_{F'(0)} \le \omega_0 \|v - u\|^2_{F'(0)}. \qquad (8.31)$$

In lieu of the cubic estimate (8.6) with (8.7) for the damped Newton method we now obtain the estimate

$$f(u + \delta u) \le f(u) + \langle F(u), \delta u \rangle + \frac{1}{2}\langle \delta u, F'(u)\delta u \rangle + \frac{1}{6}\omega_0 \|\delta u\|^3_{F'(0)}, \qquad (8.32)$$

where the Newton direction in (8.6) is replaced by some more general deviation δu.

In order to replace the Newton direction with some better correction we borrow an idea from high-dimensional, but still finite-dimensional, optimization, the so-called *truncated* PCG-method [191, 200]. Regardless of the possible nonconvexity of f we use a PCG-method as inner iteration, which would be only appropriate for symmetric positive definite linear systems. Whenever Algorithm 5.9 for the computation of the Newton correction δu encounters some q_k with negative curvature, i.e., with $\langle q_k A q_k \rangle \leq 0$, then the method terminates; there the energy functional is locally nonconvex. This information is then additionally used outside the PCG-iteration: Instead of determining a steplength for the hitherto computed iterate δu_{k-1} by minimizing the upper bound (8.29), the line search is extended to a search within a *two-dimensional search space*

$$U = \text{span}\{\delta u_{k-1}, q_k\}$$

and a correction $\delta u \in U$ is determined as minimizer of the upper bound (8.32). In close analogy to the construction of the pure Newton method the inaccessible quantity ω_0 is replaced by some algorithmically conveniently computable estimate $[\omega_0]$. The remaining part of the adaptive multigrid method can be realized in a similar way. By construction, this variant will merge into the Newton method in a neighborhood of the solution point, since there the PCG-algorithm will no longer encounter a direction with negative curvature.

For problems with large strains the above truncated PCG-method tends to lack robustness: even though the above search directions are preferable to the (preconditioned) steepest descent method with respect to the energy norm, their deformation gradients tend to be "too large" with respect to the L^∞-norm, which may quickly result in local self-penetration, i.e., $\det(I + u_x) < 0$, and thus in "too small" steps. In this case a recent *regularized* PCG-method due to A. Schiela [179] may lead to a better behavior. In this approach, the nonconvex problem is modified using the positive operator $F'(0)$ to obtain a convex problem of the kind

$$\min_{\delta u} \left(\langle F(u), \delta u \rangle + \frac{1}{2} \langle \delta u, (F'(u) + \alpha F'(0)) \delta u \rangle \right).$$

This problem can now be solved by PCG up to a prescribed relative tolerance. The parameter $\alpha \geq 0$ has to be chosen adaptively such that $\alpha = 0$, if the original problem is already convex. In contrast to solutions of the truncated PCG, a modified linear elastic problem, i.e., a physically meaningful problem, is solved. Consequently, the corrections δu tend to have increased regularity, which in turn will lead to larger steps. Using $F'(0)$ the invariance properties of the method are retained (other preconditioners may be used as well; see Exercise 8.4). Note that each step of this modified method is computationally more expensive than the truncated PCG due to the larger number of PCG steps and the possible occurrence of rejected values of α. This is outweighed by an (often significantly) smaller number of Newton steps.

An Example.   The methods described above have been applied in surgery planning. In Figure 8.5 we show a typical result for a patient, at the time 27 years old: on the left, the appearance before the operation can be seen, the center image shows the patient several weeks after an clearly successful operation. On the right, we show an overlay of:

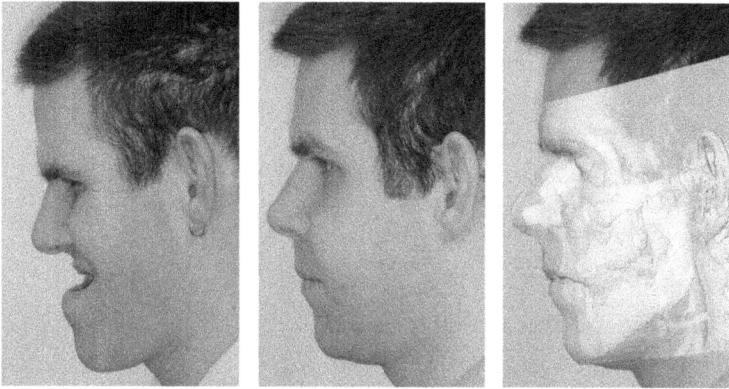

Figure 8.5. Cranio-maxillo-facial operation planning [77]. *Left*: patient *before* the operation. *Center*: patient *after* the operation. *Right*: overlay of the mathematical prognosis (before the operation) with the image of the real patient (after the operation); in addition, the underlying skeleton after operating is visible.

(a) the actual appearance;

(b) the predicted virtual appearance computed by the adaptive modified Newton-multigrid method described above;

(c) the underlying skeleton of the skull (cranium) after operating, where the cartilage below the nose is only partly correctly captured by the medical imaging.

The adaptive spatial mesh contained roughly $N = 250\,000$ nodes. The agreement of the mathematical prognosis (b) with the clinical result (a) is astounding: Even though the operation required dislocations of the jaw bones of up to 3 cm, the prognosis differed only by a maximum of 0.5 mm (above and below the nose), an accuracy highly sufficient for medical purposes.

8.4 Exercises

Exercise 8.1. Show that the hierarchical error estimation (6.36) commutes with the addition of the Newton correction δu^k, which means that the error for the systems

$$F'(u_j^k)\Delta u^k = -F(u_j^k)$$

and

$$F'(u_j^k)u_j^{k+1} = F'(u_j^k)u_j^k - F(u_j^k)$$

are the same. Why does this not hold for the residual based error estimation (6.12) and the gradient recovery (6.29)? What is the algorithmic meaning of this property?

Exercise 8.2. Let $\mathcal{T}_0$ denote a triangulation of $\Omega \subset \mathbb{R}^2$ and w_0 a given finite element function on $\mathcal{T}_0$ with $w_0|_{\partial\Omega} = 1$. Consider the finite element solution of

$$\Delta u = \Delta w_0 \quad \text{in } \Omega,$$
$$u|_{\partial\Omega} = 0 \qquad \text{on } \partial\Omega.$$

Explain which convergence rate α of the energy error $\|u - u_h\|_a \leq c h^\alpha$ is to be expected in the continued refinement of $\mathcal{T}_0$.

Exercise 8.3. Show that an energy density $\Psi : \mathbb{R}^{3\times3} \to \mathbb{R}$ that satisfies (2.21) cannot be convex.

Exercise 8.4. Let A be a symmetric indefinite matrix, and B^{-1} a symmetric positive definite preconditioner. The application of B to a vector is not usually implementable. Sometimes, however, in a nonconvex minimization context such as nonlinear hyperelasticity, it is helpful to solve the system

$$(A + \nu B)x = f,$$

where $\nu \geq 0$ is chosen in such a way that $(A + \nu B)$ is positive definite. Construct a modification of the preconditioned CG-method for this system, which works *without* the application of B. For this modification, an additional vector, but no additional applications of A and B^{-1} are needed.

Chapter 9

Adaptive Integration of Parabolic Problems

In the introductory Chapter 1 we made the acquaintence of several elementary time-dependent PDEs: in Section 1.2 the diffusion equation, in Section 1.3 the wave equation, and in Section 1.4 the Schrödinger equation. All of them can be written as *abstract Cauchy problems*, e.g., in the linear special case as

$$u' = Au, \quad u(0) = u_0, \tag{9.1}$$

where A is an unbounded spatial operator, into which the differential operator as well as the boundary conditions enter. The classical prototype of a parabolic differential equation is the diffusion equation. There the operator $-A$ is elliptic and has a spectrum on the positive real half-axis with accumulation point at infinity. For the Schrödinger equation the spectrum lies on the imaginary axis with an accumulation point at infinity, too. In a first step, we want to discuss important aspects of time discretization at these two elementary example PDEs, which are also meaningful for more general cases.

9.1 Time Discretization of Stiff Differential Equations

In Sections 9.1.1 and 9.1.2 we treat – in a crash course – the discretization of initial value problems for *ordinary* differential equations; a more extensive treatment of this topic can be found, e.g., in Volume 2. In Section 9.1.3 we will then discuss the transition to parabolic PDEs.

On the background of the above Cauchy problem (9.1) we are especially interested in "stiff" ODEs. For an explanation of this concept we study the following scalar nonautonomous example:

$$u' = g'(t) + \lambda(u - g(t)), \quad u(0) = g(0) + \epsilon_0. \tag{9.2}$$

As analytical solution we immediately get

$$u(t) = g(t) + \epsilon_0 \exp \lambda t. \tag{9.3}$$

Obviously, the function $g(t)$ is the *equilibrium solution*, which we assume to be comparatively smooth. In Figure 9.1 the two cases $\lambda \ll 0$ and $\lambda \gg 0$ are depicted for some g. If $\Re(\lambda) \ll 0$, a small deviation ϵ_0 will rapidly lead back to the equilibrium solution; this is exactly the case of "stiff" initial value problems, often just called stiff ODEs. One should keep in mind from the beginning that in the *transient phase*, i.e., the regime

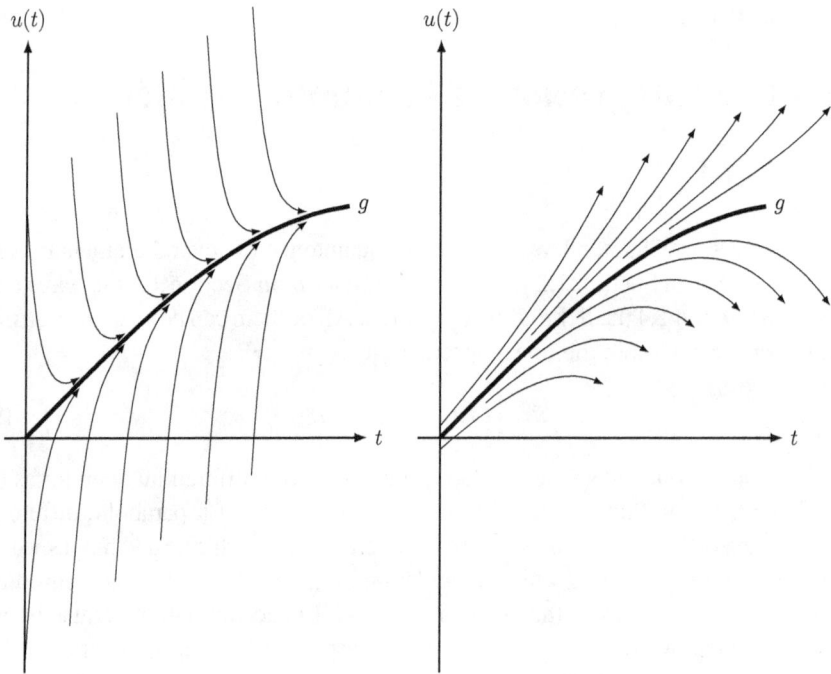

Figure 9.1. Example (9.3) for equilibrium solution $g = \sin(t), t \in [0, 1.5]$. *Left:* "stiff" problem: $\lambda = -16$, $\epsilon_0 = 1$. *Right:* inherently unstable problem: $\lambda = 3$, $\epsilon_0 = 0.05$.

before the equilibrium solution is reached, the problem is not "stiff". For $\Re(\lambda) \gg 0$ deviations are rapidly blown up, one speaks of *inherently unstable* initial value problems, often just from unstable ODEs – to be clearly distinguished from stiff ODEs (although the classical paper by C. F. Curtiss and J. O. Hirschfelder [58] from 1952, in which the term "stiff ODE" was coined, is inexact in this very point).

9.1.1 Linear Stability Theory

After the simple example we now investigate the general initial value problem for N ordinary differential equations

$$u' = F(u), \quad u(0) = u_0, \tag{9.4}$$

where we, as a preparation for later sections, already envision the nonlinear case. Now let g denote a selected nominal solution of (9.4), assumed to be sufficiently smooth, and u a different solution. Then, for the deviation

$$\delta u(t) = u(t) - g(t)$$

we have – in linearized form – the *variational equation*

$$\delta u' = F_u(g(t)) \, \delta u, \quad \delta u(0) = \delta u_0 ,$$

to prescribed "initial deviation" δu_0. In general this is a nonautonomous (i.e., the time explicitly containing) linear differential equation with time-dependent *Jacobian matrix* $J(t) = F_u(g(t))$ of dimension N. However, the analysis of nonautonomous linear systems is nearly as complex as the one of general nonlinear systems.

In order to simplify our investigation, we therefore replace the variational equation by the autonomous differential equation

$$\delta u' = J\, \delta u, \quad \delta u(0) = \delta u_0 \tag{9.5}$$

with *constant* matrix J. Here the equilibrium solution is just $\delta u(t) \equiv 0$. In order to gain some insight into the structure of the problem class, we transform (similar to Section 8.1) the variable

$$\delta u = C\, \overline{\delta u}$$

by an *arbitrary* nonsingular matrix C. Thus we obtain the transformed differential equation

$$\overline{\delta u}' = \overline{J}\, \overline{\delta u}, \quad \overline{J} = C^{-1}\, J\, C.$$

Invariants of this similarity transformation are the eigenvalues $\lambda(J)$, representative for all of the thus defined matrices is the *Jordan canonical form*. In a further step of simplification we now assume that J is diagonalizable.[1] Then the linear system (9.5) splits into N decoupled scalar differential equations of the form

$$y' = \lambda y, \quad y(0) = y_0, \quad \lambda \in \mathbb{C}. \tag{9.6}$$

Dahlquist Test Equation.   If we choose $y_0 = 1$ in (9.6), we arrive at a scalar equation suggested by G. Dahlquist in 1956 as a test equation useful to distinguish different discretization methods [59]. Let $\tau > 0$ denote a prescribed timestep. Then (9.6) can be written as

$$y(\tau) = e^z \quad \text{with } z = \lambda\tau \in \mathbb{C}.$$

To begin with, we collect several properties of the complex exponential function. The imaginary axis is a separating line in the following sense:

$$|e^z| \geq 1 \quad \Leftrightarrow \quad \Re(z) \geq 0,$$
$$|e^z| \leq 1 \quad \Leftrightarrow \quad \Re(z) \leq 0, \tag{9.7}$$
$$|e^z| = 1 \quad \Leftrightarrow \quad \Re(z) = 0.$$

For our stability analysis this property is important with respect to the question of whether a deviation from the equilibrium solution would asymptotically lead back to it. Moreover, the function possesses an *essential singularity* in the point $z = \infty$,

---

[1] The subsequent theory also holds for nondiagonalizable matrices (see Exercise 9.1).

i.e., when approaching this point, its value depends on the path of approach. In particular, we have

$$z \to \infty: \quad |e^z| \to \begin{cases} \infty & \text{for } \Re(z) > 0, \\ 0 & \text{for } \Re(z) < 0, \\ 1 & \text{for } \Re(z) = 0. \end{cases} \tag{9.8}$$

This property is predominantly important for large stepsizes $\tau \gg |\lambda|^{-1}$.

One-step Methods.   We now want to investigate what happens with these two properties by time discretization of the test equation (9.6). Among the discretizations we restrict our attention to one-step methods, since they have certain advantages with respect to time dependent PDEs (see, e.g., the more extensive exhibition in Volume 2). In one-step methods we simply have to consider one step with timestep τ; we may therefore drop the running index of the timestep and always start at u_0. As examples we select four elementary discretizations, first applied to the general differential equation $y' = f(y)$, using the short notation y_1 for the result after one step as well as $z = \lambda\tau \in \mathbb{C}$:

- explicit euler method (EE):

$$y_1 = y_0 + \tau f(y_0) \quad \Rightarrow \quad y_1 = 1 + z;$$

- implicit Euler method (IE):

$$y_1 = y_0 + \tau f(y_1) \quad \Rightarrow \quad y_1 = \frac{1}{1-z};$$

- implicit trapezoidal rule (ITR):

$$y_1 = y_0 + \tau \frac{f(y_1) + f(y_0)}{2} \quad \Rightarrow \quad y_1 = \frac{1+z/2}{1-z/2};$$

- implicit midpoint rule (IMP):

$$y_1 = y_0 + \tau f\left(\frac{y_1 + y_0}{2}\right) \quad \Rightarrow \quad y_1 = \frac{1+z/2}{1-z/2}.$$

For linear problems the methods ITR and IMP are identical. In the nonlinear case, however, the IMP preserves quadratic functionals such as the discrete *energy* (see Exercise 9.2), which is why in concrete problems it is usually preferable. From our elementary examples it becomes clear that all one-step methods can be subsumed under the general scheme

$$y_1 = R(z),$$

where R is a complex *rational* function that might have poles in $\mathbb{C}$, but no essential singularities.

Order. For "small" $z \to 0$ the concept of order p according to

$$R(z) = e^z + \mathcal{O}(z^{p+1})$$

is appropriate. It easy to verify that $p = 1$ holds for EE and IE, $p = 2$ for ITR and IMP.

A-stability. If we want to inherit the separation property (9.7) exactly into the discrete setting, from our four sample discretizations merely ITR and IMP are left. This is why G. Dahlquist [59] in 1956 has introduced the concept of A-stability, to be able to save at least part of the important property (9.7) also for other discretizations. A one-step method is called A-stable if and only if

$$|R(z)| \leq 1 \quad \text{for} \quad \Re(z) \leq 0.$$

However, this concept captures the behavior for $z \to \infty$ only insufficiently. Upon taking the fact into account that complex rational functions in $\mathbb{C}$ are unique, the comparison with (9.8) will direct us to the following discrimination:

$$z \to \infty: \quad |R(z)| \to \begin{cases} \infty & \text{for EE,} \\ 0 & \text{for IE,} \\ 1 & \text{for ITR and IMP.} \end{cases}$$

As expected, it is not possible to mimic the essential singularity $z = \infty$ of the complex exp-function by a single complex rational function. We thus have to decide which limit we prefer for a special problem. Obviously, for problems like the Schrödinger equation, with eigenvalues on the imaginary axis, we will select discretizations of the type IMP. For inherently unstable problems or also for stiff problems in the transient phase an explicit discretization like EE will be considered. For stiff problems in a neighborhood of the equilibrium solution methods like IE will be chosen.

L-stability. This concept was introduced by B. L. Ehle [83] in 1969 to select the appropriate limit especially for stiff ODEs. A one-step method is called L-stable if it is A-stable and satisfies the additional property

$$R(\infty) = 0.$$

From our four elementary candidates this additional condition selects only the implicit Euler discretization (IE). This, however, has the disadvantage that it damps too much along the imaginary axis and even in the positive complex half-plane, which is why one also speaks of undesirable *"superstability"*.

Stability Regions. The presentation so far has made clear that for an assessment of discretization methods compromises are necessary. As a useful tool one defines the stability region $\mathcal{S}$ of a one-step method by

$$\mathcal{S} = \{z \in \mathbb{C} : |R(z)| \leq 1\}.$$

Obviously, A-stability is characterized by the condition

$$\mathbb{C}^- \subset \mathcal{S}.$$

In Figure 9.2 we show the stability regions for the four elementary one-step methods.

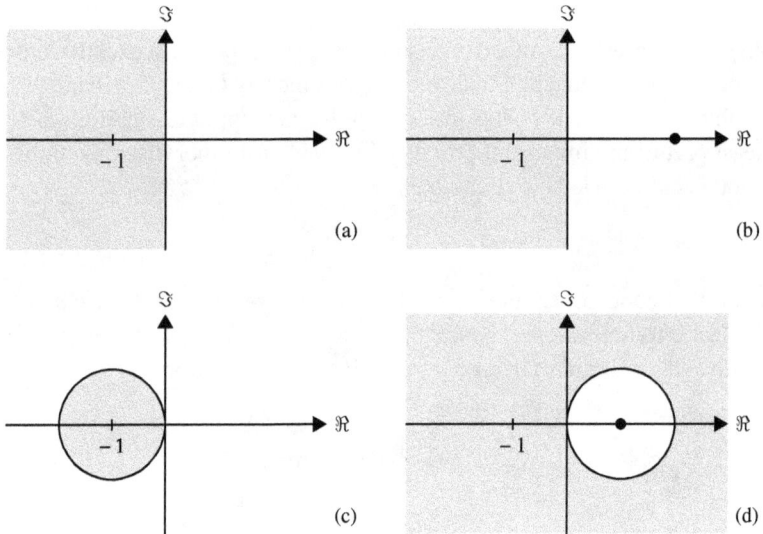

Figure 9.2. Stability regions in $\mathbb{C}$ (with poles $\bullet$) for (a) solution e^z, (b) implicit trapezoidal rule (ITR) equivalent to implicit midpoint rule (IMP), (c) explicit Euler discretization (EE), (d) implicit Euler discretization (IE).

For a better understanding of the concept we give, in Figure 9.3, a few examples of discrete solutions $y_k = (R(z))^k$ for selected points z in comparison with the continuous solution $y_k = (e^z)^k$.

A(α)- and L(α)-stability.   In the construction of methods of order $p > 1$ one strives to achieve an approximation as good as possible of the negative half-plane $\mathbb{C}^-$ by the stability region $\mathcal{S}$. In order to be able to measure the quality of the approximation in a single number, one introduces, for $\alpha \in [0, \pi/2]$, the notation

$$\overline{\mathcal{S}}(\alpha) = \{z \in \mathbb{C} : |\arg(-z)| \le \alpha\}$$

for the angular region around the negative real half-axis and extends the two concepts of A-stability and L-stability: a one-step method is called A(α)-*stable* if

$$\overline{\mathcal{S}}(\alpha) \subset \mathcal{S}.$$

Obviously, A-stability is equivalent to A($\pi/2$)-stability. In a similar way a one-step method is called L(α)-*stable* if

$$\overline{\mathcal{S}}(\alpha) \subset \mathcal{S} \quad \text{and} \quad R(\infty) = 0.$$

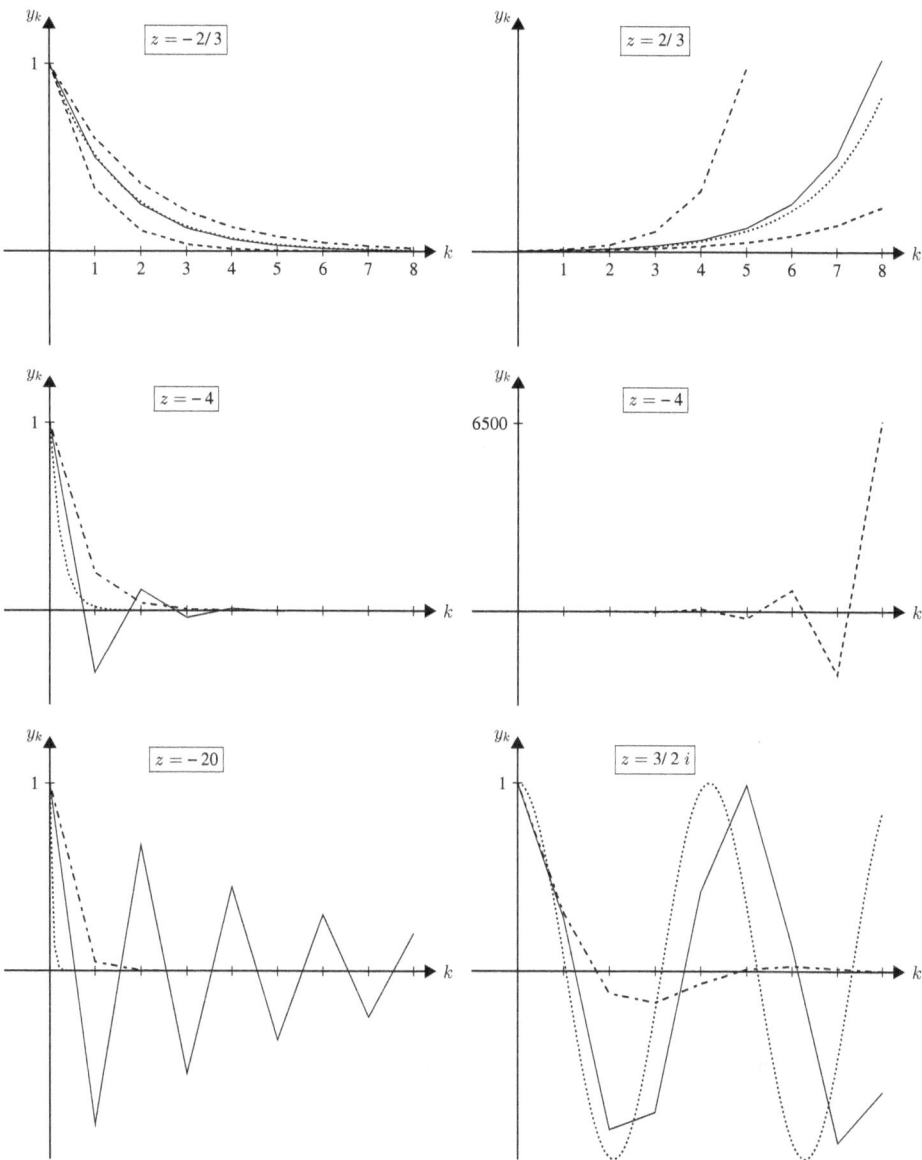

Figure 9.3. Solution $y_k = (e^z)^k$ compared with three discrete solutions $y_k = (R(z))^k$. Notations: ·········· solution, - - - - - EE, · - · - · IE, ──── IMP. *Top*: For small $z \in \mathbb{C}$ the higher order of IMP versus EE and IE pays off. *Center and bottom left*: in the "stiff" model problem only IE supplies a qualitatively correct discrete solution, IMP supplies weak, EE strong oscillations (which for $z = -20$ would leave the scale of the figure). *Bottom right*: along the imaginary axis the continuous solution exhibits undamped oscillations, which by IE are incorrectly damped ("superstability"), while IMP captures them qualitatively correctly, though with a phase shift (which, e.g., is undesirable for the Schrödinger equation; see Section 1.4).

Clearly, L-stability is equivalent to $L(\pi/2)$-stability. In both concepts one has, due to $\overline{\mathcal{S}}(\alpha_1) \subset \overline{\mathcal{S}}(\alpha_2)$ for $\alpha_1 \leq \alpha_2$: the larger the angle, the more "stable" the discretization. For a more detailed presentation, we refer to Volume 2 or the textbook of E. Hairer and G. Wanner [117].

Nonlinear Stability Concepts.   There are plenty of extensions of linear stability concepts towards nonlinear stability concepts, such as B-stability, B-convergence, algebraic stability etc. (for a comprehensive discussion of these concepts see again [117]). A different concept based on a simplified Newton method for the nonlinear initial value problem was worked out in [65] as well as in [68], Section 6.2.2. The theory for this was presented in $\mathbb{R}^N$, but applies in function space as well.

9.1.2   Linearly Implicit One-step Methods

In this section we give a brief survey on a class of adaptive stiff integrators. For a general theoretical background of this class we refer to Volume 2, Section 6.4; a more detailed treatment of this class can be found in the monograph [195] of K. Strehmel and R. Weiner.

Let us return to the initial value problem (9.4) for N ODEs, here of the form

$$Mu' = F(u), \quad u(0) = u_0, \tag{9.9}$$

where M is an invertible (N, N)-matrix. If (9.9) originated from a spatial discretization of a parabolic PDE, then, in general, N is large and the system is "stiff". For finite difference methods as spatial discretization, we usually have $M = I$, for Galerkin methods M is the symmetric positive definite mass matrix. For the Jacobian matrix $J(u) = F'(u)$ we assume a spectrum bounded from above, i.e., $v^T J v \leq \lambda_{max}|v|$. For finite element discretizations the matrices J and M are sparse. In each step of such a discretization large linear equation systems of dimension N arise in the form

$$(M - \gamma\tau J)v = w, \quad \gamma > 0. \tag{9.10}$$

The question of how to solve them numerically is postponed, for the time being.

Implicit Integrators.   Before we deal with the actual topic of this section, the linearly-implicit one-step methods, we want to quickly screen the class of purely implicit one-step and multi-step methods.

- *Implicit one-step methods.* This class contains *Runge–Kutta methods* (see, e.g., Volume 2, Sections 6.2 and 6.3), whose coefficients are determined by the solution of a set of algebraic equations. In each timestep, implicit Runge–Kutta methods require the solution of *several nonlinear* equation systems, whose number depends on the stage number s. Newton-like iteration for their solution leads to linear equation systems of the type (9.10). Particularly efficient are *Radau* methods (see Volume 2, Section 6.3), special L-stable collocation methods, realized, e.g., in the adaptive program Radau53 by E. Hairer and G. Wanner [117]. However, they have the reputation of being too costly for parabolic PDEs.

Formally speaking, *extrapolation methods* are a special subclass of Runge–Kutta methods. However, they are not generated via the solution of algebraic equation systems for the coefficients, but from a simple basis discretization by extrapolation with respect to the timestep τ. Purely implicit extrapolation methods, based on the implicit trapezoidal rule or the implicit midpoint rule, play only a minor role for parabolic PDEs, but are well used in connection with the Schrödinger equation.

- *Implicit multi-step methods.* Among them the *BDF-method* (see Volume 2, Sections 7.3.2 and 7.4.2) certainly is, for both theoretical and practical reasons, the most popular one. An adaptive implementation is realized, e.g., in the program DASSL by L. Petzold [170]. In each timestep this method requires the numerical solution of *one nonlinear* equation system, which is performed by some Newton-like iteration thus leading to linear equations of the type (9.10). Multi-step methods gain their efficiency by exploiting the "history" of the trajectory, which requires a certain smoothness over several timesteps; in our context, this is also a certain weakness, since time dependent adaptation of spatial meshes (see the subsequent Section 9.2) will perturb such a smoothness.

Structure of Linearly Implicit One-step Methods.   This type of method is based on the simple idea to subtract on both sides of (9.9) a linear autonomous term Ju, i.e.,

$$Mu' - Ju = F(u) - Ju, \qquad (9.11)$$

and discretize only the left-hand part implicitly, but the right-hand part explicitly. This means, of course, a structural simplification compared to implicit methods. For this type of methods two variants exist:

- *Methods with exact Jacobian matrix*, i.e., with $J = F'(u_0)$. This variant is realized in *Rosenbrock-Wanner methods* (in short: *ROW-methods*). The information of the exact Jacobian matrix enters into the algebraic equations for the method.

- *Methods with inexact Jacobian matrix*, i.e., with $J \approx F'(u_0)$. This variant is realized in so-called *W-methods*. Formally speaking, *linearly implicit extrapolation methods* are a subclass of such methods, even though historically they have originated independently. As the term Ju occurs on both sides of (9.11) in the same way, the choice of J is relatively free. If some detailed insight into the coupling of the various components of the ODE is available, then this class of methods permits to drop "weak" couplings in the ODE system, which arise as "small" entries in the Jacobian matrix, by setting them deliberately to zero (*deliberate sparsing*); of course, scaling needs to be carefully done, when deciding about which elements are to be regarded as "small".

If the linear equation systems (9.10) are so large that they are only accessible to an iterative numerical solution, then linearly-implicit one-step methods require only *two* nested loops (*outer loop*: time discretization, *inner loop*: iteration for linear equation system) versus *three* loops in implicit one-step of multi-step methods (*outer loop*: time discretization, *medium loop*: Newton-like iteration, *inner loop*: iteration for linear equation system). In very large systems this is an important advantage.

Linearly Implicit Runge–Kutta Methods.   This class contains both ROW- and W-methods. They are constructed by establishing N_p algebraic conditions for the coefficients of the RK-discretization scheme, depending on the order p; a more detailed examination reveals that for *exact* Jacobian matrix the number of conditions remains the same as for Runge–Kutta methods, while their structure changes considerably: The information about the Jacobian matrix directly enters into the condition equations. The number of coefficients depends on the stage number s, which is usually larger than the number of coefficients. Hence, one arrives at underdetermined systems of algebraic equations, which means that additional wishes can be fulfilled, such as: the stage number s should be as small as possible, the constructed method should be robust in the presence of inconsistent initial data, it should also work for singularly perturbed differential equations (see, e.g., Volume 2, Section 2.5) or for differential-algebraic equations up to index 1 (see again Volume 2, Section 6). Further additional equations aim at avoiding order reduction for nonlinear parabolic differential equations; see Section 9.1.3 below).

ROW-methods.   In Volume 2, Section 6.4.1 we discussed this class of linearly implicit Runge–Kutta methods. With the here introduced notation one step of this type of method can be written in the form:

(a) $J = F'(u_0)$,

(b) $i = 1, \ldots, s$:

$$(M - \gamma_{ii}\tau J)k_i = \tau \sum_{j=1}^{i-1}(\gamma_{ij} - \alpha_{ij})Jk_j + F\left(u_0 + \tau \sum_{j=1}^{i-1}\alpha_{ij}k_j\right), \qquad (9.12)$$

(c) $u_1 = u_0 + \tau \sum_{i=1}^{s} b_i k_i$.

Note that the unknowns $k_1, \ldots, k_s$ can be worked down recursively. If one sets, as is usual,

$$\gamma_{ii} = \gamma > 0, \quad i = 1, \ldots, s,$$

then all of the above matrices are identical so that a single LU-decomposition is sufficient. (Of course, in an iterative solution mode this advantage is of minor importance.)

For the determination of the coefficients the following successive order conditions occur (exemplified for $p \leq 4$)

$$p = 1: \quad \sum_{i=1}^{s} b_i = 1, \qquad\qquad p = 4: \quad \sum_{i=1}^{s} b_i \alpha_i^3 = \frac{1}{4},$$

$$p = 2: \quad \sum_{i,k=1}^{s} b_i a_{ik} = \frac{1}{2}, \qquad\qquad \sum_{i,k,l=1}^{s} b_i \alpha_{ik} a_{kl} \alpha_i^2 = \frac{1}{8},$$

$$p = 3: \quad \sum_{i=1}^{s} b_i \alpha_i^2 = \frac{1}{3}, \qquad\qquad \sum_{i,k,l,m=1}^{s} b_i a_{ik} \alpha_{kl} \alpha_{km} = \frac{1}{12},$$

$$\sum_{i,k,l=1}^{s} b_i a_{ik} a_{kl} = \frac{1}{6}, \qquad\qquad \sum_{i,k,l,m=1}^{s} b_i a_{ik} a_{kl} a_{lm} = \frac{1}{24}.$$

Here we have simplified the notation by introducing

$$\alpha_{ij} = 0, \ j \geq i, \quad \gamma_{ij} = 0, \ j > i$$

as well as the following abbreviations:

$$
\begin{aligned}
a_{ij} &= \alpha_{ij} + \gamma_{ij}, \qquad \mathcal{A} = (a_{ij})_{i,j=1}^{s}, \\
b &= (b_1, \ldots, b_s)^T, \quad e = (1, \ldots, 1)^T \in \mathbb{R}^s, \\
\alpha_i &= \sum_{j=1}^{i-1} \alpha_{ij}, \qquad \alpha^k = (\alpha_1^k, \ldots, \alpha_s^k), \qquad \gamma_i = \sum_{j=1}^{i} \gamma_{ij}.
\end{aligned}
\tag{9.13}
$$

For higher order the number of conditions increases rapidly, as shown below in the second row of Table 9.1.

In recent years, by subtle investigation of the solution structure of the order conditions, the following adaptive ROW-integrators have emerged:

- ROS3P by J. Lang and J. Verwer [141]: this method, published in 2001, has order $p = 3$ and stage number $s = 3$, is A-stable, suitable up to index 1, and does not suffer from order reduction for nonlinear parabolic differential equations (a topic that we will treat below in Section 9.1.3); in 2008 it was improved by J. Lang and D. Teleaga [140] to ROS3PL with $p = 3$ and $s = 4$, which is L-stable and robust against approximation errors of the Jacobian matrix;

- RODAS, one of the classical integrators by E. Hairer and G. Wanner [117], possesses order $p = 4$ and stage number $s = 6$, but suffers from order reduction; the method was improved in 2001 by G. Steinebach and P. Rentrop [192] to RODASP, it does not suffer from order reduction, but for linear parabolic equations only.

W-methods. None of the above mentioned linearly implicit RK-methods is a W-method, i.e., independent of the choice of the matrix J. In the context of parabolic

Table 9.1. Linearly-implicit RK-methods: number of algebraic conditions for order p.

p	1	2	3	4	5	6	7	8
N_p (ROW-methods)	1	2	4	8	17	37	85	200
N_p (W-methods)	1	3	8	21	58	166	498	1 540

PDEs, however, the "Jacobian matrix" as spatial discretization of the Fréchet-derivative will inevitably contain discretization errors, which is why W-methods seem to have a structural advantage for this problem class. Compared to ROW-methods, however, the number of algebraic conditions for the coefficients grows significantly (see Table 9.1). The theoretical reason for this stronger increase is that now terms of the kind $F'F$ and JF need to be distinguished, which leads to more "elementary differentials" and thus to more condition equations. Correspondingly, the required stage number gets higher. This background explains why this strand of methods has still been neglected in research. An alternative are the linearly implicit extrapolation methods to be presented directly after, which, by construction, can be conveniently extended to higher orders.

Timestep Control.   As in the spatial discretization in Section 6.2.1 an optimal relation of computational amount and discretization error is achieved by timesteps for which the local error contributions are *equilibrated*. For the construction of such adaptive time grids we thus need a localized error estimator and a refinement mechanism.

Error estimator. For local error estimation two discrete solutions $u_\tau, \hat{u}_\tau$ of order $p + 1$ and p are computed in parallel such that

$$u_\tau(\tau) = u(\tau) + \mathcal{O}(\tau^{p+2}), \quad \hat{u}_\tau(\tau) = u(\tau) + \mathcal{O}(\tau^{p+1}).$$

Then

$$[\hat{\epsilon}_\tau] := \|u_\tau(\tau) - \hat{u}_\tau(\tau)\| \doteq \hat{\epsilon}_\tau = \|u(\tau) - \hat{u}_\tau(\tau)\| \doteq C \, \tau^{p+1} \tag{9.14}$$

is an estimator of the actual $\hat{\epsilon}_\tau$ of $\hat{u}_\tau(\tau)$ measured in a suitable norm $\|\cdot\|$

Refinement. Unlike the case of spatial meshes, the timesteps are not determined by subdivision. It is more efficient to suggest a nearly optimal timestep for the next step in advance. Thereby one exploits the fact that the optimal timestep will change rarely.

Now let τ denote the presently selected timestep. We search for an "optimal" stepsize τ^*, for which

$$\hat{\epsilon}_{\tau^*} \leq \text{TOL}_\tau.$$

By insertion of both τ and τ^* into (9.14), we arrive at the estimation formula

$$\tau^* = \sqrt[p+1]{\frac{\rho \, \text{TOL}_\tau}{[\hat{\epsilon}_\tau]}} \, \tau \tag{9.15}$$

with a safety factor $\rho \approx 0.9$. If $[\hat{\epsilon}_\tau] \leq \text{TOL}_\tau$, then the present step is accepted and τ^* is used in the next step; otherwise the present step is rejected and repeated with timestep τ^*. In the successful case the more accurate value $u_\tau(\tau)$ will be used to start the next step.

In the context of parabolic differential equations usually *embedded* methods are realized. Therein one computes, from common intermediate values $k_1, \ldots, k_s$ due to (9.12), the two different solutions

$$u_\tau(\tau) = u_0 + \tau \sum_{i=1}^{s} b_i k_i,$$

$$\hat{u}_\tau(\tau) = u_0 + \tau \sum_{i=1}^{s} \hat{b}_i k_i.$$

Then, obviously, we get

$$[\hat{\epsilon}_\tau] = \|u_\tau(\tau) - \hat{u}_\tau(\tau)\| = \tau \left\| \sum_{i=1}^{s} (b_i - \hat{b}_i) k_i \right\|.$$

Linearly Implicit Extrapolation Methods.   The first such method was the *semiimplicit midpoint rule* (today generally called *linearly implicit midpoint rule*) suggested by G. Bader and P. Deuflhard [14] as an extension of the explicit midpoint rule. This method permits τ^2-extrapolation, but includes intermediate steps with $|R(\infty)| = 1$, which is why it is not suitable for index 1 and thus does not seem to be recommendable for parabolic PDEs. Instead the τ-extrapolation method based on the linearly implicit Euler discretization

$$(M - \tau J)(u_{k+1} - u_k) = \tau F(u_k), \quad k = 0, 1, \ldots \tag{9.16}$$

has prevailed, which realizes $R(\infty) = 0$ throughout all intermediate steps. It is implemented in the code LIMEX due to [71, 74]. If J is the exact Jacobian matrix $F'(u_k)$, then this method can be subsumed under (9.12) with

$$\gamma_{ii} = \frac{1}{n_i},$$

where $\mathscr{F} = \{n_1, n_2, \ldots\}$ is the subdivision sequence chosen for this extrapolation method. In practice, the simple harmonic sequence

$$\mathscr{F}_H = \{1, 2, 3, \ldots\}$$

has turned out to be efficient. By construction, this type of method is embedded so that a timestep control according to (9.15) can be realized in an elementary way; it can be complemented by an adaptive order control (see Volume 2, Section 6.4.2). In the integrator LIMEX a maximal order $p_{\max} = 5$ is mostly imposed.

In passing we note that in this framework W-methods of higher order p can be conveniently constructed without having to solve the highly complex (and many!) algebraic equations for the coefficients. Such methods, however, are not optimized with respect to the number of stages. But they are $L(\alpha)$-stable with $\alpha \approx \pi/2$ also for higher order and can also be applied to differential-algebraic equations, for details see [71, 74].

Exponential Integrators.   This special class of numerical integrators was suggested by M. Hochbruck and C. Lubich [121, 122] in 1997. Like for linearly implicit integrators, here one also starts from the form (9.11) of the ODE, i.e.,

$$u' - Ju = F(u) - Ju, \quad u(0) = u_0,$$

where we have set $M = I$ for simplicity. The basic idea is to integrate the left-hand side *exactly*, which for *linear F* and $J = F'$ would lead to

$$u(t) = \exp(Jt)u_0.$$

For *nonlinear F* one will make a "variation of constants" ansatz,

$$u(t) = \exp(Jt)v(t), \quad u(0) = u_0.$$

After some short calculation one then obtains the ODE

$$\exp(Jt)v' = F(\exp(Jt)v) - J\exp(Jt)v.$$

Starting from a discretization of the ODE for v and a backtransformation to u the authors construct discretization methods with respect to u. The simplest candidate is the *exponential Euler method*

$$u_1 = u_0 + \tau\varphi(J\tau)F(u_0), \qquad (9.17)$$

where

$$\varphi(z) = \frac{e^z - 1}{z}$$

and $\varphi(0) = 1$. The method is of order $p = 2$. For higher order the authors worked out ROW-methods for *exact* Jacobian matrix $J = F'(u_0)$ as well as W-methods for *inexact* Jacobian matrix $J \approx F'(u_0)$. In both cases the matrix exponential arises via the form $\varphi(\gamma\tau J)$, where the parameter γ must be suitably chosen. For (9.17) then $\gamma = 1$ will hold. In [122] the program exp4 to order $p = 4$ with stage number $s = 7$ and $\gamma = 1/3$ was suggested, into which a method of order $p = 3$ is embedded so that an adaptive timestep control is possible (see Volume 2, Section 5.4). For differential-algebraic problems (with index 1) the order reduces to $p = 3$, for inexact Jacobian with

$$J = F'(u_0) + \mathcal{O}(\tau)$$

to order $p = 2$. In total, the method is not only adaptive, but rather robust.

The knack of this approach, however, comes to bear with high dimension N, where the matrix-exponential is approximated by Krylov space methods (cf., for example, Volume 1, Section 8.3 and 8.5). It is shown in [121] that the Krylov approximation of $\varphi(\gamma J\tau)w$ (applied to some vector w) in general converges even faster than the one for $(I - \gamma\tau J)^{-1}w$, which would be required for (9.10). These methods are well-suited for "mildly stiff" problems and problems with eigenvalues along the imaginary axis. In the context of PDEs, however, one must take into account that Krylov space methods belong to *fixed* dimension N, i.e., they are structurally inconsistent with adaptive multigrid methods: By extension of the discretization spaces from some coarse to a finer mesh the information of the Krylov basis cannot be transferred – which definitely is a structural disadvantage of this type of method.

Explicit Integrators in Transient Phase.   As already exemplified in the simple test problem (9.2), the problem is not "stiff" in the transient phase, since one is not yet close to the smooth solution g; as a consequence, here even an explicit discretization is efficient, which formally may be expressed as W-method by $J = 0$. To make this a robust numerical technique, however, one would need a reliable criterion for switching between "stiff" and "nonstiff". Switching from "stiff" to "nonstiff" uses information of the Jacobian matrix, which is thus needed before; therefore this variant is not able to save very much in terms of computing time and storage, at best in the direct solution of the equation systems (9.10). Switching from "nonstiff" to "stiff" can only use information from the right-hand sides of the ODE to recognize whether the adaptively suggested timesteps are restricted by stability conditions; such strategies have been suggested several times, but do not seem to have been sufficiently robust in practical tests. Therefore, quite often the decision "stiff" or "nonstiff" is made on the basis of a priori insight into the problem to be solved. A "stiff" problem in the transient phase may then be discretized by some explicit method with small timesteps. A comparison with a stiff integrator will then have to weight the larger number of timesteps versus the reduced computational amount per timestep.

9.1.3 Order Reduction

When moving from stiff ODEs to parabolic PDEs an important restriction for the time discretization is encountered: instead of the order p of a method, one obtains an *effective* order $p^* \le p$. The reason for this is that the spectrum of the spatial differential operators is unbounded, whereas for ODEs the corresponding spectrum is bounded. For illustration, we begin with a simple test problem.

Example 9.1. Consider the scalar linear parabolic PDE

$$u_t = u_{xx} \quad \text{for } x \in \Omega =]0, \pi[\tag{9.18}$$

to homogeneous Dirichlet boundary conditions $u(0, t) = u(\pi, t) = 0$ and starting values $u(x, 0) = u_0(x)$. As elaborated in Section 1.2, the problem (9.18) may be split into independent Fourier modes

$$u(x,t) = \sum_{k=1}^{\infty} a_k e^{\lambda_k t} \sin(kx) \quad \text{to eigenvalues } \lambda_k = -k^2. \tag{9.19}$$

If we consider the problem in function space, we have the "spatial modes" (eigenfunctions) $\{\sin(kx)\} \in L^2(\Omega)$ with time-dependent coefficients $a_k e^{\lambda_k t}$. These coefficients each satisfy the Dahlquist test equation. If we discretize the problem first with respect to time by means of a (linearly) implicit Runge–Kutta method, then, due to the linearity, we get a decomposition similar to (9.19) of the form

$$u_1 = \sum_{k=1}^{\infty} a_k R(\lambda_k \tau) \sin(kx),$$

where R denotes the rational stability function of the selected one-step method (cf. Section 9.1.1). For the consistency error we obtain pointwise

$$\epsilon(\tau) = u(x, \tau) - u_1 = \sum_{k=1}^{\infty} a_k (e^{\lambda_k \tau} - R(\lambda_k \tau)) \sin(kx).$$

Now, let the one-step method be L-stable ($R(\infty) = 0$). Then the continuous function $|e^z - R(z)|$ has a maximum on the negative real axis $]-\infty, 0]$ at some point $\bar{z} < 0$, say. Thus, for stepsizes $\tau_k = \bar{z}/\lambda_k$, we have

$$\|\epsilon(\tau_k)\|^2_{L^2(\Omega)} = \frac{\pi}{2} \sum_{l=1}^{\infty} a_l^2 (e^{\lambda_l \tau_k} - R(\lambda_l \tau_k))^2 \geq a_k^2 (e^{\bar{z}} - R(\bar{z}))^2 = \mathcal{O}(a_k^2),$$

where we have used the L^2-orthogonality of the eigenfunctions and picked out one term of the sum, where the expression in brackets has its maximum. Note that this result is independent of the consistency order p of the method.

As initial condition we now choose $u_0(x) = x(\pi - x)$. For this we obtain the coefficients

$$a_k = \frac{8}{\pi k^3} = \frac{8}{\pi |\bar{z}|^{3/2}} \tau_k^{3/2} \quad \text{for odd } k,$$

as well as $a_k = 0$ for even k. Insertion then yields

$$\|\epsilon(\tau_k)\|_{L^2(\Omega)} \geq \mathcal{O}(\tau_k^{3/2}). \tag{9.20}$$

For comparison: for the consistency order of Runge–Kutta methods one has $p \geq 1$, which means that in the case of ODEs we would obtain $\mathcal{O}(\tau^{p+1})$ as consistency error. The result (9.20) thus represents an effective order $p^* = 1/2$ for the whole class of Runge–Kutta methods. It seems worth noting that this restriction of the consistency order also occurs for the very smooth initial values that we chose above. Obviously, the reason for the order reduction lies in the fact that the spectrum of the Laplace operator, in contrast to the case of ordinary differential equations, is unbounded and distributed along the whole negative real axis.

This effect is already echoed in differential-algebraic problems; but these have a large "gap" in the spectrum between the bounded differential part and the algebraic part $\lambda \to -\infty$, which is why the consistency order of L-stable methods with $e^z - R(z) \to 0$ is preserved for $z \to -\infty$ (see Volume 2, Section 6.4.2).

The phenomenon of order reduction was probably first detected by K. Strehmel and R. Weiner in the course of their B-convergence studies, and presented in 1992 in their monograph [195]; as it turs out, however, their order bounds were not sharp. This is why we give a short account of results here which were proved in 1993 by A. Ostermann and M. Roche [165] and in 1995 by C. Lubich and A. Ostermann [147, 148]. There adaptivity does not play a role, neither in space nor in time; the quoted theoretical results were derived in the framework of *uniform* spatial and temporal grids. Starting point of these analyses is the insight that in successively finer space discretization asymptotically the structure of the underlying Cauchy problem in function space will show up. As a consequence, in [147, 148, 165] only the condition equations for Runge–Kutta methods in Hilbert spaces have been investigated. At the same time it was assumed that one is already on the smooth equilibrium solution, i.e., after the transient phase (see above).

In our simple linear example (9.18) we had only considered the consistency error, i.e., the *local* error; the high frequency modes contained therein, however, are rapidly damped in the course of time and thus do not play a role in the *global* error. As a consequence, in this example the reduction of the consistency order will not end up in a reduction of the convergence order on fixed time intervals. This damping effect, however, will vanish for problems where the high frequency modes are repeatedly renewed. Such a coupling of modes occurs for linear nonautonomous as well as for nonlinear problems; the results of the corresponding theory are now briefly summarized, without going too much into technical detail.

Linear Nonautonomous Parabolic PDEs. In [165], ROW-methods (9.12) for abstract Cauchy problems of the type

$$u' = Au + f(t) \tag{9.21}$$

were considered, where the operator $-A$, which includes the boundary conditions, was assumed to be elliptic. As a prerequisite for the theory, the initial value u_0 should already be on the stationary solution, which in the reality of scientific computing need not be satisfied – cf., for example, the electrocardiological example worked out in Section 9.3. Under this assumption one obtains the effective order

$$p^* = \min\{p, s + 2 + \beta\}, \tag{9.22}$$

where s is the stage number and the quantity $\beta > 0$ can be determined from the underlying spatial elliptic problem. Due to [148] one has (with $\varepsilon > 0$ arbitrarily small)

$$\beta = \begin{cases} 3/4 - \varepsilon & \text{for homogeneous Dirichlet boundary conditions,} \\ 5/4 - \varepsilon & \text{for natural boundary conditions} \quad (d = 1), \\ 1/4 - \varepsilon & \text{for natural boundary conditions} \quad (d > 1). \end{cases} \tag{9.23}$$

It is interesting to note that for *periodic* boundary conditions no order reduction is to be expected. In the case of realistic application problems, however, there is no hope of getting hold of the value for β, so that in these cases the essential message is that $\beta > 0$.

In contrast to the adaptive control in one-step methods, where only *local* errors enter, the authors give a *global* convergence analysis for constant stepsizes τ over a fixed interval $T = n\tau$. Beyond that they work out further conditions that permit an extension of Rosenbrock methods to order $p \geq 3$. In order to break the barrier (9.22) by achieving, say, effective order $p^* = 3$ independent of the spatial regularity, the following conditions (in the notation (9.13))

$$b^T A^j (2A^2 e - \alpha^2) = 0 \quad \text{for} \quad 1 \leq j \leq s - 1 \tag{9.24}$$

must be satisfied; note that they depend on the stage number s. These conditions are identical to the ones in [195].

Example 9.2. Consider the simple example

$$u_t = u_{xx} + x\,e^{-t}, \quad x \in\ ]0, 1[, \quad t \in [0, 0.1] \tag{9.25}$$

with homogeneous Dirichlet-boundary conditions $u(0, t) = u(1, t) = 0$. In this example, Ostermann and Roche conducted numerical experiments. For that purpose they chose a spatial meshsize $h = 1/N$ for fixed $N = 1000$ and timesteps $\tau = 0.1/n$ to variable $n = 2, 4, \ldots, 128$. They generated stationary initial values by the condition $u'|_{t=0} = 0$, which leads to $u(x, 0) = \frac{1}{6}x(1 - x^2)$. For the popular Rosenbrock method GRK4T of P. Kaps and P. Rentrop [133] a reduction of the classical order $p = 4$ to $p^* = 3.25$ was achieved, whereas a method of S. Scholz [182][2], which takes extensions of the additional conditions (9.24) into account, would achieve order $p^* = p = 4$.

In Figure 9.4 we analyze the situation in more detail for differently fine spatial meshes. As in [165] we only consider the error at the final time point. Due to the damping of errors from intermediate points one expects a convergence error $\mathcal{O}(\tau^5)$ at the final point. This is actually achieved, but only for sufficiently small timesteps, corresponding to the fineness of the spatial discretization. In [165] a value of $N = 1000$ had been chosen, which would, for the selected timesteps τ, be close to the effective order $p^* = 3.25$.

---

[2] For a correction of the printing errors with respect to the coefficients, however, see [165].

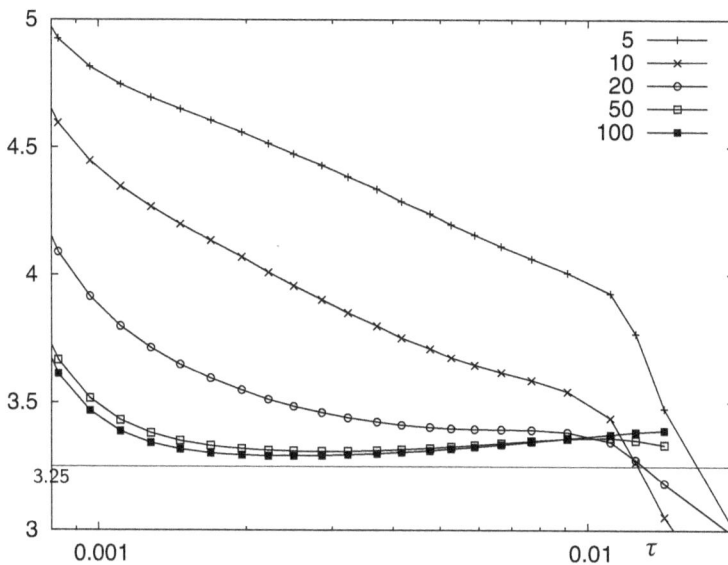

Figure 9.4. Test problem (9.25): numerically estimated convergence order of the ROW-integrator GRK4T. Spatial meshsizes $h = 1/N$ for $N = 5, \ldots, 100$, uniform timesteps $\tau = 0.1/n$ for $n = 2, 4, \ldots, 128$. Expected order $p = 5$, effective order $p^* = 3.25$.

Quasilinear Parabolic PDEs.   This problem class may again be written as an abstract Cauchy problem

$$u' = A(u)u + f(x,t), \quad u(0) = u_0, \quad x \in \Omega \subset \mathbb{R}^d, d \geq 1.$$

The nonlinear operator A again contains the boundary conditions. This nonlinear PDE is regarded as parabolic, whenever the corresponding variational equation, known to be nonautonomous, is parabolic in the sense of (9.21).

Example 9.3. Let $\Omega \subset \mathbb{R}^d$ denote a simply connected domain with smooth boundaries (mind you: no reentrant corners, no holes). Let the PDE have the form

$$u_t = \mathrm{div}(a(u)\nabla u) + f, \quad x \in \Omega, \quad 0 \leq t \leq T,$$

with symmetric diffusion matrix $a(u) \in \mathbb{R}^{d \times d}$ and associated natural boundary conditions

$$n^T a(u)\nabla u = 0, \quad x \in \partial\Omega, \quad 0 \leq t \leq T,$$

where n is again the normal unit vector of the boundary $\partial\Omega$.

Applied to this problem class for *implicit Runge–Kutta methods*, the theory [148] supplies an effective order

$$p^* = \min\{p, s + 1 + \beta\} \tag{9.26}$$

with β as in (9.23). In [147] the theory was modified for the case of *linearly implicit Runge–Kutta methods*. Here the even stronger restriction

$$p^* = \min\{p, 2 + \beta\},$$

with β as in (9.23) comes up. For *ROW-methods*, as in the linear case, an increase of the effective order is possible, if additional conditions of the kind (9.24) can be satisfied.

For *W-methods*, however, such an increase of order is *not* possible: here one essentially gets stuck at the barrier $p^* = 2$. In particular, this also holds for linearly implicit extrapolation methods on the basis of the linearly implicit Euler discretization (LIMEX). At first, this result seemed to contradict the high efficiency of the method as observed in challenging problems. A more detailed analysis in [147] revealed the surprising result that, on the one hand, the order for this method is restricted, but, on the other hand, the coefficients to second order decrease significantly with each extrapolation step. On the basis of these subtle investigations the authors suggested to replace the harmonic subdivision sequence $\mathcal{F}_H = \{1, 2, 3, \ldots\}$ by the sequence $\mathcal{F}_H^+ = \{2, 3, \ldots\}$. Extensive numerical tests, however, have shown that the thus gained reduction of coefficients is not compensated by the increased amount of computation in $\mathcal{F}_H^+$ over $\mathcal{F}_H$.

Timestep Control. In Volume 2, Section 6.4, order reduction were already theoretically described for extrapolation methods and, more general, for implicit and linearly implicit Runge–Kutta methods: it occurred in the numerical integration of singularly perturbed ODE systems of the kind

$$u' = f(u, v), \quad \epsilon v' = g(u, v) \quad \text{for} \quad \epsilon \to 0.$$

The question, there as well as here, is: Which order should be used in formula (9.15) to control the timesteps adaptively? In the practical use of extrapolation methods it has turned out that the robust controller property of the timestep control (as worked out, e.g., in Volume 2, Section 5.2) actually permits implementation of the adaptive algorithms with the highest achievable order p in an efficient way. For parabolic PDEs J. Lang [139] recommends a modification. We again start from an *error estimate* at intermediate step $t_n \to t_{n+1} = t_n + \tau_n$, written as

$$[\epsilon_\tau]_{n+1} \doteq \|u_{n+1}^{p+1} - u_{n+1}^p\| \doteq C_n \, \tau_n^{p+1}.$$

Formula (9.15) has been derived on the basis of the model assumption $C_{n+1} = C_n$. If one wants to estimate an effective order p^* therefrom, then one gets

$$p^* + 1 \doteq \log \frac{[\epsilon_\tau]_{n+1}}{[\epsilon_\tau]_n} \Big/ \log \frac{\tau_n}{\tau_{n-1}}$$

and replaces, in formula (9.15), the value p by the above p^*. Obviously, here the information of the two previous timesteps enters. An alternative strategy suggested by K. Gustafsson, M. Lundh, and G. Söderlind [109] uses the model assumption

$$\frac{C_{n+1}}{C_n} = \frac{C_n}{C_{n-1}}.$$

This yields the estimation formula for an "optimal" timestep

$$\tau^* = \frac{\tau_n}{\tau_{n-1}} \sqrt[p+1]{\frac{\rho\, \text{TOL}_\tau\, [\epsilon_\tau]_n}{[\epsilon_\tau]_{n+1}^2}}\, \tau_n$$

with a safety factor $\rho < 1$. Here, too, the information of the two previous discretization steps enters, but with slightly more freedom in the coefficients C_n.

9.2 Space-time Discretization of Parabolic PDEs

Time dependent PDEs are to be discretized with respect to both time and space. The probably most popular candidate among coupled space-time discretizations with constant spatial mesh size and fixed timesteps is the *Crank–Nicolson scheme* [57], which in time realizes an implicit trapezoidal rule, in space a simple symmetric finite difference scheme. We will not consider it any further here, since it is not adaptive; note, however, that this scheme inherits the instabilities of the trapezoidal rule (cf. Figure 9.3, bottom left), which may show up as unwanted oscillations.

If one discretizes *space first*, which is the presently most popular variant, then one obtains, in general, a large block structured *initial value problem for ODEs* (see the schematic representation in Figure 9.5, left). This approach is called *method of lines* and will be presented in Section 9.2.1. In this approach, adaptivity w.r.t. time (as timestep and order control) is nearly as easy implementable as in ODEs. But adaptivity w.r.t. space is rather restricted in its realization, in an effective way at best only in space dimension $d = 1$.

If one discretizes *time first*, then one comes up with a sequence of *boundary value problems for stationary ordinary ($d = 1$) or partial ($d > 1$) differential equations*, as depicted schematically in Figure 9.5, right. This approach is called *method of time*

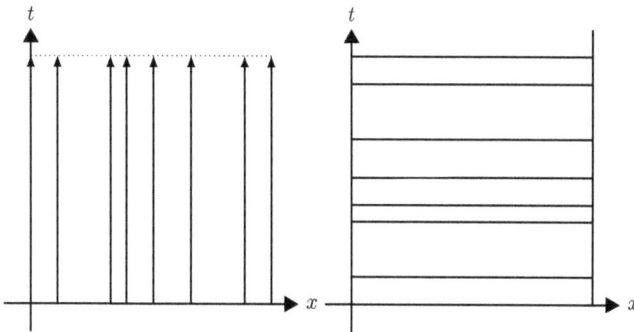

Figure 9.5. Ordering in space-time discretization. *Left: method of lines.* Initial value problem for ODEs. *Right: method of time layers*, or also *Rothe method*. Sequence of boundary value problems for stationary differential equations.

layers or also *Rothe method*. At this point we want to mention that this approach permits a natural combination of adaptive timestepping and adaptive multigrid methods. This will be worked out in Section 9.2.2. In Figure 9.6, we compare both approaches in a common diagram, where M_h again denotes the earlier introduced *mass matrix*. If one restricts the discretization to *uniformly fixed space grids and constant timesteps*, which is still the standard in the engineering world, then the diagram commutes. In *adaptive* discretizations, however, as treated in this book, the discretization ordering does play an essential role.

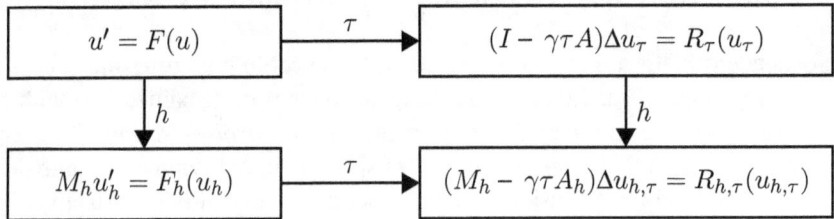

$$
\begin{array}{ccc}
\boxed{u' = F(u)} & \xrightarrow{\ \tau\ } & \boxed{(I - \gamma\tau A)\Delta u_\tau = R_\tau(u_\tau)} \\[2mm]
\Big\downarrow h & & \Big\downarrow h \\[2mm]
\boxed{M_h u'_h = F_h(u_h)} & \xrightarrow{\ \tau\ } & \boxed{(M_h - \gamma\tau A_h)\Delta u_{h,\tau} = R_{h,\tau}(u_{h,\tau})}
\end{array}
$$

Figure 9.6. *Method of lines*: first space discretization (h), then time discretization (τ), top left → bottom left → bottom right. *Method of time layers* or *Rothe method*: first time discretization (τ), then space discretization (h), top left → top right → bottom right.

9.2.1 Adaptive Method of Lines

In this approach the space discretization is performed first, in sufficiently large application problems by means of a mesh generator. In *finite difference methods* one then obtains *initial value problems* for large ODE systems of fixed dimension N (for simplicity, we drop the suffix h throughout this section)

$$u' = F(u), \quad u(0) = u_0. \tag{9.27}$$

In *Galerkin methods* one obtains large linearly implicit ODE systems

$$Mu' = F(u), \quad u(0) = u_0, \tag{9.28}$$

where M denotes the arising, usually sparse, (N, N)-mass matrix. In both cases, (9.27) was well as (9.28), the space discretization leads to block structured ODE systems. In principle they can be solved by any efficient stiff integrator with adaptive timestep control. However, for the method of lines in particular, it has become clear that linearly implicit integrators of low order are preferable (see Section 9.1.2 above; for a deeper understanding see, e.g., Volume 2, Section 6.4.2, under the keyword "method of lines").

Dynamic Regridding ($d = 1$). For problems with "moving fronts", also called *traveling waves*, a transformation onto a "moving grid" is recommended, i.e., an extension $x \rightarrow x(t)$, where the number N of nodes remains fixed and a unique correspondence of nodes over all timesteps is possible; in this framework, a change of grid will not cause any interpolation errors (in contrast to static regridding; see below). This method is often denoted as r-adaptivity (from *relocation*) – in contrast to h-adaptivity for hierarchical mesh refinement or hp-adaptivity in case of additional order adaptation. For ease of presentation, we restrict ourselves to one scalar PDE in one space dimension.

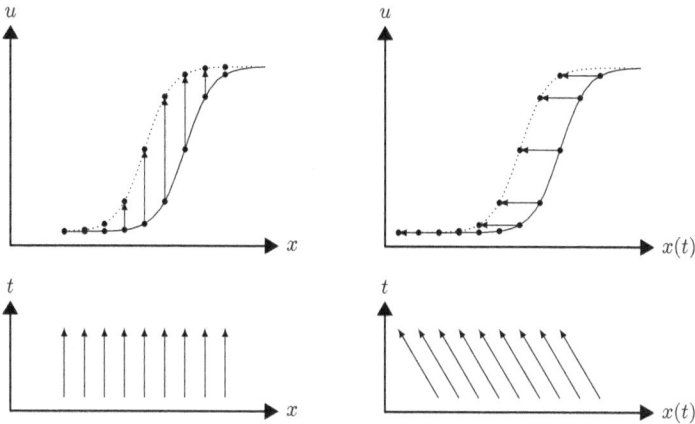

Figure 9.7. Moving front of a solution u, one time step (due to [163]). *Left*: fixed mesh, solution varies strongly. *Right*: moving mesh, solution remains nearly constant.

In Figure 9.7 the basic idea of the moving fronts is depicted graphically: When integrating on a fixed mesh, the values at all nodes in the front change, seemingly indicating some high dynamics of the problem. In moving coordinates, however, the values of the solution remain nearly constant. As a consequence, timestep control with the fixed grids will suggest "small" stepsizes, with moving grids "large" stepsizes. In moving coordinates, the solution $v(t) = u(x(t), t)$ has the time derivative

$$v' = u_x \, x' + u',$$

which gives rise to a convection term in an extended ODE (9.28) of the form

$$M(v' - u_x \, x') = F(v). \tag{9.29}$$

Obviously, here we need an additional equation for the determination of x'. For its derivation, there are quite a number of suggestions, from which we select two here.

The common basic idea in all adaptive methods of lines is to transform the "physical" space variable $x \in [a, b]$ onto a "virtual" space variable $\xi \in [0, 1]$, in which a *uniform* mesh with constant meshsize $\Delta \xi = 1/N$ is introduced. The transformation has to be chosen such that for the function $x = x(\xi)$ an inverse function $\xi = \xi(x)$ exists. Then, via backtransformation, an in general *nonuniform* mesh in x is generated. In Figure 9.8 the situation is depicted graphically, where the chosen mapping $x(\xi)$ is obviously monotone.

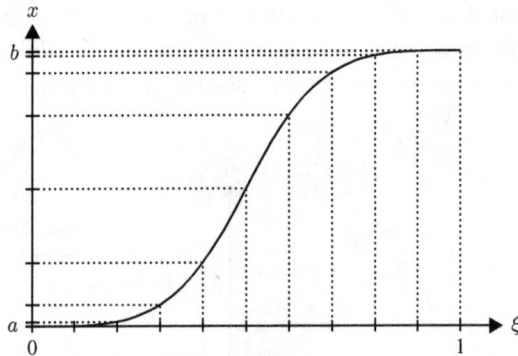

Figure 9.8. Uniform "virtual" mesh in $\xi \in [0, 1]$ transformed onto a nonuniform "physical" mesh in $x \in [a, b]$.

Monitor function. Here we follow the survey paper [45] from 2009 as well as the more recent book [125, Chapter 5] by W. Huang and R. D. Russell. The method described there extensively starts from an equilibration with respect to a *monitor function* $\mu(\xi, t)$, which leads to a condition of the form

$$\mu x_\xi = \theta(t) = \frac{\int_{\Omega_x} \mu(x, t)\, dx}{\int_{\Omega_\xi} d\xi},$$

where the domains Ω_x, Ω_ξ exist in the corresponding coordinate spaces. If one chooses $\mu > 0$ herein, then $x_\xi > 0$ holds, i.e., the mapping $x(\xi)$ is monotone and there exists a unique inverse function. Differentiation w.r.t. ξ then supplies the non-linear boundary value problem

$$\mu_\xi x_\xi^2 + \mu x_{\xi\xi} = 0, \quad x(0) = a, \quad x(1) = b. \tag{9.30}$$

The only thing left to do is the to determine of μ from some first principle. An intuitive choice is the *arc length*, i.e.,

$$\mu = \sqrt{1 + u_x^2}.$$

It is known to be not invariant under rescaling of the variable $x \to x/\sigma$, so that the ansatz is generalized according to

$$\mu = \sqrt{1 + \sigma^2 u_x^2}$$

with a scaling quantity σ given into the hands of the user. Inserting μ into the boundary value problem (9.30) one ends up with (9.29), a differential-algebraic system with non-linear algebraic equality constraint. In the context of *finite difference methods* ($M = I$ in (9.28)) the technique of C. E. Pearson [168] can be applied on the nonuniform spatial grid in x (see Section 3.3.1 above). In this way the time-dependent adaptation of the space grid is fixed.

What is missing, however, is a strategy for the choice of the virtual grid size $\Delta\xi$ or the number N of nodes, respectively, which must come from an estimate of the space discretization error. Should an estimation method for the discretization error not be available, then the interpolation error is estimated instead. However, if N is changed, then the solution values on the actual grid need to be transferred to a new grid, where again the interpolation error creeps in.

The concept presented here can in principle be carried over to higher space dimension $d > 1$ (see, e.g., [125, Chapter 4]). There and in [45] a number of articles are cited in which this has been successfully done. The examples given for the choice of a monitor function μ, however, are chosen more for illustration purposes, and so do not permit any conclusions about the efficiency in the general case.

Minimization problem. Here we present suggestions by J. M. Hyman [126] and L. R. Petzold [171] in the framework of finite difference methods, i.e., based on (9.27). In this approach, with the aim of achieving time stepsizes τ as large as possible, the requirement

$$\|v'\|_2^2 = (u' + u_x x')^T (u' + u_x x') = \min$$

is imposed. Minimization w.r.t. x' leads to the PDE

$$(u' + u_x x')^T u_x = u_x^T v' = 0.$$

Combination with (9.27) then supplies the coupled PDE system

$$\begin{aligned} M(v' - u_x \, x') &= F(v), \\ u_x^T v' &= 0. \end{aligned} \tag{9.31}$$

Space discretization due to C. E. Pearson [168] finally leads to some ODE system in $2N$ variables.

Experience has shown, however, that by this approach a *crossing of nodes* is not systematically avoided. For an illustration of this phenomenon we start from neighboring nodes $x_i < x_{i+1}$ with local velocities x_i', x_{i+1}' that can be computed from the above ODE system. Let $x_{i+1}' < x_i'$. In order to assure that in the next timestep the condition

$$x_i + \tau x_i' < x_{i+1} + \tau x_{i+1}'$$

holds as well, the maximal time stepsize

$$\tau_{\max} < \frac{x_{i+1} - x_i}{x_i' - x_{i+1}'} \tag{9.32}$$

must not be exceeded. Therefore in [156, 171] it has been suggested to couple another functional in, quasi as a "repelling term" that should work to prevent the crossing of nodes. Therefore, in view of (9.32), additionally

$$\sum_{i=2}^{N} \left\| \frac{x_i' - x_{i-1}'}{x_i - x_{i-1}} \right\|_2^2 = \min \tag{9.33}$$

is introduced. This leads to the $N - 2$ additional conditions

$$\frac{x_i' - x_{i-1}'}{(x_i - x_{i-1})^2} - \frac{x_{i+1}' - x_i'}{(x_{i+1} - x_i)^2} = 0, \quad i = 2, \dots, N - 1.$$

These equations may be interpreted as space discretization of the PDE $-x'_{xx} = 0$. If this is coupled by means of a Lagrange multiplier $\lambda > 0$, then we obtain, instead of (9.31), alternatively

$$M(v' - u_x \, x') = F(v),$$

$$u_x^T v' + \lambda x'_{xx} = 0, \quad \lambda > 0.$$

Spatial discretization then again yields a system of $2N$ coupled ODEs, which can be solved numerically by a stiff integrator. The parameter λ, however, must be chosen in each example from scratch by trial and error, a problem-independent robust choice is not known up to now. In this way the case of crossing traveling waves becomes tractable, as illustrated in Figure 9.9; for a not further specified example see [163].

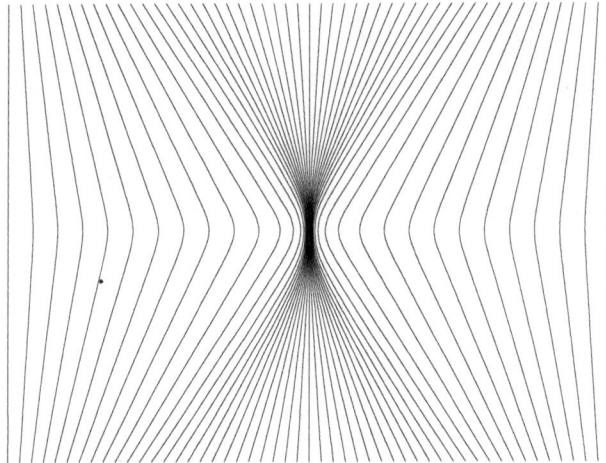

Figure 9.9. Adaptive method of lines [163]: dynamic regridding with "repelling functional" (9.33) at the example of crossing fronts. Cf. also Figure 9.13.

This approach to dynamic grid adaptation can be generalized to space dimension $d > 1$, too. For a survey on this topic we refer to the article [145]. However, as documented in [157], a grid "tangling" cannot be fully excluded, despite the addition of the "repelling functional". This might be the reason why this method has not generally prevailed yet.

Static Regridding. In this class of methods a new grid is constructed in every time-step (or also after a fixed number of timesteps). The number N of nodes is no longer constant, but varies within prescribed limits $N_{min} \leq N(t) \leq N_{max}$. There, inevitably interpolation errors occur, which must be carefully monitored apart from the discretization errors. In [163], two generally applicable strategies are suggested:

- In order to possibly avoid interpolation errors, two global grids are used in parallel, a coarse grid G with virtual grid size $\Delta\xi$ and a finer grid G^+ with grid size $\Delta\xi/2$. Thus, in local refinement, the additional value to be inserted can be just taken from G^+. For grid coarsening, G is sufficient anyway.

- In order to assure a globally monotone mapping $x(\xi)$, the cubic Hermite interpolation due to [97] is modified such that it remains locally monotone. The usual cubic Hermite interpolation (see, e.g., Volume 1, Section 7.1.2) over the interval $I_i = [x_i, x_{i+1}]$ requires as input the function values and their derivatives at the interval boundaries, i.e., within each time layer the values $u(x_i), u_x(x_i)$, $u(x_{i+1}), u_x(x_{i+1})$. In finite difference methods, however, only *difference quotients* at the boundaries are available as approximations, which reduces the approximation order from $p = 3$ to a poor $p = 1$; for compensation, superconvergence occurs at the corresponding interval midpoints $x_{i+1/2} = x(\xi_i + \Delta\xi/2)$, i.e., pointwise $p = 2$. These are the very nodes additionally held in G^+.

The goal of static regridding is to equilibrate the global discretization error by means of the choice of nodes such that in each node roughly the same local error occurs. The construction of adaptive grids is carried out stepwise by virtue of the following two algorithmic pieces:

(a) In each node x_i the discretization error is estimated; let $[\epsilon_i^x]$ denote a computationally accessible estimate of the local discretization error ϵ_i^x. If a good estimate of the discretization error is not available, it is replaced by an estimate of the interpolation error, which, in general, is easily available.

(b) Let TOL_h denote a user prescribed global upper bound and $\sigma^\pm$ local bounds for the spatial error. Then the new grid $G(t + \tau)$ is generated from the actual grid $G(t)$ according to the following *rules*:

  - if $[\epsilon_i^x] < \sigma^-$, then the node x_i is eliminated, unless it is a boundary node;
  - if $[\epsilon_i^x] \geq \sigma^-$, then the node x_i is kept;
  - ff $[\epsilon_i^x] \geq \sigma^+$, then two additional new nodes $x_{i\pm1/2}$ are inserted;
  - one ensures globally that the new grid remains *quasi-uniform*, i.e., that neighboring subintervals differ only within prescribed limits in their lengths; this leads to the recipe $h_i/h_{i+1} \in [1/\alpha, \alpha]$ for a parameter $\alpha \approx 2-3$ (to be possibly modified by the user).

With these means one hopes to be able to roughly monitor the quality of the approximations to be computed (for an illustration of these rules see Figure 9.10).

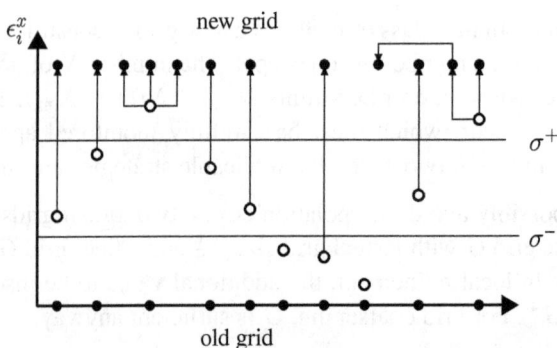

Figure 9.10. Method of lines: refinement and coarsening strategies in static regridding.

Software. In his dissertation [163] U. Nowak elaborated an efficient adaptive method of lines for parabolic PDEs in one space dimension. There he combined the above presented concepts of dynamic and static regridding by means of cleverly selected heuristics. For the estimation of the spatial error he uses a coupled extrapolation scheme in space and time that builds upon an asymptotic expansion in space and time, which, however, is mathematically not sufficiently backed up (cf. Volume 2, Sections 4.3 and 6.4.2 on extrapolation methods). In the engineering community, his program PDEX1M [164] is highly reputated, due to its efficiency (as an example see, e.g., [91]).

Example 9.4. This 1D test problem from combustion chemistry [169] describes a traveling combustion front with strongly varying velocity on a domain $\Omega = \,]{-}25, 10[$ for $t \in [0, 15]$. The PDE system

$$T_t = T_{xx} + R(T, Y),$$

$$Y_t = \frac{1}{L} Y_{xx} - R(T, Y)$$

models the evolution of the temperature T and of the fuel concentration Y with the initial and boundary conditions

$$T(x, 0) = \min(e^x, 1), \qquad T(-25, t) = 0, \quad T(10, t) = 0,$$

$$Y(x, 0) = \max(1 - e^{Lx}, 0), \quad Y(-25, t) = 1, \quad Y(10, t) = 0.$$

Therein

$$R(T, Y) = \frac{\beta^2}{2L} Y \exp\left(-\frac{\beta(1 - T)}{1 - \alpha(1 - T)}\right)$$

denotes the reaction rate and $\alpha = 0.8$ (heat release), $\beta = 20$ (activation energy) and $L = 2.0$ (Lewis number) are dimensionless parameters.

Nonadaptive methods of lineshave turned out to be unable to solve this problem with sufficient accuracy. In Figure 9.11, top and center, we illustrate the nodal flux of the

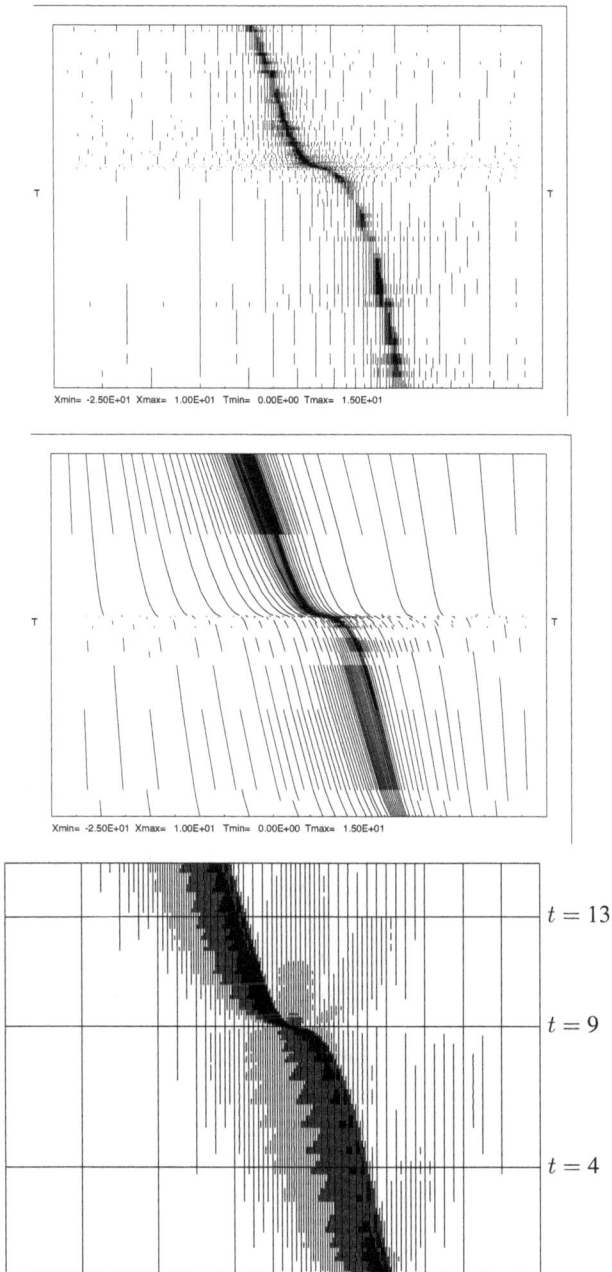

Figure 9.11. Nodal flux in Example 9.4. *Top*: adaptive method of lines [163]: static regridding. *Center*: adaptive method of lines [163]: mixed static-dynamic regridding. *Bottom*: adaptive method of time layers [72]. Temperature profiles for $t = 4, 9, 13$ see Figure 9.12.

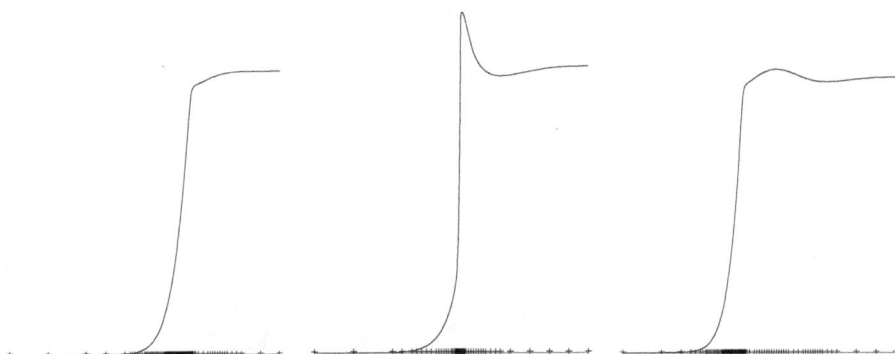

Figure 9.12. Example 9.4 (cf. Figure 9.11): temperature profiles $T(x,t)$ at the time layers $t = 4, 9, 13$ with corresponding automatic choice of nodes.

discrete solution in the adaptive method of lines due to U. Nowak [163]. For comparison, we already give at this stage the nodal flux of the method of time layers, which will be discussed below in Section 9.2.2 (see Figure 9.11, bottom). In order to give some insight into the difficulty of the problem, we additionally show the temperature profiles at critical intermediate time layers in Figure 9.11.

In [150], the combustion chemist U. Maas published numerical comparisons of simulations at difficult problems from engine combustion (see also the monograph [44]). His comparisons with respect to speed and accuracy focused on the extrapolation code LIMEX and the multistep code DASSL by L. R. Petzold [170], where the one-step method performed better by far. The reason for this performance difference is that, after a change of grid, the multistep methods must restart and build up from order $k = 1$ to the order k of the previous step (see Volume 2, Section 7.4, under the key word "startup calculation"); as an alternative, a so-called "warm restart" was suggested (see, e.g., [145]), in which, however, a sufficient monitoring of the interpolation error is lacking. As another variant, a restart by means of some one-step method of higher order has been discussed, which, however, in case of frequently changing grids leads to the nearly exclusive use of this very one-step method. For problems with high spatial-temporal dynamics, this structural property of adaptive multistep methods (like DASSL) leads to a drastic slow-down of the computations; this holds in particular for space dimension $d > 1$.

9.2.2   Adaptive Method of Time Layers

This method is also called *adaptive Rothe method*, since around 1930 E. Rothe [176] applied such a method – for uniform space and time grids – as a technique of proving existence of the solution of the diffusion equation. Around 1989 F. A. Bornemann [31] recognized that it constitutes the ideal algorithmic frame to realize *full adaptivity in space and time* for parabolic PDEs.

Preliminary Considerations.   Let us start from the situation as represented in Figure 9.6, i.e., from the abstract Cauchy problem

$$u' = F(u), \quad u(0) = u_0 \in H^1(\Omega), \quad F : H^1(\Omega) \to H^1(\Omega)^*. \tag{9.34}$$

Such a problem is regarded as parabolic whenever its linearization, i.e., the linear (in general nonautonomous) variational equation, is parabolic. If we discretize this problem first with respect to time, we obtain a sequence of linear boundary value problems with solutions $\{u_\tau(t), t = t_0, t_1, \ldots\} \subset H^1(\Omega)$.

As worked out in Section 9.1.2, in the finite dimensional case of PDEs the situation is as follows. Whenever the exact Jacobian matrix $J = F'(u_0)$ can be evaluated, then ROW-methods may be constructed, otherwise W-methods or linearly implicit extrapolation methods. In the infinite dimensional case of interest here, however, an exact functional derivative $F'(u_0)$ is not available, but at best, in the context of FE-methods, a projection on the FE-subspace S_h or an approximation of it. Thus, if we interpret the time integration within the method of time layers as in Section 9.1.3 as *function space method*, we need to consider either W-methods or ROW-methods, where in the latter the quantities to be computed can be approximated *to sufficient accuracy* such that the order of the time integration method is essentially preserved. From (9.26) we know that in W-methods we can expect an effective order of maximally $p^* = 2$. This is why, up to now, in the developments of the method of time layers ROW-methods dominate, where adaptive mesh refinement realizes the sufficiently accurate approximation of $F'(u_0)$. Upon satisfying the additional conditions (9.24) the order reduction elaborated in Section 9.1.3 can be avoided so that time integrators of high effective order can be implemented.

Timestep Control.   From (9.15) one obtains an optimal timestep via the formula

$$\tau^* = \sqrt[p+1]{\frac{\rho \, \text{TOL}_\tau}{[\hat{\epsilon}_\tau]}} \, \tau \tag{9.35}$$

with $\rho < 1$ and

$$[\hat{\epsilon}_\tau] = \|u_\tau(\tau) - \hat{u}_\tau(\tau)\| \doteq C_p \, \tau^{p+1},$$

where in our context now $u_\tau, \hat{u}_\tau \in H_0^1(\Omega)$. For the time being, the norm $\| \cdot \|$ may remain unspecified. The associated accuracy monitor would then read

$$[\hat{\epsilon}_\tau] \le \text{TOL}_\tau. \tag{9.36}$$

Obviously, the formula as such cannot be directly evaluated. At best, the norm of the difference $[\hat{\epsilon}_\tau]$ can be approximated by

$$[\hat{\epsilon}_\tau]_h = \|u_{h,\tau}(\tau) - \hat{u}_{h,\tau}(\tau)\|, \quad u_{h,\tau} \in S_h \subset H_0^1(\Omega), \tag{9.37}$$

i.e., by the difference of discretizations $u_{h,\tau}$ and $\hat{u}_{h,\tau}$. In order to realize adaptivity of
the algorithm w.r.t. time in a reliable way, one must assure that

$$[\hat{\epsilon}_\tau] \approx [\hat{\epsilon}_\tau]_h \le \text{TOL}_\tau.$$

This can be done in connection with adaptive multigrid methods, as will be demonstrated below in Section 6.1.

Already at this stage we want to indicate that in the explicit evaluation of $[\hat{\epsilon}_\tau]_h$ special attention must be paid to the possible occurrence of *numerical extinction of leading digits*. For example, if one computes u_τ and $\hat{u}_\tau$ independently from each other on
different meshes, then the evaluation of the above difference indeed induces a massive
blow-up of the various discretization errors. In order to cope with this problem, two
different strategies have been developed, which we are going to present next.

Linear problems

F. Bornemann worked out the Rothe method for linear nonautonomous scalar parabolic PDEs, first in one space dimension [31, 32], later in two space dimensions [33].
That is why we consider this special case first and start from the abstract linear Cauchy
problem

$$u' = Au + f(t), \quad u(0) = u_0 \tag{9.38}$$

with $u \in H_0^1(\Omega)$ and $-A(\cdot)$ an elliptic operator. As time discretization he began with
the extrapolated linearly implicit Euler method (code LIMEX; see Section 9.1.2), but
did not couple the grids for the discrete solutions associated with neighboring orders
w.r.t. τ and thus had to compensate large error amplification factors. Nevertheless,
already in this first realization, the principal advantage of an adaptive method of time
layers could be seen: in contrast to the method of lines, the space discretization is
not frozen over all time layers, so that a crossing of traveling fronts does not exhibit
any problem. For illustration, see a test problem suggested by M. Bietermann and
I. Babuška [28].

Example 9.5. In this example the solution u consists of two components

$$u = u^{(1)} + u^{(2)},$$

where

$$u^{(1)}(x,t) = 0.25\Big(1 + \tanh\big(100(x - 10t)\big)\Big),$$
$$u^{(2)}(x,t) = 0.25\Big(1 + \tanh\big(80(1 - x - 30t)\big)\Big).$$

Obviously, the two solution parts model two "countertraveling" waves with different
velocities (10 and 30). In the PDE

$$u_t = u_{xx} + f(t), \quad t > 0, \ x \in [0, 1]$$

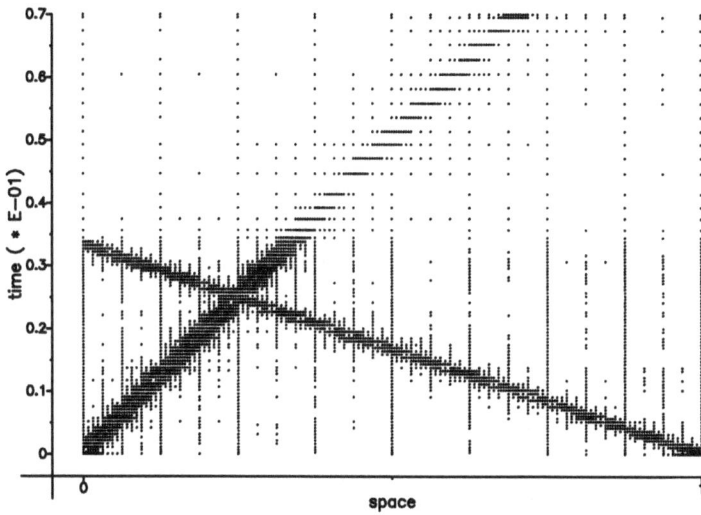

Figure 9.13. Adaptive method of time layers [31]: nodal flux for two crossing traveling fronts. Cf. also Figure 9.9.

Dirichlet boundary conditions and the function $f(t)$ are chosen such that the above superposition ends up as the unique solution. In Figure 9.13 we show the nodal flux obtained when applying the adaptive method of time layers based on LIMEX (see Section 9.1.2); in each time layer an adaptive multigrid method has been applied.

In order to circumvent the above problem of unwanted amplification of space discretization errors in $[\epsilon_\tau]_h$ (see (9.37)), F. A. Bornemann [32, 33] developed two specific classes of time discretizations, first only for *autonomous* source terms f: Let, for increasing order p, solutions $u_\tau^p \in H^1(\Omega)$ be defined according to

$$u_\tau^{p+1} = u_\tau^p + \Delta u_\tau^p, \quad p = 0, 1, \ldots.$$

The special idea here is to compute the differences $\Delta u_\tau^0, \Delta u_\tau^1, \ldots$ *multiplicatively* from one another.

In a linear stability model for stiff ODEs (see above Section 9.1.1) this idea corresponds to the rational functions $r_p(z)$, recursively computed according to

$$r_{p+1}(z) = r_p(z) + \rho_p(z), \quad \rho_{p+1}(z) = \gamma_p \frac{z}{1-z} \rho_p(z), \quad p = 0, 1, \ldots,$$

where $z = \lambda \tau \in \mathbb{C}$ belongs to eigenvalues $\lambda(A)$. As corresponding starting values we have $r_0(z) = \rho_0(z) = 1$. In our context we are interested in the special case $\Re(z) < 0$, which is why the methods are started with an implicit Euler step

$$r_1(z) = \frac{1}{1-z}.$$

A more accurate analysis shows that for the choice of ρ_1 only two reasonable possibilities exist,

$$\rho_1^A(z) = -\frac{1}{2}\frac{z^2}{(1-z)^2} \quad \text{and} \quad \rho_1^L(z) = -\frac{1}{2}\frac{z^2}{(1-z)^3},$$

which lead to $A(\alpha)$- and $L(\alpha)$-stable methods. Apart from automatically avoiding the error amplification when computing the difference (9.37) the small stage number $s = p$ is attractive.

A transfer to the linear *nonautonomous* case, i.e., with $f(t)$ in (9.38), is worked out in [32], there, however, only for the first two steps of the A-stable discretization, which means up to $s = p = 2$. (Note that the nonautonomous case is included in the general nonlinear case if the usual autonomization trick is applied; see Volume 2).

Remark 9.6. For the numerical solution of countable differential equations (roughly: discrete PDEs) the method of time layers is also crucial in the context of *discrete* Galerkin methods. For this problem class, M. Wulkow [220] suggested a transfer of the above mentioned method from the nonautonomous case to the general *nonlinear* case. This method has proved to be successful in practice (see, e.g., [221]); however, for the general case there is still no consistency theory yet, which is why we do not pursue it any further here.

Nonlinear problems

In 2001, J. Lang [139] managed to extend the adaptive method of times layers efficiently to nonlinear parabolic PDEs (9.34). He employed ROW-methods as linearly implicit time discretizations, avoiding order reduction (see Section 9.1.3) by satisfying the additional conditions (9.24).

Due to (9.12) ROW-methods require the solution of a formal linear system in each time step, here written as operator equation

$$i = 1, \ldots, s:$$
$$(I - \gamma\tau A)k_i = \tau \sum_{j=1}^{i-1}(\gamma_{ij} - \alpha_{ij})Ak_j + F\left(u_0 + \tau \sum_{j=1}^{i-1}\alpha_{ij}k_j\right). \tag{9.39}$$

There the eigenvalues λ_i of the operator $-A = -F'(u_0)$ are restricted according to $\Re(\lambda_i) \leq C$ for some $C > 0$; due to $\gamma\tau > 0$ the operator $I - \gamma\tau A$ is then elliptic for sufficiently small timesteps $\tau < 1/\gamma C$. Because of the special structure of ROW-methods the above system (9.39) is recursive, i.e., the $k_1, \ldots, k_s \in H_0^1(\Omega)$ can be computed one after the other by solving *linear elliptic boundary value problems*. This is why for nonlinear parabolic PDEs predominantly linearly implicit time integrators are applied.

In an *embedded* ROW-method (see Section 9.1.2) two discrete solutions for neighboring orders can be computed from these intermediate quantities as follows:

$$u_\tau(\tau) = u_0 + \tau \sum_{i=1}^{s} b_i k_i \in H_0^1(\Omega), \quad \text{order } p+1,$$

$$\hat{u}_\tau(\tau) = u_0 + \tau \sum_{i=1}^{s} \hat{b}_i k_i \in H_0^1(\Omega), \quad \text{order } p.$$

As error estimator w.r.t. time one obtains, purely formally,

$$[\hat{\epsilon}_\tau] = \|u_\tau(\tau) - \hat{u}_\tau(\tau)\| = \tau \left\| \sum_{i=1}^{s} (b_i - \hat{b}_i) k_i \right\| \doteq C_p \, \tau^{p+1}.$$

None of these formulas can be directly evaluated numerically. Instead, sufficiently accurate spatial approximations must be computed, e.g., by adaptive multilevel methods.

For this reason, we turn to the weak formulation in a finite element space $S_h \subset H_0^1(\Omega)$. We obtain a recursive system for the approximations $k_{h,1}, \ldots, k_{h,s} \in S_h$ of the form

$$i = 1, \ldots, s:$$
$$\langle (I - \gamma\tau A) k_{h,i}, v \rangle \tag{9.40}$$

$$= \tau \sum_{j=1}^{i-1} (\gamma_{ij} - \alpha_{ij}) \langle A k_{h,j}, v \rangle + \left\langle F\left(u_0 + \tau \sum_{j=1}^{i-1} \alpha_{ij} k_{h,j} \right), v \right\rangle, \quad v \in S_h.$$

In this FE-approximation we now have

$$u_{h,\tau}(\tau) = u_0 + \tau \sum_{i=1}^{s} b_i k_{h,i} \in S_h,$$

$$\hat{u}_{h,\tau}(\tau) = u_0 + \tau \sum_{i=1}^{s} \hat{b}_i k_{h,i} \in S_h.$$

The space discretization error can be approximately determined by a hierarchical extension $S_h^+ = S_h \oplus S_h^\oplus$, $S_h^\oplus = \text{span}(\varphi_l)_{l=1,\ldots,n}$ defining

$$k_{h,i}^\oplus = \sum_l \eta_{h,i,l} \varphi_l \in S_h^\oplus$$

via a hierarchical error estimator (e.g., the DLY-estimator; see Section 6.1.4):

$$\langle (I - \gamma\tau A) \varphi_l, \varphi_l \rangle \eta_{h,i,l} = \tau \sum_{j=1}^{i-1} (\gamma_{ij} - \alpha_{ij}) \langle A k_{h,j}^+, \varphi_l \rangle \tag{9.41}$$

$$+ \left\langle F\left(u_0 + \tau \sum_{j=1}^{i-1} \alpha_{ij} k_{h,j}^+ \right), \varphi_l \right\rangle - \langle (I - \gamma\tau A) k_{h,i}, \varphi_l \rangle.$$

With this, the terms $k_{h,i}^+ = k_{h,i} + k_{h,i}^\oplus$ and

$$u_{h,\tau}^+(\tau) = u_0 + \tau \sum_{i=1}^{s} b_i k_{h,i}^+ \in S_h^+,$$

$$\hat{u}_{h,\tau}^+(\tau) = u_0 + \tau \sum_{i=1}^{s} \hat{b}_i k_{h,i}^+ \in S_h^+$$

may be defined. For ease of reading, all four notations are arranged in Table 9.2.

Table 9.2. Four different approximations of the solution $u(\tau) \in H_0^1(\Omega)$.

h ＼ τ	order p	order $p+1$
S_h	$\hat{u}_{h,\tau}$	$u_{h,\tau}$
S_h^+	$\hat{u}_{h,\tau}^+$	$u_{h,\tau}^+$

For the acceptance test (9.36) we replace the unavailable estimate $[\hat{\epsilon}_\tau]$ by the most accurate available spatial estimate

$$[\hat{\epsilon}_\tau]_h^+ = \|u_{h,\tau}^+ - \hat{u}_{h,\tau}^+\| = \tau \left\| \sum_{i=1}^{s} (b_i - \hat{b}_i) k_{h,i}^+ \right\| \le \mathrm{TOL}_\tau, \qquad (9.42)$$

which we also insert into the timestep control (9.35).

In addition to the pure time discretization error $\hat{\epsilon}_\tau$ we obtain the space discretization error

$$\hat{\epsilon}_h = \|\hat{u}_\tau - \hat{u}_{h,\tau}\| = \tau \left\| \sum_{i=1}^{s} \hat{b}_i (k_i - k_{h,i}) \right\|$$

and bound the suitable estimator by virtue of

$$[\hat{\epsilon}_h] = \|\hat{u}_{h,\tau}^+ - \hat{u}_{h,\tau}\| = \tau \left\| \sum_{i=1}^{s} \hat{b}_i k_{h,i}^\oplus \right\| \le \mathrm{TOL}_h. \qquad (9.43)$$

For the estimation of the total error we replace $u(\tau) \in H_0^1(\Omega)$ by its best available approximation $u_{h,\tau}^+ \in S_h^+$ according to

$$\hat{\epsilon}_{h,\tau} = \|u(\tau) - \hat{u}_{h,\tau}\| \approx \|u_{h,\tau}^+ - \hat{u}_{h,\tau}\| =: [\hat{\epsilon}_{h,\tau}]$$

and thus obtain the desired bound

$$[\hat{\epsilon}_{h,\tau}] = \|u_{h,\tau}^+ - \hat{u}_{h,\tau}\|$$
$$\le \|u_{h,\tau}^+ - \hat{u}_{h,\tau}^+\| + \|\hat{u}_{h,\tau}^+ - \hat{u}_{h,\tau}\| = [\hat{\epsilon}_\tau]_h^+ + [\hat{\epsilon}_h] \le \mathrm{TOL}_\tau + \mathrm{TOL}_h \le \mathrm{TOL}$$

with a user prescribed tolerance TOL.

Computational Complexity Model. Following [33] we split the required accuracy into its temporal and spatial parts

$$\text{TOL}_\tau = \sigma\text{TOL}, \quad \text{TOL}_h = (1 - \sigma)\text{TOL}$$

with a parameter $0 < \sigma < 1$ to be determined. The choice of σ does not influence the size of the total error, but it does very much so the computational complexity. In particular, a reduction of the spatial error mostly goes with a significantly larger computational amount than the corresponding reduction of the temporal error.

For a more detailed analysis we use the following simple model. We start from the situation that the local timesteps have been selected by the timestep control. In Volume 1, Section 9.5.3, we derived the local timestep control (in the simple example of numerical quadrature) on the basis of a model in which the *local amount per unit timestep* has been minimized and thus the error globally equilibrated. This model is applicable to every evolution problem, i.e., also in the present case. When using an optimal multigrid solver, the local computational amount W per timestep τ depends linearly on the number $N_h = \dim S_h$ of degrees of freedom:

$$W \sim N_h/\tau.$$

Note that also in the *time adaptive* case, this result is equivalent to the model

$$W \sim N_h N_\tau \quad \text{with} \quad N_\tau \sim 1/\tau.$$

For *space adaptive* meshes we have, asymptotically just as in the quasi-uniform special case,

$$\text{TOL}_h \approx \hat{\epsilon}_h \sim N_h^{-q/d} \|\hat{u}_\tau - \hat{u}_{h,\tau}\|_A,$$

where the convergence order q of the finite elements depends both on the ansatz order of the space S_h and on the chosen norm, i.e., in linear finite elements $q = 1$ for the energy norm (Theorem 4.19), $q = 2$ for the L^2-norm (Theorem 4.21). Because of $\hat{u}_0 - \hat{u}_{h,0} = 0$ we have $\|\hat{u}_\tau - \hat{u}_{h,\tau}\|_A \sim \tau$, from which we conclude

$$(1 - \sigma)\,\text{TOL} = \text{TOL}_h \sim N_h^{-q/d}\tau \quad \Rightarrow \quad N_h \sim \left(\frac{\tau}{(1 - \sigma)\,\text{TOL}}\right)^{d/q}.$$

A more detailed consideration leads to the insight that TOL should not be regarded as independent of the timestep τ. In fact, when integrating over a fixed time interval of length T (not discussed further here) the local errors of an order of magnitude of TOL will give rise to a final accuracy $\epsilon(T)$ as discussed extensively in Volume 2, Section 5.5. In parabolic problems, which are known to be dissipative, the local errors will be damped in the stationary phase, so that then $\epsilon(T) \approx \text{TOL}$ arises. In the transient phase, however, the timestep control will, for *fixed final accuracy*, establish the relation

$$\epsilon(T) \approx N_\tau\,\text{TOL} \quad \Rightarrow \quad \text{TOL} \sim \tau.$$

For the local timesteps one gets

$$\tau^{p+1} \sim \text{TOL}_\tau = \sigma\,\text{TOL} \sim \sigma\tau \quad \Rightarrow \quad \tau \sim \sigma^{1/p}.$$

This finally supplies the functional dependence

$$W \sim (1-\sigma)^{-d/q}\sigma^{-1/p} =: \varphi(\sigma).$$

As $\varphi(0)$ and $\varphi(1)$ are unbounded positive, the condition $\varphi'(\sigma) = 0$ yields the minimum in

$$\sigma_{\min} = \frac{q}{q+dp}. \tag{9.44}$$

As expected, the result (9.44) produces a smaller σ for larger space dimension, higher order in time and lower order in space, and thus permits a relatively larger spatial part in the total error.

Let us exemplify the case for linear finite elements, i.e., $q = 2$ in the L^2-norm or $q = 1$ in the energy norm, respectively. For the implicit Euler discretization ($p = 1$) one then gets $\sigma_{\min} = 1/(1 + d/2)$ (nearly the result in [33]) or $\sigma_{\min} = 1/(1 + d)$, respectively. For ROW-methods of order $p = 4$ and in space dimension $d = 3$ the formula (9.44) supplies the value $\sigma_{\min} = 1/13$ and thus $\text{TOL}_\tau \approx \text{TOL}_h/12$. This low value expresses the fact that a reduction of the spatial error causes a significantly larger amount of work than a reduction of the temporal error. Facing this accuracy discrepancy, it may be asked whether the hierarchical spatial error estimator used in (9.42) will at all permit a reliable determination of the temporal error – after all, it still depends, by a not too large factor, on the parameter β of the saturation property (see Definition 6.7). Therefore, in practice, the factor σ should be chosen sufficiently large, e.g., by the restriction $\sigma \geq 1/4$. The considerations so far lead us to the following algorithm.

Algorithm 9.7. *Adaptive method of time layers.*
$(u_{h,\tau}^+, \tau, \tau^*) := \text{AMOT}(u, \tau^*, \text{TOL}, \sigma)$

    do
        $\tau := \tau^*$
        for $i := 1$ **to** s **do**
            determine $k_{h,i}$ and $k_{h,i}^\oplus$ according to (9.40) and (9.41)
        end for
        $[\hat{e}_h] := \tau \sum_{i=1}^s \hat{b}_i k_{h,i}^\oplus$
        if $\|[\hat{e}_h]\| \geq \text{TOL}_h$ **then**
            refine the mesh, where $\|[\hat{e}_h]\|$ is large
        end if
        compute τ^* according to (9.35)
    until $[\hat{e}_\tau]_h^+ \leq \text{TOL}_\tau$ and $[\hat{e}_h] \leq \text{TOL}_h$

Note that *all* stage vectors $k_{h,i}$ and $k_{h,i}^{\oplus}$ need to be recomputed on a finer mesh if (9.43) is violated. Fortunately, this costly case occurred rather rarely in our numerical tests.

Mesh Coarsening. Up to now we have discussed only mesh *refinement*. However, especially for nonlinear time-dependent equations a shift of the spatial solution structure in time is typical. Algorithm 9.7 merely realizes mesh refinement in case of insufficient accuracy so that a large part of the domain will be gradually filled with small elements that were only necessary at some earlier time layer. This phenomenon does not affect the accuracy of the solution, but but it doese affect the computational efficiency. For this reason, a *coarsening* of meshes, where small elements are no longer needed, will be an essential piece of an efficient algorithm. In the context of nonlinear parabolic PDEs two types of coarsening methods have evolved.

A priori coarsening. Here the mesh is coarsened before a timestep is computed, either globally by one mesh level as a whole or only locally where the error estimator in the previous timestep supplied particularly small error indicators $[\epsilon_T]$. The adaptive mesh refinement in the next timestep then generates a mesh corresponding to the required accuracy. In order to avoid losing information and thus accuracy by the coarsening, the initial value u of the step on the finer mesh is kept stored, which, however, complicates the implementation significantly.

A posteriori coarsening. Here the tolerance is slightly reduced and the thusly gained scope afterwards used to coarsen the solution. In this coarsening the complete information is available, so that the thusly generated error can be exactly determined in advance and be used to control the coarsening process. In this variant the arising systems to be solved become slightly larger.

9.2.3 Goal-oriented Error Estimation

As in Section 6.1.5, for parabolic problems, goal-oriented error estimators for quantities of interest $J(u) = \langle j, u \rangle$ can also be expressed via weight functions z that are computed as solutions of the adjoint problem. In contrast to elliptic equations, the information flow here is directed in time, which leads to an interesting structure of the goal-oriented error estimation.

For ease of a simpler derivation of the weight function, we consider the model problem

$$u' = \Delta u + f(u) + r \quad \text{on } \Omega \times [0, T], \qquad u|_{t=0} = u_0, \quad u|_{\partial\Omega} = 0 \qquad (9.45)$$

in abstract notation $c(u, r) = 0$, where r is again the residual. To each r there belongs a solution $u(r)$ with $c(u(r), r) \equiv 0$. Taking the derivative w.r.t. r leads to

$$c_u\big(u(r), r\big)u_r(r) + I = 0, \quad \text{i.e.,} \quad u_r = -c_u^{-1}.$$

Let $u = u(0)$ denote the exact solution and $u_h = u(r)$ an approximation. Then, in first order approximation,

$$\epsilon_h = \langle j, u_h - u \rangle \approx \langle j, u_r(r) \, r \rangle = \langle -c_u(u_h)^{-*} j, r \rangle.$$

Thus the weight function $z = -c_u(u_h)^{-*} j$ is again a solution of the adjoint problem

$$c_u(u_h)^* z = -j, \tag{9.46}$$

which can be written as a parabolic equation backward in time (see Exercise 9.5):

$$-z' = \Delta z + f'(u_h) z - j \quad \text{on } \Omega \times [0, T], \qquad z|_{t=T} = 0, \quad z|_{\partial \Omega} = 0. \tag{9.47}$$

In accordance with the sign of the time derivative, a final value is prescribed here (see Section 1.2). The goal-oriented error estimator is then defined in analogy to the elliptic case as

$$[\epsilon_h] = \langle z, r \rangle, \quad [\epsilon_T] = \langle z|_T, r \rangle. \tag{9.48}$$

For point evaluations $J(u) = u(\hat{x}, \hat{t})$ and a simple diffusion with $f'(u_h) = 0$ the weight function is just the fundamental solution (A.18), but in backward temporal direction. In particular we have $z(x, t) \equiv 0$ for $t > \hat{t}$ – in agreement with the causality interpretation that errors arising after $\hat{t}$ cannot have an influence on the value $u(\hat{x}, \hat{t})$.

Other than in the case of pure diffusion, in *nonlinear* parabolic equations the weight function z corresponding to a point evaluation J need not decay rapidly, as shown by the following example.

Example 9.8. Consider the equation

$$u_t = 10^{-3} u_{xx} + af(u) \quad \text{on } \Omega \times [0, T] = \,]0, 1[\times [0, 10] \tag{9.49}$$

with $f(u) = u(u-0.1)(1-u)$ and initial value $u_0 = \max(0, 1-2x)$ to homogeneous Neumann boundary conditions. For $a = 0$ we obtain the linear diffusion equation familiar from Section 1.2. For $a = 10$, however, equation (9.49) is a simple model for excitable media. The reaction term $f(u)$ has two stable fixed points, 0 (the "state in rest") and 1 (the "excited state"), as well as an unstable fixed point 0.1. By diffusion, domains in rest are lifted by neighboring excited domains over the unstable fixed point and, due to the reaction, aspire to the excited state. In this way the excitation spreads through the domain in the form of a traveling wave – from left to right for this starting value. In an unbounded domain, solutions of the form $u(x, t) = w(x - vt)$ would come up, which will occur as well on bounded domains, as long as boundary effects do not play a dominant role. The *shape* of the solution is stable in the sense that small perturbations, in particular those of high frequencies, are rapidly damped. Not stable, however, is the *position* of the solution. Position errors remain permanent, since both $u(x, t) = w(x - vt)$ and the shifted variant $u(x, t) = w(x - vt + \epsilon)$ are solutions of (9.49). Whenever perturbations have an influence on the position of the solution, then this is not damped. This is clearly recognized at the weight function z in Figure 9.14, left, where residuals along the whole wave front in the space-time domain have an essentially constant influence on the point evaluation $u(0.9, T)$. The reason for this behavior is that for $0.05 \leq u \leq 0.68$ the reaction derivative $f'(u)$ is positive, which implies that the differential operator $10^{-3}\Delta + f'(u)$ in the adjoint equation (9.47) has positive eigenvalues.

In contrast to this behavior, the influence of perturbations in the case of pure diffusion (Figure 9.14, center) is restricted to spatially and temporally nearby points. This holds all the more for nonlinear problems if the solution to be evaluated is in a stable fixed point of the reaction (Figure 9.14, right), where perturbations are additionally damped.

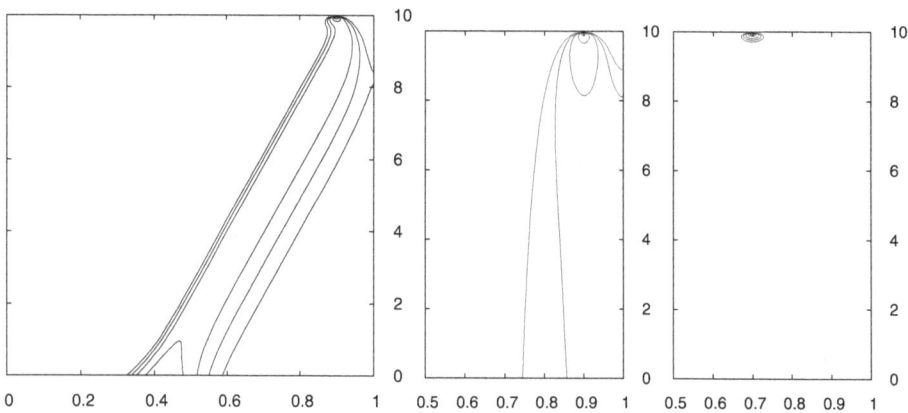

Figure 9.14. Isolines of the weight function $z(x, t)$ for the equation (9.49) for point evaluation. The heights of the isolines are 50, 100, 200, 500, 1000. *Left:* $a = 10$, $J = u(0.9, T)$. *Center:* $a = 0$, $J = u(0.9, T)$. *Right:* $a = 10$, $J = u(0.8, T)$.

As illustrated by this example, nonlinear parabolic problems often exhibit a significant global error transport. Unlike elliptic problems with strongly localized effects, here mesh refinement, due to a local residual or the locally estimated error, generally does not lead to meshes optimally adapted to the problem. Therefore, if we want to apply (9.48) for mesh refinement, we must compute the weight function z from (9.47), where the solution u_h in (9.47) and the corresponding residual r in (9.48) enter. This implies that u_h must be simultaneously available on the whole space-time domain $\Omega \times [0, T]$, which for $d = 3$ may involve an enormous amount of storage. In order to reduce this amount, several approaches like checkpointing [107] or state trajectory compression [214] have been developed.

9.3   Electrical Excitation of the Heart Muscle

In the last part of this chapter we present a parabolic problem from a medical application, the excitation of the excitation of the heart muscle. As will turn out, its complexity by far exceeds that of the simple model problems for illustration given so far. However, it nicely indicates how the adaptive algorithmic pieces suggested in this book affect the efficiency of numerical computation.

9.3.1   Mathematical Models

In order to pump enough blood through the body, a uniform coherent contraction of the heart muscle is required, which is imposed by an interaction of elastomechanical and electrical processes. Here we skip the elastomechanical part and only look at the electrocardiology.

Physiological model. While normal skeleton muscles are excited individually, the muscle cells of the heart are excited collectively. The electrical excitation runs through the muscle tissue in the form of a depolarization wave. In the course of depolarization the transmembrane voltage between the interior of the cell and its exterior changes from a rest value −85 mV to an excitation potential of 20 mV. This is followed by a discharge of Ca^{2+}, which causes a contraction of the muscle cells. In the healthy heartbeat this wave is triggered by some regularly repeated electrical impulse from the Sine node (see Figure 9.15, left). In some kinds of arrhythmia, out-of-phase impulses arise which totally change the spatio-temporal pattern (see Figure 9.15, right). The propagation of the depolarization front is based on an interplay of passive ion diffusion on the one hand and active ion exchange between cell interior and exterior on the other hand. Ion diffusion takes place both outside the cells and inside, as well as in between. The ion transport between the cells, i.e., through the cell membrane, runs through the ion channels. These may, depending on the transmembrane voltage and the interior states, be open or closed. By an action of the chemical substance adenosintriphosphat (ATP), ions can be transported even against the potential difference, and different kinds of ions can be exchanged.

To avoid a description of individual cells, *continuous* mathematical models are established. The essential quantities there are:

- the electric potentials ϕ_i in the cell interior and ϕ_e in the cell exterior;
- the transmembrane voltage $u = \phi_i - \phi_e$;
- the interior states w of the ion channels.

Membrane Models.   The ion current I_{ion} through the channels as well as the temporal evolution R of the internal states is described by one of the numerous membrane models (see, e.g., [161]). Physiologically-oriented membrane models contain descriptions of the dynamics of real ion channels as well as the concentrations of the various ions (Na^+, K^+, Ca^{2+}). The mathematical formulation of these connections leads to a system of PDEs plus between one and 100 ODEs, which act at each point of the volume.

Bidomain Model. This model is the most elaborate one among the established electrocardiological models. It consists of two PDEs and a set of ODEs:

$$\chi C_m u' = \operatorname{div}(\sigma_i \nabla(u + \phi_e)) - \chi I_{\text{ion}}(u, w), \tag{9.50}$$
$$0 = \operatorname{div}(\sigma_i \nabla u) + \operatorname{div}((\sigma_i + \sigma_e)\nabla\phi_e) + I_e, \tag{9.51}$$
$$w' = R(u, w).$$

The physical quantities used above have the following meaning:

- χ: membrane surface per volume;
- C_m: electrical capacity per membrane surface;
- $\sigma_i, \sigma_e \in \mathbb{R}^{d \times d}$: conductivities of the intra- and the extracellular medium, anisotropic because of the fiber structure of the heart muscle (and therefore tensors);
- I_e: external excitation current.

The elliptic equation (9.51) obviously has the function of an algebraic equality constraint, so that a differential-algebraic system of index 1 arises (cf. Volume 2, Section 2.6). In most cases one applies isolating boundary conditions

$$n^T \sigma_i \nabla u = -n^T \sigma_i \phi_e,$$
$$n^T \sigma_e \nabla \phi_e = 0.$$

Because of the high computational complexity of the bidomain model, simpler models are often applied (the limitations of which one should keep in mind).

Monodomain Model. If the diffusion tensors σ_i and σ_e are proportional to each other, then the second equation (9.51) may be inserted into (9.50), which produces the differential equation system:

$$\chi C_m u' = \operatorname{div}(\sigma_m \nabla u) - \chi I_{\text{ion}}(u, w),$$
$$w' = R(u, w).$$

Even though the assumption underlying this simplification does not hold in general, this model is taken as a good approximation for a number of questions; moreover, it is also popular due to its lower computational complexity.

9.3.2 Numerical Simulation

Changes in the normal excitation propagation due to genetic defects or local tissue damages as a consequence of infarcts, electric shocks, or drug abuse may perturb the coherent contraction pattern and lead to severe diseases like cardiac insufficiency or tachycardia, as well as to fatal fibrillation. In the future, reliable numerical simulations of the excitation propagation will play an increasing role for the fundamental understanding of the functioning and the prediction of therapeutic effects.

The numerical solution of the above mathematical model equations is a real challenge (see also the monograph [196]). Main reasons for that are:

- *Strong spreading of spatial scales.* While a human heart has a diameter of about 10 cm, the width of a depolarization front lies around 1 mm. In order to obtain a roughly acceptable accuracy of the representation, the spatial resolution at the front position must be less than 0.5 mm.

- *Strong spreading of time scales.* A healthy heart beat takes about 1 s, whereas the transmembrane voltage changes from the rest value to the excitation potential within the order of 2 ms. For a roughly acceptable accuracy an equidistant timestep of 1 ms should not be exceeded.

- *Large systems.* Recent membrane models take into account various ions and their concentrations in several organelles within the cells as well as a large selection of ion channels. Such models may include up to 100 variables (in w) with rather different dynamics.

- *Complex geometry and anisotropy.* The lengthy form of the muscle cells and their arrangement in layers lead to a marked spatially dependent and rather diverse anisotropy of the conductivities σ_i and σ_e. In addition, the geometry of the heart is rather complex: it is pervaded by blood vessels that do not propagate the excitation impulse; vice versa, the heart ventricles are pervaded by fiber-like structures that well conduct the depolarization front. Special conduction fibers (His-bundles and the Purkinje system) exhibit a much faster propagation of stimuli.

The numerical simulation of this obviously difficult problem has been approached in two stages. Since hitherto only uniform spatial meshes had been used, the effect of spatial and temporal adaptivity in the solution of the problem was first tested on a quadrilateral piece of tissue (see [53]). This gave rise to 200 adaptive timesteps vs. 10 000 equidistant timesteps that would have been necessary with constant timestep τ. In the space meshes, a factor of approximately 200 could be saved. These results were seen as an encouragement for attacking the problem with realistic geometry (see [69]): the 3D coarse mesh of the heart muscle was taken from [142] consisting of 11 306 nodes for 56 581 tetrahedra. The simulation was done by the adaptive finite element code KARDOS due to J. Lang [139], which realizes an adaptive method of time layers. For time integration, the recently developed linearly implicit integrator ROS3PL [140] was used (see Section 9.1.2). For space discretization, linear finite elements were selected. The adaptive meshes were constructed by means of the hierarchical DLY-error estimator (see Section 6.1.4). The numerical solution of the arising huge systems of linear equations was performed by the iterative solver BI-CGSTAB from [207] with

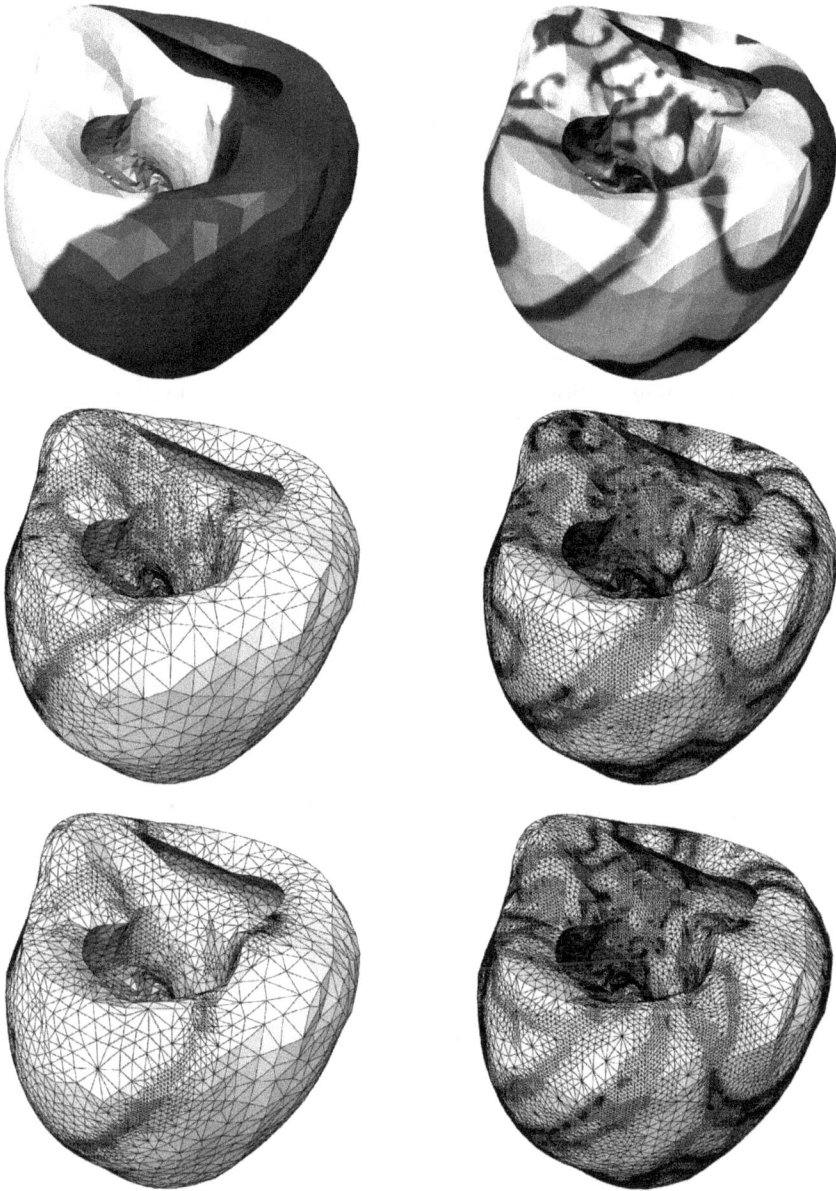

Figure 9.15. Adaptive method of time layers: Traveling depolarization fronts in the process of the electrical excitation of the heart. *Left*: normal heart beat: *top*: visualization of the solution at timepoint $t = 40$ ms, *center*: associated spatial mesh with front, *bottom*: spatial mesh at $t = 55$ ms. *Right*: Fibrillation: *top*: visualization of the solution at $t = 710$ ms, *center*: associated spatial mesh with multiple fronts, *bottom*: spatial mesh at $t = 725$ ms.

ILU-preconditioning[3] (see Remark 5.11). This preconditioning is an unsymmetric extension of the incomplete Cholesky decomposition presented in Section 5.3.2; the efficiency of ILU-preconditioners, however, has not yet been proven.

In Figure 9.15, typical transmembrane voltages from the monodomain equations with an Aliev–Panfilov membrane model are shown for one healthy heart beat (left) and for the case of cardiac fibrillation (right). The simple Aliev–Panfilov model was particularly developed for a mathematical description of cardiac arrhythmia. One may clearly observe the local refinement of the meshes along the fronts. When the fronts move through the domain, the position of the mesh refinement also moves. In the case of cardiac fibrillation, with mesh refinements at several parts of the domain, up to 2.1 million nodes were necessary at intermediate timepoints. An equidistant mesh with the same local resolution would have required around 310 million nodes and thus 150 times more degrees of freedom for spatial discretization.

Despite the significant reduction of the number of degrees of freedom by adaptive methods the simulation of the electrical excitation of the heart remains a real challenge. This comes not the least from the fact that the adaptive methods require an additional amount of computing compared to methods with fixed meshes: the computation of the error estimators, the mesh manipulation as such, and in particular the assembly of the mass and stiffness matrices, which need to be repeated after each mesh adaptation.

9.4   Exercises

Exercise 9.1. The concept of *A-stability* will be examined for the case of nondiagonalizable matrices. For this purpose we consider the linear ODE $y' = Ay$ and a corresponding one-step method $y_{k+1} = R(\tau A)y_k$ with stability function $R(z), z \in \mathbb{C}$. Let

$$|R(z)| \leq 1 \quad \text{for} \quad \Re(z) \leq 0.$$

Show that if the ODE is asymptotically stable, then the sequence $\{y_k\}$ is a null sequence.

    Hint: Transform A to Jordan canonical form; for the individual Jordan blocks, construct a similarity transformation depending on some sufficiently small parameter ε such that the off-diagonal elements become arbitrarily small.

Exercise 9.2. Let $\mathcal{E}$ denote a first integral of the ODE $u' = f(u)$, i.e., let

$$\mathcal{E}\big(u(t)\big) = \mathcal{E}\big(u(0)\big).$$

With the notations $u_0 = u(0)$, $u_k = u_\tau(k\tau)$, $k = 0, 1, \ldots$ consider the following one-step methods:

---

[3] This may not be the last word on the subject, from the point of view of multigrid methods.

- implicit midpoint rule: $u_{k+1} = u_k + \tau f(\frac{1}{2}(u_k + u_{k+1}))$;
- implicit trapezoidal rule: $u_{k+1} = u_k + \frac{\tau}{2}(f(u_k) + f(u_{k+1}))$.

Show that:

1. for the invariant the following result must hold: grad $\mathcal{E}(u(t))^T f(u(t)) = 0$;

2. the relation $\mathcal{E}(u_\tau(\tau)) = \mathcal{E}(u_0)$ holds for the implicit midpoint rule, but not for the implicit trapezoidal rule.

   Hint: Introduce the quantity $g(\theta) := \mathcal{E}(u_\tau(\theta\tau))$.

Exercise 9.3. For $z = i\sigma$ in the complex plane, represent the continuous solution of the Dahlquist test equation and the corresponding discrete solutions for the one-step methods EE, IE, and IMP.

Exercise 9.4. Consider a parabolic PDE in the abstract Cauchy form

$$u' = F(u) = -f_u(u), \quad u(0) = u_0,$$

where f is a strictly convex function. Show that f decreases monotonely along the trajectory $u(t)$.

Exercise 9.5. Show that, under the assumption of sufficient regularity, the solution z of the problem (9.46) adjoint to (9.45) satisfies just the parabolic equation (9.47).

Exercise 9.6. Consider the case that in an adaptive method of lines in each timestep merely the *local* error is monitored by some prescribed tolerance TOL independent of τ. Derive a decomposition of the time and the space part of the total error.

A

Appendix

A.1 Fourier Analysis and Fourier Transform

Fourier analysis imparts the representation of a function by its values depending on space and time and its representation as an superposition of sin- and cos-functions of various frequencies. Thereby periodic functions can be represented by discrete frequencies, while non-periodic functions in general lead to a continuous frequency spectrum.

The basis for a compact notation is the Euler identity

$$e^{iy} = \cos y + i \sin y,$$

which is why Fourier analysis is mostly formulated in complex numbers.

Fourier Series. Given a periodic function $f : [0, T] \to \mathbb{C}$, *Fourier analysis* defines Fourier coefficients as

$$\hat{f}_k = \frac{1}{\sqrt{2\pi}} \int_{[0,T]} f(t)\, e^{-i2\pi kt/T}\, dt, \quad k \in \mathbb{Z}.$$

Vice versa, given the coefficients $\hat{f}_k$, the function f may be recovered by *Fourier synthesis* as a Fourier series

$$f(t) = \frac{\sqrt{2\pi}}{T} \sum_{k \in \mathbb{Z}} \hat{f}_k\, e^{i2\pi kt/T}.$$

For real-valued functions one has, due to $e^{-iy} = \overline{e^{iy}}$, also $\hat{f}_{-k} = \overline{\hat{f}_k}$, so that a restriction to $k \in \mathbb{N}$ is possible.

Continuous Fourier Transform. Given a function $f : \mathbb{R}^d \to \mathbb{C}$, possibly non-periodic, then its *Fourier transform* is defined by

$$\hat{f}(\omega) = \frac{1}{\sqrt{2\pi}} \int_{\mathbb{R}^d} f(x)\, e^{-i\omega^T x}\, dx.$$

For $d > 1$ the frequency ω is a frequency vector, which permits different frequencies in different space directions. The *inverse Fourier transform*

$$f(x) = \frac{1}{\sqrt{2\pi}} \int_{\mathbb{R}^d} \hat{f}(\omega)\, e^{-i\omega^T x}\, d\omega$$

recovers the original function.

The Fourier transform is a linear isometry of the Lebesgue space $L^2(\mathbb{R}^d)$, but can also be extended to ampler spaces such as the space of distributions.

A.2 Differential Operators in $\mathbb{R}^3$

In the world of PDEs from science and engineering a series of extremely useful abbreviations for differential operators have evolved historically, which we are going to introduce here briefly. Their suggestive names are not already self-explaining by their definitions, but through their connections via integral theorems that we we will present in Appendix A.3 below.

Derivative and Gradient. For scalar functions $u : \mathbb{R}^d \to \mathbb{R}$ the derivative u_x associates a linearization of u to each point, i.e., a mapping $u_x : \mathbb{R}^d \to (\mathbb{R}^d)^*$ into the dual space $(\mathbb{R}^d)^*$ of $\mathbb{R}^d$, such that $u(x+\delta x) = u(x) + \langle u_x(x), \delta x \rangle + o(|\delta x|)$. The representation of functionals from $(\mathbb{R}^d)^*$ is most conveniently done as *row vector* (or matrix from $\mathbb{R}^{1 \times d}$), so that the dual pairing can be written as matrix multiplication: $\langle u_x(x), \delta x \rangle = u_x(x)\delta x$.

Often it is useful to interpret the derivative not as a linear functional in $(\mathbb{R}^d)^*$, but as a *column vector* in $\mathbb{R}^d$. As $\mathbb{R}^d$ equipped with the Euclidean scalar product is a Hilbert space, it can be identified with its dual space via the Riesz isomorphism. In this sense one defines the *gradient* $\nabla u(x) \in \mathbb{R}^d$ as adjoint to the derivative. The reformulation is especially simple here, since

$$\nabla u = u_x^T = \begin{bmatrix} u_{x_1} \\ \vdots \\ u_{x_d} \end{bmatrix}.$$

The gradient operator ∇ is also called *nabla operator*.

Geometric interpretation: The gradient is orthogonal (or normal) with respect to the Euclidean scalar product on all *level surfaces* $u = $ const.

Proof. Consider an arbitrary curve $x(s)$ within a level surface (parametrized by a variable s). For this the relation

$$u\,(x(s)) \equiv \text{const}$$

must hold, which implies

$$u_s = u_x x_s = \nabla u^T x_s = 0.$$

Note that the derivative x_s naturally is a column vector and an arbitrary tangential vector at the level surface, from which the geometric interpretation follows. □

The gradient points in the direction of increasing values of u. The *unit normal vector* normalized to length 1 is obtained as

$$n = \frac{\nabla u}{|\nabla u|},$$ (A.1)

where $|\cdot|$ is the Euclidean norm.

In $\mathbb{R}^d$, the distinction between derivative and gradient mostly serves for notational clarity, whereas in infinite dimensional function spaces it is crucial. In particular, in contrast to the derivative, the gradient therein depends on the selected scalar product.

Divergence. For vector fields $q : \mathbb{R}^d \to \mathbb{R}^d$ the divergence is defined as

$$\text{div}\, q = \sum_{i=1}^{d} \frac{\partial q_i}{\partial x_i},$$

sometimes written in sloppy form as "scalar product" $\nabla \cdot q$.

Curl. For vector fields $q : \mathbb{R}^3 \to \mathbb{R}^3$, in Euclidean coordinates

$$q(x, y, z) = \begin{bmatrix} q_1(x, y, z) \\ q_2(x, y, z) \\ q_3(x, y, z) \end{bmatrix},$$

the "curl" is defined by

$$\text{curl}\, q = \begin{bmatrix} q_{3y} - q_{2z} \\ q_{1z} - q_{3x} \\ q_{2x} - q_{1y} \end{bmatrix}.$$

An extension to more than three space dimension is nontrivial. In two space dimensions (x, y) the curl can be defined by embedding it into $\mathbb{R}^3$ with $z = 0$ as vector $\text{curl}\, q = [0, 0, q_{2x} - q_{1y}]$ orthogonal to the embedding plane. Because of its trivial first two components, the curl is mostly regarded as scalar.

Laplace Operator. By means of the differential operators defined above, the Laplace operator, again in Euclidean coordinates, can be represented as

$$\Delta u = \text{div}\, \nabla u = \sum_{i=1}^{d} u_{x_i x_i},$$

which can be verified by simple calculation using the above definitions.

Connections. The introduced differential operators satisfy certain elementary relations, which runs like a thread through the analysis of PDEs in science and engineering:

Lemma A.1. *For sufficiently smooth vector fields q and scalar functions u the following results hold:*

$$\operatorname{div}\operatorname{curl} q = 0, \quad \operatorname{curl} \nabla u = 0. \tag{A.2}$$

The properties (A.2) are easily verified by insertion of the definitions.

Let in particular u be a physical "potential" with an associated "potential field" $q = \nabla u$. Then the above assertions may be interpreted as:

Curl fields are source-free; potential fields are curl-free.

Useful Reformulation.   In the context of this book we make repeatedly use of the identity

$$\operatorname{curl}\operatorname{curl} H = \nabla \operatorname{div} H - \Delta H, \quad H : \mathbb{R}^3 \to \mathbb{R}^3, \tag{A.3}$$

where the Laplace operator Δ is understood to act componentwise.

A.3   Integral Theorems

In this section we collect some classical integral theorems, which are of crucial importance in the mathematical analysis of PDEs and therefore applied at many places throughout this textbook.

Definition A.2. *Domain.* A set $\Omega \subset \mathbb{R}^d$ is called *domain*, if it is open, nonempty, and connected.

Theorem A.3 (Theorem of Gauss). *Let $\Omega \subset \mathbb{R}^d$ denote a bounded domain with piecewise smooth boundary. Then the following result holds for continuously differentiable vector fields $q : \Omega \to \mathbb{R}^d$,*

$$\int_\Omega \operatorname{div} q \, dx = \int_{\Gamma = \partial\Omega} q^T n \, ds, \tag{A.4}$$

where n is the unit normal vector, i.e., the vector orthogonal to the domain boundary $\partial\Omega$, pointing outward and normalized to length 1.

Before we begin with this proof, we want to give a physically motivated interpretation with the aim of better understanding the assertion. Let the vector field q describe a mass flux, then div q indicates the existence of local sources or drains (see Figure A.1). Thus the left-hand side of (A.4) is the total mass added by sources or subtracted by drains to the domain Ω. The mass flux through the boundary $\partial\Omega$ is described by $q^T n$. The right-hand side corresponds to the total mass taken out from Ω via the flux through the boundary. Hence, the Theorem of Gauss is nothing else than mass conservation.

Proof. In order to avoid confusion, we here write $d\Omega$ for the differential volume element and $d\Gamma$ for the differential area element. For the time being, let the boundary

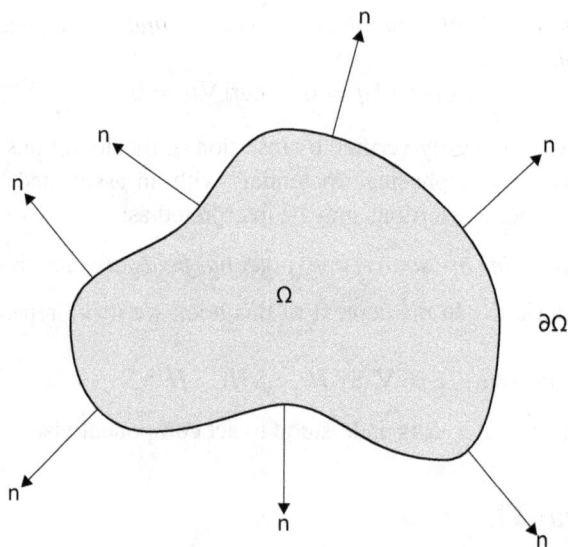

Figure A.1. The divergence as indicator of sources and drains of a vector field.

$\partial\Omega$ be intersected by any line parallel to the x_d-coordinate axis in at most two points. Thus the area $\partial\Omega$ consists of two subareas, an "upper" subarea

$$\overline{\partial\Omega} = \{x \in \partial\Omega : n^T e_d \geq 0\}, \tag{A.5}$$

and a lower subarea

$$\underline{\partial\Omega} = \{x \in \partial\Omega : n^T e_d < 0\}, \tag{A.6}$$

where e_d is the unit vector in the x_d-coordinate direction. We still need to define the "basic area"

$$\Omega_0 = \{x \in \mathbb{R}^{d-1} : \exists z \in \mathbb{R} : (x, z) \in \Omega\}$$

and the "height" $\overline{z}(x_0) = \max\{z \in \mathbb{R} : (x_0, z) \in \partial\Omega\}$ of the upper subarea as well as $\underline{z}(x_0) = \min\{z \in \mathbb{R} : (x_0, z) \in \partial\Omega\}$ of the lower subarea. A geometric representation with associated notations is given in Figure A.2.

With this notation, the Theorem of Fubini and the main theorem of differential and integral calculus imply

$$\int_\Omega \frac{\partial q_d}{\partial x_d} d\Omega = \int_{\Omega_0} \int_{\underline{z}(x_0)}^{\overline{z}(x_0)} \frac{\partial q_d}{\partial x_d} dz d\Omega_0 = \int_{\Omega_0} \left(q_d\left(x_0, \overline{z}(x_0)\right) - q_d\left(x_0, \underline{z}(x_0)\right) \right) d\Omega_0.$$

The associated area elements $d\Omega_0$ each are projections of the surface element $d\Gamma$, so that

$$d\Omega_0 = \cos\gamma \, d\Gamma = |n^T e_n| \, d\Gamma,$$

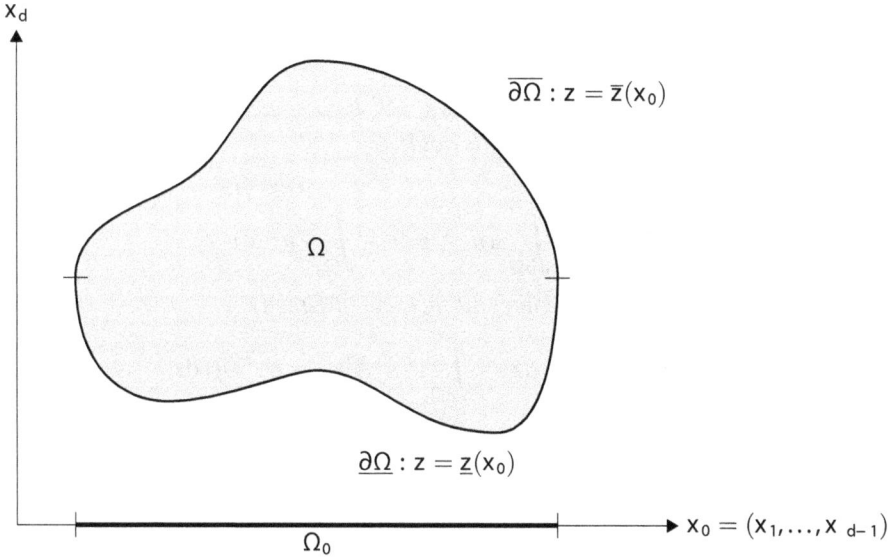

Figure A.2. Projection of the surface element $d\Gamma$ onto the upper and the lower subarea.

where γ is the angle between the normal vector n and the x_d-axis. With the sign structure of (A.5) and (A.6) we immediately get

$$\int_{\Omega} \frac{\partial q_d}{\partial x_d}\, d\Omega = \int_{\overline{\partial\Omega}} q_d |n^T e_d|\, d\Gamma - \int_{\underline{\partial\Omega}} q_d |n^T e_d|\, d\Gamma = \int_{\partial\Omega} q_d n^T e_d\, d\Gamma.$$

An analogous consideration for the coordinates $x_1, \ldots, x_{d-1}$ and summation over all coordinates finally supplies

$$\int_{\Omega} \operatorname{div} q\, d\Omega = \sum_{i=1}^{d} \int_{\Omega} \frac{\partial q_i}{\partial x_i}\, d\Omega = \sum_{i=1}^{d} \int_{\Gamma} q_i n^T e_i\, d\Gamma = \int_{\Gamma} n^T q\, d\Gamma.$$

For the sake of completeness we want to mention: if $\partial\Omega$ must be split into more than two subareas, then a partition into several subintegrals will be necessary, which finally leads to the same result.   □

Remark A.4. The Theorem of Gauss permits an alternative definition of the divergence:

$$\operatorname{div} q := \lim_{r \to 0} \frac{1}{|B(x;r)|} \int_{\partial B(x;r)} n^T q\, ds.$$

This version is used in numerical mathematics as an idea for a spatial discretization (finite volume method). It plays a particular role in fluid dynamics, which, however, will not be further pursued in this book.

By a special choice of the vector field q in the above Theorem of Gauss one is led to the following theorem:

Theorem A.5 (Theorems of Green). *Let* $\Omega \subset \mathbb{R}^d$ *be bounded. Then, for scalar functions* $u, v \in C^2(\Omega)$, *the following results hold:*
First integral theorem of Green:

$$\int_\Omega u \Delta v \, dx = \int_{\partial\Omega} u n^T \nabla v \, ds - \int_\Omega \nabla u^T \nabla v \, dx, \tag{A.7}$$

Second theorem of Green (also integral theorem of Green):

$$\int_\Omega (u \Delta v - v \Delta u) \, dx = \int_{\partial\Omega} (u n^T \nabla v - v n^T \nabla u) \, ds. \tag{A.8}$$

Proof. We choose particularly $q = u \operatorname{grad} v$ in (A.4) with scalar functions u, v and obtain

$$\operatorname{div}(u \nabla v) = \sum_{i=1}^d (u v_{x_i})_{x_i} = \sum_{i=1}^d \left(u_{x_i} v_{x_i} + u v_{x_i x_i} \right) = \nabla u^T \nabla v + u \Delta v.$$

This leads to the intermediate result

$$\int_\Omega u \Delta v \, dx = \int_\Omega \operatorname{div}(u \nabla v) \, dx - \int_\Omega \nabla u^T \nabla v \, dx.$$

Application of the Theorem of Gauss to the first of the two right-hand side terms supplies

$$\int_\Omega \operatorname{div}(u \nabla v) \, dx = \int_{\partial\Omega} u \nabla v^T n \, ds.$$

Insertion of this term finally leads us to the result (A.7).

Next we exchange u and v in (A.7) and subtract the two results: thereby the symmetric second term drops out, which immediately supplies statement (A.8) of the theorem. □

From Green's first theorem an auxiliary result follows immediately, which is helpful in proving uniqueness of the solution of the Poisson equation.

Theorem A.6 (Energy identity). *For every scalar function* $u \in C^2(\Omega)$ *one has*

$$\int_\Omega u \Delta u \, dx + \int_\Omega (\nabla u)^2 \, dx = \int_{\partial\Omega} u \frac{\partial u}{\partial n} \, ds, \tag{A.9}$$

Proof. We start from Green's first integral theorem (A.7) and set $v = u$, which brings us directly to the above identity. □

In the last one among the classical integral theorems the role of the differential operator curl becomes clear.

Theorem A.7 (Theorem of Stokes). *Let* Ω *denote a two-dimensional area in* $\mathbb{R}^3$, *bounded by a closed, piecewise* C^1*-curve* Γ. *Let* t *be the piecewise continuous tangential vector to* Γ, *oriented such that* t *circulates the area in the positive sense, and* n *normal unit vector to* Ω. *Then*

$$\int_\Omega n^T \, \text{curl} \, q \, dx = \int_\Gamma q^T t \, ds.$$

This theorem is particularly important in electrodynamics.

Remark A.8. Integral theorems like the ones of Gauss and of Stokes also hold in higher space dimensions. A comprehensive approach to this topic is supplied by the *external differential calculus* invented by H. Cartan, for which we refer to the special literature. In this elegant calculus the two theorems are even more similar than in the formulation with the differential operators curl, div.

A.4 Delta-Distribution and Green's Functions

Green's functions are solutions of linear differential equations to pointwise data in $x_0 \in \mathbb{R}^d$; they are particularly important when considering local perturbations of boundary data or source terms. We here restrict ourselves to translation invariant problems on the whole $\mathbb{R}^d$ and thus set $x_0 = 0$ w.l.o.g., which means we just drop x_0.

Delta-distribution.   With local function we mean a function ϕ with compact support. We are interested in the limit case of locality, where the support consists of only one point. For this purpose we consider the Dirac sequence

$$\phi_\epsilon(x) = \begin{cases} \frac{1}{\epsilon^d V_d}, & \|x\| < \epsilon, \\ 0, & \|x\| \geq \epsilon. \end{cases}$$

where $V_d = \pi^{d/2}/\Gamma(d/2 + 1)$ is the volume of the d-dimensional unit sphere S^d (compare [229, 0.1.6]) and Γ the Gamma-function. Obviously, for each $f \in C(\mathbb{R}^d)$ we have

$$\int_{\mathbb{R}^d} \phi_\epsilon(x) f(x) \, dx = \frac{1}{\epsilon^d V_d} \int_{\epsilon S^d} f(x) \, dx \rightarrow f(0).$$

Mind the nesting in Figure A.3. Therefore the limit

$$\delta = \lim_{\epsilon \to 0} \phi_\epsilon$$

in the dual space $C(\mathbb{R}^d)^*$ of the continuous functions, i.e., the measures, is given via the dual pairing $\langle \cdot, \cdot \rangle$ as

$$\langle \delta, f \rangle = \int_{\mathbb{R}^d} f \, d\delta = f(0).$$

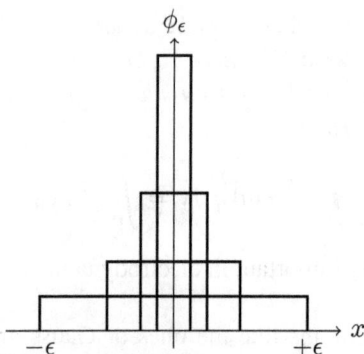

Figure A.3. Nested sharpening of characteristic functions $\phi_\epsilon(x)$ for $\epsilon \to 0$.

We will extend the concept of integration in such a way that we may write

$$\int_{\mathbb{R}^d} f(x)\, \delta(x - x_0)\, dx = f(x_0).$$

In particular, we obtain that the Fourier transform of the δ-distribution is constant,

$$k \mapsto \int_{\mathbb{R}^d} e^{\pm ik^T x}\, \delta(x)\, dx = 1,$$

which means that all frequencies k occur with equal weight ("white noise").

Poisson Equation.   The Green's function G_P of the Laplace operator in $\mathbb{R}^d$ is de-fined as the solution of the Poisson equation

$$\Delta u(x) = \delta(x) \quad \text{on} \quad \mathbb{R}^d,$$

where we, in view of the radial symmetry, are merely interested in radially symmetric solutions $u(x) = u(r)$ with $r = \|x\|$. The representation of the Laplace operator in radial coordinates (compare Exercises 1.1 and 1.2) leads, since $\delta(r) = 0$ for all $r > 0$, to

$$\frac{1}{r^{d-1}} \frac{\partial}{\partial r} \left(r^{d-1} \frac{\partial}{\partial r} u(r) \right) = 0, \quad r > 0$$

and, after multiplication by r^{d-1}, to

$$r^{d-1} \frac{\partial}{\partial r} u(r) = c_d \in \mathbb{R}. \tag{A.10}$$

Back in Cartesian coordinates we obtain for the gradient of u

$$\nabla u(x) = \frac{c_d}{r^d} x.$$

Application of the Theorem of Gauss A.3 on the unit ball $\Omega = S^d$ leads to

$$1 = \int_\Omega \delta(x)\,dx = \int_\Omega \Delta u(x)\,dx = \int_{\partial\Omega} \frac{c_d}{r^d} n^T x\,ds.$$

In fact, we have only shown the Theorem of Gauss for continuously differentiable vector fields, but it can be generalized to less smooth functions and in particular to distributions. For $x \in \partial S^d$ we now have $n^T x = 1$ and $r = 1$, which gives us

$$1 = c_n \int_{\partial\Omega} ds = c_d O_d. \tag{A.11}$$

The surface O_d of S^d can be expressed by the Gamma-function Γ (cf., for example, [229, 0.1.6]):

$$O_d = \frac{2\pi^{d/2}}{\Gamma(d/2)}. \tag{A.12}$$

Thus, from (A.11) and (A.12) we obtain the normalizing factors

$$c_1 = \frac{1}{2}, \quad c_2 = \frac{1}{2\pi}, \quad c_3 = \frac{1}{4\pi}.$$

Finally, the integration of (A.10) yields, up to an additive constant, the solutions

$$G_P(x) = \begin{cases} r/2, & d = 1 \\ \frac{\ln r}{2\pi}, & d = 2 \\ \frac{r^{2-d}}{4\pi(2-d)}, & d \geq 3, \end{cases} \quad \text{with} \quad r = \|x\|. \tag{A.13}$$

Diffusion Equation. In the diffusion equation with constant scalar diffusion $\sigma \in \mathbb{R}$ we are interested in point perturbations of the initial values and thus define the Green's function as the solution of

$$(G_D)_t = \sigma \Delta G_D \quad \text{in } \mathbb{R} \times \mathbb{R}_+, \qquad G_D(x) = \delta(x) \quad \text{for } t = 0. \tag{A.14}$$

As in Section 1.2 we represent G_D by a Fourier series with time dependent coefficients, where we, because of the nonperiodicity, apply the continuous inverse Fourier transform:

$$G_D(x,t) = \frac{1}{\sqrt{2\pi}} \int_\mathbb{R} A(\omega,t) e^{i\omega x}\,d\omega. \tag{A.15}$$

As before, the exponential functions are eigenfunctions of the Laplace operator,

$$(e^{i\omega x})_{xx} = -\omega^2 e^{i\omega x},$$

so that (A.14) leads to

$$\frac{1}{\sqrt{2\pi}} \int_\mathbb{R} A_t(\omega,t) e^{i\omega x}\,d\omega = -\frac{\sigma}{\sqrt{2\pi}} \int_\mathbb{R} A(\omega,t)\omega^2 e^{i\omega x}\,d\omega.$$

Because of the orthogonality of the ansatz functions we have $A_t(\omega, t) = -\omega^2 \sigma A$ (ω, t) for all $\omega \in \mathbb{R}$ and $t \in \mathbb{R}_+$. The solution of this linear differential equations supplies

$$A(\omega, t) = e^{-\omega^2 \sigma t} A(\omega, 0). \tag{A.16}$$

We are left with calculating the Fourier transform of the initial value δ as

$$A(\omega, 0) = \frac{1}{\sqrt{2\pi}} \int_{\mathbb{R}} \delta(x) e^{-i\omega x}\, dx = \frac{1}{\sqrt{2\pi}}. \tag{A.17}$$

Insertion of (A.17) and (A.15) into (A.16) yields

$$G_D(x, t) = \frac{1}{2\pi} \int_{\mathbb{R}} e^{-\sigma \omega^2 t} e^{i\omega x}\, d\omega = \frac{1}{\sqrt{4\pi \sigma t}} \frac{1}{\sqrt{2\pi}} \int_{\mathbb{R}} \sqrt{2\sigma t}\, e^{-\sigma t \omega^2} e^{i\omega x}\, d\omega.$$

The Fourier transform of the Gaussian bell-shaped curve $e^{-x^2/4a}$ is just $\sqrt{2a}\, e^{-x^2/a}$, from which we may directly read off the Green's function. Analogously, for all space dimensions d one obtains the general form

$$G_D(x, t) = \frac{e^{-\frac{x^2}{4\sigma t}}}{\sqrt{2\pi}(2\sigma t)^{d/2}}. \tag{A.18}$$

Helmholtz Equation.   The Green's function G_H for the Helmholtz equation is defined as the solution of

$$\Delta G_H(x) + k^2 G_H(x) = \delta(x) \quad \text{on } \mathbb{R}^d, \tag{A.19}$$

where k can even be complex. Similar to the procedure for the Poisson equation we make a radially symmetric ansatz $G_H(x) = u(\|x\|)$ and get

$$\frac{1}{r^{d-1}} \frac{\partial}{\partial r} \left(r^{d-1} \frac{\partial}{\partial r} u(r) \right) + k^2 u(r) = 0, \quad r > 0. \tag{A.20}$$

The solution of this ODE, however, is slightly more difficult than in the case of the Poisson equation. Therefore we consider the dimensions $d = 1, 2, 3$ separately. As before, the normalizing constants are determined by insertion into (A.19). Here we obtain

$$1 = \int_{\epsilon S^d} \delta(x)\, dx = \int_{\epsilon S^d} (\Delta G_H + k^2 G_H)\, dx$$

$$= \int_{\epsilon \partial S^d} n^T \nabla G_H\, ds + k^2 \int_{\epsilon S^d} G_H\, dx$$

for all $\epsilon > 0$. In particular, we have

$$\int_{\epsilon S^d} G_H\, dx \to 0 \quad \text{for } \epsilon \to 0$$

and thus

$$1 = \lim_{\epsilon \to 0} \int_{\epsilon \partial S^n} n^T \nabla G_H \, ds = \lim_{\epsilon \to 0} \epsilon^{d-1} O_d u_r(\epsilon). \tag{A.21}$$

In the one-dimensional case (A.20) reduces to $u_{rr} + k^2 u = 0$. For the Fourier ansatz $u(r) = c_s \sin kr - c_c \cos kr$ the equation (A.21) enforces $c_s = 1/k$, whereas c_c can be freely chosen, since $\cos kr$ solves the homogeneous Helmholtz equation. We are interested in solutions as local as possible, especially in those that do not increase for $r \to \infty$. In Section 1.5 only wave numbers k with $\Im(k^2) \le 0$ appeared, which is why we restrict our consideration to $-\pi/2 \le \arg k \le 0$. For real-valued wave numbers we simply choose $c_c = 0$ and obtain the real-valued Green's function

$$G_H(x) = \frac{\sin k \|x\|}{k} \quad \text{for } d = 1, k > 0.$$

Otherwise, due to $\Re(ik) > 0$,

$$\sin z = \frac{e^{iz} - e^{-iz}}{2i} \quad \text{and} \quad \cos z = \frac{e^{iz} + e^{-iz}}{2}$$

the term e^{ikr} in the sum $c_s \sin kr + c_c \cos kr$ must vanish. We thus choose $c_c = -ic_s$ and obtain

$$G_H(x) = \frac{1}{k} (\sin k \|x\| + i \cos k \|x\|) \quad \text{for } d = 1, \Im(k) < 0.$$

For $d = 2$ and $d = 3$ the ODE (A.20) requires the individual terms to be differentiated and multiplied by r^2. Scaling of the radial coordinate as $\rho = kr$ then leads to the homogeneous Bessel differential equation

$$\rho^2 u_{\rho\rho} + (d-1)\rho u_\rho + \rho^2 u = 0.$$

Its linearly independent solutions are the Bessel function $J_0(\rho)$ and the Neumann function $Y_0(\rho)$ for $d = 2$ and the spherical Bessel function $j_0(\rho) = \sin(\rho)/\rho$ and the spherical Neumann function $y_0(\rho) = -\cos(\rho)/\rho$ for $d = 3$, see [1]. Thus the Green's function is defined by

$$u(r) = c_J J_0(kr) + c_Y Y_0(kr) \quad \text{for } d = 2 \quad \text{and}$$
$$u(r) = c_j j_0(kr) + c_y y_0(kr) \quad \text{for } d = 3.$$

Here J_0 or j_0, respectively, satisfy the homogeneous Helmholtz equation so that the normalizing factors $c_Y = 1/4$ and $c_y = k/(4\pi)$ are determined by (A.21). The freedom of choice for c_J and c_j can again be used to filter out those solutions that asymptotically decrease for $r \to \infty$. For real wave numbers we will again choose $c_J = c_j = 0$ and obtain the real-valued Green's functions

$$G_H(x) = \frac{Y_0(kr)}{4} \quad \text{for } d = 2, k > 0,$$

$$G_H(x) = -\frac{\cos kr}{4\pi r} \quad \text{for } d = 3, k > 0.$$

Otherwise the terms increasing for $r \to \infty$ vanish by the choices

$$G_H(x) = \frac{Y_0(kr) + i J_0(kr)}{4} \quad \text{for } d = 2, \, \Im(k) < 0,$$

$$G_H(x) = -\frac{\cos kr - i \sin kr}{4\pi r} \quad \text{for } d = 3, \, \Im(k) < 0.$$

A.5 Sobolev Spaces

The analysis of PDEs, in particular of elliptic and parabolic type, is in large parts based on Sobolev spaces. Here we collect, with all due brevity, the functional analytic basic concepts.

Definition A.9. A vector space V which is complete with respect to a norm $\| \cdot \|$ defined thereon is called *Banach space*. If, in addition, the norm is induced by some scalar product $(\cdot, \cdot)$, then the space is a *Hilbert space*.

Important Banach spaces of functions over a domain $\Omega \subset \mathbb{R}^d$ are the Lebesgue spaces $L^p(\Omega) = \{u : \Omega \to \mathbb{R} : \|u\|_{L^p(\Omega)} < \infty\}$, which are defined via norms

$$\|u\|_{L^p(\Omega)}^p = \int_\Omega u^p \, dx, \quad \text{for } p \in [1, \infty[\quad \text{and} \quad \|u\|_{L^\infty(\Omega)} = \operatorname*{ess\,sup}_{x \in \Omega} |u(x)|.$$

As in the integration sets of measure zero do not play a role, the Lebesgue spaces consist, strictly speaking, of equivalence classes of functions that differ merely by sets of measure zero. For $p = 2$ one obtains the Hilbert space $L^2(\Omega)$.

Of special importance for the theory of elliptic PDEs is that they can be studied in Hilbert spaces in terms of symmetric, positive bilinear forms.

Definition A.10. A symmetric bilinear form $a : V \times V \to \mathbb{R}$ on a Hilbert space V is called *elliptic*, if constants $0 < \alpha_a \le C_a < \infty$ exist such that a is both continuous

$$|a(u, v)| \le C_a \|u\| \, \|v\| \quad \text{for all } u, v \in V$$

and coercive (stable)

$$a(u, u) \ge \alpha_a \|u\|^2 \quad \text{for all } u \in V.$$

From a geometrical standpoint, C_a is the circumcircle radius and α_a the in-circle radius of the level ellipsoid $a(u, u) = 1$. In terms of theory, the essential property of elliptic bilinear forms is the equivalence of the energy norm $\|v\|_a^2 := a(v, v)$ induced by the bilinear form and the norm of the space V.

An elliptic bilinear form a induces a continuous operator $A : V \to V^*$ via

$$\langle Av, w \rangle = a(v, w) \quad \text{for all } w \in V,$$

which is symmetric and positive definite. Note that a Banach space, on which a continuous symmetric and positive definite operator is defined, is always also a Hilbert space.

Theorem A.11 (Theorem of Lax-Milgram). *Let* $a : V \times V \to \mathbb{R}$ *be elliptic on a Hilbert space* V *and* $b \in V^*$. *Then there exists a unique minimizer* u *of the functional*

$$J(v) = \tfrac{1}{2} a(v, v) - \langle b, v \rangle.$$

This satisfies the weak equation

$$a(u, v) = \langle b, v \rangle \quad \text{for all } v \in V \tag{A.22}$$

and is bounded by

$$\|u\|_V \leq \frac{\|b\|_{V^*}}{\alpha_a}. \tag{A.23}$$

Proof. Due to $J(v) \geq \alpha_a \|v\|^2 - \|b\| \, \|v\|$ the functional J is bounded from below. Let $(v_n)_{n \in \mathbb{N}}$ denote a minimizing sequence with $\lim_{n \to \infty} J(v_n) = J_{\min} = \inf_{v \in V} J(v)$. From the strong convexity of J we now conclude that $(v_n)_n$ is a Cauchy sequence. First, similar to the scalar equation $(x - y)^2 = x^2 + y^2 - 2xy = 2x^2 + 2y^2 - (x + y)^2$ for $v, w \in V$ we have

$$\alpha \|v - w\|^2 \leq a(v - w, v - w) = 2a(v, v) + 2a(w, w) - a(v + w, v + w)$$

$$= 2a(v, v) + 2a(w, w) - 4a \left(\frac{v + w}{2}, \frac{v + w}{2} \right).$$

Addition of the linear terms $\langle b, \cdot \rangle$ then leads to

$$\alpha \|v - w\|^2 \leq 4J(v) + 4J(w) - 8J \left(\frac{v + w}{2} \right) \leq 4J(v) + 4J(w) - 8J_{\min}. \tag{A.24}$$

For the minimizing sequence $(v_n)_n$ this implies $\|v_n - v_m\| \to 0$ for $n, m \to \infty$, so that the limit $u = \lim_{n \to \infty} v_n \in V$ exists. From the continuity of J we conclude $J(u) = J_{\min}$. The uniqueness follows from the strong convexity: Let u_1 and u_2 be minimizers, then (A.24) implies $u_1 = u_2$.

As u is the minimizer, necessarily all directional derivatives vanish, $0 = \langle J'(u), v \rangle = a(u, v) - \langle b, v \rangle$, so that (A.22) is satisfied. The choice $v = u$ leads to

$$\alpha_a \|u\|_V^2 \leq a(u, u) = \langle b, u \rangle \leq \|b\|_{V^*} \|u\|_V$$

and thus to the norm estimate (A.23). $\qquad \square$

The Theorem of Lax–Milgram is an essential motivation to build the theory and numerics of elliptic PDEs on the weak formulation and not to restrict them to the classical solutions $u \in C^2(\Omega) \cap C(\bar{\Omega})$. The Laplace operator Δ as a prototype of an elliptic operator is not elliptic on $C^2(\Omega) \cap C(\bar{\Omega})$, so that the existence of classical solutions is not guaranteed. In particular, for domains with re-entrant corners, discontinuous coefficients in the differential operator, and less smooth right-hand sides classical solutions do not exist.

From the requirement of ellipticity a suitable function space can be directly constructed. If we define a scalar product $(u, v)_a = a(u, v)$, the ellipticity is trivially given. The associated function space is obtained by completion with respect to the induced energy norm:

$$V = \overline{\{v \in C^\infty(\Omega) : \|v\|_a < \infty\}}^{\|\cdot\|_a}.$$

The operator $\Delta + I$ gives rise to the bilinear form

$$a(u, v) = \int_\Omega (\nabla u^T \nabla v + uv) \, dx,$$

which thus generates the *Sobolev space* $H^1(\Omega)$. More general and also for higher order derivatives one may define Sobolev spaces in the simplest way recursively.

Definition A.12. For $m \geq 0$ and $1 \leq p < \infty$, the *Sobolev norms* are given by

$$\|u\|_{W^{m,p}(\Omega)}^p = \|\nabla u\|_{W^{m-1,p}(\Omega)}^p + \|u\|_{L^p(\Omega)}^p, \quad \|u\|_{W^{0,p}(\Omega)} = \|u\|_{L^p(\Omega)}.$$

The induced Banach spaces

$$W^{m,p}(\Omega) = \overline{\{v \in C^\infty(\Omega) : \|v\|_{W^{m,p}(\Omega)} < \infty\}}^{\|\cdot\|_{W^{m,p}(\Omega)}}$$

are called *Sobolev spaces*. For $p = 2$ they are Hilbert spaces denoted by $H^m(\Omega) = W^{m,2}(\Omega)$.

In approximation estimates for finite element methods often certain *semi-norms* $|u|_{H^m} = \|\nabla^m u\|_{L^2(\Omega)}$ are used. One may directly recognize that $H^1(\Omega)$ contains significantly less smooth functions than the classical solution space $C^2(\Omega) \cap C(\bar{\Omega})$, since the gradients are allowed to be discontinuous.

For quite a lot of statements about Sobolev spaces some regularity of the domain Ω must be assumed, which fortunately is almost always satisfied in practically relevant situations. Such an assumption is the following cone condition.

Definition A.13. *Cone condition.* A domain $\Omega \subset \mathbb{R}^d$ satisfies the *cone condition*, if an opening angle $\alpha > 0$ and a length $l > 0$ and for each $x \in \partial\Omega$ a $\xi(x) \in \mathbb{R}^d$ exist such that for all $y \in \mathbb{R}^d$ with $0 < \|y\| < l$ the following relation holds:

$$y^T \xi \geq \cos\alpha \|y\| \|\xi\| \Rightarrow x + y \in \Omega.$$

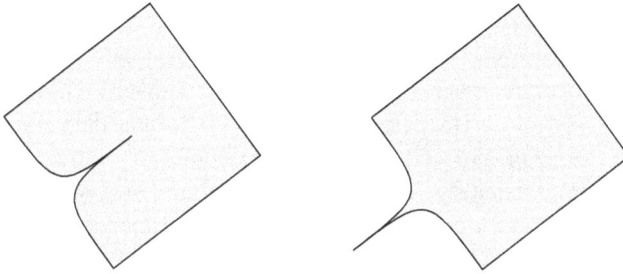

Figure A.4. Illustration of the cone condition A.13. The left domain satisfies the cone condition, the right one does not.

Clearly, the cone condition states that a fixed open cone exists that with its tip can be moved to each point on the boundary without sticking out of the domain (see Figure A.4).

On the smoothness of functions contained in Sobolev spaces H^m the following theorem supplies some information (see [2]).

Theorem A.14 (Sobolev embedding theorem). *Let $\Omega \subset \mathbb{R}^d$ denote a domain that satisfies the cone condition and $mp > d$. Then*

$$W^{m,p}(\Omega) \subset C(\Omega) \cap L^{\infty}(\Omega).$$

If one imposes homogeneous Dirichlet boundary conditions $u|_{\partial\Omega} = 0$ or Robin boundary conditions $n^T \nabla u + \alpha u = \beta$, then the values of u on the boundary must be well-defined even for possibly discontinuous $u \in H^1(\Omega)$. Fortunately, for functions in $H^1(\Omega)$ the restriction to the boundary is continuous in some integral sense, even though not pointwise (see, e.g., [39]).

Theorem A.15 (Trace theorem). *Let $\Omega \subset \mathbb{R}^d$ be a bounded domain with piecewise smooth boundary that satisfies the cone condition A.13. Then there exists a constant $C_S < \infty$ dependent on the shape of the domain, but independent of its diameter h such that*

$$\|u\|_{L^2(\partial\Omega)} \le C_S \sqrt{h} \, \|u\|_{H^1(\Omega)}$$

for all $u \in H^1(\Omega)$.

In Robin boundary conditions the boundary integrals

$$\int_{\partial\Omega} \beta v \, dx \quad \text{and} \quad \int_{\partial\Omega} \alpha u v \, dx, \quad v \in H^1(\Omega),$$

arise in their weak form. These boundary conditions are therefore well-defined at least for $\alpha \in L^{\infty}(\partial\Omega)$ and $\beta \in L^2(\partial\Omega)$. In fact, Theorem A.15 can be sharpened such that less regular boundary data α, β are sufficient, which leads to Sobolev spaces with rational indices.

For Dirichlet boundary conditions we will merely impose $u|_{\partial\Omega} = 0$ nearly everywhere, which corresponds to equality in $L^2(\partial\Omega)$. The boundary conditions are enforced by restriction of the space. For this purpose one defines, in the sense of the trace theorem, $H_0^1(\Omega) = \{u \in H^1(\Omega) : u|_{\partial\Omega} = 0\}$ and formulates the variational problem on H_0^1 instead of H^1. One proceeds similarly, if Dirichlet conditions are only imposed on a part $\partial\Omega_D$ of the boundary. Then the variational problem will be formulated on $H_D^1(\Omega) = \{u \in H^1(\Omega) : u|_{\partial\Omega_D} = 0\}$. The fact that these energy norms constituted by the Laplace operator are equivalent to the H^1-norm is stated by the inequalities of Poincaré (A.25) and of Friedrichs (A.26) below.

Theorem A.16 (Poincaré and Friedrichs inequality). *Let a bounded domain $\Omega \subset \mathbb{R}^d$ satisfy the interior cone condition. Then there exist constants C_P, C_F dependent on the shape of Ω, but not on its diameter h such that*

$$\|u - \bar{u}\|_{L^2(\Omega)} \le C_P h |u|_{H^1(\Omega)}, \quad \bar{u} = \frac{1}{|\Omega|} \int_\Omega u \, \mathrm{d}x \qquad (A.25)$$

and

$$\|u\|_{L^2(\Omega)} \le C_F h |u|_{H^1(\Omega)} \quad \text{for all } u \in H_D^1(\Omega). \qquad (A.26)$$

In the case $\partial\Omega_D = \partial\Omega$ one has $C_F = 1$.

Both statements are consequences of the following theorem, the proof of which may be found in [4, 5.15].

Theorem A.17 (General Poincaré inequality). *Let $\Omega \subset \mathbb{R}^d$ denote a bounded domain with Lipschitz boundary and $M \subset W^{1,p}(\Omega)$ convex and closed with $1 < p < \infty$. Let further exist some $u_0 \in M$ and $C_0 < \infty$ such that for all $\xi \in \mathbb{R}$ the following result holds:*

$$u_0 + \xi \in M \Rightarrow |\xi| \le C_0.$$

Then there exists a constant $C_{GP} < \infty$ such that

$$\|u\|_{L^p(\Omega)} \le C_{GP}(\|\nabla u\|_{L^p(\Omega)} + 1)$$

for all $u \in M$. If, in addition, M is a cone (i.e., $\mathbb{R}_+ M = M$), then

$$\|u\|_{L^p(\Omega)} \le C_{GP} \|\nabla u\|_{L^p(\Omega)}. \qquad (A.27)$$

While Sobolev spaces and their norms constitute an excellent framework for the analysis of elliptic PDEs, they are less suitable for the numerical solution of concrete problems. They do not distinguish between Robin and Neumann boundary conditions and do not specifically value space dependent coefficients. In particular they are, different from the energy norm, not scaling invariant: The norms change when the domains are described in other length units.

An essential consequence of the difference between Sobolev norm and energy norm is that Galerkin solutions in Sobolev spaces no longer represent, in general, the best approximations, but only quasi-optimal ones, as shown in the following lemma.

Lemma A.18 (Céa Lemma). *Let a denote a V-elliptic bilinear form with $V = H^m(\Omega)$ or $V = H_0^m(\Omega)$. Furthermore let u or u_h, respectively, denote the solutions of the corresponding variational problems in V or $S_h \subset V$, respectively. Then*

$$\|u - u_h\|_m \leq \frac{C_a}{\alpha_a} \inf_{v_h \in S_h} \|u - v_h\|_m .$$

Proof. Due to the ellipticity of a we have for all $v_h \in S_h$

$$
\begin{aligned}
\alpha_a \|u - u_h\|_m^2 &\leq a(u - u_h, u - u_h) \\
&= a(u - u_h, u - v_h) + \underbrace{a(u - u_h, v_h - u_h)}_{=0} \\
&\leq C_a \|u - u_h\|_m \|u - v_h\|_m
\end{aligned}
$$

and thus $\alpha_a \|u - u_h\|_m \leq C_a \|u - v_h\|_m$. $\square$

The fact that Sobolev spaces nevertheless play an essential role stems from the property that the interpolation error estimates from Section 4.4 can be generally formulated in terms of Sobolev norms.

A.6   Optimality Conditions

The solutions x of constrained optimization problems of the form

$$\text{minimize } J(\xi) \quad \text{subject to } c(\xi) = 0$$

satisfy under certain regularity assumptions on the cost functional J and the equality constraint c the Karush–Kuhn–Tucker conditions (briefly also: KKT conditions).

Theorem A.19. *Let X and Y be Hilbert spaces, $J : X \to \mathbb{R}$ differentiable and $c : X \to Y$ Fréchet-differentiable with surjective derivative c'. If $c(x) = 0$ and $J(x) \leq J(\xi)$ for all ξ with $c(\xi) = 0$, then there exists a Lagrange multiplier $\lambda \in Y^*$ such that the adjoint equation holds:*

$$J'(x) - c'(x)^* \lambda = 0.$$

B

Software

For the adaptive algorithms described in this book, there are public domain software packages available, most of which are highly defeloped, and which can be downloaded from the internet. The complexity of these software systems, due to the required data structures and the large class of covered problems, should not be underestimated – an intensive advance phase of learning and preparation will in any case be necessary. Here an incomplete (!) list:

B.1 Adaptive Finite Element Codes

ALBERTA an adaptive FE-package implemented in C for simplicial meshes with refinement by residual based error estimators and bisection.
http://www.alberta-fem.de/

deal.II an adaptive FE-package for nonconformal hexahedral meshes with hanging nodes.
http://www.dealii.org/

DUNE a C++ library, which offers the basic pieces for FE-codes (interface to various mesh implementations, shape functions, quadrature, iterative solvers); supports distributed computing.
http://www.dune-project.org/

KARDOS a fully time and space adaptive FE-code implemented in C for parabolic problems on the basis of former KASKADE versions.
http://www.zib.de/de/numerik/software/kardos.html

KASKADE a family of adaptive FE-codes for the solution of PDEs; the present version 7 is a flexible FE-package implemented in C++ on the basis of DUNE for the solution of linear and nonlinear, stationary and instationary problems.
http://www.zib.de/en/numerik/software/kaskade-7.html

PLTMG an adaptive FE-package for $d = 2$ implemented in FORTRAN with triangular meshes and red-green refinement.
http://ccom.ucsd.edu/~reb/software.html

UG an adaptive FE-package implemented in C for general unstructured meshes with red-green refinement; supports distributed computing.
http://atlas.gcsc.uni-frankfurt.de/~ug/

Stanford University Unstructured (SU^2) a modularized C++-package for the adaptive solution of PDEs and shape optimization, particularly for fluid dynamics.
http://adl.stanford.edu/docs/display/SUSQUARED/SU2+Home

Commercial software such as ABAQUS (`http://www.simulia.com/`), ANSYS (`http://www.ansys.com/`), and NASTRAN (`http://www.mscsoftware.com/`) are generally marked by their voluminous pre- and post-processing and their adaptation to many concrete problems as well as highly sophisticated numerical algorithms.

B.2   Direct Solvers

MUMPS multifrontal method   acronym for **MU**ltifrontal **M**assively **P**arallel **S**olver), see, e.g., [7].

    `http://mumps.enseeiht.fr/`

PARDISO a robust program package for large symmetric and unsymmetric systems on shared-memory multiprocessors [178].

    `http://www.pardiso-project.org/`

SuperLU for large sparse nonsymmetric systems on HPC-computers [64], in C.

    `http://crd.lbl.gov/~xiaoye/SuperLU/`

TAUCS for large sparse symmetric and nonsymmetric systems including incomplete factorizations, in C.

    `http://www.tau.ac.il/~stoledo/taucs/`

UMFPACK for sparse nonsymmetric systems [62], in C.

    `http://www.cise.ufl.edu/research/sparse/umfpack/`

B.3   Nonlinear Solvers

NewtonLib library with many Newton-type methods for the solution of a large class of nonlinear problems, contains in particular the methods mentioned in Section 8, i.e., NLEQ-OPT and GIANT-PCG.

    `http://www.zib.de/de/numerik/software/newtonlib.html`

KASKADE 7 FE-package (see above), contains, among others, the Newton-multigrid method from Section 8.2.

Bibliography

[1] M. Abramowitz and I. A. Stegun, editors. *Handbook of mathematical functions with formulas, graphs, and mathematical tables.* Wiley, 1972.

[2] R. A. Adams and J. J. F. Fournier. *Sobolev spaces.* Academic Press, 2. edition, 2003.

[3] M. Ainsworth and J. T. Oden. A unified approach to a posteriori error estimation using element residual methods. *Numer. Math.*, 65:23–50, 1993.

[4] H. W. Alt. *Lineare Funktionalanalysis.* Springer, 5. edition, 2008.

[5] P. R. Amestoy, A. Buttari, I. S. Duff, A. Guermouche, J.-Y. L'Excellent, and B. Uçar. The Multifrontal Method. In D. Padua, editor, *Encyclopedia of Parallel Computing.* Springer, New York, 2011.

[6] P. R. Amestoy, A. Buttari, I. S. Duff, A. Guermouche, J.-Y. L'Excellent, and B. Uçar. MUMPS. In D. Padua, editor, *Encyclopedia of Parallel Computing.* Springer, New York, 2011.

[7] P. R. Amestoy, I. S. Duff, J. Koster, and J.-Y. L'Excellent. A fully asynchronous multifrontal solver using distributed dynamic scheduling. *SIAM Journal on Matrix Analysis and Applications*, 23(1):15–41, 2001.

[8] J. P. Aubin. Behaviour of the error of the approximate solution of boundary value problems for linear elliptic operators by Galerkin's and finite difference methods. *Ann. Scuola Norm. Sup. Pisa*, 21:599–637, 1967.

[9] I. Babuška and A. K. Aziz. Survey lectures on the mathematical foundations of the finite element method. In A.K Aziz, editor, *Mathematical Foundations of the Finite Element Method with Applications to Partial Differential Equations*, pages 1–359. Academic Press, New York, 1972.

[10] I. Babuška and A. K. Aziz. On the angle condition in the finite element method. *SIAM J. Numer. Anal.*, 13:214–226, 1976.

[11] I. Babuška and A. Miller. A feedback finite element method with a posteriori error estimation: Part I. The finite element method and some basic properties of the a posteriori error estimator. *Comp. Meth. Appl. Mech. Eng.*, 61(3):1–40, 1987.

[12] I. Babuška and J. E. Osborn. Estimates for the errors in eigenvalue and eigenvector approximation by Galerkin methods, with particular attention to the case of multiple eigenvalues. *SIAM J. Numer. Anal.*, 24:1249–1276, 1987.

[13] I. Babuška and W. C. Rheinboldt. Error estimates for adaptive finite element computations. *SIAM J. Numer. Anal.*, 15:736–754, 1978.

[14] G. Bader and P. Deuflhard. A semi-implicit mid-point rule for stiff systems of ordinary differential equations. *Numer. Math.*, 41:373–398, 1983.

[15] W. Bangerth and R. Rannacher. *Adaptive Finite Element Methods for Differential Equations*. Lectures in Mathematics. Birkhäuser, 2003.

[16] R. E. Bank. *PLTMG: A Software Package for Solving Elliptic Partial Differential Equations. Users' Guide 8.0*. Frontiers in Applied Mathematics. SIAM, 1998.

[17] R. E. Bank, T. Dupont, and H. Yserentant. The hierarchical basis multigrid method. *Numer. Math.*, 52(4):427–458, 1988.

[18] R. E. Bank, A. H. Sherman, and A. Weiser. Some refinement algorithms and data structures for regular local mesh refinement. In R. S. Stepleman, editor, *Scientific Computing. Applications of Mathematics and Computing to the Physical Sciences*, pages 3–17, Amsterdam, 1983. North-Holland.

[19] R. E. Bank and A. Weiser. Some a posteriori error estimators for elliptic partial differential equations. *Math. Comp.*, 44(170):283–301, 1985.

[20] R. E. Bank and J. Xu. Asymptotically exact a posteriori error estimators. I: Grids with superconvergence. *SIAM J. Numer. Anal.*, 41(6):2294–2312, 2003.

[21] R. E. Bank and J. Xu. Asymptotically exact a posteriori error estimators. II: General unstructured grids. *SIAM J. Numer. Anal.*, 41(6):2313–2332, 2003.

[22] P. Bastian, W. Hackbusch, and G Wittum. Additive and multiplicative multigrid – a comparison. *Computing*, 60:345–368, 1998.

[23] R. Becker, M. Braack, and R. Rannacher. Adaptive finite element methods for flow problems. In R. A. DeVore, A. Iserles, and E. Süli, editors, *Foundations of Computational Mathematics*, pages 21–44. Cambridge University Press, 2001.

[24] R. Becker, H. Kapp, and R. Rannacher. Adaptive finite element methods for optimal control of partial differential equations: basic concepts. *SIAM J. Control Optim.*, 39:113–132, 2000.

[25] R. Becker and R. Rannacher. An optimal control approach to a posteriori error estimation in finite element methods. *Acta Numerica*, 10:1–102, 2001.

[26] P. Berini. Plasmon-polariton waves guided by thin lossy metal films of finite width: Bound modes of symmetric structures. *Phys. Rev. B*, 61(15):10484–10503, 2000.

[27] J. Bey. Tetrahedral grid refinement. *Computing*, 55(4):355–378, 1995.

[28] M. Bietermann and I. Babuška. An adaptive method of lines with error control for parabolic equations of the reaction-diffusion type. *J. Comp. Phys.*, 63(1):33–66, 1986.

[29] P. Binev, W. Dahmen, and R. DeVore. Adaptive finite element methods with convergence rates. *Numer. Math.*, 97(2):219–268, 2004.

[30] L. Bonaventura and G. Rosatti. A cascadic conjugate gradient algorithm for mass conservative, semi-implicit discretization of the shallow water equations on locally refined structured grids. *Int. J.Numer. Meth. Fluids*, 40(1-2):217–230, 2002.

[31] F. A. Bornemann. An Adaptive Multilevel Approach to Parabolic Equations. I. General Theory and 1D Implementation. *IMPACT Comp. Sci. Engin.*, 2(4):279–317, 1990.

[32] F. A. Bornemann. An Adaptive Multilevel Approach to Parabolic Equations. II. Variable-Order Time Discretization Based on a Multiplicative Error Correction. *IMPACT Comp. Sci. Engin.*, 3(2):93–122, 1991.

[33] F. A. Bornemann. *An Adaptive Multilevel Approach to Parabolic Equations in Two Space Dimensions*. Dissertation, Freie Universität Berlin, Institut für Mathematik, 1991.

[34] F. A. Bornemann. Die Maximalwinkelbedingung für Finite Elemente. http://www-m3.ma.tum.de/Allgemeines/FolkmarBornemannPublications, 1993.

[35] F. A. Bornemann and P. Deuflhard. The cascadic multigrid method for elliptic problems. *Numer. Math.*, 75(2):135–152, 1996.

[36] F. A. Bornemann, B. Erdmann, and R. Kornhuber. Adaptive multilevel methods in three space dimensions. *Int. J. Numer. Meth. Engin.*, 36:3187–3203, 1993.

[37] F. A. Bornemann, B. Erdmann, and R. Kornhuber. A posteriori error estimates for elliptic problems in two and three space dimensions. *SIAM J. Numer. Anal.*, 33:1188–1204, 1996.

[38] F. A. Bornemann and R. Krause. Classical and cascadic multigrid – a methodological comparison. In P. Bjørstad, M. Espedal, and D. Keyes, editors, *Proc. 9th Int. Conference on Domain Decomposition Methods 1996*, pages 64–71, Ullensvang, Norway, 1998. Domain Decomposition Press.

[39] D. Braess. *Finite Elements: Theory, Fast Solvers and Applications in Solid Mechanics*. Cambridge University Press, 2007.

[40] D. Braess and W. Hackbusch. A new convergence proof for the multigrid method including the V-cycle. *SIAM J. Numer. Anal.*, 20(5):967–975, 1983.

[41] H. Brakhage. Über die numerische Behandlung von Integralgleichungen nach der Quadraturformelmethode. *Numer. Math.*, 2:183–196, 1960.

[42] J. H. Bramble, J. E. Pasciak, and J. Xu. Parallel multilevel preconditioners. *Math. Comput.*, 55(191):1–22, 1990.

[43] A. Brandt. Multi-level adaptive technique (MLAT) for fast numerical solution to boundary value problems. In H. Cabannes and R. Temam, editors, *Proc. of the Third Int. Conf. on Numerical Methods in Fluid Mechanics*, volume 18 of *Lecture Notes in Physics*, pages 82–89. Springer, Berlin, 1973.

[44] K. E. Brenan, S. L. Campbell, and L. R. Petzold. *Numerical Solution of Initial-Value Problems in Differential-Algebraic Equations*, volume 14 of *Classics in Applied Mathematics*. SIAM, 1995.

[45] C. J. Budd, W. Huang, and R. D. Russell. Adaptivity with moving grids. *Acta Numerica*, 18:111–241, 2009.

[46] S. Burger, L. Zschiedrich, J. Pomplun, F. Schmidt, B. Kettner, and D. Lockau. 3D Finite-Element Simulations of Enhanced Light Transmission Through Arrays of Holes in Metal Films. In *Numerical Methods in Optical Metrology, Proc. SPIE*, volume 7390, page 73900H, 2009.

[47] C. Canuto, M. Y. Hussaini, A. Quarteroni, and T. A. Zang. *Spectral Methods: Fundamentals in Single Domains*. Springer, 2006.

[48] C. Canuto, M. Y. Hussaini, A. Quarteroni, and T. A. Zang. *Spectral Methods: Evolution to Complex Goemetries and Applications to Fluid Dynamics*. Springer, 2007.

[49] C. Carstensen. Reliable and efficient averaging techniques as universal tool for a posteriori finite element error control on unstructured grids. *Int. J. Num. Anal. Model.*, 3(3):333–347, 2006.

[50] A. J. Chorin and J. E. Marsden. *A Mathematical Introduction to Fluid Mechanics*. Springer, 3. edition, 1993.

[51] P. G. Ciarlet. *The Finite Element Method for Elliptic Problems*. SIAM, 2002.

[52] P. G. Ciarlet and P.-A. Raviart. General Lagrange and Hermite interpolation in $\mathbb{R}^n$ with applications to finite element methods. *Arch. Rat. Mech. Anal.*, 46(3):177–199, 1972.

[53] P. Colli Franzone, P. Deuflhard, B. Erdmann, J. Lang, and L. Pavarino. Adaptivity in space and time for reaction-diffusion systems in electrocardiology. *SIAM J. Sci. Comput.*, 28(3):942–962, 2006.

[54] R. Cools. Constructing cubature formulae: The science behind the art. *Acta Numerica*, 6:1–54, 1997.

[55] R. Courant. Variational methods for the solution of problems of equilibrium and vibrations. *Bull. Amer. Math. Soc.*, 49:1–23, 1943.

[56] R. Courant and D. Hilbert. *Methods of Mathematical Physics*, volume 2. Wiley-VCH, 1968.

[57] J. Crank and P. Nicolson. A practical method for numerical evaluation of solutions of partial differential equations of the heat-conduction type. *Proc. Cambridge Philos. Soc.*, 43:50–67, 1947.

[58] C. F. Curtiss and J. O. Hirschfelder. Integration of stiff equations. *Proc. Nat. Acad. Sci. USA*, 38:235–243, 1952.

[59] G. Dahlquist. Convergence and stability in the numerical integration of ordinary differential equations. *Math. Scand.*, 4:33–53, 1956.

[60] W. Dahmen and A. Kunoth. Multilevel preconditioning. *Numer. Math.*, 63:315–344, 1992.

[61] H. Darcy. *Les Fontaines Publiques de la Ville de Dijon*, Appendix D, pages 559–603. Dalmont, Paris, 1856.

[62] T. A. Davis. A column pre-ordering strategy for the unsymmetric-pattern multifrontal method. *ACM Transactions on Mathematical Software*, 30(2):165–195, 2004.

[63] T. A. Davis. *Direct Methods for Sparse Linear Systems*. SIAM, 2006.

[64] J. W. Demmel, S. C. Eisenstat, J. R. Gilbert, X. S. Li, and J. W. H. Liu. A supernodal approach to sparse partial pivoting. *SIAM J. Matrix Analysis and Applications*, 20(3):720–755, 1999.

[65]   P. Deuflhard. Uniqueness theorems for stiff ODE initial value problems. In D. F. Grif-
       fiths and G. A. Watson, editors, *Proceedings 13th Biennial Conference on Numerical
       Analysis 1989*, pages 74–88, Harlow, Essex, UK, 1990. Longman.

[66]   P. Deuflhard. Cascadic conjugate gradient methods for elliptic partial differential equa-
       tions: Algorithm and numerical results. *Contemporary Mathematics*, 180:29–42, 1994.

[67]   P. Deuflhard. Differential equations in technology and medicine: Computational con-
       cepts, adaptive algorithms, and virtual labs. In R. Burkard, P. Deuflhard, A. Jameson,
       J.-L. Lions, and G. Strang, editors, *Computational Mathematics Driven by Industrial
       Problems*, pages 69–125, Berlin, Heidelberg, New York, 2000. Springer.

[68]   P. Deuflhard. *Newton Methods for Nonlinear Problems. Affine Invariance and Adaptive
       Algorithms*. Springer, 2. edition, 2006.

[69]   P. Deuflhard, B. Erdmann, R. Roitzsch, and G. T. Lines. Adaptive finite element sim-
       ulation of ventricular fibrillation dynamics. *Comput. Vis. Sci.*, 12:201–205, 2009.

[70]   P. Deuflhard, T. Friese, F. Schmidt, R. März, and H.-P. Nolting. Effiziente Eigen-
       modenberechnung für den Entwurf integriert-optischer Chips. In K.-H. Hoffmann,
       W. Jäger, T. Lohmann, and H. Schunk, editors, *Mathematik. Schlüsseltechnologie für
       die Zukunft*, pages 267–279. Springer, 1997.

[71]   P. Deuflhard, E. Hairer, and J. Zugck. One–step and extrapolation methods for differ-
       ential–algebraic systems. *Numer. Math.*, 51:501–516, 1987.

[72]   P. Deuflhard, J. Lang, and U. Nowak. Adaptive Algorithms in Dynamical Process
       Simulation. In H. Neunzert, editor, *Progress in Industrial Mathematics at ECMI 94*,
       pages 122–137, Chichester, Stuttgart, 1996. Wiley, Teubner.

[73]   P. Deuflhard, P. Leinen, and H. Yserentant. Concepts of an adaptive hierarchical finite
       element code. *IMPACT Comp. Sci. Engin.*, 1:3–35, 1989.

[74]   P. Deuflhard and U. Nowak. Extrapolation integrators for quasilinear implicit ODEs. In
       P. Deuflhard and B. Engquist, editors, *Large Scale Scientific Computing*, pages 37–50,
       Boston, Basel, Stuttgart, 1987. Birkhäuser.

[75]   P. Deuflhard and M. Weiser. Local inexact Newton multilevel FEM for nonlinear
       elliptic problems. In M.-O. Bristeau, G. Etgen, W. Fitzgibbon, J-L. Lions, J. Periaux,
       and M. Wheeler, editors, *Computational Science for the 21st Century*, pages 129–138.
       Wiley-Interscience-Europe, 1997.

[76]   P. Deuflhard and M. Weiser. Global inexact Newton multilevel FEM for nonlinear el-
       liptic problems. In W. Hackbusch and G. Wittum, editors, *Multigrid Methods*, volume 3
       of *Lecture Notes in Computational Science and Engineering*, pages 71–89. Springer
       International, 1998.

[77]   P. Deuflhard, M. Weiser, and S. Zachow. Mathematics in facial surgery. *Notices Amer.
       Math. Soc.*, 53:1012–1016, 2006.

[78]   B. Döhler. Ein neues Gradientenverfahren zur simultanen Berechnung der kleinsten
       oder größten Eigenwerte des allgemeinen Eigenwertproblems. *Numer. Math.*, 40:79–
       91, 1982.

[79] W. Dörfler. A convergent adaptive algorithm for Poisson's equation. *SIAM J. Numer. Anal.*, 33(1106–1124), 1996.

[80] W. Dörfler and R. H. Nochetto. Small data oscillation implies the saturation assumption. *Numer. Math.*, 91:1–12, 2002.

[81] I. S. Duff and J. A. Scott. A frontal code for the solution of sparse positive-definite symmetric systems arising from finite-element applications. *ACM Trans. Math. Sofware*, 25(4):404–424, 1999.

[82] G. Dziuk. *Theorie und Numerik partieller Differentialgleichungen*. De Gruyter, 2010.

[83] B. L. Ehle. On Padé approximations to the exponential function and A-stable methods for the numerical solution of initial value problems. Research Report CSRR 2010, Dept. AACS, Univ. Waterloo, Ontario, Canada, 1969.

[84] I. Ekeland and R. Témam. *Convex Analysis and Variational Problems*. Number 28 in Classics in Applied Mathematics. SIAM, 1999.

[85] D. Farina, Y. Jiang, and O. Dössel. Acceleration of FEM-based transfer matrix computation for forward and inverse problems of electrocardiography. *Medical and Biological Engineering and Computing*, 47(12):1229–1236, 2009.

[86] R. P. Fedorenko. The speed of convergence of an iterative process (russian). *USSR Comput. Math. and Math. Phys.*, 4.3:227–235, 1964.

[87] R. P. Fedorenko. Iterative methods for elliptic difference equations. *Russian Mathematical Surveys*, 28(2):129–195, 1973.

[88] K. Feng. Difference schemes based on variational principle (chinese). *J. Appl. Comp. Math.*, 2(4):238–262, 1965.

[89] R. P. Feynman, R. B. Leighton, and M. Sands. *The Feynman Lectures on Physics. The Definitive and Extended Edition*, volume 2. Benjamin-Cummings, 2 edition, 2005.

[90] L. Fox. Solution by relaxation methods of plane potential problems with mixed boundary conditions. *Q. Appl. Math.*, 2:251–257, 1944.

[91] J. Frauhammer, H. Klein, G. Eigenberger, and U. Nowak. Solving moving boundary problems with an adaptive moving grid method: Rotary heat exchangers with condensation and evaporation. *Chemical Engineering Science*, 53(19):3393–3411, 1998.

[92] H. Freudenthal. Simplizialzerlegungen von beschränkter Flachheit. *Ann. Math.*, 43(3):580–582, 1942.

[93] P. J. Frey and P.-L. George. *Mesh Generation*. Wiley, 2. edition, 2008.

[94] T. Friese. Eine adaptive Spektralmethode zur Berechnung periodischer Orbits. Master's thesis, Freie Universität Berlin, 1994.

[95] T. Friese. *Eine Mehrgitter-Methode zur Lösung des Eigenwertproblems der komplexen Helmholtz-Gleichung*. Dissertation, Freie Universität Berlin, Institut für Mathematik, 1998.

[96] T. Friese, P. Deuflhard, and F. Schmidt. A multigrid method for the complex Helmholtz eigenvalue problem. In *Eighth International Conference on Domain Decomposition*, pages 18–26. DDM.org, 1999.

[97] F. N. Fritsch and J. Butland. A method for constructing local monotone piecewise cubic interpolants. *SIAM J. Sci. Comput.*, 5:300–304, 1984.

[98] Y. C. Fung. *Biomechanics: Mechanical Properties of Living Tissues*. Springer, 1993.

[99] A. Gentry. The origins of lift. `http://www.arvelgentry.com/techs/origins_of_lift.pdf`, 2006.

[100] A. George and J. W.-H. Liu. *Computer Solution of Large Sparse Positive Definite Systems*. Prentice-Hall, 1981.

[101] D. Gilbarg and N. S. Trudinger. *Elliptic partial differential equations of second order*. Springer, 2 edition, 2001.

[102] M. E. Go Ong. *Hierachical basis preconditioners for second order elliptic problems in three dimensions*. Dissertation, Washington University, Seattle, USA, 1989.

[103] S. K. Godunov and G. P. Prokopov. On the solution of the Laplace difference equation (Russian). *Zh. vychisl. Mat. mat. Fiz*, 9:462–468, 1969.

[104] G. H. Golub and C. F. van Loan. *Matrix Computations*. The Johns Hopkins University Press, 1996.

[105] D. Gottlieb, M. Y. Hussaini, and S. Orszag. Introduction: Theory and applications of spectral methods. In R. G. Voigt, D. Gottlieb, and M. Y. Hussaini, editors, *Spectral Methods for Partial Differential Equations*, pages 1–54. SIAM, 1984.

[106] C. Gräser and R. Kornhuber. Multigrid methods for obstacle problems. *J. Comput. Math.*, 27(1):1–44, 2009.

[107] A. Griewank and A. Walther. *Evaluating Derivatives*. SIAM, 2 edition, 2008.

[108] P. Grisvard. *Elliptic problems in nonsmooth domains*, volume 24 of *Monographs and Studies in Mathematics*. Pitman, 1985.

[109] K. Gustafsson, M. Lundh, and G. Söderlind. A PI stepsize control for the numerical solution of ordinary differential equations. *BIT*, 28:270–287, 1988.

[110] S. Götschel, M. Weiser, and A. Schiela. Solving Optimal Control Problems with the Kaskade 7 Finite Element Toolbox. In A. Dedner, B. Flemisch, and R. Klöfkorn, editors, *Proceedings of the first DUNE User's workshop*. Springer, 2012.

[111] W. Hackbusch. Ein iteratives Verfahren zur schnellen Auflösung elliptischer Randwertprobleme. Report 76–12, Universität zu Köln, 1976.

[112] W. Hackbusch. On the computation of approximate eigenvalues and eigenfunctions of elliptic operators by means of a multigrid method. *SIAM J. Numer. Anal.*, 16:201–215, 1979.

[113] W. Hackbusch. Survey of convergence proofs for multi-grid iterations. In J. Frehse, D. Pallaschke, and U. Trottenberg, editors, *Special Topics of Applied Mathematics, Proc. Bonn, Oct. 1979*, volume 18, pages 151–164. North-Holland, Amsterdam, 1980.

[114] W. Hackbusch. *Multi-Grid Methods and Applications*, volume 4 of *Computational Mathematics*. Springer, 1985.

[115] W. Hackbusch. *Theorie und Numerik elliptischer Differentialgleichungen*. Teubner, 1986.

[116] W. Hackbusch and U. Trottenberg. *Multigrid Methods*. Springer, 1982.

[117] E. Hairer and G. Wanner. *Solving Ordinary Differential Equations II. Stiff and Differential–Algebraic Problems*. Springer, 2nd edition, 1996.

[118] O. Heaviside. On the forces, stresses and fluxes of energy in the electromagnetic field. *Phil. Trans. Roy. Soc.*, 183A:423ff., 1892.

[119] P. Henrici. *Applied and computational analysis*, volume 3. J. Wiley, New York, 1986.

[120] M. R. Hestenes and E. Stiefel. Methods of conjugate gradients for solving linear systems. *J. Res. Nat. Bur. Stand.*, 49:409–436, 1952.

[121] M. Hochbruck and C. Lubich. On Krylov subspace approximations to the matrix exponential operator. *SIAM J. Numer. Anal.*, 34(5):1911–1925, 1997.

[122] M. Hochbruck, C. Lubich, and H. Selhofer. Exponential integrators for large systems of differential equations. *SIAM J. Sci. Comp.*, 19(5):1552–1574, 1998.

[123] G. A. Holzapfel. *Nonlinear Solid Mechanics*. Wiley, 2000.

[124] L.-C. Hsu and C. Mavriplis. Adaptive meshes for the spectral element method. In P. E. Bjørstad, M. S. Espedal, and D. E. Keyes, editors, *9th International Conference on Domain Decomposition Methods*, pages 374–381. DDM.org, 1998.

[125] W. Huang and R. D. Russell. *Adaptive Moving Mesh Methods*. Applied Mathematical Sciences. Springer, 2011.

[126] J. M. Hyman. Moving Mesh Methods for Partial Differential Equations. In L. Goldstein, S. Rosencrans, and G. Sod, editors, *Mathematics Applied to Science*, pages 129–153. Academic Press, 1988.

[127] L. Hörmander. *The Analysis of Linear Partial Differential Operators I*, volume 256 of *Grundl. Math. Wissenschaft*. Springer, 2003.

[128] V. P. Il'yin. Some estimates for conjugate gradient methods. *USSR Comput. Math. and Math. Phys.*, 16:22–30, 1976.

[129] B. Jakobsen and F. Rosendahl. The Sleipner Platform Accident. *Structural Engineering International*, 4(3):190–193, 1994.

[130] P. Jamet. Estimations d'erreur pour des elements finis droits presque degeneres. *R. A. I. R. O. Anal. Numér.*, 10:43–60, 1976.

[131] E. W. Jenkins, C. E. Kees, C. T. Kelley, and C. T. Miller. An aggregation-based domain decomposition preconditioner for groundwater flow. *SIAM J. Sci. Comput.*, 23(2):430–441, 2001.

[132] F. John. *Partial Differential Equations*. Springer, 1982.

[133] P. Kaps and P. Rentrop. Generalized Runge-Kutta methods of order four with stepsize control for stiff ordinary differential equations. *Numer. Math.*, 33:55–68, 1979.

[134] R. Kornhuber. Nonlinear Multigrid Techniques. In J. F. Blowey, J. P. Coleman, and A. W. Craig, editors, *Theory and Numerics of Differential Equations*, pages 179–229. Springer Universitext, Heidelberg, New York, 2001.

[135] M. Křížek. On the maximum angle condition for linear tetrahedral elements. *SIAM J. Numer. Anal.*, 29(2):513–520, 1992.

[136] T. Kröger and T. Preusser. Stability of the 8-tetrahedra shortest-interior-edge partitioning method. *Numer. Math.*, 109(3):435–457, 2008.

[137] D. Kröner. *Numerical Schemes for Conservation Laws*. J. Wiley and Teubner, 1997.

[138] D. Kröner, M. Ohlberger, and C. Rohde. *An Introduction to Recent Developments in Theory and Numerics for Conservation Laws*, volume 5 of *Lecture Notes in Computational Science and Engineering*. Springer, 1999.

[139] J. Lang. *Adaptive Multilevel Solution of Nonlinear Parabolic PDE Systems*, volume 16 of *Lecture Notes in Computational Science and Engineering*. Springer, 2001.

[140] J. Lang and D. Teleaga. Towards a Fully Space-Time Adaptive FEM for Magnetoquasistatics. *IEEE Trans. Magn.*, 44(6):1238–1241, 2008.

[141] J. Lang and J. Verwer. ROS3P – an accurate third-order Rosenbrock solver designed for parabolic problems. *BIT*, 41:730–737, 2001.

[142] I. LeGrice, P. Hunter, A. Young, and B. Smaill. The architecture of the heart: a database model. *Phil. Trans. of the Royal Society London A*, 359(1783):1217–1232, 2001.

[143] R. B. Lehoucq and D. C. Sorensen. Deflation Techniques For An Implicitly Re-Started Arnoldi Iteration. *SIAM J. Matrix Anal. Appl*, 17:789–821, 1996.

[144] R. J. LeVeque. *Finite Volume Methods for Hyperbolic Problems*. Cambridge University Press, 2002.

[145] S. Li, L. Petzold, and J. M. Hyman. Solution Adapted Mesh Refinement and Sensitivity Analysis for Parabolic Partial Differential Equation Systems. In L. T. Biegler, O. Ghattas, M. Heinkenschloss, and B. van Bloemen Waanders, editors, *Large-Scale PDE-Constrained Optimization*, volume 30 of *Lecture Notes in Computational Science and Engineering*, Heidelberg, 2003. Springer.

[146] D. E. Longsine and S. F. McCormick. Simultaneous Rayleigh-Quotient Minimization Methods for $Ax = \lambda Bx$. *Lin. Alg. Appl.*, 34:195–234, 1980.

[147] C. Lubich and A. Ostermann. Linearly implicit time discretization of non-linear parabolic equations. *IMA J. Numer. Anal.*, 15(4):555–583, 1995.

[148] C. Lubich and A. Ostermann. Runge-Kutta approximation of quasi-linear parabolic equations. *Math. Comp.*, 64:601–627, 1995.

[149] J. N. Lyness and R. Cools. A survey of numerical cubature over triangles. In W. Gautschi, editor, *Mathematics of computation 1943-1993: a half- century of computational mathematics. Mathematics of computation 50th anniversary symposium*, volume 48 of *American Mathematical Society. Proc. Symp. Appl. Math.*, pages 127–150, 1994.

[150] U. Maas and J. Warnatz. Ignition processes in carbon monoxide-hydrogen-oxygen mixtures. *Proc. Comb. Inst.*, 22(1):1695–1704, 1989.

[151] J. Mandel and S. F. McCormick. A multilevel variational method for $Au = \lambda Bu$ on composite grids. *J. Comp. Phys.*, 80:442–452, 1989.

[152] C. Mavriplis. A posteriori error estimators for adaptive spectral element techniques. In P. Wesseling, editor, *Proc. 8th GAMM-Conference on Numerical Methods in Fluid Mechanics*, volume 29, pages 333–342. Vieweg, Braunschweig, 1990.

[153] J. C. Maxwell. A dynamical theory of the electromagnetic field. *Roy. Soc. Trans.*, 155:459–512, 1865.

[154] S. F. McCormick. *Multilevel Projection Methods for Partial Differential Equations*, volume 62 of *CBMS-NSF*. SIAM, Philadelphia, 1992.

[155] J. A. Meijerink and H. A. van der Vorst. An iterative solution method for linear systems of which the coefficient matrix is a symmetric M-matrix. *Math. Comp.*, 31:148–162, 1977.

[156] K. Miller. Moving finite elements ii. *SIAM J. Numer. Anal.*, 18(6):1033–1057, 1981.

[157] K. Miller and R. N. Miller. Moving finite elements i. *SIAM J. Numer. Anal.*, 18(6): 1019–1032, 1981.

[158] J. C. Nédélec. Mixed finite elements in $\mathbb{R}^3$. *Numer. Math.*, 35:315–341, 1980.

[159] N. Neuss. V-cycle convergence with unsymmetric smoothers and application to an anisotropic model problem. *SIAM J. Numer. Anal.*, 35(3):1201–1212, 1998.

[160] J. A. Nitsche. Ein Kriterium für die Quasioptimalität des Ritzschen Verfahrens. *Numer. Math.*, 11:346–348, 1968.

[161] D. Noble and Y. Rudy. Models of cardiac ventricular action potentials: iterative interaction between experiment and simulation. *Phil. Trans. Royal Society London A*, 359:1127–1142, 2001.

[162] R. H. Nochetto, K. G. Siebert, and A. Veeser. Theory of adaptive finite element methods: An introduction. In R. A. DeVore and A. Kunoth, editors, *Multiscale, Nonlinear and Adaptive Approximation*, pages 409–542. Springer, 2009.

[163] U. Nowak. *Adaptive Linienmethoden für nichtlineare parabolische Systeme in einer Raumdimension*. Dissertation, Freie Universität Berlin, Institut für Mathematik, 1993.

[164] U. Nowak. PDEX1M – a software package for the numerical simulation of parabolic systems in one space dimension. In F. Keil, W. Mackens, H. Voß, and J. Werther, editors, *Scientific Computing in Chemical Engineering*, pages 163–169. Springer, 1996.

[165] A. Ostermann and M. Roche. Rosenbrock methods for partial differential equations and fractional order of convergence. *SIAM J. Numer. Anal.*, 30(4):1084–1098, 1993.

[166] P. Oswald. On discrete norm estimates related to multilevel preconditioners in the finite element method. In *Proc. Conf. Constructive Theory of Functions*, Varna, 1991.

[167] L. F. Pavarino. Additive Schwarz methods for the *p*-version finite element method. *Numer. Math.*, 66(1):493–515, 1994.

[168] C. E. Pearson. On a differential equation of boundary layer type. *J. Math. Phys.*, 47:134–154, 1968.

[169] N. Peters and J. Warnatz. *Numerical Methods in Laminar Flame Propagation*, volume 6 of *Notes in Numerical Fluid Dynamics*. Vieweg, 1982.

[170] L. R. Petzold. A description of DASSL: A differential-algebraic system solver. In R. S. Stepleman et al., editor, *Scientific Computing*, pages 65–68, Amsterdam, 1983. North-Holland.

[171] L. R. Petzold. Observations on an adaptive moving grid method for one-dimensional systems of partial differential equations. *Appl. Numer. Math.*, 3:347–360, 1987.

[172] L. Prandtl. Über Flüssigkeitsbewegung bei sehr kleiner Reibung. In *Verhandlungen des dritten Internationalen Mathematiker-Kongresses*, pages 484–491, Heidelberg, 1904.

[173] A. Quarteroni and A. Valli. *Domain Decomposition Methods for Partial Differential Equations*. Oxford University Press, 1999.

[174] R. Rannacher. On the convergence of the Newton-Raphson method for strongly non-linear finite element equations. In P. Wriggers and W. Wagner, editors, *Nonlinear Computational Mechanics – State of the Art*, pages 11–30. Springer, 1991.

[175] M.-C. Rivara. Mesh refinement process based on the generalized bisection of simplices. *SIAM J. Numer. Anal.*, 21(3):604–613, 1984.

[176] E. Rothe. Zweidimensionale parabolische Randwertaufgabe als Grenzfall eindimensionaler Randwertaufgaben. *Math. Ann.*, 102:650–670, 1930.

[177] S. Sauter and G. Wittum. A Multigrid Method for the Computation of Eigenmodes of Closed Water Basins. *Impact Comp. Sci. Eng.*, 4:124–152, 1992.

[178] O. Schenk, M. Bollhöfer, and R. Römer. On large-scale diagonalization techniques for the Anderson model of localization. *SIAM Review*, 50(1):91–112, 2008.

[179] A. Schiela. A cubic regularization algorithm for nonconvex optimization in function space. ZIB Report 12-16, Zuse Institute Berlin, 2012.

[180] A. Schmidt and K. Siebert. ALBERT – Software for Scientific Computations and Applications. *Acta Math. Univ. Comenianae, Proc. Algoritmy 2000*, LXX:105–122, 2001.

[181] J. Schöberl, J. M. Melenk, C. Pechstein, and S. Zaglmayr. Additive Schwarz preconditioning for p-version triangular and tetrahedral finite elements. *IMA J. Num. Anal.*, 28(1):1–24, 2008.

[182] S. Scholz. Order barriers for the B-convergence of ROW methods. *Computing*, 41(3):219–235, 1989.

[183] C. Schwab. *p- and hp- Finite Element Methods. Theory and Applications in Solid and Fluid Mechanics*. Oxford University Press, 1998.

[184] H. A. Schwarz. Über einen Grenzübergang durch alternierendes Verfahren. *Vierteljahrsschrift der Naturforschenden Gesellschaft in Zürich*, 15:272–286, 1870. In: Gesammelte Mathematische Abhandlungen, Band 2, 133-134, Springer 1890.

[185] H. R. Schwarz. *Methode der finiten Elemente*. Teubner, 3. edition, 2000.

[186] L. R. Scott and S. Zhang. Finite element interpolation of non-smooth functions satisfying boundary conditions. *Math. Comp.*, 54:483–493, 1990.

[187] V. V. Shaidurov. Some estimates of the rate of convergence for the cascadic conjugate-gradient method. *Comp. & Math. Appl.*, 31(4-5):161–171, 1996.

[188] J. R. Shewchuk. Triangle: engineering a 2D quality mesh generator and delaunay triangulator. In M. C. Lin and D. Manocha, editors, *Applied Computational Geometry: Towards Geometric Engineering*, volume 1148 of *Lecture Notes in Computer Science*, pages 203–222. Springer, 1996.

[189] P. Šolín, K. Segeth, and I. Doležel. *Higher-Order Finite Element Methods*. Studies in Advanced Mathematics. Chapman & Hall, 2004.

[190] D. Stalling, M. Westerhoff, and H.-C. Hege. Amira: A highly interactive system for visual data analysis. In C. D. Hansen and C. R. Johnson, editors, *The Visualization Handbook*, chapter 38, pages 749–767. Elsevier, 2005.

[191] T. Steihaug. The conjugate gradient method and trust regions in large scale optimization. *SIAM J. Numer. Anal.*, 20(3):626–637, 1983.

[192] G. Steinebach and P. Rentrop. An adaptive method of lines approach for modelling flow and transport in rivers. In A. V. Wouver, P. Saucez, and W. E. Schiesser, editors, *Adaptive Method of Lines*, pages 181–205, Boca Raton, FL, 2001. Chapman Hall / CRC.

[193] R. Stevenson. Optimality of a standard adaptive finite element method. *Found. Comput. Math.*, 7(2):245–269, 2007.

[194] G. Strang and G. J. Fix. *An Analysis of the Finite Element Method*. Cambridge University Press, 2. edition, 2008.

[195] K. Strehmel and R. Weiner. *Linear-implizite Runge-Kutta-Methoden und ihre Anwendung*. Teubner, Stuttgart, 1992.

[196] J. Sundnes, G. T. Lines, X. Cai, B. F. Nielsen, K.-A. Mardal, and A. Tveito. *Computing the Electrical Activity in the Heart*, volume 1 of *Monographs in Computational Science and Engineering*. Springer, 2006.

[197] J. L. Synge. *The Hypercircle in Mathematical Physics*. Cambridge University Press, 1957.

[198] M. A. Taylor, B. A. Wingate, and R. E. Vincent. An algorithm for computing Fekete points in the triangle. *SIAM J. Numer. Anal.*, 38(5):1707–1720, 2000.

[199] F. Tisseur and K. Meerbergen. The quadratic eigenvalue problem. *SIAM Rev.*, 43(2):235–286, 2001.

[200] P. L. Toint. Towards an efficient sparsity exploiting Newton method for minimization. In I. S. Duff, editor, *Sparse Matrices and Their Uses*, pages 57–88. Academic Press, 1981.

[201] A. Toselli and O. B. Widlund. *Domain Decomposition Methods – Algorithms and Theory*, volume 34 of *Computational Mathematics*. Springer, 2005.

[202] L. N. Trefethen and M. Embree. *Spectra and Pseudospectra: The Behavior of Nonnormal Matrices and Operators*. Princeton University Press, 2005.

[203] L. N. Trefethen, A. E. Trefethen, S. C. Reddy, and T. A. Driscoll. Hydrodynamic stability without eigenvalues. *Science*, 261:578–584, 1993.

[204] U. Trottenberg, C. Oosterlee, and A. Schüller. *Multigrid*. Academic Press, 2001.

[205] M. J. Turner, R. M. Clough, H. C. Martin, and L. J. Topp. Stiffness and deflection anal-
       ysis of complex structures. *J. Aeron. Sci.*, 23:805–823, 854, 1956.

[206] G. M. Vainikko. Asymptotic error estimates for projective methods in the eigenvalue
       problem (Russian). *Zh. Vychisl, Mat.*, 4:404–425, 1964.

[207] H. A. van der Vorst. BI-CGSTAB: a fast and smoothly converging variant of BI-CG
       for the solution of nonsymmetric linear systems. *SIAM J. Sci. Stat.*, 13(2):631–644,
       1992.

[208] R. Varga. *Matrix Iterative Analysis*. Prentice Hall, 1962.

[209] R. Verfürth. *A Review of A-Posteriori Error Estimation and Adaptive Mesh-Refinement
       Techniques*. Wiley-Teubner, New York, Stuttgart, 1996.

[210] T. Warburton. An explicit construction of interpolation nodes on the simplex. *J. Eng.
       Math.*, 56(3):247–262, 2006.

[211] L. K. Waters, G. J. Fix, and C. L. Cox. The method of Glowinski and Pironneau for the
       unsteady Stokes problem. *Computers Math. Appl.*, 48:1191–1211, 2004.

[212] M. Weiser. On Goal-Oriented Adaptivity for Elliptic Optimal Control Problems. *Op-
       tim. Meth. Softw.*, accepted.

[213] M. Weiser, P. Deuflhard, and B. Erdmann. Affine conjugate adaptive Newton methods
       for nonlinear elastomechanics. *OMS*, 22(3):413–431, 2007.

[214] M. Weiser and S. Götschel. State Trajectory Compression for Optimal Control with
       Parabolic PDEs. *SIAM J. Sci. Comput.*, 34(1):A161–A184, 2012.

[215] M. Weiser, A. Schiela, and P. Deuflhard. Asymptotic mesh independence of Newton's
       method revisited. *SIAM J. Numer. Anal.*, 42(5):1830–1845, 2005.

[216] M. Weiser, S. Zachow, and P. Deuflhard. Craniofacial surgery planning based on vir-
       tual patient models. *Information Technology*, 52(5):258–263, 2010.

[217] D. Werner. *Funktionalanalysis*. Springer, 5. edition, 2007.

[218] G. Wittum. Multi-grid methods for Stokes and Navier-Stokes equations with trans-
       forming smoothers: algorithms and numerical results. *Numer. Math.*, 54:543–563,
       1989.

[219] G. Wittum. On the convergence of multi-grid methods with transforming smoothers.
       Numer. Math., 57:15–38, 1990.

[220] M. Wulkow. *Numerical treatment of countable systems of ordinary differential equa-
       tions*. Dissertation, Freie Universität Berlin, Institut für Mathematik, 1990.

[221] M. Wulkow. The simulation of molecular weight distributions in polyreaction kinetics
       by discrete Galerkin methods. *Macromol. Theory Simul.*, 5(3):393–416, 1996.

[222] J. Xu. *Theory of Multilevel Methods*. Dissertation, Pennsylvania State University,
       University Park, USA, 1989.

[223] J. Xu. Iterative methods by space decomposition and subspace correction. *SIAM Re-
       view*, 34(4):581–613, 1992.

[224] H. Yang. Conjugate gradient methods for the Rayleigh quotient minimization of gen-
       eralized eigenvalue problems. *Computing*, 51(1):79–94, 1993.

[225] D. M. Young. *Iterative Solution of Large Linear Systems*. Computer Science and
       Applied Mathematics. New York: Academic Press, 1971.

[226] H. Yserentant. On the multi-level splitting of finite element spaces. *Numer. Math.*,
       49(4):379–412, 1986.

[227] H. Yserentant. Two preconditioners based on the multilevel splitting of finite element
       spaces. *Numer. Math.*, 58:163–184, 1990.

[228] H. Yserentant. Old and new convergence proofs for multigrid methods. *Acta Numerica*,
       2:285–326, 1993.

[229] E. Zeidler, editor. *Teubner-Taschenbuch der Mathematik*. Teubner, 2003.

[230] S. Zhang. Successive subdivisions of tetrahedra and multigrid methods on tetrahedral
       meshes. *Houston J. Math.*, 21(3):541–556, 1995.

[231] O. C. Zienkiewicz, R. L. Taylor, and J. Z. Zhu. *The Finite Element Method. Its Basis
       & Fundamentals*, volume 1. Elsevier, 5. edition, 2005.

[232] O. C. Zienkiewicz and J. Z. Zhu. A simple error estimator and adaptive procedure for
       practical engineering analysis. *Int. J. Numer. Meth. Engrg.*, 24(2):337–357, 1987.

[233] M. Zöckler, D. Stalling, and H.-C. Hege. Interactive visualization of 3D-vector fields
       using illuminated streamlines. In *IEEE Visualization 1996*, pages 107–113, 1996.

[234] F. Zöllner. *Leonardo da Vinci. Sämtliche Gemälde und Zeichnungen*. Verlag Taschen,
       Köln, 2003.

[235] G. Zumbusch. Symmetric hierarchical polynomials and the adaptive h-p-version. In
       A. V. Ilin and L. R. Scott, editors, *Proc. of the Third Int. Conf. on Spectral and High
       Order Methods, ICOSAHOM '95*, Houston Journal of Mathematics, pages 529–540,
       1996.

Index